Bivalve Filter Feeders

in Estuarine and Coastal Ecosystem Processes

NATO ASI Series

Advanced Science Institutes Series

A series presenting the results of activities sponsored by the NATO Science Committee, which aims at the dissemination of advanced scientific and technological knowledge, with a view to strengthening links between scientific communities.

The Series is published by an international board of publishers in conjunction with the NATO Scientific Affairs Division

A Life Sciences B Physics	Plenum Publishing Corporation London and New York
C Mathematical and Physical Sciences D Behavioural and Social Sciences E Applied Sciences	Kluwer Academic Publishers Dordrecht, Boston and London
F Computer and Systems Sciences G Ecological Sciences H Cell Biology I Global Environmental Change	Springer-Verlag Berlin Heidelberg New York London Paris Tokyo Hong Kong Barcelona Budapest

NATO-PCO DATABASE

The electronic index to the NATO ASI Series provides full bibliographical references (with keywords and/or abstracts) to more than 30000 contributions from international scientists published in all sections of the NATO ASI Series. Access to the NATO-PCO DATABASE compiled by the NATO Publication Coordination Office is possible in two ways:

- via online FILE 128 (NATO-PCO DATABASE) hosted by ESRIN, Via Galileo Galilei, I-00044 Frascati, Italy.

- via CD-ROM "NATO Science & Technology Disk" with user-friendly retrieval software in English, French and German (© WTV GmbH and DATAWARE Technologies Inc. 1992).

The CD-ROM can be ordered through any member of the Board of Publishers or through NATO-PCO, Overijse, Belgium.

Series G: Ecological Sciences, Vol. 33

Bivalve Filter Feeders

in Estuarine and Coastal Ecosystem Processes

Edited by

Richard F. Dame

Marine Science Department
Coastal Carolina University
P. O. Box 1954
Conway, SC 29526, USA

Springer-Verlag
Berlin Heidelberg New York London Paris Tokyo
Hong Kong Barcelona Budapest
Published in cooperation with NATO Scientific Affairs Division

Proceedings of the NATO Advanced Research Workshop on The Role of Bivalve Filter Feeders in Marine Ecosystem Processes, held at Renesse, The Netherlands, November 30 to December 4, 1992

Library of Congress Cataloging-in-Publication Data. Bivalve filter feeders in estuarine and coastal ecosystem processes / edited by Richard F. Dame. p. cm. – (Nato ASI series. Series G, Ecological sciences; vol. 33) "Proceedings of the NATO Advanced Research Workshop on the Role of Bivalve Filter Feeders in Marine Ecosystem Processes, held at Renesse, The Netherlands, November 30 to December 4, 1992" – T.p. verso. Includes bibliographical references (p.) and index.
ISBN-13: 978-3-642-78355-5 e-ISBN-13: 978-3-642-78353-1
DOI: 10.1007/ 978-3-642-78353-1
1. Bivalve–Ecology–Congresses. 2. Bivalve–Food–Congresses. 3. Estuarine ecology–Congresses. 4. Coastal ecology–Congresses. I. Dame, Richard F. II. NATO Advance Research Workshop on the Role of Bivalve Filter Feeders in Marine Ecosystem Processes (1992: Renesse, Netherlands) III. Series: NATO ASI series. Series G, Ecological sciences; no. 33. QL430.6.B57 1993 594'.11045–dc20 93-28010

Softcover reprint of the hardcover 1st edition 1993

Typesetting: Camera ready by authors
31/3145 - 5 4 3 2 1 0 - Printed on acid-free paper

TABLE OF CONTENTS

ACKNOWLEDGMENTS: The editor is grateful for the efforts of the steering committee of B. Bayne, N. Dankers, J. Grant, M. Heral, and A. Smaal in preparing the workshop proposal and identifying speakers and participants. Aat Smaal and Theo Prins were particularly helpful with the local arrangements at Slot Moermond, Renesse, The Netherlands. Finally, I wish to thank Amanda M. Dame for her untiring support of this project and the supervision of the final indexing.

FEEDING PHYSIOLOGY OF BIVALVES: TIME-DEPENDENCE AND COMPENSATION FOR CHANGES IN FOOD AVAILABILITY

Brian L Bayne,
Plymouth Marine Laboratory,
Prospect Place, West Hoe,
Plymouth PL1 3DH
United Kingdom

INTRODUCTION

In spite of considerable experimental and observational study over many years, controversies and uncertainties still exist concerning fundamental features of feeding behaviour in suspension-feeding bivalves (Bayne and Newell, 1983; Griffiths and Griffiths, 1987; Jørgensen, 1991). These include uncertainties over the mechanisms of particle capture (Jørgensen, 1983; Silvester and Sleigh, 1984; Shimeta and Jumars, 1991; Ward et al., 1991) and controversy over the role of physiological processes in determining feeding behaviour (Bayne et al., 1988; Jørgensen, 1991; Iglesias et al., 1992). This paper addresses the second of these topics and suggests that physiological and behavioural compensations for changes in the food environment are important elements for a full understanding of suspension-feeding. This is not to deny the significance of physical constraints on feeding, such as the viscocity of the medium, or frictional forces influencing water flows within the mantle cavity, but rather to argue that behavioural and physiological flexibility in aspects of particle capture, sorting, digestion and absorption are also important. Since physiological traits are inherently time-dependent, it follows that feeding in bivalves should be viewed as a linked series of processes with different time-constants, coupled to relevant time-scales of change in the environmental availability of food.

Early studies of feeding in bivalves concentrated upon the effects of key environmental variables (temperature, salinity, the concentration of suspended particles) and of body size on feeding rate (Winter, 1978). More recent investigations have

NATO ASI Series, Vol. G 33
Bivalve Filter Feeders
in Estuarine and Coastal Ecosystem Processes
Edited by Richard F. Dame

become concerned with the role of particle type (or "food quality") and of past feeding history on interpretations of present feeding behaviour (Bayne et al., 1987). The result is an extensive but empirical data base on feeding response to environmental change. In parallel with such studies have been attempts to model feeding behaviour, either as statistical fits to experimental data (eg Ross and Nisbet, 1990), or in terms of postulated physical mechanisms of feeding (Shimata and Jumars, 1991) or of optimality models based on assumptions of "trade off" amongst physiological traits (Willows, 1992). All such models have a role to play in an analysis of feeding behaviour, but those which are to be most effective will recognise the existence of both physical and physiological influences on feeding behaviour, set in the context of response to the considerable spatial and temporal variability within the natural food environment.

I will consider the results of some recent studies on feeding behaviour in relation to three scales of temporal variability, and one of spatial variability, in food availability: short-term changes typical of tidal effects and patchiness; "event-scale" changes typical of the initiation and decline of plankton blooms; seasonal changes, as between winter and summer in temperate environments; and differences between habitats in the composition of the seston. It is the extent to which the time constants of changes in feeding behaviour are themselves linked to the time scales of environmental change that are of most interest. When these processes are viewed in their totality, it is the coupling of physiological and morphological processes that can be seen to comprise and to control integrated feeding behaviour.

SHORT-TERM CHANGES IN FOOD QUANTITY AND QUALITY

These are the changes typical of tidal effects within estuaries and coastal seas, brought about by tidal resuspension and deposition of sediments, particulate detritus, benthic diatoms and the like, patchiness in plankton abundance within the water column, and the activities of populations of bivalves themselves, causing local depletion of food. Variability at this scale may be considerable, seston concentrations fluctuating over a range in excess of the seasonal cycle in mean concentration, with large changes also in particle quality, measured as organic content, or concentration of chlorophyll *a*.

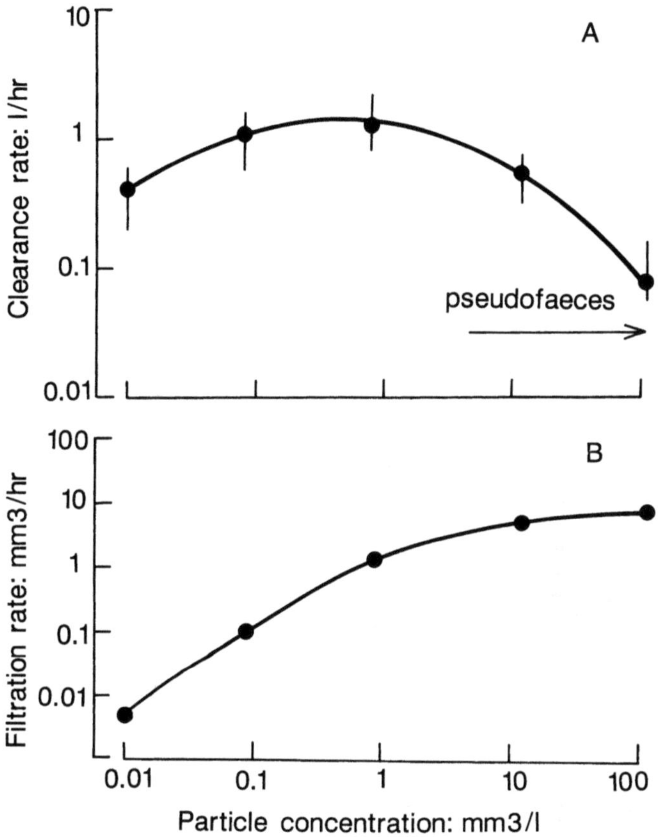

Fig. 1. A: Rates of clearance of *Phaeodactylum tricornutum* cells at different concentrations by *Mytilus edulis*; the concentration range at which pseudofaeces were produced is indicated by an arrow. Note logarithmic scales; error bars are 2 x standard errors. B: Rates of filtration calculated as clearance x particle concentration. Mussels were 39.1±7.6 mg dry flesh weight, at 9.5±0.8C. Given appropriate scaling for body size, 1 l h^{-1} clearance rate by these animals is equivalent to 8.8 l h^{-1} per gram dry flesh weight.

Figure 1A illustrates the relationship between clearance rate (the volume of water cleared of particles in unit time) and particle concentration; in Figure 1B the filtration rate (or grazing rate; the volume of particles captured per unit time) is derived as clearance rate x particle volume concentration to illustrate the commonly observed functional feeding response.

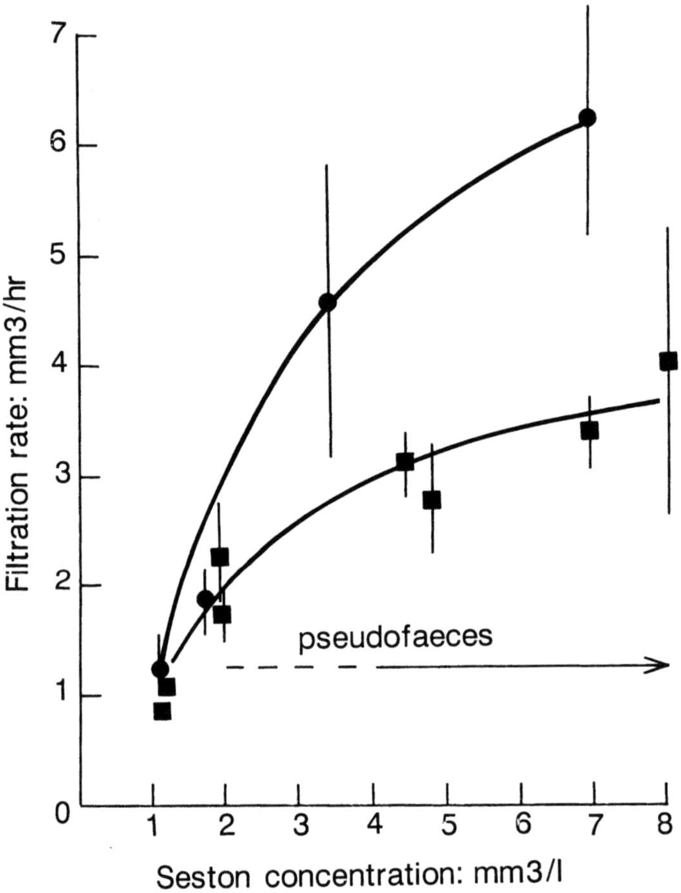

Fig. 2. Rates of filtration by *Mytilus edulis* feeding on different concentrations of seston made up as a mixture of algal cells, detritus and sediment particles. The concentration range at which pseudofaeces were produced is indicated by an arrow. Means±2 x standard errors. Circles: seston organic content was 0.41±0.04; Squares: 0.68±0.03 organic content. From Bayne et al. (1992).

The simple functional response of Fig. 1B is rendered more complex when the available particles (the "seston") are more representative of natural diets, in comprising mixtures of algal cells, detritus and particulate inorganic matter. In Figure 2 are shown results from experiments in which mussels were exposed over a period of two days to mixtures of silt particles and two algal species; the functional response is illustrated for two such mixtures representing organic contents of 68% and 41% of total mass. Filtration rates increased sharply on the diet of lower organic content, particularly at the higher seston concentrations.

In addition to these behavioural differences in response to what may be termed the 'bulk' properties of the seston (concentration, organic content), mussels and other bivalves have been shown to be capable of selective feeding on particles of different size and on different particles of similar size. Stenton-Dozey and Brown (1992) record short-term changes in the size fraction of available particulates filtered by *Venerupis corrugatus* which "...maximised filtration and retention efficiency of those particles that would yield the most food material per litre of water" (p. 182). Newell et al. (1989) observed for mussels, using flow cytometry, an ability to select flourescent from non-flourescent particles of similar size; this response to particle type showed a threshold below which "...the mussels lost their ability to selectively filter out phytoplankton from mixed particle assemblages" (p. 193).

It is clear that mussels and other bivalves (Shumway et al., 1985) are able to select amongst particle types trapped by the gills and, in addition, vary their rates of feeding in response to composition of the seston.

Under natural environmental conditions mussels living within estuarine and shallow coastal habitats are frequently exposed to levels of seston above the threshold at which some material filtered from suspension is rejected from the mantle cavity (as pseudofaeces). The proportion of filtered material rejected in this way varies both with the quantity and the organic content of the seston (Figure 3A). In the process of rejection, a degree of sorting is effected by which material relatively enriched in organic content is selected for ingestion (Kiørboe and Møhlenberg, 1981). The efficiency of this selection varies with the organic content of the suspended particulates, as illustrated in Figure 3B.

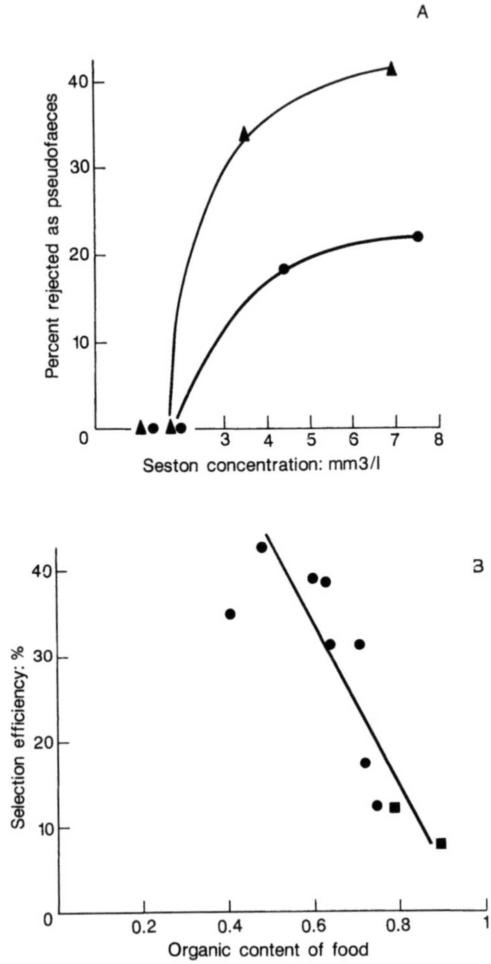

Fig. 3. A. The percentage of filtered material rejected as pseudofaeces by *Mytilus edulis* as a function of seston concentration. Note the "pseudofaeces threshold" at ca. 2 mm^3 seston l^{-1}. Triangles: seston organic content was 0.41±0.04; Circles: 0.68±0.03 organic content. B. Selection efficiency related to the organic content of the seston. The different symbols refer to different experiments under otherwise standard conditions. Based on Bayne et al. (1992).

The net result of these physiological/behavioural responses to short-term changes in the diet is to maintain the ingestion rate of organic matter in a simple relationship with the concentration of seston (Figure 4). This represents compensation for differences in diet quality, evidenced as no significant difference between ingestion rates (for total organics) at similar concentrations of particles which differ in their organic matter by mass. Bayne et al. (1992) estimated a 35% increase in the ingestion rate for particulate organic matter over values which would be expected if no compensation for reduced diet quality occurred. At higher seston concentrations, Prins et al. (1991) observed that *M. edulis* may double the ingestion of chlorophyll *a* over values to be expected in the absence of any selection.

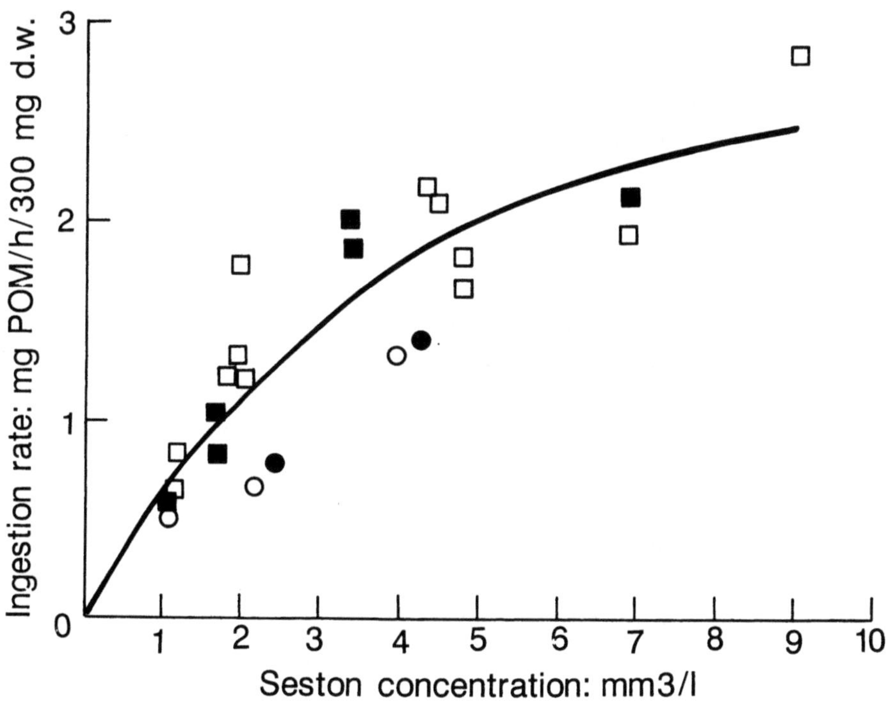

Fig. 4. Ingestion rate by *Mytilus edulis* feeding on different types of seston, at different concentrations. Open circles and squares, seston ca. 68% organic matter by mass: Closed circles and squares, seston 41% organic. Circles, 1986 experiments: Squares, 1987 experiments. Based on Bayne *et al.* (1987; 1992).

To quote Navarro et al. (1992, p. 84): "...regulating ingestion rate by means of pseudofaeces production therefore appears beneficial, helping to keep absorption rate nearly proportional to the rate of POM filtration".

We do not know what mechanisms effect these changes, though they presumably reside in ciliary processes on the ctenidia and labial palps (Shimata and Jumars, 1991). Some recent evidence indicates a role for chemical cues in facilitating selection between particles of similar size (Ward and Targett, 1989). There is empirical evidence to suggest that the rate of rejection of filtered particles as pseudofaeces may depend largely on the volume of material filtered (Widdows *et al* 1979); a limit to the rate of ingestion is set by the volume capacity of the digestive system. Even less is known of the processes of particle sorting by the labial palps (Newell and Jordan, 1983). Direct observation by endoscopy (Ward et al., 1991) probably represents the best current means available to determine how filtered particles are treated within the mantle cavity for differential filtration, rejection or ingestion.

"EVENT SCALE" COMPENSATIONS

Variability at this scale, over days to a few weeks, includes the growth and decline of algal blooms, and changes in the food environment due to the spring/neap tidal cycle.

In a series of experiments we have held mussels on similar diets for up to two weeks and measured absorption efficiency (AE) after 2 and 12-14 days. Some results are shown in Figure 5; in all cases there has been a significant increase in AE over the period of the experiment. More recently we have shown that this increase in AE over time occured not just for total organics in the diet, but also for the individual biochemical components, carbohydrates, proteins and lipids (Bayne et al., 1992).

The net result of these changes in absorption efficiency is an increase in the rate of absorption of organic matter (POM) over two weeks, when mussels are held on equivalent diets

(Figure 6). Evidently there has been a change in digestive physiology which increases the absorption of material from ingested food. Certainly, the null hypothesis of no change in rate of absorption of organics over time has not been supported in these experiments.

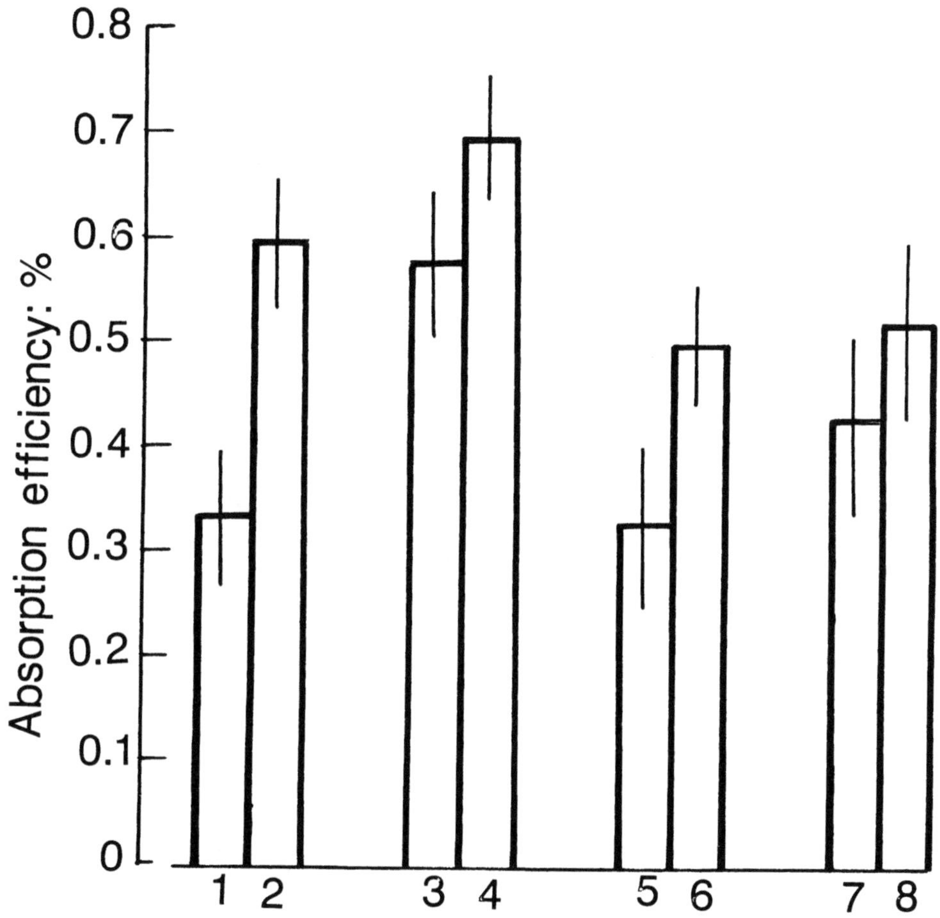

Fig. 5. Absorption efficiency (AE) by *Mytilus edulis* feeding on different types of seston, to show increased AE over time. Values are means±2 x standard errors. 1, seston 0.34 mg POM mm^{-3} at 2 days: 2, as for 1 but at 14 days: 3, 0.43 mg POM mm^{-3} at 2 days: 4, as for 3 but at 14 days: 5, 0.70 mg POM mm^{-3} at 2 days: 6, as for 5 but at 12 days: 7, 0.72 mg POM mm^{-3} at 2 days: 8, as for 7 but at 12 days. Data from Bayne et al. (1987, 1992).

Here also, as with changes evidenced over shorter time scales, we remain uncertain as to the mechanisms involved. There are three possibilities for supporting an increase in absorption efficiency at constant ingestion rate; changes in the volume of material processed within the gut (ie. in the volume of ingested material relative to the total gut volume), differential gut passage times for different types of particles (cp. constant gut passage times for the bulk diet), and changes in the production of digestive enzymes. Although these are analysed seperately, they will be functionally integrated within the whole organism.

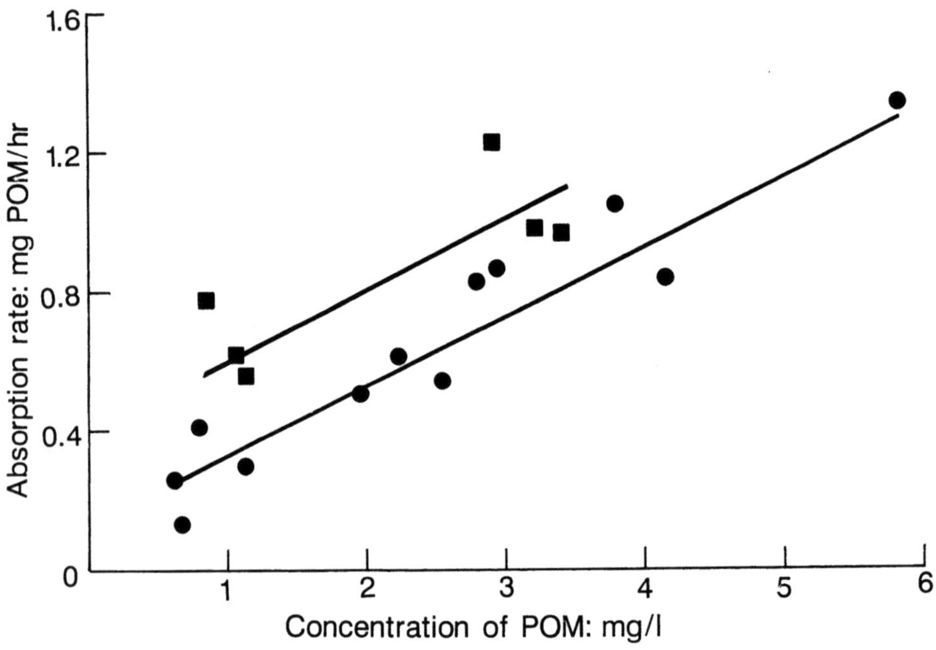

Fig. 6. Rates of absorption of organic matter by *Mytilus edulis* related to the concentration of organic matter (POM) in suspension. Circles, at 2 days: Squares, at 12 days. From Bayne et al. (1992).

Bayne et al. (1987) and Navarro et al. (1992) have suggested that the proportion of the gut volume which is occupied by ingesta may vary, facilitating increases in absorption efficiency with little change in the mean gut passage time. There may be selective mechanisms at work within the stomach to effect organic enrichment in the particle stream entering the digestive tubules. Equally, organic particles that are ingested may be reduced in volume more rapidly following acclimation, facilitating an increase in the amount of material capable of being processed within the gut. Shumway et al. (1985) have demonstrated "post-ingestive selection" of particles within the bivalve gut, whereby a cryptomonad flagellate was preferentially absorbed compared with a dinoflagellate and a diatom of similar size. We have observed (Hawkins and Bayne, unpublished data) differential gut passage times for particles of silt, a diatom and a naked flagellate, which point to a capacity for selection amongst particle types within the gut, effecting a higher rate of absorption for some particles over others of similar size.

There is more evidence for bivalves varying the production of digestive enzymes in response to changes in the diet. Seiderer and Newell (1979) observed adjustments of the activity of *a*-amylase from the crystalline style of *Choromytilus meridionalis* in response to changes in temperature, which coincided with changes in the proportion of phytoplankton cells and detritus in the diet. Brock (1989) recorded changes in the kinetics of cellulase in *Crassostrea gigas* coincident with changes in the food, and Hawkins and Bayne (1992) have reported changes in *a*-amylase and laminarinase activities in *Mytilus edulis* held on high and low levels of ration.

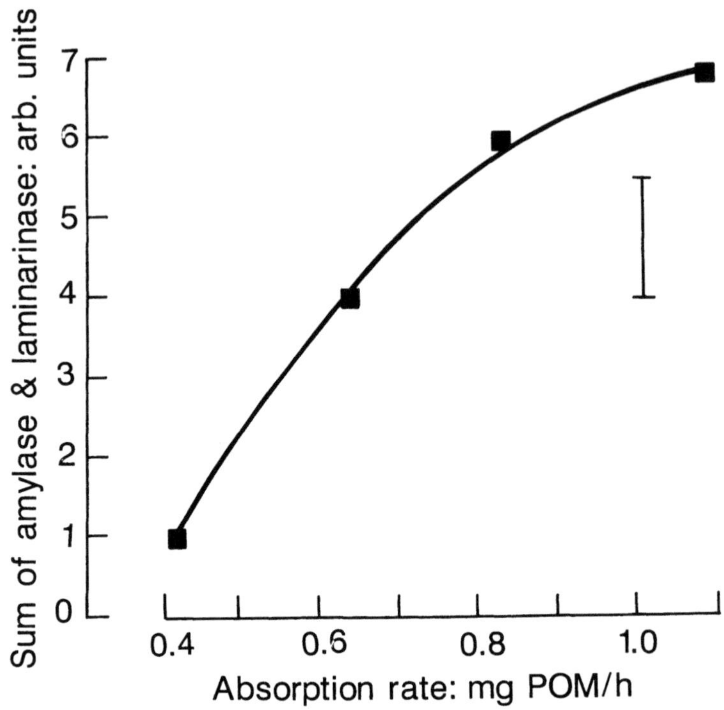

Fig. 7. Protein-specific activities of digestive enzymes (sum of *a*-amylase and laminarinase, arbitrary units) related to rates of absorption of organic matter, in *Mytilus edulis* held in the laboratory on a mixed seston diet for 15 days. The vertical bar represents an estimate of the Least Significant Difference ($P<0.05$) amongst the measures of activity. Courtesy of J-F Samain, unpublished data.

Jean-Francois Samain (pers. comm; see Figure 7) has observed a positive correlation between the sum of *a*-amylase and laminarinase activity and rate of absorption of organics in *M. edulis* fed a mixed diet of silt and algal cells, and found that the mussels required up to 15 days to show such correlated changes in the protein-specific activity of digestive enzymes.

Whatever the mechanism, the net result is to enhance the rate of energy acquisition by the mussel, and its scope for growth, over a period of a few days to two weeks, in response to changes in both the amount and the quality of available food. Some of these changes may be feasible on shorter time scales (eg. sorting of particles within the gut) but others, such as the induction and synthesis of new digestive enzymes, presumably require longer periods on the order of days.

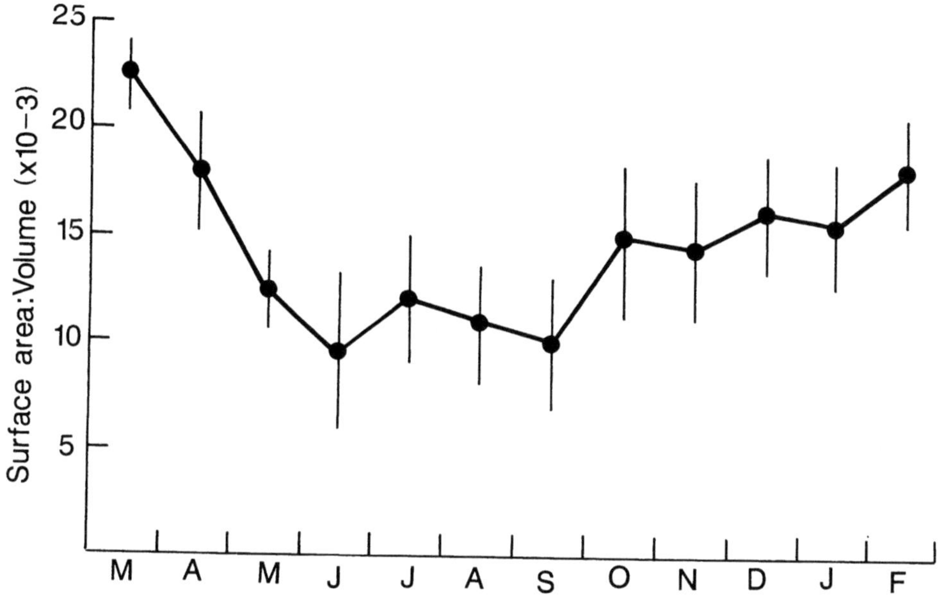

Fig. 8. The surface area:volume ratios for the epithelium of the digestive tubules of *Mytilus edulis* in different months of the year. Values are means±95% confidence intervals. Courtesy of D Lowe, unpublished data.

SEASONAL CHANGES

Seasonal changes in the food environment in north temperate habitats comprise the winter/summer cycle of changes in both total concentration of suspended particles and in relative organic content of the seston (eg Widdows et al., 1979). Such environments are to be compared, for example, with the Cape peninsular of South Africa, where tidal and "event scale" variability in food resources for suspension feeders may predominate over seasonal changes (Fielding and Davis, 1989).

Figure 8 shows the results of a seasonal survey of the morphology of the digestive tubules of *M. edulis* from a population at Whitsand, SW England (unpublished data, courtesy of David Lowe, Plymouth Marine Laboratory). Specifically, within transversely sectioned tubules, the area and the volume of the digestive epithelium were measured on eight microscope fields for each of a random sample of five mussels, using computer image analysis as described by Lowe and Clarke (1989). In the winter and early spring, the volume of the digestive cells is reduced, and surface area:volume ratio increased, relative to the summer condition of much increased cell volume, and lower surface area:volume ratio. These two conditions are similar to the "holding state" and "digestive or absorptive state", respectively, of the digestive diverticula of bivalves, as reviewed by Morton (1983), and constitute a seasonal equivalent to the commonly reported tidal or diurnal rhythmicity. As Morton (loc. cit., p. 116) observes, "...the assimilative capacity of a bivalve (which is reflected in the morphology of the digestive tubules) acts as an endogenous rhythm influencing filtration rate in a given concentration of food".

These observations are consistent with physiological observations by Hawkins and Bayne (1984). Mussels were held under similar conditions of temperature and food in the laboratory in March and June, but showed significant differences in aspects of feeding behaviour, including a higher gut content, but lower absorption efficiency, in March:

	March	June
% Organic gut content	4.1 - 8.5	2.4 - 4.4
Absorption Efficiency	0.27 - 0.38	0.52 - 0.57

In addition, mussels in March showed reduced rates of ingestion and extended gut passage times, when offered an organic rich diet. "Time averaged" (Doyle, 1979) responses to natural seasonal changes in food availability (low in March, high in June) were indicated.

A full interpretation of seasonal data of this type requires an analysis also of other seasonally variable features of the physiology of the individual, including growth, reproductive condition, the cycle of stored nutrients (Gabbott, 1983) and the consequences of these for the energy or nutrient demands of the animal. For the Whitsand population of mussels indicated here, the winter is a time of reduced availability of food, coupled with the utilisation of stored energy reserves to support gametogenesis. In the summer, the metabolic demand for food is high, in order to support growth at a time of reduced storage reserves.

In such circumstances a seasonal shift in feeding and digestion potential is consistent with the overall adaptive strategy, setting reduced 'investment' costs in digestive tissue and digestive enzymes, for example, against the consequent relative inability to respond in the short term to any changes in the normal seasonal cycle of food availability. Hence the interpretation of a time-averaged feeding behaviour which responds, in certain traits at least, to the longer wave length of seasonal variation rather than the shorter wave lengths of daily or tidal changes in food availability. Where the seasonal signal is less significant, such as the Cape peninsular, South Africa, time averaging in feeding behaviour may be less evident (Stenton-Dozey and Brown, 1992).

DIFFERENCES BETWEEN HABITATS

In view of the plasticity in feeding behaviour indicated above, it is hardly surprising that many studies have demonstrated differences in feeding rates and other traits between mussels from different locations (Bayne and Widdows, 1978). Newell et al. (1989), for example, compared clearance rates by mussels from three sites ("lease areas") and found significant differences (x1.5 between lowest and highest) due to enhanced clearance of flourescent particles (=phytoplankton) at their "Webb Cove" site. Indeed, higher clearance rates on organic as compared to inorganic particles may be an important factor explaining other recorded differences in feeding rates between mussels in different populations. In other cases differences between sites are apparent even when the food environment is standardised within laboratory experiments. In Table 1 some unpublished data (courtesy of John Widdows, Plymouth Marine Laboratory) illustrate the wide range in clearance rates that may be observed. In spite of suggestions to the contrary (Jørgensen, 1991), such differences are real and not the result of methodological artefacts, and they indicate a degree of persistence in extrinsically forced changes in feeding behaviour.

Table 1. Clearance rates for mussels, *Mytilus edulis*, from different populations within the UK sampled within a few days of each other and measured in the laboratory, under standardised conditions of temperature (15C), salinity (33ppt), food concentration (8×10^3 cells *Phaeodactylum tricornutum* ml^{-1}), body size (0.5 g dry flesh) and season (summer); sample size was 16 animals per location. Courtesy of J Widdows, Plymouth Marine Laboratory.

Location	**Clearance rate: litres h^{-1}±s.e.**
Orkney	4.02±0.19
Shetland	3.60±0.15
Ythan	2.44±0.14
Holy Island	3.23±0.18
Hunstanton	1.65±0.15
Walberswick	3.15±0.14
Lynher	1.37±0.15

Other, morphological, differences have been recorded which correlate potential feeding traits with the concentrations of seston (or suspended particulate matter, SPM) typical for different habitats (Theissen, 1977, 1982). Some differences, such as gill surface area, may be quantitatively small though statistically significant (Jones et al., 1992). However, Essink et al. (1989) recorded significant differences in surface areas of gills and labial palps, relative to shell area, for mussels from the Waddensea and North Sea which related to ambient concentrations of SPM and for which a phenotypic plasticity was reported following reciprocal transplantations for 3-4 months. Such morphological differences are known to affect feeding behaviour and sorting efficiency (Kiørboe and Møhlenberg, 1981). Amongst different sites within the Waddensea, however, no consistent differences in relative gill and palp sizes were evident. Essink et al. (1989) conclude that this "...is probably caused by the large variance of SPM concentrations at any Waddensea location" (p. 48); the mean concentration is of less consequence than the variance in such highly variable habitats.

MODELS OF FEEDING BEHAVIOUR

The model of feeding behaviour proposed by Jørgensen *et al*(1986) considers that filtration rates are determined solely by the physical properties of the "bivalve pump" and by the effects of animal disturbance influencing valve gape and consequently the frictional resistance to water flows. This model fails to explain many recorded features of feeding behaviour, such as temperature acclimation of filtration rate (Widdows, 1976), changes in filtration rate in response to increased seston concentrations (Winter, 1978), seasonal variability in temperature-specific ventilation rates (Jørgensen et al., 1990), and many of the features of variable physiological response discussed in this paper. Whatever the true physical constraints on the ciliary filtration process may be (Shimeta and Jumars, 1991), a model is needed which also incorporates elements of physiological regulation.

Such a model has been proposed by Willows (1992), based on principles of optimality. This model reproduces many of the features of feeding in bivalves which have been described, and it offers insights into likely responses to changes in both the quantity and the quality of the food. The model formalises the metabolic investment that the individual makes to effect digestion and absorption of food, and makes the important prediction that when particle abundance is low, it is the flexibility in this digestive investment that is important in energy maximisation. Above critical particle concentrations, and depending only on the abundance of the particles of high food quality, the mechanisms of particle selection become significant, and at the highest concentrations, when all poor-quality particles are rejected, flexibility comprises the efficient use of effective gut capacity "...brought about through an increased rate of digestive investment as greater volumes of (high quality) particles are collected" (p. 846). At intermediate particle concentrations, where selection is operating but not fully effective, flexibility in net energy gain is achieved by increasing filtration rate whilst simultaneously excluding poorer components of the available diet. In this analysis, then, the costs of increasing metabolic investment in digestion are incurred at high and low particle concentrations; the costs of filtration, which can be a significant proportion of ingested energy at low particle concentrations, are incurred at medium and at high particle abundance. Although these are features of feeding behaviour that have not yet been tested rigorously by experiment, they are consistent with recent empirical data on *Cerastoderma edule* and *Mytilus edulis* by Iglesias et al. (1992) and Bayne et al. (1992), respectively. For example, Bayne et al. (loc. cit.) record a greater acclimation increase in absorption efficiency for proteins and lipids at low than at high seston concentrations, indicative of a greater metabolic investment in digestion.

DISCUSSION

Mussels respond in different ways to changes in the food environment (food quantity and quality), dependent at least in part on the time scale of the environmental variability (Table 2). Long-term, habitat-specific differences in seston concentration are reflected in morphological features such as relative sizes of the main organs of particle capture (the gills) and sorting for rejection or ingestion (the palps). Other morphological differences, within the digestive gland, are expressed on seasonal time scales. Responses to "event scale" variation in food, over periods of days to a few weeks, include not only changes in feeding rate, but also processes within the digestive system, such as the induction and synthesis of digestive enzymes. Navarro and Iglesias (this volume) have shown how changes in digestion, gut passage times etc will have natural consequences for changes in feeding behaviour itself; the component traits are integrated as a means of effecting physiological compensations for changes in diet. And on shorter time scales other mechanisms are expressed to maximise net energy gain, including the production of pseudofaeces, coupled with selection processes both within the gills and on the labial palps. Of course, all of these processes will function within the physical constraints of the feeding apparatus (ciliary action, the secretion of mucous) and the systems for digestion and absorption.

When considering such time-dependent features of feeding behaviour, two different properties should be recognised; the time required for a physiological or morphological trait to change, and the persistence of this response in the context of further environmental change. We know very little of this latter aspect of bivalve feeding.

Table 2. Time scales of response by physiological and morphological traits in mussels to changes in the ambient food environment (see text).

Physiological/Morphological Trait	**Time-scale of response**
Clearance rate	Minutes
Pseudofaeces production (inc. selection for organics)	Minutes to Hours
Digestive "investment" (eg. enzymes)	Days
Morphology of digestive tubules	Weeks
Relative sizes of gills and labial palps	Months

This view of a series of physiological traits which are linked to environmental changes on different time scales, and which affect feeding behaviour via the feeding history of the individual, is essentially similar to current interpretations of feeding in marine copepods (Donaghay, 1988; Huntley, 1988). The copepod *Calanus pacificus*, for example, "...adapts behaviourally, biochemically and energetically to food-limiting conditions; however, the direction and extent of these adaptations vary on different temporal scales" (Landry and Hassett, 1985, p. 209). There is a contrast in this species between a seasonal response which minimises rates of energy utilisation, and responses at shorter time scales which maximise the rate of energy gain. In both copepods and mussels there is an adaptive fitting of various traits to changes in food availability, including time-averaged responses which may represent a trade-off between long term energy balance considerations and the potentiality to respond to short-term fluctuations in resources. Huntley (1988) has suggested that a primacy should be afforded to feeding history as affecting observed feeding behaviour, in order to become predictive of copepod feeding rates. The same can be argued for bivalves.

Models of feeding behaviour based on the principles of optimality amongst traits, and of energy maximisation, are proving effective in integrating our information and setting hypotheses for experimental testing. The model of Willows (1992) explicitly includes a consideration of the relationships between particle type, ingestion rate and absorption efficiency, but assumes that the organism acclimates instantaneously to any change in feeding conditions. Such models should be enhanced by incorporating the effects of previous feeding history on the functional/digestion response to food availability, and potential trade-offs which are expressed as temporal differences in the capacity of the digestive system to process food. Equally, such models pose a challenge to better experimental evaluation of the properties and processes of "digestive investment" that contribute to flexibility in the feeding response.

A clearer understanding of the physiological elements of feeding by bivalves is essential also for ecological models of the impacts of suspension feeding on, for example, particle flux and nutrient dynamics. The challenge is to represent integrated relationships (eg. Figure 4) whilst recognising explicitly the constraints on the mechanisms upon which such integrations are based.

ACKNOWLEDGEMENTS

I am grateful to Drs J-F Samain, D Lowe and J Widdows for permission to use some of their unpublished results, and to Drs A Hawkins, E Navarro, I P Iglesias, R Willows and R Harris for many stimulating discussions on these topics.

REFERENCES

Bayne BL Newell RC (1983) Physiological energetics of marine molluscs. In Wilbur KM Saleuddin AS (eds) The Mollusca. Vol 4 Academic Press New York pp 407-515

Bayne BL Widdows J (1978) The physiological ecology of two populations of *Mytlus edulis* L. Oecologia 37:137-162

Bayne BL Hawkins AJS Navarro E (1987) Feeding and digestion by the common mussel *Mytilus edulis* L. (Bivalvia: Mollusca) in mixtures of silt and algal cells at low concentrations. J Exp Mar Biol Ecol 111:1-22

Bayne BL Hawkins AJS Navarro E Iglesias JIP (1989) The effects of seston concentration on feeding, digestion and growth of the mussel, *Mytilus edulis* L. Mar Ecol Prog Ser 55:47-54

Bayne BL Iglesias JIP Hawkins AJS Navarro E Heral M Deslous-Pauli JM (1992) Feeding behaviour of the mussel *Mytilus edulis* L: responses to variations in both quantity and organic content of seston. J Mar Biol Ass UK In Press

Brock V (1989) *Crassostrea gigas* (Thunberg) hepatopancreas: cellulase kinetics and cellulolysis of living monocellular algae with cellulose walls. J Exp Mar Biol Ecol 128:157-164

Donaghay PL (1988) Role of temporal scales of acclimation, food quality and trophic dominance in controlling the evolution of copepod feeding behaviour. Bull Mar Sci 43:469-485

Doyle RW (1979) Ingestion rate of a selective deposit feeder in a complex mixture of particles: testing the energy-optimisation hypothesis. Limnol Oceanog 24:867-874

Essink K Tydeman P de Koning F Kleef HL (1989) On the adaptation of the mussel *Mytilus edulis* L to different environmental suspended matter concentrations. In Klekowski RZ et al (eds) Proc 21st EMBS Gdansk Ossolineumn Warsaw pp 41-51

Fielding PJ Davis CL (1989) Carbon and nitrogen resources available to kelp bed filter feeders in an upwelling environment. Mar Ecol Prog Ser 55:181-189

Gabbott PA (1983) Development and seasonal metabolic activities in marine molluscs. In Hochachka PW (ed) The Mollusca. Vol 2 Academic Press New York pp 165-217

Griffiths CL Griffiths RJ (1987) Bivalvia. In Pandian TJ Vernberg FJ (eds) Animal Energetics. Vol 2 Academic Press New York pp 1-88

Hawkins AJS Bayne BL (1984) Seasonal variation in the balance between physiological mechanisms of feeding and digestion in *Mytilus edulis* (Bivalvia: Mollusca) Mar Biol 82:233-240

Hawkins AJS Bayne BL (1992) Physiological interrelations, and the regulation of production. In Gosling E (ed) The mussel *Mytilus*: ecology, physiology, genetics and culture. Elsevier Amsterdam pp 171-222

Huntley M (1988) Feeding biology of *Calanus*: a new perspective. In Boxshall GA Schminke HK (eds) Biology of Copepods. Hydrobiologia 167/168:83-99

Iglesias JIP Navarro E Alvarez Jorna P Armentia I (1992) Feeding, particle selection and absorption in cockles *Cerastoderma edule* exposed to variable conditions of food concentration and quality. J Exp Mar Biol Ecol In Press

Jones HD Richards OG Southern TA (1992) Gill dimensions, water pumping rate and body size in the mussel *Mytilus edulis* L. J Exp Mar Biol Ecol 155:213-237

Jørgensen CB (1981) A hydrodynamic principle for particle retention in *Mytilus edulis* and other ciliary suspension feeders. Mar Biol 61:277-282

Jørgensen CB (1991) Bivalve filter feeding: hydrodynamics, bioenergetics, physiology and ecology. Olsen and Olsen Fredensborg

Jørgensen CB Larsen PS Riisgard HU (1990) Effects of temperature on the mussel pump. Mar Ecol Prog Ser 64:89-97

Kiørboe T Møhlenberg F (1981) Particle selection in suspension- feeding bivalves. Mar Ecol Prog Ser 5:291-296

Landry MR Hassett RP (1985) Time scales in behavioural, biochemical and energetic adaptations to food-limiting conditions by a marine copepod. Arch Hydrobiol Beih 21:209-221

Lowe DM, Clarke KR (1989) Contaminant-induced changes in the structure of the digestive epithelium of *Mytilus edulis.* Aq Toxicol 15:345-358

Morton B (1983) Feeding and digestion in Bivalvia. In Saleuddin ASM Wilbur KM (eds) The Mollusca. Vol 5 Academic Press New York pp 65-147

Navarro E Iglesias JIP Ortega M (1992) Natural sediment as a food source for the cockle *Cerastoderma edule* (L): effect of variable particle concentration on feeding, digestion and the scope for growth. J Exp Mar Biol Ecol 156:69-87

Newell CR Shumway SE Cucci TL Selvin R (1989) The effects of natural seston particle size and type on feeding rates, feeding selectivity and food resource availability for the mussel *Mytilus edulis* Linnaeus, 1758 at bottom culture sites in Maine. J Shellfish Res 8:187-196

Newell RIE Jordan SJ (1983) Preferential ingestion of organic material by the american oyster *Crassostrea virginica*. Mar Ecol Prog Ser 13:47-53

Prins TC Smaal AC Pouwer AJ (1991) Selective ingestion of phytoplankton by the bivalves *Mytilus edulis* L and *Cerastoderma edule* (L). Hydrobiol Bull 25:93-100

Ross AH Nisbet RM (1990) Dynamic models of growth and reproduction of the mussel *Mytilus edulis* L. Funct Ecol 4:777-787.

Seiderer LJ Newell RC (1979) Adjustments of the activity of a-amylase extracted from the style of the black mussel *Choromytilus meridionalis* (Krauss) in response to thermal acclimation. J Exp Mar Biol Ecol 39:79-86

Silvester NR Sleigh MA (1984) Hydrodynamic aspects of particle capture by *Mytilus*. J Mar Biol Ass UK 64:859-879

Shimeta J Jumars PA (1991) Physical mechanisms and rates of particle capture by suspension-feeders. Oceanog Mar Biol Ann Revs 29:191-259

Shumway SE Cucci TL Newell RC Yentsch CM (1985) Particle selection, ingestion and absorption in filter-feeding bivalves. J Exp Mar Biol Ecol 91:77-92

Stenton-Dozey JME Brown AC (1992) Clearance and retention efficiency of natural suspended particles by the rock-pool bivalve *Venerupis corrugatus* in relation to tidal availability. Mar Ecol Prog Ser 82:175-186

Theissen BF (1977) Filter feeding of *Mytilus edulis* L (Bivalvia) from different parts of Danish water of different turbidity. Ophelia 16:221-232

Theissen BF (1982) Variation in size of gills, labial palps and adductor muscle in *Mytilus edulis* L (Bivalvia) from Danish waters. Ophelia 21:49-63

Ward JE Targett NM (1989) Influence of marine microalgal metabolites on the feeding behaviour of the blue mussel *Mytilus edulis*. Mar Biol 101:313-321

Ward JE Beninger PG MacDonald BA Thompson RJ (1991) Direct observations of feeding structures and mechanisms in bivalve molluscs using endoscopic examination and video image analysis. Mar Biol 111:287-291

Widdows J (1976) Physiological adaptations of *Mytilus edulis* to cyclic temperatures. J Comp Physiol 195:115-128

Widdows J Fieth P Worrall CM (1979) Relations between seston, available food and feeding activity in the common mussel *Mytilus edulis*. Mar Biol 50:195-207

Willows RI (1992) Optimal digestive investment: a model for filter feeders experiencing variable diets. Limnol Oceanog 37:829-847

Winter JE (1978) A review of the knowledge of suspension feeding in lamellibranchiate bivalves, with special reference to artificial aquaculture systems. Aquaculture 13:1-33

INFAUNAL FILTER-FEEDING BIVALVES AND THE PHYSIOLOGICAL RESPONSE TO SHORT-TERM FLUCTUATIONS IN FOOD AVAILABILITY AND COMPOSITION.

Enrique Navarro and Juan I. P. Iglesias
Departamento de Biologla Animal y Genetica,
Universidad del Pals Vasco,
Aptado 644, 48080 Bilbao, Spain.

INTRODUCTION

An evaluation of nutritional determinants of growth in natural populations of bivalve molluscs involves two considerations:(a) The discontinuity of food environment, characterized by broad fluctuations in both the concentration and composition of the particulate complex that is accessible to filtration systems of animals.(b) The time-course of this discontinuity, ranging from seasonal to diurnal or tidal time scales.

In shallow estuaries, actual seston composition appears strongly determined by hydrodynamic factors (tide and wind-driven water movements) which cause resuspension of bottom material. Incorporation to the water column of such materials,(essentially inorganic but comprising also particles of food value: benthic microflora, bacteria, detritus, etc.; Grant et al 1986 1987, Wainright 1990) has a complex effect: While increasing the total concentration of particulate organic matter(POM) it normally produces a dramatic decrease in particle quality measured in terms of the organic or energy content of suspension (Widdows et al 1979; Berg and Newell 1986; Smaal et al 1986).

There have been many attempts to address the nutritional effects of such resuspension phenomenon, this problem being of great relevance as regards growth rates of species of bivalves.

NATO ASI Series, Vol. G 33
Bivalve Filter Feeders
in Estuarine and Coastal Ecosystem Processes
Edited by Richard F. Dame

Some experiments that used mixed diets of silt and phytoplankton to simulate natural seston concluded positive influence of silt (Winter 1976; Kiørboe et al 1980, 1981): Enhanced growth essentially resulted from high rates of pseudofeces production coupled to pre-ingestive selection for organics that acted to counterbalance the potential "diluting" effect of sediments of low organic content. This would allow taking advantage of extra organic matter brought in suspension with these sediments. Even at low seston concentration (below the threshold for pseudofeces production), the notion of organic dilution of phytoplankton with sedimentary particles is ambiguous as being based on the organic content by weight. Bayne et al (1987) have discussed the relevance of size-distribution of particles while contributing to precise the concept of food quality: When this is expressed as energy (or POM) per unit of particulate volume -volume appearing as a key feature given the nature of ingestive/digestive constraints-, silt particles may exhibit quality values compatible with moderate positive absorption efficiencies.

Above considerations are essential in determining the potential use of suspended sediments by bivalves as a source of assimilable organic matter, capable under certain circumstances of promoting growth in populations where phytoplankton may be limiting (Grant et al 1990; Frechette and Grant 1991). In this respect, differences in silt composition might partly explain reported variability concerning the effects of resuspension of bottom deposits on energetics and growth of species of bivalves (Wildish and Kristmanson 1984 1985; Grizzle and Lutz 1989; Grizzle and Morin 1989; Frechette and Grant 1991; Grant et al 1990; Grant and Thorpe 1991).

Additional attention should be accorded to inter or intraspecific (populational) differences in feeding behaviour (Theissen 1977; Bayne et al,]984 1987; Smaal et al 1986), that would introduce another source of variability acting to enlarge the range of physiological response to particular feeding conditions. Microhabitat characteristics of the food environment(e.g., gradients associated with vertical positioning of species) might provide an environmental correlate for such differences in physiology that, accordingly, would indicate the existence of patterns of response to prevailing conditions of food quantity and quality. It is in this context where contrasting feeding behaviors that emerge from some comparisons involving infaunal and epifaunal species of bivalves could be interpreted.

CONTRASTING STRATEGIES OF FEEDING AND DIGESTION

In a comparative study, Hawkins et al (1990) demonstrated great differences in feeding and digestive parameters between epifaunal (*Mytilus edulis*) and infaunal (*Cerastoderma edule*) species. As an example, cockles had much larger gut contents (per unit dry tissue weight) which enabled the maintenance of higher ingestion rates coupled with longer retention times of food particles inside the gut. Consequences of these differences are clearly illustrated by considering the effects of food quality on the absorption efficiency (AE): In both species, high values of AE measured with organically rich suspensions tend to exponentially decrease with decreasing quality of food; however, a lower limit for positive net AE has been estimated to correspond with qualities around 0.09 mg POM mm^{-3} particulate matter in cockles (Navarro et al 1993), while in species of *Mytilus* AE would take negative values below 0.15 to 0.25 mg POM mm^{-3} (Bayne et al 1987, 1988; Navarro et al 1991). Above characteristics would enable cockles to maximize absorption from organically poor particles: In fact, positive growth (measured as SFG) has been reported for *C. edule* fed resuspended sediment of low food value (~ 12 % organic content) (Navarro et al 1992).

Full appreciation of implications of the above differences in the potential for adaptation to changing dietary characteristics requires the temporal variable to be considered. Absorption rates of sedimentary organics were measured as a function of POM concentration in cockles (*C. edule*) exposed for 2 or 10 days to diets of low organic content (Table I). Net rates of absorption were positive in all the experimental range of particle concentration, confirming the ability of the feeding-digestive system of cockles for actively obtaining energy at such reduced qualities; nevertheless, acclimation did not improve absorptive capabilities, as indicated by lack of significance for differences between lines for 2 and 10 days, either in slope ($t = 0.55$; $p > 0.50$) or elevation ($t = 0.23$; $p>0.50$). This finding contrasts with results from a similar experiment performed on mussels (*M. edulis*) where net rates of absorption from different diets increased significantly between 2 and 12 days, this increase being essentially based on a consistent rise in AE (Bayne et al 1993). Indeed, the most relevant compensatory phenomenon documented in mussels held on organically poor diets consisted in a rather dramatic change in AE over time (10 days), leading from negative AE to values as high as 50-60 % (Bayne et al 1987 1988).

Table I. Parameters of lineal regression equations relating - on logarithmic scales- the absorption rate of sedimentary organics (A: J h^{-1}) to the concentration of particulate organic matter (mg POM l^{-1}), in cockles (*C. edule*) exposed for 2 and 10 days to variable loads of suspended sediment

	a ± 95% CI	b ± 95% CI	r^2	p
2d	0.771 ± 0.056	0.675 ± 0.278	0.824	< 0.001
10d	0.708 ± 0.167	0.942 ± 0.841	0.708	0.03

This brief comparative analysis suggests that the main differences between both species of filter feeders refer to time-course of the response, rather than to features of feeding behavior, and this might well correlate with microhabitat differences. Infaunal specimens are most likely exposed to larger and more frequent fluctuations in particle concentration and associated changes in food quality, due to resuspension pulses which would reach the position of epifaunal bivalves comparatively attenuated. Now, for functional value to be accorded to observed adjustments in components of feeding and digestion, some coupling should exist between time-course of physiological response and actual patterns of fluctuation in food parameters.

M. edulis is a frequent non-siphonate inhabitant of phytoplankton dominated waters. Compensatory digestive adjustments for changes in food quality have been demonstrated in this species but such responses appear to require days (or weeks) and are only appropriate on a seasonal time scale, consistent with the observation that energy optimization by mussels is only conceivable in a time-averaged sense (Hawkins et al 1985. Conversely, feeding behaviour of the infaunal *C. edule* appears best suited to deal with highly concentrated suspensions of low quality, characteristic from the water-sediment interface. This would include the possibility of short-term adjustments of components of feeding and digestion to changes both in the availability and composition of food. Physiological mechanisms underlying these responses were analyzed in cockles fed experimental diets where nutritionally-relevant factors of food quantity and quality were independently varied. Feeding and digestive parameters -including rates and efficiencies- were determined and relationship between them analyzed by means of empirically based models which have been reported elsewhere (Iglesias et al 1992; Navarro et al 1993). For the present aim, these models have been developed to explore aspects of feeding behaviour of cockles in a range of values of food availability and composition representative of natural suspensions in the near-bottom area.

Food processing, from handling of particles by the filtration apparatus to final absorption of digested products is integrated by two sequential steps: (a) The pre-ingestive processes, including particle clearing from the suspension by the ctenidia followed -when particle concentration exceeds a certain, threshold- by sorting prior to rejection of excess material as pseudofeces; and (b) the digestive processes, involving the participation of separate compartments for extra and intracellular digestion and absorption.

A relevant feature of feeding behaviour of cockles is represented by the existence of controls on feeding rate that are induced by characteristics of food environment. Pre-ingestive processes are primarily involved in such controls, but the final explanation for these particular feeding responses comes from the consideration of possibilities given and limits set by digestive processing of ingested food. Consequently, functional relationships inherent in digestion and absorption will be considered first, in order to provide a basis for the posterior analysis of feeding behavior.

DIGESTIVE PROCESSES

Factors limiting the energy uptake by the digestive system

Absorption of nutrients across the gut wall represents energy gain from feeding and digestive processes standing out as the term to be optimized. Such energetic consideration strictly applies to the net rate of absorption, defined as the result from a balance between ingestion and defecation of organics. Nevertheless, in a detailed description of the various components involved in digestion it is extremely useful the quantification of rates of absorption of exogenous (food derived) organics, also termed gross absorption. The difference between gross and net rates of absorption constitutes endogenous losses inherent to digestion, so called Metabolic fecal losses (MFL; Hawkins and Bayne 1985; Bayne and Hawkins, 1990); these represent digestive investments that are not reabsorbed. In the experimental practice with bivalves, both rates of absorption are obtained as the product of the ingestion rate and the corresponding absorption efficiencies. Net AE is usually determined by the Conover method (Conover 1966); gross AE can be obtained as the ratio of absorbed (ingested - defecated) to ingested ^{14}C following the ingestion of ^{14}C labeled algal food (Hawkins and Bayne 1984).

When analyzing the mechanisms by which absorption can be optimized, account should be taken of digestive trade-offs. In experiments with *C. edule*, Navarro et al (1993) made measurements of net absorption efficiency (Conover method) under different conditions of food concentration and quality, and developed an empirical model providing for the simultaneous dependence of AE on the ingestion rate and quality of ingested food. Curves shown in Fig. 1 have been derived from that model in a realistic range of ingestion rates and food qualities (0.25 to 0.45 mg POM mm^{-3}) representative of diets ingested by filter feeders from the infaunal environment.

Theoretical importance of these curves resides in the graphical illustration of one fundamental digestive trade-off: The negative dependence between net AE and ingestion rate. We have tried to substantiate such a dependence by quantitatively analyzing the consequences of digestive constraints responsible for some basic relationships between parameters of digestion and absorption. Two types of constraints can be considered: (1) Morphological constraints imposed by limited gut capacities and (2) Functional constraints represented by the indirect costs of digestion (MFL).

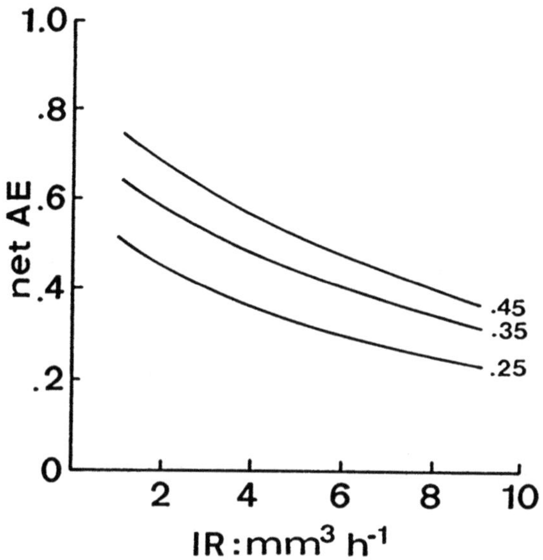

Fig. 1. Net absorption efficiency of cockles as a function of the ingestion rate (IR) at various qualities of ingested food (q= mg POM mm^{-3}). Curves derived from model of Navarro et al 1993)

$$AE = e^{-0.087IR}\left[1 - e^{-4.763(q-0.099)}\right]$$

Occurrence of volumetric digestive limits determines a negative relationship between the amount of food that may be processed per unit time (ingestion rate) and the time it takes the food to traverse the digestive tube (gut passage time = GPT). That relationship can be more precisely formulated by writing:

gut contents = ingestion rate x gut passage time,

gut contents representing the amount of food held within the gut at one time. From the assumption that gut is always full, gut contents constituting a constant, a simple proportional decrease in GPT can be predicted with an increase in the ingestion rate. Such a negative dependence has been empirically demonstrated by using independent estimations of both these feeding parameters (e.g. ^{14}C labeled food to determine GPT) in several species of bivalves; however, the form of the relationship corresponds to a negative exponential, indicating that decrease of GPT with IR increase occurs at a reducing rate (Bayne et al 1984, 1989; Navarro et al 1992).

Departure from a simple lineal model suggests the possibility of variable gut contents. In fact, gut contents of cockles fed different ration levels have been found to increase with ingestion rate (see Table II) according to a hyperbolic saturating function, which appears to involve variable degrees of packing of food particles. Possibility of adjusting the amount of material held by the gut is an important mechanism by which the effects of ingestion, imposing a faster passage of food across the digestive tract, may be attenuated.

Table II. Gut contents (mm^3) of 200 mg dry weight cockles, corresponding to different rates of ingestion and food qualities. Values were estimated from an hyperbolic model fitted to experimental data (from Navarro et al 1993).

Quality	Ingestion rate (mm^3 h^{-1})				
(mg POM mm^{-3})	2	4	6	8	10
0.25	4.07	6.31	7.73	8.71	9.42
0.35	5.11	7.49	8.87	9.77	10.41
0.45	5.95	8.36	9.66	10.48	11.04

Benefits of an extended GPT as regards the digestive yield are obvious since this parameter would represent the effective time for the enzymatic attack and absorption of digestion products. Fig. 2 shows the relationship between gross AE and GPT established for *C. edule* fed diets differing in concentration and composition. Potential utilization of food becomes saturated, in the sense that the rate of change in AE with GPT rapidly decreases with longer permanence of food inside the gut.

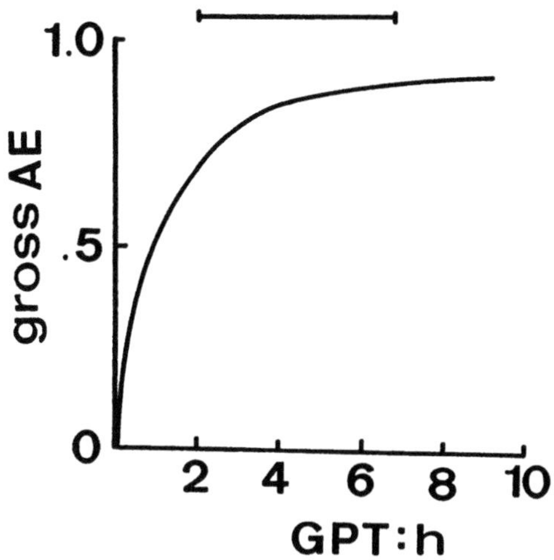

Fig. 2. Gross AE as a function of gut passage time(GPT). A range of GPT corresponding to extreme feeding conditions in the natural environment is indicated.

We can assume 1 and 10 mm^3 particulate matter per hour as broad limits to the ingestion rate of medium sized cockles (= 200 mg tissues dry weight), and qualities of ingested particles varying from 0.25 to 0.45 mg POM mm^{-3}, in order to estimate an interval for GPTs representative of extreme food conditions in the environment. That interval has been represented in Fig. 2 to illustrate the change in gross AE that would result from fluctuations within these extremes, and particularly to evaluate the effects of increasing ingestion up to the maximum value considered. As it can be appreciated, GPTs are mostly displayed in a sort of optimal interval characterized by high values of AE that are subject to minor variations; still, these GPT are not much extended and compatible with relatively high ingestion rates.

Relationships underlying that constancy in gross AE deserve special attention: In the lower range of ingestion rates GPT would be expected to rapidly rise, moving into the saturating region, where no benefits derive from a longer permanence of food inside the gut. On the other hand, with moderate to high values of ingestion GPT varies at much lower rates as partly caused by adjustments of gut contents, and so preventing GPT to be reduced to values (i.e., below 2h) which would cause a dramatic decrease in gross AE.

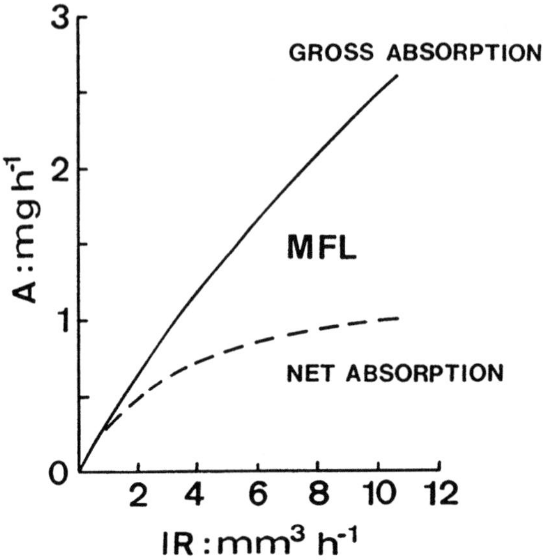

Fig. 3. Gross and net rates of absorption (A) as a function of the ingestion rate of mid-quality food (q = 0.35 mg POM mm^{-3}).

The above mechanisms provide for rates of gross absorption that are nearly proportional to ingestion rates (Fig. 3). Conversely, the curve for net absorption tends to level-off, suggesting saturation of the functional capacity of the digestive system as regards the net uptake of nutrients; in other words, indirect costs of digestion represented by metabolic fecal losses -the area between lines in Fig.3- exponentially rises with ingestion rate, imposing severe limits to the amount of food that can be processed per unit time. In consequence, adjustments of volumetric gut capacity in order to accommodate more food when ingestion rises achieves minimal restrictions on absorption efficiency coming from morphological constraints; rather, limits would arise from functional constraints represented by disproportionate increases of endogenous losses inherent in digestive processes. Consideration of these limits makes plain the energetic benefits implicit in a strict regulation of ingestion rate.

Evidences of digestive food selection

Natural assemblages of particles entering the gut are heterogeneous, creating an opportunity for digestive food selection. Very neat differences in food value among particles provide for a functional basis of that selection. Sorting devices at points where ducts leading to digestive diverticula open into the stomach have been reported at the anatomical level (reviewed by Morton 1983), but its role in digestive selection is only conjectural up to date. However, existence of such a phenomenon would help to understand the digestive adjustments underlying energy maximization processes found in cockles exposed to diets of increasing quality (Navarro et al 1993).

Bivalves, among other molluscs, are characterized by a biphasic digestion process, involving the stomach and digestive gland as compartments for extra and intracellular digestion, respectively (Purchon 1968). Accordingly, two components of food processing have been acknowledged: A fast" component includes the fraction of food subjected to extracellular digestion in the stomach followed by intestinal absorption and is voided as "intestinal faces", while a slow component comprises the food which undergo intracellular digestion and absorption within the digestive gland, giving rise to the so-called glandular faces van Weel 1961; Calow 1975; Widdows et al 1979; Bricelj and Malouf 1984). Both components to the defecation process can eventually be identified by using isotopically labeled food (Hawkins and Bayne 1984; Decho and Luoma 1991). It is broadly accepted that filling of the digestive gland proceeds up to satiation and increasing ingestion beyond this point results in excess material being egested as intestinal faces (Widdows et al 1979). Increased proportion of particles by-passing the digestive ducts (fast component) subjected so to weaker absorptive processes would specifically explain reductions in both GPT and AE associated with higher rates of food processing.

Recent evidence indicates that a re-elaboration of the above scheme may be opportune on account of digestive food selection. We have undertaken an indirect approach to this phenomenon by comparing two different estimates of gut passage time available from reported research on cockles. GPT can be indirectly calculated as the ratio of gut contents to ingestion rate; when food materials are given as packed volumes of particles and gut contents are corrected for volume changes upon digestion, resulting GPT would represent an average time for the whole ingested material to travel across the digestive tract (GPT_A). Quite differently, GPT determinations based on time-courses of defecation for ^{14}C following the ingestion of labeled algal food would in fact represent the retention time of the organic fraction of the ,diet (GPT_O). An analysis of the relationship between both these estimates (Fig. 4) lead to the following conclusions:

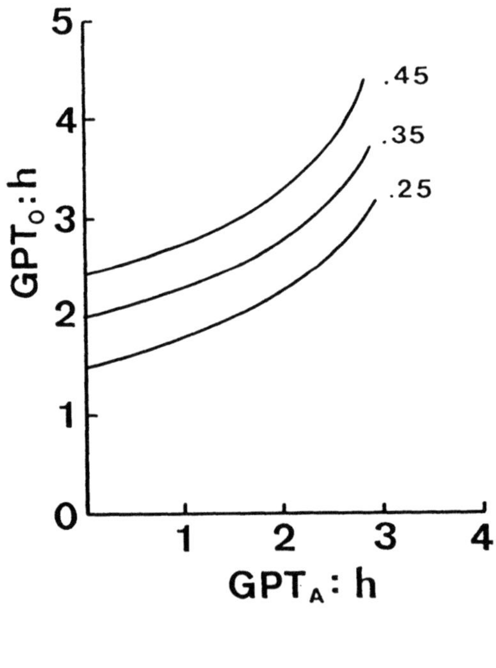

Fig. 4. GPT_O as a function of GTP_A, at three food qualities (mg POM mm^{-3})

(1) Longer GPT_O relative to GPT_A are observed, that would be explained by the existence of positive selection for organics. This effect appears to be dependent on rates of ingestion as the slope of lines rises with increasing retention times (i.e. IR tending to 0). On the other hand, intercepts of curves in Fig. 4 represent the asymptotic minima that GPTo tends to reach with increasing ingestion. (2) For a given GPT_A, the retention time for organics increases with increasing quality of food ingested (v.gr. intercepts are a linear function of quality). So, permanence of food within the gut appears to occur in direct proportion to the abundance of high quality particles.

This last result suggests that particle incorporation to the digestive gland might be subjected to active regulation, rather than being determined by filling to satiation. If such regulation exists, interestingly, the ability to vary the allocation of ingested matter between functionally separate parts of the gut might well be related to the selection process itself: Selection of particles with high food value would be based on some "recognition" at the ciliary areas conforming the different sorting devices that have been described within the stomach of bivalves (Morton 1983). Probability that nutritionally desirable particles interact with specific ciliary tracts conveying food to the digestive diverticula would increase with increasing abundance of these particles in the diet i.e., quality) and so, the amount of material taken up by the digestive gland could become positively dependent on food quality through this sort of stochastic effect. On the other hand, this hypothesis implies that the ratio between the slow and fast components would tend to decrease as food passes faster the sorting area, thus accounting for a negative relationship between GPT and ingestion rate.

How does digestive selection effect adjustments involved in maximization of rates of absorption

Benefits associated with a greater availability of organics per unit of ingested matter, brought about by improving food quality, are enhanced through digestive adjustments to finally result in maximization of absorptive rates. In the range of values here used for illustration, organic matter available for digestion increases x1.8 as the sole consequence of changing food quality from 0.25 to 0.45; whereas, energy return from digestive processes measured as net rates of absorption increases by a factor considerably higher, which in turn varies with the rate of ingestion (i.e., from 2.58 to 2.85 for IR ranging from 1 to 10 mm^3 h^{-1}).

Digestive selection appears to contribute to the above adjustments through varying the assignment of food to the compartment for intracellular digestion, with two important consequences: (1) Volumetric constraints are expected to be modulated by food quality as enlarged gut capacities would result from a greater incorporation to the digestive gland of high quality particles per unit ingestion. Such an effect, that has been measured in the form of increased gut contents (see Table II), would enable the maintenance of rates of ingestion with extended gut retention times and increased efficiencies of absorption. Resulting effects may be quantified in terms of gross rates of absorption and explain a proportion, variable according to IR, of the quality dependent increments in net AE. (2) Digestive investments in the form of enzymes and other endogenous products contributed to digestion are expected to increase, especially within the digestive gland, to match the quality-induced increase in gut contents. This could constitute a mechanism for maintaining constant the time-specific AE (i.e. per unit of GPT) with rising quality. As an independent confirmation of this assumption, activity of digestive enzymes - particularly carbohydrases- in the digestive gland of cockles have been found to increase in response to enhanced quality of the diet (Table III), while no such induction was apparent for carbohydrases of the crystalline style involved in extracellular digestion (Ibarrola, unpublished results). Since MFL represent the remaining of these digestive investments that are not reabsorbed, a positive dependence of these losses on food quality, as noticed in *C. edule* (Navarro et al 1993), is also implicit in the above statements.

Table III. Carbohydrase activities (A: amylase; C: cellulase; L: laminarinase) measured in the digestive gland of cockles (C. edule), under different food conditions: a and b, low quality (0.31 mg POM mm^{-3}); c and d, high quality (0.52 mg POM mm^{-3}); a and c, low ration (1.5 mm3 l^{-1}); b and d, high ration (3.0 mm3 l^{-1}). Activities are given as mg. maltose released per hour and per standard digestive gland of 200 mg dry weight cockles (from Ibarrola, unpublished data).

Diets	A	C	L
a	4.08 ± 0.69	1.28 ± 0.49	1.69 ± 0.59
b	3.79 ± 0.42	1.44 ± 0.33	1.52 ± 0.41
c	4.79 ± 0.72	2.26 ± 0.63	2.32 ± 0.67
d	5.21 ± 0.16	2.33 ± 0.81	1.71 ± 0.52

Differences between qualities (:i.e., a + b vs c + d) were found significant for amylase ($p = 0.02$) and cellulase ($p = 0.001$)

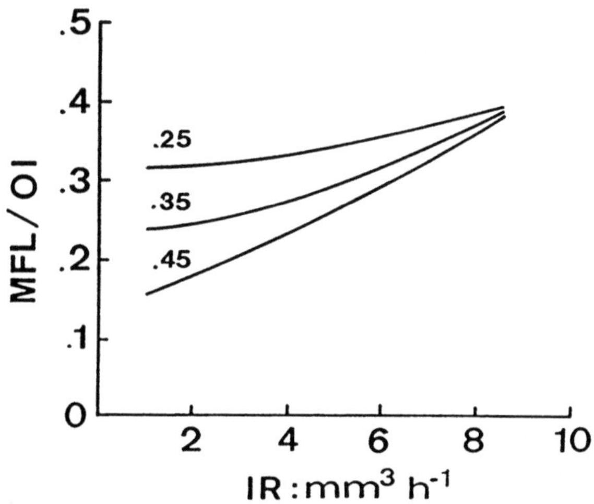

Fig. 5. Metabolic fecal losses (MFL) per unit of organic ingestion, as a function of the total ingestion rate, at different food qualities (mg POM mm^{-3}).

In spite of having increased the rate of digestive investments, MFL per unit organic ingestion decrease with quality (Fig. 5), meaning that costs of digestion tend to decline in relative terms. Since MFL /OI represents the gap between gross and net efficiencies of absorption, this behavior of MFL explains the remaining of the above referred quality-dependent increments in the energy absorbed per unit of ingested energy. As also shown in Fig. 5, these effects tend to be overcome at high rates of ingestion, perhaps reflecting a less efficient reabsorption of endogenous digestive products as greater proportions of food are egested without entering the digestive gland. In summary, at low rates of ingestion, increased quality of food result in an increment of AE that is essentially based on reducing endogenous losses per unit organic ingestion. As ingestion rises, however, maintenance of higher AE would turn to depend more on digestive adjustments leading to enlarged gut contents and associated longer permanence of food inside the digestive system. Quick enzyme induction (as reported for carbohydrases of the digestive gland) would constitute the particular expression of increasing digestive investments that may be required under these circumstances in order to guarantee "an efficient use of the effective gut capacity" (Willows, 1992). Short-term adjustments of this kind would enable cockles to optimize energy intake under a food regime that characteristically includes rapid fluctuations in food quality.

Briefly, knowledge of the! relationships inherent in functioning of the digestive system of cockles leads to two important conclusions: (a) Net rates of absorption are constrained by disproportionate increments of MFL with rising rates of food processing, setting the basis for a trade-off between ingestion rate and absorption efficiency. Optimization would then demand an efficient regulation of ingestion rate at high particle concentrations. (b) As food quality rises, the increment in organics supplied per unit of volume ingested combines with higher absorption efficiencies to considerably enhance absorption rates; consequently, any organic enrichment of food upon ingestion would be highly beneficial in terms of energy gain. We will consider next the extent to which the above requirements are fulfilled through adjustments in feeding behaviour.

BEHAVIORAL AND PHYSIOLOGICAL COMPONENTS OF FEEDING

Behavioral and physiological mechanisms involved in the regulation of ingestion rate have two components: Control on pumping or clearance rates and control on the level of pseudofeces production.

Early work by Winter (1978) called the attention on the ability exhibited by several species of bivalves to adjust the rate of pumping in response to a rise in the particulate food concentration. An exponential decrease above a certain threshold concentration is usually involved, resulting in rates of filtration that typically increase with rising concentration though at a decreasing rate, until a maximum and almost constant value is attained at high loads of particles. Further development of the concept of regulation comes from various authors (Foster-Smith 1975; Widdows et al 1979) who have considered a second behavioral component in the control of ingestion: The rejection of variable amounts of filtered material as pseudofeces before the remaining is processed by the digestive system. More or less implicit in these studies is the assumption that food is conveyed through the gut at a standard rate; resorting to rejection of pseudofeces would constitute then an overflow mechanism (see also Jørgensen 1990) which, above a certain concentration level, avoids saturation of the feeding system by excess material. On the other hand, Foster-Smith (1975) characterized the feeding behaviour of species of bivalves (particularly, mussels *M. edulis*, cockles *C. edule* and clams *Venerupis pullastra*) according to their preingestive responses to changing particle concentration: Mussels, for example, were observed to maintain approximately constant pumping rates, regulation being based on increasing production of pseudofeces; under the same circumstances, cockles and clams did not rely as much on pseudofeces production but controlled the rate of ingestion by adjusting their pumping rates.

To our present state of knowledge, these views appear somehow stereotyped as they disregard essential aspects of the plasticity these regulatory mechanisms may exhibit in response to variable food conditions. Particularly, recent research that included food composition as a variable has provided some insights into the following aspects: (a) Regulation leads to maximum values of ingestion, but these maxima may be set at different levels according to food quality. (b) The particular mechanism of regulation used by a filter-feeder (i.e., control on pumping rate or pseudofeces production) is largely dependent on food composition (quality), rather than being some sort of species-specific mechanism. In this respect, an understanding of the choice an organism accomplishes requires taking account of a second aspect of feeding behaviour: The possibility of preferentially rejecting particles of low or no food value based on selection abilities of the pallial organs. Although denied or ignored at first (Foster-Smith 1975; Widdows et al 1979), strong evidence in favor of this phenomenon has come from posterior studies of different species of bivalves (Kiørboe and Mohlenberg 1981; Newell and Jordan 1983; Bayne et al 1989; Prins et al 1991; Navarro et al 1992).

As already stated, recently obtained data for cockles (Iglesias et al 1992 and unpublished results) have been used to analyze relationships between different components of feeding at the preingestive level (filtration rates, pseudofeces production and food selection) under realistic feeding conditions so identified in the course of field experiments. Eventually, data relative to infaunal clams (*Tapes philippinarum*) obtained in our laboratory (Alvarez-Jorna, unpublished results) have been added to the discussion for the sake of a broader perspective. Functional value of feeding traits will finally be analyzed by reference to the requirements for optimal absorption rates.

Filtration rates

Quantitative relationships between filtration rates and parameters of food suspensions have been analyzed by various authors (Kiørboe et al 1980; Kiørboe and Møhlenberg 1981; Prins et al 1991). Experimental approaches used in these studies relied in adding increasing amounts of silt to a fixed algal concentration; this procedure results in simultaneous variation of both particle concentration and organic content among diets, precluding the independent consideration of these food descriptors. By using mixed diets of different quality dosed at different rates, Iglesias et al (1992) studied the influence of food quantity and quality on rates of filtration of cockles independently, to conclude that interactive effects between these two variables constituted a fundamental feature of feeding behaviour. These interactions can be summarized as follows: when cockles are exposed to increasing particle concentrations, filtration rates remain almost constant with high quality diets (= regulation) but become progressively dependent on concentration with decreasing food quality. Essentially the same behaviour has been observed for the clam *T. philippinarum* under similar conditions (Alvarez-Jorna, unpublished results). For these two infaunal species then, at low seston loads higher filtration rates appear associated with high food quality, while the opposite occurs as particle concentration rises

To quantify the above interactive effects, we have chosen power functions (on account of the best-fit criterion) to describe relationships between filtration rate and particle concentration (TPM), adding specific terms to account for quality. Data from two sources (Iglesias et al 1992 plus a recent set of unpublished measurements at much higher seston loads) on filtration rates of cockles were fitted with these models. Results for a naturally-occurring range of organic contents of food (f = 0.1, 0.3 and 0.5) are shown in Fig 6A.

Similar trends with seston quality has been reported for mussels, *M. edulis* (Bayne et al 1993), where dependence of filtration rate on seston concentration appears generally higher. As an example, at two organic contents of food (f = 0.41 and 0.68), estimated power b values were:

f	cockles	mussels
0.41	0.72	0.98
0.68	0.55	0.75

Consequently, once discarded the effects of quality, results of this comparison lend support to the idea of intraspecific differences in regulatory abilities attributed to these two species by Foster-Smith (1975).

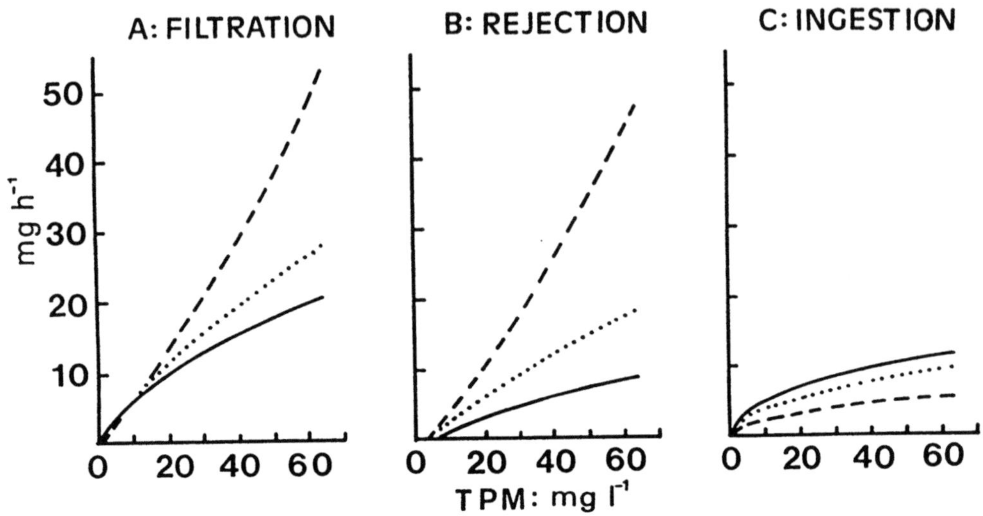

Fig. 6. (A) Relationships between filtration rate(FR: mg h-1) and seston concentration (mg TPM 1-1), at three organic contents of seston (f = 0.1: dashed line, 0.3: dotted line and 0.5: solid line). Lines were obtained from the following general expression,

$$FR = 0.38 + 0.42 \log TPM + 0.77 f - 0.75 \log TPM \times f$$

fitted to experimental data by multiple regression analysis. (B) Relationships between rates of rejection (RR: mg h-1) and seston concentration (mg TPM 1^{-1}), for f = 0.1, 0.3 and 0.5. Lines were derived from a combination of the equation above (FR vs. TPM) and the following linear equation relating RR to the filtration rate (FR) and organic content of seston

$$RR = -2.001 + 1.033\ FR - 1.004\ FT \times f$$

This expression was obtained after fitting data for cockles (Iglesias et al 1992 and own unpublished data) with a multiple regression model. (C) Ingestion rates (obtained as Filtration - Rejection) as a function of seston concentration (mg TPM 1^{-1})

Pseudofeces production and rates of ingestion

In the analysis of factors affecting pseudofeces production we have taken Kiørboe et al (1980) approach expressing the rates of rejection as a linear function of the filtration rate. When interactive effects of food quality are accounted for by means of specific terms (Fig. 6 B), the resulting model establishes that, for a given rate of filtration, rejection rate constitutes a negative, linear function of the organic content of food; in other words, the proportion of filtered matter that is ingested increases with rising food quality. These differences in rates of rejection induced by food composition are large enough to overcome differences of the same sign in filtration (Fig 6, A vs. B); consequently, the ingestion rate -calculated as filtration minus rejection- becomes positively dependent on food quality at any TPM concentration (Fig. 6 C).

In spite of absolute differences in the rate of ingestion, a similar degree of regulation is achieved with diets of different quality; the specific mechanism of regulation, however, is determined by food composition. As exemplified by the series of curves plotted in Fig. 6, at high concentration of high quality particles, regulation is mainly based on controlling rates of filtration through adjustments of pumping activity; conversely, no such adjustments appear associated with low quality suspensions and ingestion becomes regulated by high rates of pseudofecal production that acts to counterbalance the high unrestrained filtration rates. Possibilities given by pre-ingestive sorting and selection of particles with different food value may provide a functional interpretation for these contrasting behaviors.

Sorting and selection

In order to evaluate the efficiency in selecting the organic fraction of food when pseudofeces are being produced we used an index of selection efficiency (SE) which is defined by Bayne et al (1989) as:

$$\mathbf{SE = 1 - (p / f)}$$

for p and f representing the organic contents of pseudofeces and food, respectively. This version of SE constitutes a transformation of the index defined by Kiørboe et al (1980) and has the merit of being independent on the ratio of ingestion to filtration.

Values of SE obtained with this formula are plotted against the organic content of food (f) in Fig 7 (dotted line) for diets obtained by mixing phytoplankton and fresh silt (Iglesias et al 1992). Although SE is expected to decline to 0 at both ends of the organic content axis (no selection is possible with pure organic or inorganic particles), these restrictions give no account of the observed symmetry of the curve. In this respect, two additional factors may be considered to explain a regular decline in SE for f < 0.5:

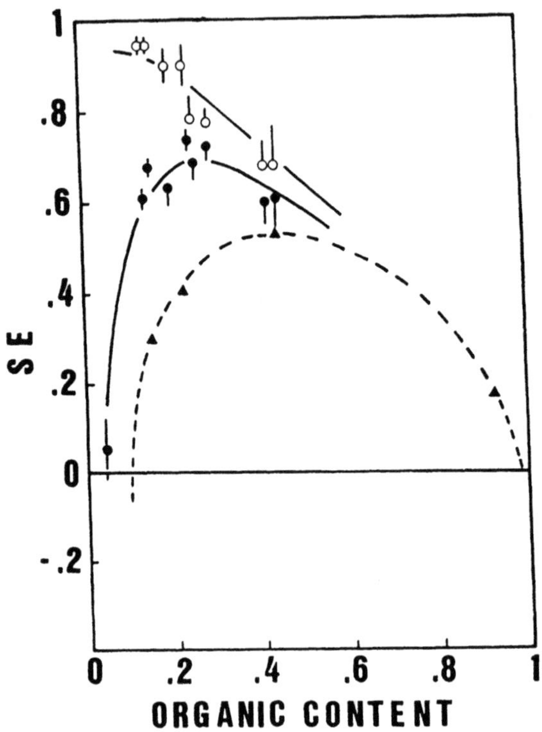

Fig.7. Selection efficiencies (SE) of cockles based on total organics (solid symbols) or Chl *a* (hollow symbols), as a function of food organic content (f). Triangles: Diets consisting in mixtures of phytoplankton (*Tetraselmis suecica*) and fresh silt (Iglesias et al 1992); circles: fresh silt was substituted by ashed (organic-free) silt (Urrutia, unpublished results). Lines, fitted by eye, are only indicative.

(a) Given the actual composition, of diets, there is a reduction in the proportion of living algal cells inherent in decreasing food organic content; consequently, lower SE would be expected for low quality diets if we assume that algae are selected in preference to other forms of particulate organic matter (i.e., detritus). Newell and Jordan (1983) put forward this possibility to explain higher SE for the particulate N compared with C, in oysters.

(b) If the amount of mucus secreted per unit of dry mass of pseudofeces produced were a constant, then the relative enrichment of rejected material with endogenous organics would rise with the decreasing quality of food, leading to reduced SE values (Iglesias et al 1992).

Both these possibilities were tested in our laboratory by performing experiments where cockles were fed mixed suspensions of the algae *Tetraselmis suecica* and ashed (organic-free) silt in a range of 0 to 50 % organic content (Urrutia, unpublished results). Higher SE for these diets -where living cells constitute the only organic component- (full circles, solid line in Fig. 7), compared with values previously reported by Iglesias et al ~1992) (i.e. triangles, dotted line), involves the preferential ingestion of the algae, confirming hypothesis a. On the other hand, hypothesis b may be tested by using Chl a contents instead of organic contents, in order to compute an index of SE that is not affected by mucous secretion (hollow circles, upper solid line in Fig. 7). The difference between the two solid lines represents the effect of an endogenous enrichment of pseudofeces with, mucus that, accordingly, could be estimated to constitute ~ 3 % of total dry mass of pseudofeces.

Therefore, selection of phytoplankton within the pallial organs of cockles seems to occur with increasing efficiency as its abundance in the seston decreases, although under conditions resembling natural food availability characteristic of turbid environments, this feature appears obscured by both the presence of other forms of POM and the mucous losses associated with the rejection process.

Evaluation of energetic benefits of food selection

While constituting a good indicator of the degree of efficiency achieved in the sorting process, the SE index does not provide a proper estimation of the associated energetic benefits. To quantify such benefits, Prins et al (1991) proposed the index ES:

$$ES = (Chl\ \mathbf{a}\ I - Chl\ \mathbf{a}\ I^*) / Chl\ \mathbf{a}\ I$$

where Chl *a* I is the actual ingestion rate of Chl *a* and Chl *a* I* the ingestion rate of Chl *a* in the absence of pre- ingestive food selection. Obviously, ingestion of Chl *a* was assumed to represent the ingestion rate of phytoplankton.

While taking this approach, we have introduced some modifications in order to obviate two limitations we appreciate in this ES index: (1) The use of Chl *a* as a marker implies only phytoplanktonic matter is taken into account, effects of other organic components of the diet or derived from mucous losses being ignored. Since algal cells are selected with maxima efficiencies, an overestimation of the true benefits would be expected. (2) Both terms of ingestion are calculated on the basis of total dry mass ingested while ingestion in volume units appears as a preferable reference (most probably, ingestion is limited by volumetric constraints; Bayne et al 1987). This consideration might have quantitative importance in cases where sorting result in a significant change of the specific density of ingested food. Our Benefit Ratio (BR) takes then the form:

$$BR = (OI - OI^*) / OI$$

for OI and OI* being the actual organic ingestion and organic ingestion without preingestive sorting, respectively, both calculated for reference to a same rate of total ingested volume (VI). Since:

$$OI = VI \times q_I \quad \text{and} \quad VI^* = VI \times q_F$$

for q_I and q_F representing the volumetric quality (mg POM mm^{-3}) of ingested and filtered material, respectively, then,

$$BR = 1- (q_I / q_F)\ \text{(see Iqlesias et al 1992)}.$$

Thus, BR represents also the relative increment in food quality achieved after filtered material has undergone the process of selection.

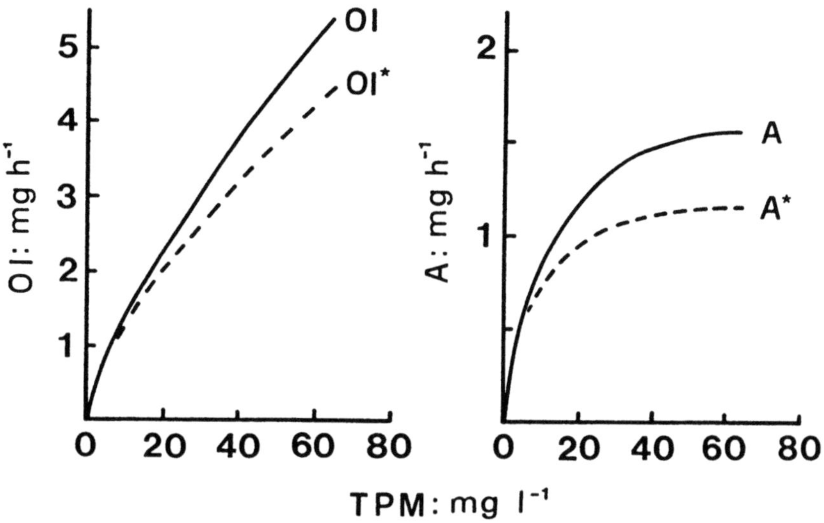

Fig. 8. Rates of organic ingestion (OI) and absorption (A) as a function of seston concentration (mg TPM l^{-1}) at mid organic content (f = 0.3). Solid lines: actual rates; dashed lines: rates under hypothetical conditions of no selection.

Both these versions of BR index provide complementary information relative to the benefits of food selection: (1) Improvement of organic ingestion following pseudofeces production -as given by the difference (OI - OI*)- is a rising function of seston concentration (Fig. 8A), which represents the advantages of increasing rejection; in fact, BR presents an hyperbolic, positive dependence on the rejection/filtration ratio (Iglesias et al 1992). (2) In terms of rates of absorption, the improvement of quality of food ingested relative to filtered material -also quantified by BR index- represents an additional benefit derived from its effects on absorption efficiency. By using the appropriate AE values computed with the equation of Fig. 1 rates of absorption can be calculated from rates of organic ingestion, for actual conditions (A) and under conditions of no-selection (A*) (Fig. 8B). For comparison with curves for organic ingestion, differences between both estimates of absorption appear clearly increased due to these AE-mediated quality effects.

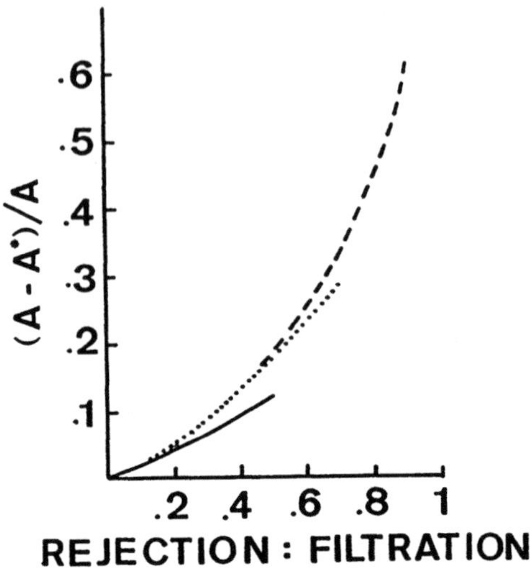

Fig. 9. Benefit ratio based on absorption rates as a function of the relative rejection (= rejection:filtration). Dashed line: f = 0.1; dotted line: f = 0.3; solid line: f = 0.5.

A benefit ratio based on rates of absorption (i.e.: A-A*/A) might be of interest as it includes consideration of both effects of selection. When this index is plotted against the relative rejection rate for diets of variable organic content (Fig. 9), two conclusions are evident: 1) Benefits rapidly increase with the intensity of rejection. 2) These benefits are comparatively higher when selection acts on diets of poor quality, and this in spite of a reduced selection efficiency.

The above relationships allow analyzing the specific feeding responses to changing food quantity and quality under an energy-optimization approach. Since benefits of selection are high with suspensions of low organic content, an efficient-cost strategy would rely on the maintenance of high filtration rates and high rates of pseudofeces production in order to achieve a considerable organic enrichment of food upon ingestion. With high quality suspensions, benefits of selection are comparatively less. Even if metabolic costs of filtration and rejection are low (Jørgensen et al 1986; Widdows and Hawkins, 1989), they may still counterbalance small advantages of selection; consequently, adjustments of the rate of pumping in order to achieve moderate rates of ingestion would appear as the selected regulatory behavior.

IMPLICATIONS OF FEATURES OF FEEDING BEHAVIOR UNDER ENVIRONMENTAL CONDITIONS OF FOOD AVAILABILITY

Particular feeding conditions of discrete nature have been assayed in the laboratory to quantitatively analyze relationships between nutritional factors and physiological components of growth. Extrapolation of results from these studies to predict growth rates in the field would be essentially limited by the lack of a precise knowledge about seston composition in the near-bottom, particularly concerning its temporal dynamics. By assuming some generalizations, however, a range of realistic feeding conditions can be obtained where the appropriate physiological relationships might be applied to compute the scope for growth. By this simulation the study of growth trends could be approached as strictly dependent on gross nutritional factors.

As already mentioned, resuspension of material from the bottom causes increments in total seston availability together with a reduction in the food value of the suspension. Such a phenomenon can be formalized by using relationships between POM and TPM (see Berg and Newell 1986). For the purposes of the present simulation, first-hand data from estuaries extremely different in seston composition (i.e., low TPM, high organic content: Ria de Arosa; Navarro et al 1991, and high TPM low organic content: Marennes-Oleron Bay: Goulletquer et al 1989 and own unpublished data) were fitted with a power equation of the form:

$$POM = 0.5\ TPM^{0.6}$$

TPM values can be substituted in equations of Fig. 6, together with corresponding f (organic content) values derived from the above equation, to obtain rates of filtration and ingestion. Organic ingestion was then calculated from the organic filtration by using the equation:

$$OI - OF\ (a + b\ I/F)$$

where a is SE (selection efficiency), b = 1-a, and I/F the ratio of total ingestion to filtration (see Iglesias et al 1992). For comparison, we also computed organic ingestion in the absence of selection (OI*), by setting a = 0 in the above expression . Absorption was derived from organic ingestion by applying the appropriate AE values (see before) and rates of metabolism (R) were calculated from rates of absorption by using a previous relationship (Navarro et al 1993); from these, SFG was computed as A - R. SFG values plotted in Fig. 10 were obtained assuming a 30 % reduction in energy gain and considering rates of aerial metabolism to be 67 % of the aquatic rate (Iglesias 1986) in order to reproduce mid intertidal conditions.

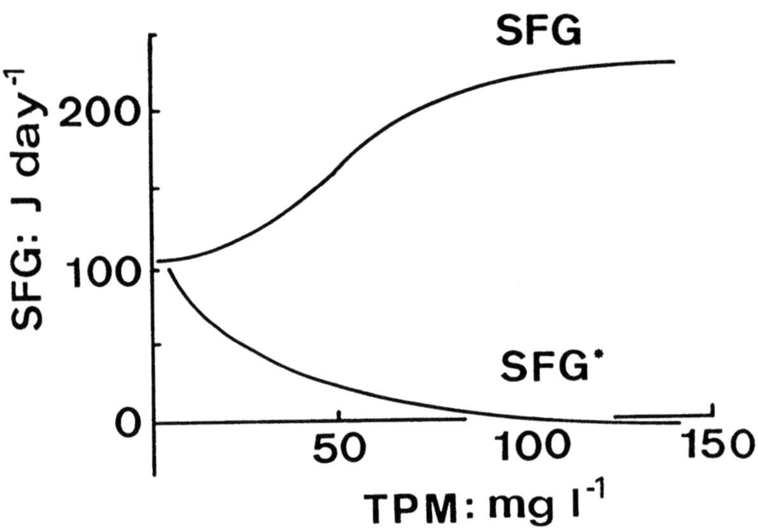

Fig. 10. Scope for growth as a function of seston concentration (mg TPM l^{-1}), under simulated environmental conditions. Actual values (SFG) are compared with the expected trends without selection (SFG*).

According to this simulation (Fig. 10) growth rates would tend to continuously increase with rising seston loads. Steeper dependence occurs at TPM values between 20 to 60 mg l^{-1}, when high availability of food of moderate quality allows maximum benefit to be obtained from selective processes. At higher particle concentration however, extremely reduced quality determines very low efficiencies of selection that are scarcely compensated for by increased POM availability; consequently, growth potential tends to level off. Importance of pre-ingestive selection is clearly illustrated by the behaviour of SFG under hypothetical conditions of regulation without organic enrichment of food upon ingestion. In that case, growth would continuously decrease with increasing concentration to reach negative values at moderate to high seston loads, a tendency that essentially reflects the effect of decreasing quality on the AE. This last observation enlightens the importance of digestive constraints in conditioning energy gain from the environment. As regards cockles, and possibly other infaunal species of bivalves, maxima energy yields in turbid environments appear to depend on high rates of pseudofeces production and selection efficiencies that, when appropriately compared, are considerably higher than found in epifaunal species such as mussels (see Bayne et al, 1993). A precise coupling should exist, however, between these selective abilities at the pre-ingestive level and digestive adjustments involved in optimizing rates of absorption, in order to take maximum advantage of quality improvements of the ingested food.

ACKNOWLEDGMENTS

We are grateful to P Alvarez Jorna, I Ibarrola and MB Urrutia for allowing us the use of original research data, and to BL Bayne (Plymouth Marine Laboratory, UK) and TC Prins (Delta Institute for Hydrobiological Research, The Netherlands), for valuable suggestions. This work was founded by CICYT (Project MAR89-0185).

REFERENCES

Bayne BL Hawkins AJS (1990) Filter feeding in bivalve molluscs: controls on energy balance. In: Mellinger J, Truchot JP, Lahlou B (eds) Animal nutrition and transport processes. Vol. 1. Nutrition in wild and domestic animals. Comparative Physiology. Vol 5. Karger, Basel, p. 60

Bayne BL Hawkins AJS Nalarro E (1987) Feeding and digestion by the mussel *Mytilus edulis* L. (Bivalvia: Mollusca) in mixtures of silt and algal cells at low concentrations. J Exp Mar Biol Ecol 111:1-22

Bayne BL Hawkins AJS Navarro E (1988) Feeding and digestion in suspension-feeding bivalve molluscs: The relevance of physiological compensations. Am Zool 28: 147-159

Bayne BL Hawkins AJS Navarro E Iglesias JIP (1989) Effects of seston concentration on feeding, digestion and growth in the mussel *Mytilus edulis*. Mar Ecol Prog Ser 55: 47-54

Bayne BL Iglesias JIP Hawkins AJS Navarro E Heral M Deslous-Paoli JM (1993) Feeding behaviour of the mussel, *Mytilus edulis* L.; responses to variations in both quantity and organic content of seston. J Mar Biol Ass U K (in press)

Bayne BL Klumpp DW Clarke KR (1984) Aspects of feeding, including estimates of gut residence time, in three mytilid species (Bivalvia,Mollusca) at two contrasting sites in the Cape Peninsula, South Africa. Oecologia (Berl.) 64: 26-33

Berg JA Newell RIE (1986) Temporal and spatial variations in the composition of seston available to the suspension feeder *Crassostrea virqinica*. Est Coast Shelf Sci 23: 375-386

Bricelj VM Malouf RE (1984) Influence of algal and suspended sediment concentrations on the feeding physiology of the hard clam *Mercenaria mercenaria*. Mar Biol 84: 155-165

Calow P (1975) The feeding strategies of two freshwater gastropods, *Ancylus fluviatilis* Mull. and *Planorbis contortus* Linn. (Pulmonata), in terms of ingestion rates and absorption efficiencies. Oecologia (Berl.) 20: 33-49

Conover RJ (1966) Assimilation of organic matter by zooplankton. Limnol Oceanogr 11: 338-345

Decho AW Luoma SN (1991) Time course in the retention of food material in the bivalves *Potamocorbula amurensis* and *Macoma balthica*: significance to the absorption of carbon and chromium. Mar Ecol Prog Ser 78: 303-314

Foster-Smith RL (1975) The effect of concentration of suspension on the filtration rates and pseudofaecal production for *Mytilus edulis* L., *Cerastoderma edule* (L.) and *Venerupis pullastra* (Montagu). J Exp Mar Biol Ecol 17: 1-22

Frechette M Grant J (1991) An in situ estimation of the effect of wind-driven resuspension on the growth of the mussel *Mytilus edulis* L. J Exp Mar Biol Ecol 148: 201-213

Goulletquer P Heral M Deslous-Paoli JM Prou J Razet D Boromthanarat W (1989) Ecophysiologie et bilan energetique de la palourde japonaise d elevage *Ruditapes philippinarum*. J Exp Mar Biol Ecol 132: 85-108

Grant J Bathman W Mills EL (1986) The interaction between benthic diatom films and sediment transport. Est Coast Shelf Sci 23: 225-238

Grant J Enright CT Griswold A (1990) Resuspension and growth of *Ostrea edulis*: a field experiment. Mar Biol 104: 51-59

Grant J Thorpe B (1991) Effects of suspended sediment on growth, respiration and excretion of the soft-shell clam (*Mya arenaria*). Can J Fish Aquat Sci 48: 1285-1292

Grant J Volckaert F Roberts-Regan DL (1987) Resuspendable organic matter in Nova Scotian shelf and slope sediments. Contin Shelf Res 7: 1123-1138

Grizzle RE Lutz RA (1983) A statistical model relating horizontal seston fluxes and bottom sediment

characteristics to growth of *Mercenaria mercenaria*. Mar Biol 102: 95-105
Grizzle RE Morin PJ (1989) Effect of tidal currents, seston, and bottom sediments on growth of Mercenaria mercenaria: results of a field experiment. Mar Biol 102: 85-93
Hawkins AJS Bayne BL (1984) Seasonal variation in the balance between physiological mechanisms of feeding and digestion in *Mytilus edulis* (Bivalvia:Mollusca). Mar Biol 82: 233-240
Hawkins AJS Bayne BL (1985) Seasonal variation in the relative utilisation of carbon and nitrogen by the mussel *Mytilus edulis*: Budgets, conversion efficiencies and maintenance requirements. Mar Ecol Prog Ser 25: 181-188
Hawkins AJS Navarro E Iglesias JIP (1990) Comparative allometries of gut content, gut passage time and metabolic faecal loss in *Mytilus edulis* and *Cerastoderma edule*. Mar Biol 105: 197-204
Hawkins AJS Salkeld PN Bayne BL Gnaiger E Lowe DM (1985) Feeding and resource allocation in the mussel *Mytilus edulis*: evidence for time-averaged optimization. Mar Ecol Prog Ser 20: 273-287
Iglesias JIP (1986) Bases energeticas del crecimiento y la reproduccion en *Cerastoderma edule* (Bivalvia:Cardiidae) U npublished PhD
Iglesias JIP Navarro E Alvarez-Jorna P Armentia I (1992) Feeding, particle selection and absorption in cockles *Cerastoderma edule* exposed to variable conditions of food concentration and quality. J Exp Mar Biol Ecol 162: 177-198
Jørgensen CB (1990) Bivalve filter feeding: Hydrodynamics, bioenergetics, physiology and ecology, Olsen and Olsen,
Fredensborg Jørgensen CB Møhlenberg F Sten-Knudsen O (1986) Nature of relation between ventilation and oxygen consumption in filter feeders. Mar Ecol Prog Ser 29: 73-88
Kiørboe T Møhlenberg F (1981) Particle selection in suspension-feeding bivalves. Mar Ecol Prog Ser 5: 291-296
Kiørboe T Møhlenberg F Nøhr O (1980) Feeding, particle selection and carbon absorption in Mytilus edulis in different mixtures of algae and resuspended bottom material. Ophelia 19: 193-205
Kiørboe T Møhlenberg F Nøhr O (1981) Effect of suspended botom material on growth and en,ergetics in MYtilus edulis. Mar Biol 61: 283-288
Morton BS (1983) Feeding and digestion in Bivalvia. In Saleuddin ASM, Wilbur KM (eds) The Mollusca, Vol 5: Physiology, Part 2. Academic Press, New York, p.65
Navarro E Iglesias JIP Ortega M (1992) Natural sediment as a food source for the cockle *Cerastoderma edule* (L.): effect of variable particle concentration on feeding, digestion and the scope for growth. J Exp Mar Biol Ecol 156: 69-87
Navarro E Iglesias JIP Ortega M Larretxea X (1993) Interrelations between processes of feeding and digestion in cockles *Cerastoderma edule* (Bivalvia, Cardiidae): The basis for a functional response to variable food quantity a nd quality (In prep.)
Navarro E Iglesias JIP Perez Camacho A Labarta U Beiras R (1991) The physiological energetics of mussels (*Mytilus galloprovincialis* Lmk.) from different cultivation rafts in the Rla de Arosa (Galicia, N.W.Spain). Aquaculture 94: 197-212

Newell RIE Jordan SE (1983) Preferential ingestion of organic material by the American oyster *Crassostrea virqinica*. Mar Ecol Prog Ser 13: 47-53

Prins TC Smaal AC Pouwer AJ (1991) Selective ingestion of phytoplankton by the bivalves Mytilus edulis L. and *Cerastoderma edule* (L.). Hydrobiol Bull 25: 93-100

Purchon RD (1968) The Biology of the Mollusca. Pergamon Press Oxford England

Smaal AC Verhagen JGH Coosen J Haas HA (1986) Interaction betwen seston quantity and quality and benthic suspension feeders in the Oosterschelde, The Netherlands. Ophelia 26: 385-399

Theisen BF (1977) Feeding rate of *Mytilus edulis* L. (Bivalvia) from different parts of Danish waters in water of different turbidity. Ophelia 16: 221-232

Van Weel PB (1961) The comparative physiology of digestion in molluscs. ~m Zool 1: 245-252

Wainright SC (1990) Sediment-to-water fluxes of particulate material and microbes by resuspension and their contribution to the planktonic food web. Mar Ecol Prog Ser 62: 271-281

Widdows J Fieth P Worrall CM (1979) Relationships between seston, available food and feeding activity in the common mussel *Mytilus edulis*. Mar Biol 50: 195-207

Widdows J Hawkins AJS (1989) Partitioning of rate of heat dissipation by *Mytilus edulis* into maintenance, feeding and growth components. Physiol Zool 62: 764-784

Wildish DJ Kristmanson DD (1984) Importance to mussels of the benthic boundary layer. Can J Fish Aquat Sci 41: 1618-1625

Wildish DJ Kristmanson DD (1985) Control of suspension feeding bivalve production by current speed. Helgolander Meeresunters 3g: 237-243

Willows RI (1992) Optimal digestive investment: A model for filter feeders experiencing variable diets. Limnol Oceanogr 37: 829-847

Winter JE (1976) Feeding experiments with *Mytilus edulis* L. at small laboratory scale. II. The influence of suspended silt in addition to algal suspensions on growth. In: Persoone G, Jaspers E, (eds) Proc 10th Europ Mar Biol Symp Vol 1, Universa Press, Wetteren, p.583

Winter JE (1978) A critical review on some aspects of filter-feeding in lamellibranchiate bivalves. Haliotis 7: 71-87.

PHYTOPLANKTON-MUSSEL BED INTERACTIONS IN INTERTIDAL ECOSYSTEMS

Harald Asmus and Ragnhild M. Asmus
Biologische Anstalt Helgoland, Wattenmeerstation Sylt, D-2282 List

INTRODUCTION

Dense aggregations of filter-feeding molluscs characterize the benthic fauna of many estuarine and shallow coastal regions. These regions meet the light and nutrient requirements for intensive phytoplankton growth which is the basis of food resources for bivalves. Tidal currents transport water masses from a large surrounding area to the sites where banks of bivalve filter feeders occur and enable a continuous input of fresh phytoplankton. Comparing the volume of tidal water resident on a tidal flat during a tidal cycle to the water volume filtered by the entire mussel or oyster community, it is evident that the concentration of phytoplankton is distinctly influenced by the bivalve aggregations. Both grazing on phytoplankton and fertilization of the water by bivalves has to be considered, when describing the interactions between mussel beds and phytoplankton. We will review here recent publications and experiments focussed on these bivalve filter feeder-phytoplankton interactions.

THE ROLE OF PHYTOPLANKTON AS A FOOD SOURCE

Phytoplankton constitute only a minor portion of the particulate organic matter fraction of seston (Hickel 1984). The greater portion is formed by detritus and a further, minor part by bacteria. Zooplankton may also contribute to particulate organic matter but its concentration is small.

Many authors show phytoplankton playing a dominant role in the food of various bivalve species. Grant and Cranford (1991)

NATO ASI Series, Vol. G 33
Bivalve Filter Feeders
in Estuarine and Coastal Ecosystem Processes
Edited by Richard F. Dame

revealed that resuspended sediments and kelp detritus could not act as sole sources in the diet of sea scallops. The nutritional quality for sea scallops was ranked as follows: Aged kelp debris > phytoplankton > fresh kelp debris > resuspended sediments (Cranford and Grant 1990). *Mytilus edulis* has shown poor growth when supplied only with nonphytoplankton food sources in laboratory experiments (Winter 1974; Williams 1981). Langdon and Newell (1990) showed that utilization of bacteria as a food source could make a significant contribution to the nitrogen requirements of the oyster *Crassostrea virginica* and to the carbon and nitrogen requirements of the mussel *Geukensia demissa* during the summer. However, cellulosic detritus and bacteria do not appear to meet fully the requirements of these bivalve species for carbon and nitrogen. Utilization of other food sources such as phytoplankton, nanozooplankton or non-cellulosic particulate and dissolved organic matter is required. Deslous-Paoli et al. (1990) and Navarro et al. (1991) showed an increasing absorption efficiency with an increasing organic fraction of the food. The former authors found a significant relationship between the assimilated energy and the production of a mussel population, considering the quality of the food, expressed as percentage of algal pigments of total seston. These results show the importance of phytoplankton blooms as high quality food for the production of mussels, as Incze et al. (1980) pointed out for *M. edulis* and Deslous-Paoli et al.(1982) for *Crassostrea gigas*. Even if organic matter of detrital origin is abundant in estuarine ecosystems throughout the year, it is less important for growth than spring and autumnal phytoplankton blooms. Thus, phytoplankton plays a crucial role within the organic fraction of seston as food for mussels.

In many estuarine and coastal environments, high concentrations of organic matter are coupled with a large fraction of inorganic material such as clay, silt and sand particles.

Some recent studies showed that these components will also influence the filtration and ingestion of phytoplankton by mussels (Kiørboe et al. 1980; 1981; Newell and Jordan 1983; Bayne et al. 1987). The food deluting effect of inorganic particles is effectively counteracted by particle selection, by an increased ingestion rate and, apparently, by utilization of some of the organic matter that originates from silt (Kiørboe et al. 1980). There may be an adaptive strategy to utilization of phytoplankton at different concentrations of inorganic seston on both the population and the species level (Jørgensen 1990). The central role phytoplankton plays in the food budgets of different bivalve species leads to the question: Is phytoplankton the key factor regulating the uptake of seston by mussel beds? Mussels appear to react with different ingestion, filtration or growth rates, when different species of phytoplankton are dominant in the seston they feed upon. This may be mainly a consequence of different sizes and shapes of the phytoplankton (Winter 1978), but it is also possible that the chemical composition of algae is important (Ward and Targett 1989). When mussels are exposed to toxic algae, growth is reduced (Nielsen and Strømgren 1991) but also reduced filtration rates are reported (Shumway 1990). Clams are able to sort different algal species in their passage through the gut. The gut passage time in *Mercenaria mercenaria* of *Pseudoisochrysis paradoxa* was significantly higher than that of the chlorophytes *Nannochloris atomus* and *Stichococcus* and Cyanobacteria. The latter three forms were ineffectively used by *Mercenaria mercenaria*. Control of gut clearance rates, through more rapid elimination of those algal species which are also poorly utilized, may contribute to the species adaptive strategy (Bricelj et al. 1984). Growth experiments also showed low absorption efficiencies and no tissue growth of clams when fed with the former small chlorophytes and cyanobacteria (Bass et al. 1990). We have to consider changes in quantity as well as in quality of phytoplankton when we look upon the interactions between phytoplankton and intertidal mussel beds.

THE IMPACT OF MUSSEL BEDS ON THE AMOUNT OF PHYTOPLANKTON

Many authors report on a distinct uptake of seston by bivalves in laboratory experiments (Widdows et al. 1979; Riisgard 1988; Cole et al. 1992) and in the field (Dame andDankers 1988; Dame et al. 1991). Uptake rates measured in Benthic Ecosystem Tunnels ranged between 4.4 and 7.3 g dw m^{-2} h^{-}

^{1} for seston or in terms of organic matter between 0.8 and 1.1 g m^{-2} h^{-1} (Dame and Dankers 1988). Few authors have measured phytoplankton uptake in field experiments. Fréchette et al (1989) measured fluorescence of phytoplankton over an intertidal mussel bed in the St. Lawrence River estuary. These measurements indicate a significant reduction in phytoplankton concentration close to the bed. Phytoplankton carbon concentration over an intertidal mussel bed in the German Wadden Sea ranged between 3 and 291 µg C l^{-1} during field experiments with the Sylt flume. The mussel bed in the flume removed phytoplankton by mean rates over tidal cycles of -0.02 to -0.52 g C m^{-2} h^{-1}. Regarding the biomass of 1561 g AFDW m^{-2} in the flume, these rates are equivalent to -0.01 to -0.33 mg C g^{-1} AFDW m^{-2} h^{-1}. These rates are based on phytoplankton cell counts. Phytoplankton uptake in these experiments also occurred when at the same time there was a net release of total seston from the mussel bed, showing the independence of seston flux and phytoplankton flux to and from the mussel bed.

Seston flux is influenced by uptake (filtration by bivalves, sedimentation of particles) as well as release processes (defaecation, pseudofaeces, resuspension) whereas phytoplankton flux is dominated by mussel filtration during the flume passage time. Phytoplankton is transported by the tidal currents passing over the mussel banks. After passage over an intertidal mussel bed 20 m wide, the amount of phytoplankton was reduced on average to 37 $\pm$ 20 % (n= 22) in summer 1986 (Asmus and Asmus 1991) which was confirmed during a second summer (35 $\pm$ 27 %, n=14) (1989). Phytoplankton uptake by the mussel bed was positively correlated with the amount of phytoplankton available (Asmus and Asmus 1991). Adding to these data from 1986, new data from 1989 confirmed a linear regression ($y= 0.15 x + 41.88$, $r= 0.69$, $P< 0.01$) (Fig. 1). In other areas, a depletion of phytoplankton over dense communities of bivalves has also been observed (Wright et al. 1982; Carlson et al. 1984; Cohen et al. 1984; Nichols et al 1985; Josefsen and Schlüter 1993). Riemann et al (1988) used large plastic bag enclosures to assess the influence of *Mytilus edulis* on phytoplankton, zooplankton and nutrients on the water column. Addition of *M. edulis* reduced the phytoplankton biomass to 10-59 % compared to control experiments.

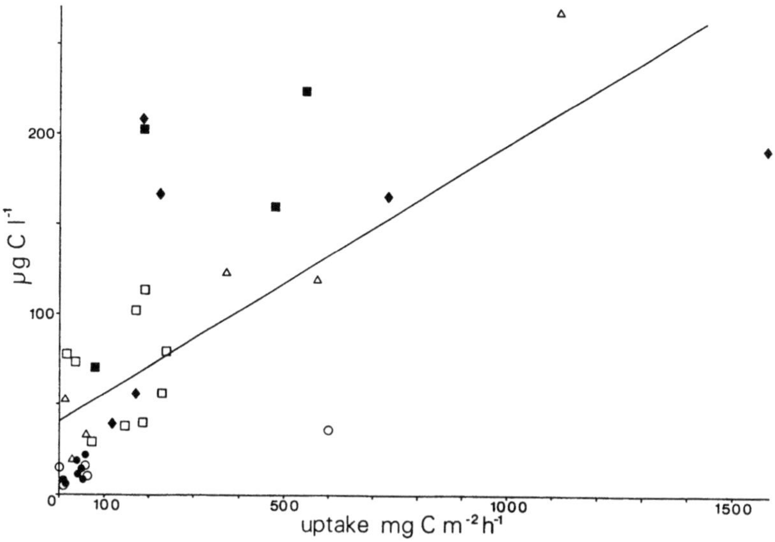

Fig. 1. Correlation between phytoplankton biomass in tidal waters flowing into the Sylt flume and the uptake rate by the enclosed mussel bed (in 1986: 21 July (•), 1 August (Δ), 1 September (□), in 1989: 1 June (■), 14. August (O), 30 August (◆))

The most important combination of factors controlling phytoplankton biomass was the balance between nutrient input, phytoplankton size structure and the physiological state of the mussels. Phytoplankton biomass is a food source characterized by a fluctuating or pulsating availability to which mussels must adapt. They are able to counteract low phytoplankton availability by increasing ingestion rates, absorption efficiency and digestion capacity (Hawkins and Bayne 1984; Bayne et al. 1987; Hawkins et al. 1990). Mussel beds mainly occur in places where the ambient particle concentration is high and thus preingestive algal selection constitutes the main compensatory mechanism for low phytoplankton availability (Newell and Jordan 1983; Prins et al. 1991). High productivity of phytoplankton results in sudden bursts of food and enables large benthic populations to grow. The predictability is low and both temporal and spatial variations are high. In the course of the year, these influences of phytoplankton are changing. The net effect is to produce a more or less seasonal or regular variability (e.g. phytoplankton bloom in spring and autumn) to which filter-feeders can adapt (Levinton 1972). Growth and growth efficiencies of *M. edulis* were measured at different phytoplankton levels by Riisgard and Randlov (1981). Filtration rate was independent of algal concentrations between about 1.5 x 10^6 and 30 x 10^6 cells l^{-1} *Phaeodactylum tricornutum*. At algal concentrations below 1.5 x 10^6 cells l^{-1}, closing of mussels occurred. The authors interpreted this as a possible physiological adaptation to suspension feeding in extremely meagre phytoplankton suspensions. Comparing these phytoplankton concentrations to field values, it becomes evident that the former are as high as a phytoplankton bloom formed by small species. However, in coastal waters, other seston constituents such as detritus and inorganic particles shift the base line for starting filtration by mussels to lower phytoplankton densities. Doering and Oviatt (1986) tested different filtration rate models to compare these to filtration of *Mercenaria mercenaria* in mesocosm experiments. Models of filtration rates based on dyes or algal monocultures yielded poor predictions to generate data and overestimated gross sedimentation due to bivalves by up to an order of magnitude. Such overestimation may exaggerate the role of bivalves in enhancing sedimentation and controlling phytoplankton biomass in shallow waters. Direct measurements of ATP-uptake over intertidal oyster beds showed that the net flow of microbial carbon into the oyster reef was 20 times higher than reported estimates of particulate carbon flow from the entire marsh system to the ocean (Chrzanowski et al. 1986).

EFFECTS OF PHYTOPLANKTON SPECIES COMPOSITION ON BIVALVE FILTER-FEEDER COMMUNITIES

In natural waters, phytoplankton occur in mixed populations of different species. Dominant species of phytoplankton show seasonal variations and a natural succession between phytoplankton groups. Phytoplankton blooms are accompanied by increased growth, production and an improved condition index of bivalves (Beukema and Cadée 1991; Hickman et al. 1991). Although the role of phytoplankton as a main constituent in the diet of bivalves is confirmed by many authors (Walne 1970; Officer et al.1982; Webb and Chu 1983; Epifanio 1983; Nichols 1985; Enright et al. 1986; Lucas et al. 1987; Tiu et al. 1989; Cranford and Grant 1990), some investigations show a different nutritive value of algal species revealing a distinct impact on bivalve growth, development and condition index (Laing et al. 1987; Tiu et al. 1989), food uptake and respiration rate (Davids 1964; Laing et al. 1987; Cranford and Grant 1990).

The food value of algae has been assessed for the culture of oysters (Walne 1970; Tenore and Dunstan 1973; Enright et al. 1986). Nutritional value of algae tested was in the order *Isochrysis cf. galbana, Skeletonema costatum* > *Chaetoceros calcitrans, Chroomonas salina, Thalassiosira pseudonana* > *Tetraselmis suecica* > *Phaeodactylum tricornutum* > *Chlamydomonas coccoides*, when the same weight of these species was given in the diet to 3 clam species (Laing et al. 1987). Walne (1970) found high growth rates of *Mercenaria mercenaria* fed with *Skeletonema costatum*, low growth rates with *P. tricornutum* and very little growth with *C. coccoides*. Epifanio (1979) showed that *T. suecia* promoted moderate growth of *Mercenaria* juveniles. Other molluscan species fed with these diets show similar relative growth rates (Peirson 1983; Laing and Millican 1986; Enright et al. 1986). These authors have suggested that one of the more important factors determining nutritional value of algae is their content of polyunsaturated fatty acids. Algae containing special forms of fatty acids in relatively high amounts generally support much better growth than species without either (Langdon and Waldock 1981; Webb and Chu 1983). The food value appears to be different on the species level rather than on the level of taxonomic algal groups as diatoms, dinoflagellates etc.. *Skeletonema costatum* is a diatom of high food value for growth of bivalves (Laing et al. 1987), whereas *Phaeodactylum tricornutum*, also a diatom, is indigestible to many bivalves (Epifanio 1983).

Bivalves are capable to exploit the ambient phytoplankton

by ingesting preferent species and by rejecting others (Shumway et al. 1985; 1987; 1990; Cucci et al. 1985). This selection seems to be a process occurring mainly during the feeding process by preingestive selection mechanisms (eg. forming pseudofaeces) (Shumway et al. 1987; Iglesias et al. 1992;) or postingestive mechanisms (eg. different digestability or different gut passage times) (Bricelj and Malouf 1984; Bricelj et al. 1984; Shumway et al. 1987).

In the field, Beukema and Cadée (1991) compared the patterns of annual growth and condition of *Macoma balthica* with the patterns of abundance of phytoplankton components over 15 years. Growth fluctuations in *M. balthica* were found to parallel fluctuations of planktonic diatoms but not those of *Phaeocystis pouchetii* (Prymnesiophyceae). It is concluded that *M. balthica* responded by faster growth and higher condition to better feeding conditions in years of high diatom abundance. Diatoms rather than *P. pouchetii* or other flagellates will be the most important food source for *M. balthica*.

In contrast to investigations on the influence of different phytoplankton species on growth of bivalves, low evidence is found on the influence of different phytoplankton species on the filtration or water processing of bivalves. Negative effects on filtration and growth were observed when toxic phytoplankton were offered to mussels which were not previously exposed to toxic algae (Shumway 1990; Nielsen and Strømgren 1991).

Field studies on natural phytoplankton uptake by mussels are mainly focussed on the amount of total phytoplankton carbon removed from the water (Smaal et al. 1986; Dame and Dankers 1988; Prins and Smaal 1990). Few studies considered the species composition of phytoplankton in measuring phytoplankton uptake by mussels (Asmus and Asmus 1991). Flume measurements on a mussel bed showed that uptake was mainly dependent on the amount of phytoplankton available rather than on species composition. Phytoplankton dominated by diatoms was taken up during a tidal cycle with low phytoplankton biomass in much lower but still significant rates than during a bloom of *Chaetoceros socialis* (Fig. 2).

High rates occur around high water when the water coming from the deep tidal gully and the adjacent North Sea is generally rich in phytoplankton. During 2 blooms of *Phaeocystis globosa* (September 1986 and June 1989) an intensive uptake by the mussel bed occurred in the flume (Fig. 3.). The effect of the mussel bed on the overflowing phytoplankton became particularly evident by introducing a control lane (in 1989) where mussels had been removed and only the bare sand was left (Asmus et al. 1992).

Phytoplankton cells over the whole natural size range were taken up by the mussel bed (Asmus and Asmus 1991). This is in accordance with laboratory findings revealing that *M. edulis* take up cells from 4 µm cell diameter onwards with 100% efficiency (Riisgard 1988; Jørgensen 1990). A large phytoplankton species *Ceratium furca* (up to 300 x 30 µm) occurred in low and in high abundance during 2 tidal cycles of flume measurements (Fig. 4). Uptake rates were very high in the mussel lane whereas varying rates with no clear tendency were measured in the control lane. As shown by these different field situations, uptake of phytoplankton from the water column is more dependent on phytoplankton concentration than on phytoplankton composition. Individual clearance rates estimated from these experiments indicate a decreasing trend when the amount of phytoplankton carbon and the percentage phytoplankton carbon of total particulate organic carbon increase. Phytoplankton species composition seems to have no influence on the water volume processed by a mussel bed, but the different nutritive value of phytoplankton species affects the growth and production of bivalves.

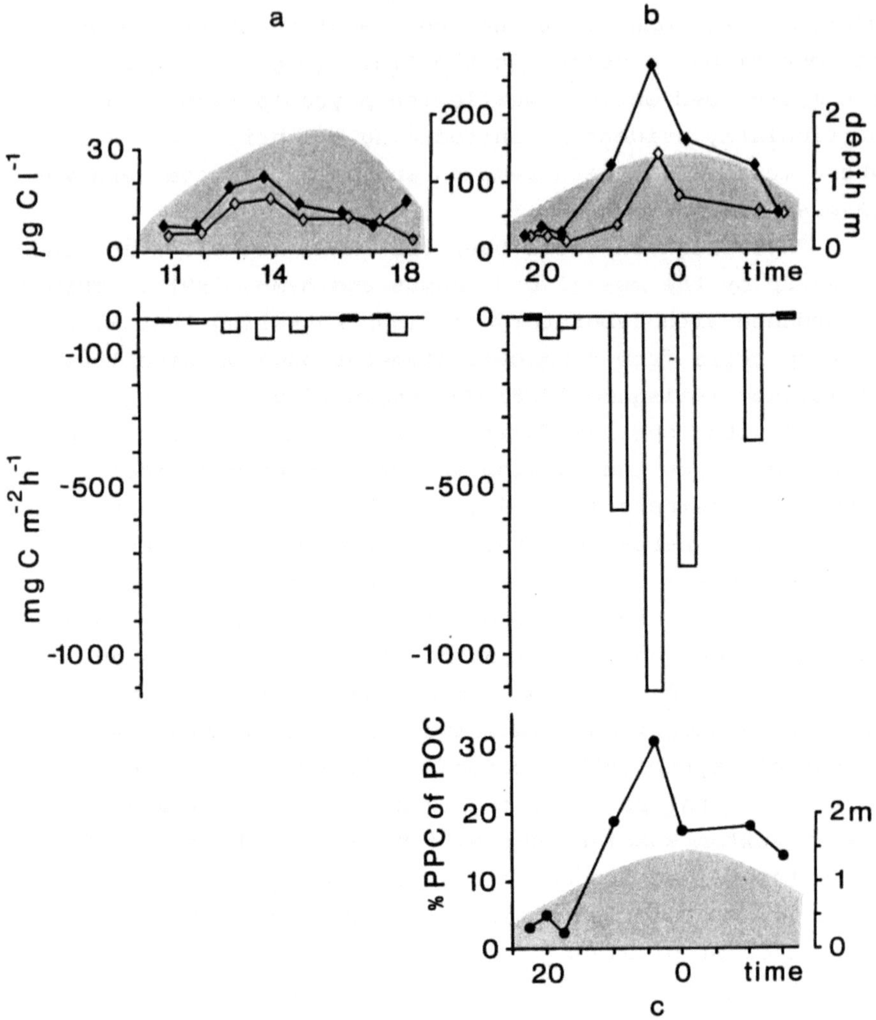

Fig. 2. Phytoplankton biomass in inflow (♦) and outflow (◊) of the Sylt flume during a tidal cycle with a mixed algal composition (a) (21 July 1986) and during a bloom of *Chaetoceros socialis* (b) (1 August 1986). Uptake of phytoplankton by the mussel bed is indicated by negative values. Water depth is given (right scale, shading). (c) Percentage of phytoplankton carbon in particulate organic carbon on 1 August .

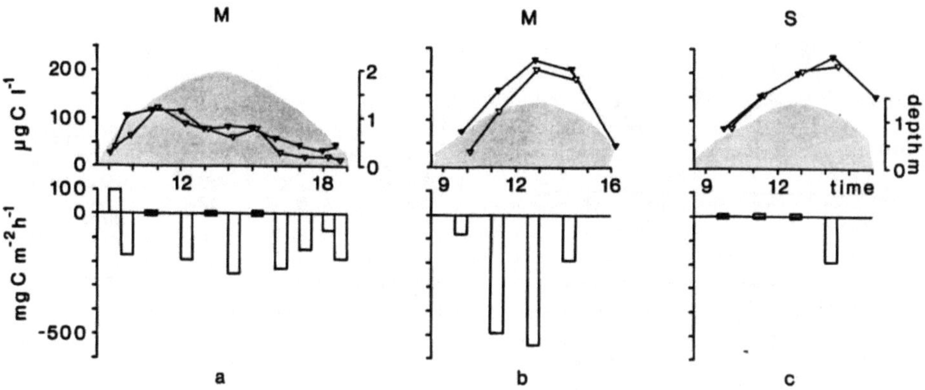

Fig. 3. Phytoplankton biomass in inflowing (▲) and outflowing water (Δ) in the Sylt flume during tidal cycles when *Phaeocystis globosa* dominated (1 September 1986 (a), 1 June 1989 (b, c) (redrawn from Asmus et al. 1992). M= mussel lane, S= sand lane of the flume. Water depth is given (right scale, shading).

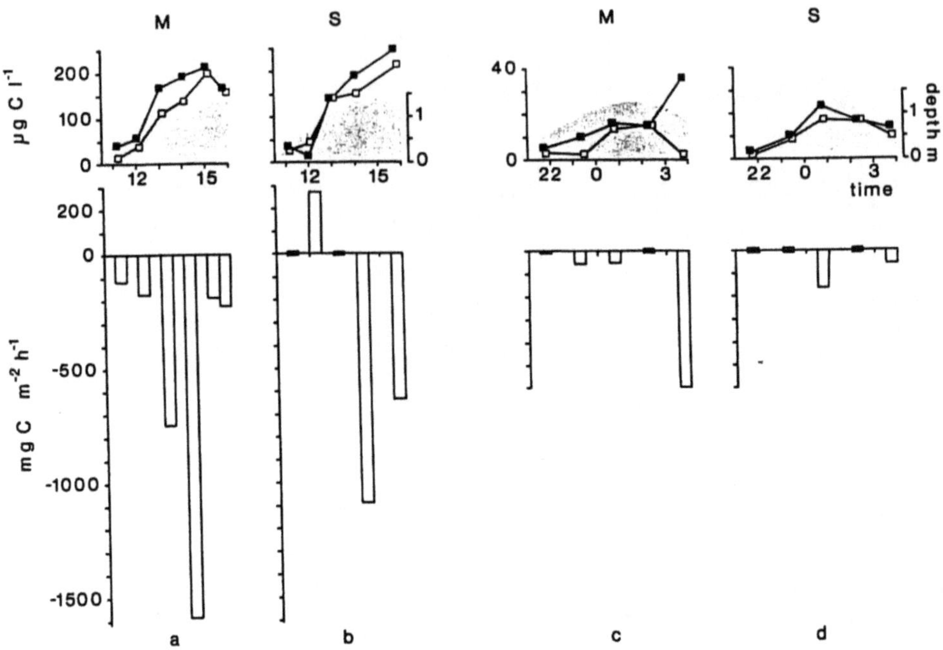

Fig. 4. Phytoplankton concentration in water flowing into (■) and out of (□) the Sylt flume at times of dominance of the dinoflagellate *Ceratium furca*. (a, b) Rates in mussel lane (M) and sand lane (S) during the bloom. (c, d) Rates in mussel and sand lane when inflow concentrations were low. Water depth is indicated (right scale, shading).

CONTRIBUTION OF RESUSPENDED DIATOMS TO THE FOOD OF MUSSEL BEDS

Since benthic microalgae are very abundant on shallow sediments, and nonadhering species may be easily resuspended, it could be assumed that benthic microalgae might be an important food source for filter-feeding organisms. Shumway et al. (1987) found in gut analysis of *Placopecten magellanicus* benthic diatoms as well as planktonic ones. This opportunistic filter-feeder exploits pelagic as well as benthic organisms especially by its ability to stir up the "fluff" layer. Also Newell et al. (1989) identified many benthic algal species from gut contents of *Mytilus edulis*. However, Fréchette and Grant (1991) suggest that decreased mussel growth was primarily caused by phytoplankton depletion which could not be ameliorated by resuspended food supplies.

In the tidal waters of our study area in the northern Wadden Sea, benthic diatoms constitute on average only about 3 % of the pelagic microalgal biomass in carbon units (3.42 $\pm$ 3.33 %, n= 98) in summer. This figure is based on cell counts during flume studies. Benthic diatoms may only be an important food source for filter feeders during times of very low phytoplankton biomass, between blooms, or at winter time. When benthic diatoms are abundant in the water column, they are taken up by mussel beds. The diatom *Pleurosigma spp.* formed the greatest part of the resuspended diatoms in our study. The uptake rates of benthic diatom biomass by the mussel bed enclosed in the Sylt flume were distinct but very low compared to the high rates of phytoplankton (Fig. 5). These findings may be influenced by local environmental conditons. However, the more phytoplankton is available to filter-feeder the smaller will be the importance of resuspended benthic organisms in the diet.

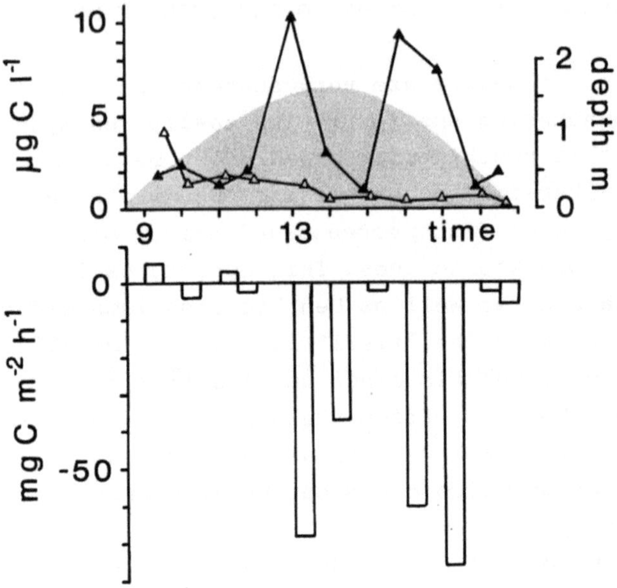

Fig. 5. Biomass of resuspended diatoms, dominated by *Pleurosigma spp.* in the inflow (Δ) and outflow (Δ) of the Sylt flume, and the resulting flux rates (+ discharge, - uptake/sedimentation)

THE IMPORTANCE OF PHYTOPLANKTON/ MUSSEL BED INTERACTIONS IN CARBON AND NITROGEN CYCLES

The percentage of phytoplankton comprising total organic carbon (POC) in tidal waters of estuaries and intertidal coastal areas varies both with season and tide. In the Wadden Sea, large phytoplankton shares in total POC are mainly a consequence of high absolute values of phytoplankton. A share of phytoplankton of up to 20 % of POC (averaged over a tidal cycle) is formed by phytoplankton during a bloom. Phytoplankton concentration increases towards high tide and decreases during ebbing tide. The elevated concentrations at high tide indicate an input of phytoplankton from deeper water originating outside the tidal flats. The amount of total POC is high at the beginning of inundation, when particles from the sediment surface are resuspended easily. At low ambient phytoplankton concentrations, its share in POC is about 1 % and increases to

2.9 % at high tide decreasing again during ebbing tide. During a phytoplankton bloom, the share of phytoplankton in POC attains values of up to 36 % during high tide (Fig. 2c).

Carbon input from phytoplankton to an intertidal mussel bed varies with the tide parallel to increasing availability. Detrital components are dominant in organic seston which form a major part of the carbon uptake by a mussel bed, although the capacity of water processed may be directed by the phytoplankton concentration. In spite of a larger uptake of detrital carbon compared to phytoplankton carbon, carbon flux of detrital components reveals a variable direction, changing the function of mussel beds from a sink to a source of organic material due to weather conditions. Considering phytoplankton carbon, the mussel bed is acting permanently as a sink. Phytoplankton does not dominate the carbon flux over an intertidal mussel bed quantitatively, but it is essential to the mussel bed as a source of assimilated carbon. This can be explained by the low absorption efficiencies for detrital carbon relative to phytoplankton carbon (Crosby et al. 1989). Langdon and Newell (1990) found that oysters absorb refractory cellulosic material with an efficiency of only 3 %. Other bivalves which are more adapted to detrital food sources like *Geukensia demissa* show absorption efficiencies of 9-14 %. Detritus contributes up to 0.7 % of the metabolic requirements of oysters and up to 8.6 % of those of *Geukensia*. The contribution of bacteria is 5.5 % of the metabolic requirements for carbon in the diet of oysters and up to 31 % of required carbon in the diet of *Geukensia*. In the discussion on the role of detritus as a food source for mussels it seems necessary to differentiate between refractory organic matter and organic matter of higher food value (e.g. aged kelp debris). Depending on the relative contribution of refractory and nonrefractory matter to the particulate organic matter in suspension, the amount of material absorbed by mussels will be highly variable in space and time.

Table 1. Mean phytoplankton carbon value (PPC) (µg C l^{-1}) with standard deviation in waters over mussel beds in the Königshafen Bay. Percentage of PPC of particulate organic carbon (POC) and percentage of phytoplankton nitrogen (PPN) of particulate nitrogen (PN) (mean values over tidal cycles and standard deviation).

	PPC	% PPC of POC	% PPN of PN
high phytoplankton biomass:			
1 August 1986	110 ± 84	15.9 ± 10.1	28.3 ± 25.3
1 September 1986	75 ± 31	9.6 ± 2.8	13.0 ± 5.2
1 June 1989	171 ± 68	19.3 ± 5.1	24.1 ± 6.3
30 August 1989	155 ± 75	17.5 ± 12.0	24.6 ± 16.7
low phytoplankton biomass:			
21 July 1986	14 ± 6	3.3 ± 1.6	3.7 ± 2.0
14 August 1989	15 ± 8	1.5 ± 0.7	1.9 ± 0.9

The share of phytoplankton carbon in total POC is lower than the share of phytoplankton nitrogen in particulate nitrogen (Table 1). Phytoplankton nitrogen was calculated after Redfield et al. (1963) considering the conversion from atomic to weight-specific ratio. Phytoplankton appears to be a dominant constituent in the nitrogen budget of bivalve food. However, this may be dependent on the bivalve species and its adaptation to food resources. Only 27 % of the metabolic nitrogen requirements of oysters (*Crassostrea virginica*) are met by bacterial nitrogen, the remaining 73 % must be met by other food sources, mainly phytoplankton. In contrast, *Geukensia demissa* is able to utilize mainly bacterial nitrogen contributing 70.6 %

of the metabolic requirements of this species (Langdon and Newell 1990). For sea scallops phytoplankton provided at least six times greater carbon and nitrogen balance than detrital sources. Sediment allowed a relatively favourable nitrogen balance due to C/N values but infavourable carbon balance due to high inorganic content. In contrast, detritus (kelp derivates) provided comparatively better carbon than nitrogen nutrition (Grant and Cranford 1991). However, dietary protein and lipid must be present in sufficient quantities to support rapid growth in clams (Wikfors et al. 1992). Growth of bivalves increased with dietary protein content up to a maximum level, beyond which growth decreased at higher dietary protein contents (Hawkins and Bayne 1991).

Asmus et al. (1990) showed that the intensity of release and uptake of particulate nitrogen (PN) by an intertidal mussel bed displayed a trend different from that of POC. It was assumed that the PN- flux over a mussel bed reflected feeding activities of the mussels more than POC or total seston, because high uptake rates of PN coincided with high release of dissolved inorganic nitrogen. These experiments also showed the highest rates of uptake of phytoplankton carbon (PPC) and PN coupled with a consistent increase in CN-ratio during water passage through the flume. Accordingly we may assume that this increase in CN-ratios is an effect of bivalve filter-feeding on phytoplankton. Grant and Cranford (1991) argued that nitrogen limitation provides a satisfactory explanation for the inadequacy of the diet derived from macroalgal detritus. It supports the classical paradigm of detrital-based food webs: Despite abundant particulate organic carbon particulate nitrogen can improve limits on secondary production (Mann 1988).

In summary we conclude that mussel beds change the carbon and nitrogen budget of the overflowing tidal water by filtering phytoplankton . Phytoplankton nitrogen and carbon may be the primary subject to storage in tissue of mussels and for enhanced remineralisation by digestion, whereas detrital components may be mainly deposited .

FEED BACK MECHANISMS BETWEEN MUSSEL BEDS AND PHYTOPLANKTON

The impact mussel beds have on the total estuarine and shallow coastal ecosystem by filtering phytoplankton is mainly determined by the capacity of water processing. The higher the volume of water processed by bivalve communities compared to total water exchange between the coastal ecosystem and adjacent waters, the more intense will be the control of phytoplankton assemblages by the bivalves. Maximal clearance rates for mussel beds range between 6 and 12 $m^3\ m^{-2}\ h^{-1}$ were reported (for summary see Jørgensen 1990). Filtration capacities of an estuarine bay in France were between 7.55 and 6.36 $m^3\ m^{-2}\ d^{-1}$ (Hily 1991). The benthic community in that bay in France consisted mainly of filter-feeding organisms. This community is capable of filtering the water of the bay with a time constant shorter than the hydrodynamic residence time and comparable with the time constant for phytoplankton growth. Clearance rates measured in flumes and tunnels over intertidal mussel beds in the northern Wadden Sea (Königshafen area) are 2.1 to 2.9 $m^3\ m^{-2}\ h^{-1}$ (Asmus et al. 1992). Assuming that 10 % of the Königshafen Bay is covered by mussel beds, a value of 6 $m^3\ m^{-2}\ d^{-1}$ would indicate the filtering capacity of the mussels of the whole bay, a value comparable to that found by Hily (1991).

In mesocosms Doering et al. (1986) did not find a limitation of phytoplankton biomass in the water column through filtration, although clams enhanced sedimentation. He concluded that a lack of changes in water column concentrations of particles do not indicate a lack of filter-feeder influence on material cycles, because clams also enhance the production of phytoplankton. Mesocosm experiments last for days or weeks and represent net results of processes integrated over a longer period. Flow-through measurements in tunnels or flumes last only hours over several tidal cycles and reveal short-term effects of filter-feeders on the water column. Thus, reduction of phytoplankton is a short-term effect, whereas enhancing phytoplankton productivity may balance this reduction on a longer time scale under certain conditions (nutrient limitation, light conditions, stability of water column).

Beside the removal of phytoplankton together with other

fractions of seston from the overflowing waters, mussel banks and oyster beds release simultaneously inorganic nutrients as ammonium, phosphate and silicate in the highest rates ever measured over natural benthic communities. This results in distinctly reduced phytoplankton biomass and increased nutrient concentrations in waters passing bivalve filter-feeder assemblages. Some years ago, before results from flow-through devices for the measurement of benthic-pelagic exchange processes were available, great attention was payed to the potential of benthic filter-feeders to reduce phytoplankton biomass in coastal ecosystems (Cloern 1982; Officer et al. 1982; Alpine and Cloern 1992). During the last few years , the awareness of a possible importance of a feed-back mechanism exerted by mussel and oyster beds on phytoplankton increased due to strong experimental evidence (Loo and Rosenberg 1989; Prins and Smaal 1990; Dame et al. 1991; Smaal 1991). The remineralization of filtered material is rapid and intense, either by direct excretion by filter-feeders or via bacterial processes in the underlying sediments. In summer, nitrogen and silicate concentrations may become limiting for phytoplankton in shallow coastal areas. In a situation like this, the release of nutrients by filter-feeder communities may promote phytoplankton productivity while phytoplankton biomass is reduced. This is possible because filter-feeders not only feed upon phytoplankton but seston which includes a variety of components. In an area of the northern Wadden Sea, it was shown that the potentially induced phytoplankton primary production by the ammonium release of a mussel bed may be even higher than the actual uptake of phytoplankton by the same mussel bed (Asmus and Asmus 1991). Realization of this potential depends upon various environmental conditions. At ambient nutrient levels of strong eutrophication, a further input of nutrients by filter-feeders may be ineffective. In a situation of nutrient limitation, however, the input of nutrients from filter-feeder communities may be rapidly incorporated into algal biomass. This is not necessarily only phytoplankton. Brown-green- or red-algae at times grow on natural mussel or oyster beds. These macroalgae benefit from the nutrients released by a mussel bed. It is also possible that benthic microalgae or adjacent seagrass beds may take up released nutrients. The fate of the nutrients discharged by filter-feeder communities will depend on the local ecological features.

CONCLUSIONS

The interaction between mussel beds and phytoplankton is a key process within the material exchange between the benthic and pelagic subsystems. The amount, as well as the composition of phytoplankton triggers the removal of particles from the water column by the benthos. Selective feeding by bivalves directs

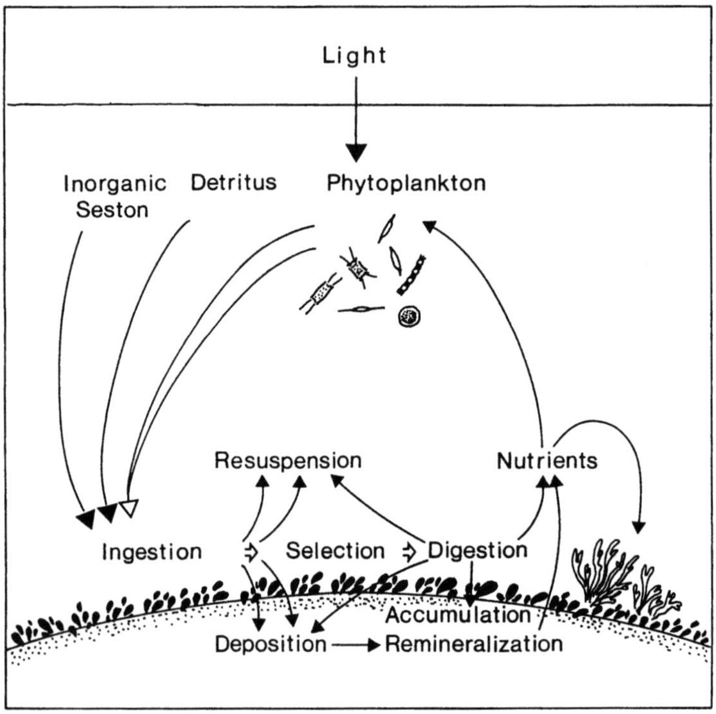

Fig. 6. Simplified diagram showing the interactions between a mussel bed, inorganic seston, detritus (black arrows) and phytoplankton (white arrows) as well as the biotic processes involved.

the carbon and nitrogen flux originating from phytoplankton primary production to growth and secondary production as well as to fast remineralization. Remineralization of phytoplankton carbon and nitrogen by mussels enlarged by additional sources (detritus, bacteria) gives rise to a distinct feedback to primary production. Mussel beds enhance the turnover of material by grazing upon phytoplankton, they are important accelerators for the ecological circuits of materials in shallow water ecosystems.

Acknowledgements. We are indebted to Birgit Maaß who counted the plankton samples of 1989 with great care and patience. We like to thank all our friends who helped us in the field work.

REFERENCES

Alpine AE Cloern JE (1992) Trophic interactions and direct physical effects control phytoplankton biomass and production in an estuary. Limnol Oceanogr 37 (5):946-955

Asmus H Asmus RM Reise K (1990) Exchange processes in an intertidal mussel bed: a Sylt-flume study in the Wadden Sea. Ber Biol Anst Helgoland 6:1-79

Asmus RM, Asmus H (1991) Mussel beds: limiting or promoting phytoplankton ? J Exp Mar Biol Ecol 148:215-232

Asmus H Asmus RM Prins TC Dankers N Francés G Maaß B Reise K (1992) Benthic-pelagic flux rates on mussel beds: tunnel and tidal flume methodoloy compared. Helgoländer Meeresunters. 46:341 -361

Bass AE Malouf RE Shumway SE (1990) Growth of northern quahogs (*Mercenaria mercenaria*)(Linneaeus, 1758)) fed on picoplankton. J Shellfish Res 9(2):299-307

Bayne BL Hawkins AJS Navarro E (1987) Feeding and digestion by the mussel *Mytilus edulis* L. (Bivalvia: Mollusca) in mixtures of silt and algal cells at low concentrations. J Exp Mar Biol Ecol 111:1-22

Beukema JJ Cadée GC (1991) Growth rates of the bivalve *Macoma balthica* in the Wadden Sea during a period of eutrophication: relationships with concentrations of pelagic diatoms and flagellates. Mar Ecol Prog Ser 68: 249- 256

Bricelj VM Bass AE Lopez GR (1984) Absorption and gut passage time of microalgae in a suspension feeder: an evaluation of the 51Cr: 14C twin tracer technique. Mar Ecol Prog Ser 17:57-63

Bricelj VM Malouf RE de Quillfeldt C (1984) Growth of juvenile *Mercenaria mercenaria* and the effect of resuspended bottom sediments. Mar Biol 84:167-173

Carlson DJ, Townsend DW, Hilyard AL, Eaton JF (1984) Effect of an intertidal mudflat on plankton of the overlying water column. Can J Fish Aquat Sci 41: 1523-1528

Chrzanowski TH, Spurrier JD, Dame RF, Zingmark RG (1986) Processing of microbial biomass by an intertidal reef community. Mar Ecol Prog Ser 30: 181-189

Cloern JE (1982) Does the benthos control phytoplankton biomass in South San Francisco Bay ? Mar Ecol Prog Ser 9:191-202

Cohen RRH Dresler PV Phillips EJP Cory RL (1984) The effect of the Asiatic clam, *Corbicula fluminea*, on phytoplankton of the Potomac River, Maryland. Limnol Oceanogr 29(1): 170-180

Cole BE Thompson JK Cloern JE (1992) Measurement of filtration rates by infaunal bivalves in a recirculating flume. Mar Biol 113:219-225

Cranford PJ Grant J (1990) Particle clearance and absorption of phytoplankton and detritus by the sea scallop *Placopecten magellanicus* (Gmelin). J Exp Mar Biol Ecol 137:105-121

Crosby MP Langdon CJ Newell RIE (1989) Importance of refractory plant material to the carbon budget of the oyster *Crassostrea virginica*. Mar Biol 100:343-352

Cucci TL Shumway SE Newell RC Selvin R Guillard RRL Yentsch CM (1985) Flow cytometry: a new method for characterization of differential ingestion, digestion and egestion by suspension feeders. Mar Ecol Prog Ser 24:201-204

Dame RF Dankers N (1988) Uptake and release of materials by a Wadden Sea mussel bed. J Exp Mar Biol Ecol 118:207-216

Dame RF Spurrier JD Williams TM Kjerfve B Zingmark RG

Wolaver TG Chrzanowski TH McKellar HN Vernberg FJ (1991) Annual material processing by a salt marsh-estuarine basin in South Carolina, USA. Mar Ecol Prog Ser 72:153-166

Davids C (1964) The influence of suspensions of microorganisms of different concentrations on the pumping and retention of food by the mussel (*Mytilus edulis L.*). Neth J Sea Res 2(2):233-249

Deslous-Paoli JM Héral M Zanette Y (1982) Problèmes posés par l'analyse des relations trophiques hûitre-milieu. Journées du GABIM, 18-20. Nov 1981, Publ. CNEXO (Actes Colloq.) 14: 335-340

Deslous-Paoli JM Boromthanarat S Heral M Boromthanarat W Razet D (1990) Energy budget of a *Mytilus edulis L.* population during its first year on bouchonts in the bay of Marennes-Oléron. Aquaculture 91:49-63

Doering PH Oviatt CA (1986) Application of filtration rate models to field populations of bivalves: an assessment using experimental mesocosms. Mar Ecol Prog Ser 31:265-275

Doering PH Oviatt CA Kelly JR (1986) The effects of the filter-feeding clam *Mercenaria mercenaria* on carbon cycling in experimental marine mesocosms. J Mar Res 44: 839-861

Enright CT Newkirk CF Craigie JS Castell JD (1986) Evalulation of phytoplankton as diets for juvenile *Ostrea edulis L.*. J Exp Mar Biol Ecol 96:1-13

Epifanio CE (1979) Growth in bivalve molluscs: Nutritional effects of two or more species of algae in diets fed to the American oyster *Crassostrea virginica* (Gmelin) and the hard clam *Mercenaria mercenaria* (L.). Aquaculture 18:1-12

Epifanio CE (1983) Phytoplankton and yeast as food for juvenile bivalves: a review of research at the University of Delaware. In: Pruder GD, Langdon C, Conklin D (eds.) Proc Sec Int Conf Aquaculture Nutrition. Biochemical and physiological approaches to shellfish nutrition. Louisiana State University, Baton Rouge, Louisiana:292-304

Fréchette M Butman CA Geyer WR (1989). The importance of boundary-layer flows in supplying phytoplankton to the benthic suspension feeder, *Mytilus edulis* L.. Limnol.

Oceanogr 34:19-36

Fréchette M Grant J (1991) An in situ estimation of the effect of wind-driven resuspension on the growth of the mussel *Mytilus edulis* L.. J Exp Mar Biol Ecol 148:201-213

Grant J Cranford PJ (1991) Carbon and nitrogen scope for growth as a function of diet in the sea scallop *Placopecten magellanicus*. J mar biol Ass UK 71:437-450

Hawkins AJS Bayne BL (1984) Seasonal variation in the balance between physiological mechanisms of feeding and digestion in *Mytilus edulis* (Bivalvia: Mollusca). Mar Biol 82: 233-240

Hawkins AJS, Bayne BL (1991) Nutrition of marine mussels: factors influencing the relative utilizations of protein and energy. Aquaculture 94:177-196

Hawkins AJS Navarro E Iglesias JIP (1990) Comparative allometries of gut-passage time, gut content and metabolic faecal loss in *Mytilus edulis* and *Cerastoderma edule*. Mar Biol 105:197-204

Hickel W (1984) Seston in the Wadden Sea of Sylt (German Bight, North Sea). Neth Inst Sea Res-Publ Ser 10:113-131

Hickman RW Waite RP Illingworth J Meredyth-Young JL Payne G (1991) The relationship between farmed mussels, *Perna canaliculus*, and available food in Pelorus-Kenepuru Sound, New Zealand, 1983-1985. Aquaculture 99:49-68

Hily C (1991) Is the activity of benthic suspension feeders a factor controlling water quality in the Bay of Brest ? Mar Ecol Prog Ser 69:179- 188

Iglesias JIP Navarro E Alvarez Jorna P Armentia I (1992) Feeding, particle selection and absorption in cockles *Cerastoderma edule* (L.) exposed to variable conditions of food concentration and quality. J Exp Mar Biol Ecol 162: 177-198

Incze LS Lutz RA Watling L (1980) Relationships between effects of environmental temperature and seston on growth and mortalilty of *Mytilus edulis* in a temperate northern estuary. Mar Biol 57:147-156

Jørgensen CB (1990) Bivalve filter feeding: Hydrodynamics, Bioenergetics, Physiology and Ecology. Olsen and Olsen Fredensborg Denmark

Josefsen SB Schlüter L (1993) The influence of an intertidal mussel bed (*Mytilus edulis* L.) on nutrient fluxes in the Kerteminde fjord, Denmark; a flume study. In: Dyer K (ed.) Changes in fluxes in estuaries: implications from science to management. Proc ECSA 22 Symposium 1992, Olsen and Olsen Fredensborg Denmark (in press)

Kiørboe T Møhlenberg F Nøhr O (1980) Feeding, particle selection and carbon absorption in *Mytilus edulis* in different mixtures of algae and resuspended bottom material. OPHELIA 19(2):193-205

Kiørboe T Møhlenberg F Nøhr O (1981) Effect of suspended bottom material on growth and energetics in *Mytilus edulis*. Mar Biol 61: 282-288

Laing I, Millican PF (1986) Relative growth and growth efficiency of *Ostrea edulis* L. spat fed various algal diets. Aquaculture 54:245-262

Laing I Utting SD Kilada RWS (1987) Interactive effect of diet and temperature on the growth of juvenile clams. J Exp Mar Biol Ecol 113:23-38

Langdon CJ Waldock MJ (1981) The effect of algal and artificial diets on the growth and fatty acid composition of *Crassostrea gigas* spat. J mar biol Ass UK 61:431-448

Langdon CJ Newell RIE (1990) Utilization of detritus and bacteria as food sources by two bivalve suspension-feeders, the oyster *Crassostrea virginica* and the mussel *Geukensia demissa*. Mar Ecol Prog Ser 58:299-310

Levinton J (1972) Stability and trophic structure in deposit-feeding and suspension-feeding communities. The American Naturalist 106:472-486

Loo L-O Rosenberg R (1989) Bivalve suspension-feeding dynamics and benthic-pelagic coupling in an eutrophicated marine bay. J Exp Mar Biol Ecol 130:253- 276

Lucas MI Newell RC Shumway SE Seiderer LJ Bally R (1987) Particle clearance and yield in relation to bacterioplankton and suspended particulate availability in estuarine and open coast populations of the mussel *Mytilus edulis*. Mar Ecol Prog Ser 36:215-224

Mann KH (1988) Production and use of detritus in various freshwater, estuarine and coastal marine ecosystems. Limnol Oceanogr 33:910-930

Navarro E Iglesias JIP Perez Camacho A Labarta U Beiras R

(1991) The physiological energetics of mussels (*Mytilus galloprovincialis* Lmk) from different cultivation rafts in the Ria de Arosa (Galicia, N.W. Spain). Aquaculture 94: 197-212

Newell CR Shumway SE Cucci TL Selvin R (1989) The effects of natural seston particle size and type on feeding rates, feeding selectivity and food resource availability for the mussel *Mytilus edulis* Linnaeus, 1758 at bottom culture sites in Maine. J Shellfish Res 8(1):187-196

Newell RIE Jordan SJ (1983) Preferential ingestion of organic material by the American oyster *Crassostrea virginica*. Mar Ecol Prog Ser 13:47-53

Nichols FH (1985) Increased benthic grazing: an alternative explanation for low phytoplankton biomass in Northern San Francisco Bay during the 1976- 1977 drought. Estuarine Coastal Shelf Sci 21:379-388

Nielsen MV Strømgren T (1991) Shell growth response of mussels (*Mytilus edulis*) exposed to toxic microalgae. Mar Biol 108:263-267

Officer CB Smayda TJ Mann R (1982) Benthic filter feeding: a natural eutrophication control. Mar Ecol Prog Ser 9:203-210

Peirson WM (1983) Utilization of eight algal species by the scallop *Argopecten irradians concentricus* (Say.). J Exp Mar Biol Ecol 68:1-11

Prins TC Smaal AC (1990) Benthic-pelagic coupling: The release of inorganic nutrients by an intertidal bed of *Mytilus edulis*. In: Barnes M, Gibson RN (eds) Trophic Relationships in the Marine Environment Aberdeen University Press: 89-103

Prins TC Smaal AC Pouwer AJ (1991) Selective ingestion of phytoplankton by the bivalves *Mytilus edulis* L. and *Cerastoderma edule* (L.). Hydrobiol. Bull. 25:93-100

Redfield AC Ketchum BH Richards FA (1963) The influence of organisms on the composition of seawater. In: Hill MN (ed.) The Sea Vol 2 The composition of sea-water. Interscience Publishers (Wiley and Sons) New York: 26-77

Riemann B Nielsen TG Horsted SJ Bjørnsen PK Pock-Steen J (1988) Regulation of phytoplankton biomass in estuarine enclosures. Mar Ecol Prog Ser 48:205-215

Riisgard HU (1988) Efficiency of particle retention and filtration rate in 6 species of Northeast American bivalves. Mar Ecol Prog Ser 45:217-223

Riisgard HU Randløv A (1981) Energy budgets, growth and filtration rates in *Mytilus edulis* at different algal concentrations. Mar Biol 61:227-234

Shumway SE (1990) A review of the algal blooms on shellfish and aquaculture. J World Aquaculture Soc 21(2):65-104

Shumway SE Cucci TL Newell RC Yentsch CM (1985) Particle selection, ingestion, and absorption in filter-feeding bivalves. J Exp Mar Biol Ecol 91:77-92

Shumway SE Selvin R Schick DF (1987) Food resources related to habitat in the scallop *Placopecten magellanicus* (Gmelin, 1791): a qualitative study. J Shellfish Res 6(2): 89- 95

Shumway SE Newell RC Crisp DJ Cucci TL (1990) Particle selection in filter-feeding bivalve molluscs: a new technique on an old theme. In: Morton B (ed.) The Bivalvia-Proc Mem Symp in Honour Sir Charles Maurice Yonge Edinburgh Hong Kong University Press Hong Kong p. 151-165

Smaal AC (1991) The ecology and cultivation of mussels: new advances. Aquaculture 94:245-261

Smaal AC Verhagen JHG Coosen J Haas HA (1986) Interaction between seston quantity and quality and benthic suspension feeders in the Oosterschelde, The Netherlands. OPHELIA 26: 385- 399

Tenore KR Dunstan WM (1973) Comparison of rates of feeding and biodeposition of the Amercan oyster, *Crassostrea virginica* Gmelin fed different species of phytoplankton. J Exp Mar Biol Ecol 12:19-26

Tiu AT Vaughan D Chiles T Bird K (1989) Food value of eurytopic microalgae to bivalve larvae of *Cryptopleura costata* (Linnaeus, 1758), *Crassostrea virginica* (Gmelin, 1791) and *Mercenaria mercenaria* (Linnaeus, 1758). J Shellfish Res 8:399-405

Walne PR (1970) Studies on the food value of nineteen genera of algae to juvenile bivalves of the genera *Ostrea, Crassostrea, Mercenaria* and *Mytilus* . Fish Invest London Ser II 5:1-62

Ward JE Targett NM (1989) Influence of marine microalgal

metabolites on the feeding behavior of the blue mussel *Mytilus edulis*. Mar Biol 101:313-321

Webb KL Chu FLE (1983) Phytoplankton as a food source for bivalve larvae. In: Pruder GD, Langdon C, Conklin D (eds) Proc Sec International Conf Aquaculture Nutrition. Biochemical and physiological approaches to shellfish nutrition. Louisiana State University, Baton Rouge Louisiana: 272-291

Widdows J Fieth P Worrall CM (1979) Relationships between seston, available food and feeding activity in the common mussel *Mytilus edulis*. Mar Biol 50:195-207

Wikfors GH Ferris GE Smith BC (1992) The relationship between gross biochemical composition of cultured algal foods and growth of the hard clam, *Mercenaria mercenaria* (L.). Aquaculture 108:135-154

Williams P (1981) Detritus utilization by *Mytilus edulis*. Estuarine Coastal Shelf Sci 12:739-746

Winter JE (1974) Growth in *Mytilus edulis* using different types of food. Ber dtsch Wiss Kommn Meeresforsch 23:360-375

Winter JE (1978) A review on the knowledge of suspension-feeding in lamellibranchiate bivalves, with special reference to artificial aquaculture systems. Aquaculture 13:1-33

Wright RT Coffin RB Ersing CP Pearson D (1982) Field and laboratory measurements of bivavle filtration of natural marine bacterioplankton. Limnol Oceanogr 27(1):91- 98

GRAZING OF NATURAL PARTICULATES BY BIVALVE MOLLUSCS: A SPATIAL AND TEMPORAL PERSPECTIVE

Carter R. Newell
Great Eastern Mussel Farms, Inc.
Tenants Harbor, Maine 04860 U.S.A.
and
Sandra E. Shumway
Bigelow Laboratory for Ocean Sciences
McKown Point, West Boothbay Harbor, Maine 04575 U.S.A.

INTRODUCTION

The production of suspension feeders may be limited by processes ranging from cm to km scales, and time periods from seasonal changes in temperature to tidal cycle changes in the seston flux. Field observations have shown that the processing of phytoplankton, organic detritus and inorganic particulates by shellfish beds results in kilometer scale effects on particle concentration which can be simulated by computer modelling. A four-year study (1986-1990) (Newell 1991) at four commercial mussel bottom lease sites in Maine, USA, examined oceanographic and biological processes in a variety of field experiments. The data were used to build and test mussel production models for determination of carrying capacity and optimal seeding density. Seven key factors in the benthic-pelagic coupling were studied:

- Height of the mussel feeding zone
- Fine vertical gradients (0-50 cm) in the food supply
- Variations in mussel pumping rates over tidal cycles
- Differential feeding of mussels on phytoplankton and silt particles
- Clearance of different phytoplankton species and effects of food quality on growth.
- Density-dependent growth of mussels within patches
- Spatial variations in water volume flux, and benthic boundary layer particle flux, as determined with

NATO ASI Series, Vol. G 33
Bivalve Filter Feeders
in Estuarine and Coastal Ecosystem Processes
Edited by Richard F. Dame

physical flow models forced by tide, and influenced by local bathymetry and boundary layer roughness

In addition, studies on remote sensing (LANDSAT, airborne video at 400 m) along with sea-truthing transects in a Maine estuary system suggest the estuary-scale processes of tidally-supplied diatom blooms to be a major forcing function of production of benthic filter-feeders.

In shallow, nutrient-rich bays where the hydrodynamic residence time is on the order of the water recycling time of the benthic filter-feeding biomass (Officer et al 1982), an increasing quantity of literature has suggested that benthic filter-feeders play a major role in controlling eutrophication and stabilizing estuarine ecosystems (Herman and Scholten 1990). In regions of high shellfish biomass, [e.g. 100-1500 g $ADW{\cdot}m^{-2}$ in intertidal wild mussel populations (Smaal 1991); 450-1900 g $ADW{\cdot}m^{-2}$ in Maine bottom-cultured mussels (Newell, 1990); and 800-2200 g $ADW{\cdot}m^{-2}$ in the Netherlands (Prins and Smaal 1990)], filtration rates of over 50 mg chl *a* m^{-2} h^{-1} may be observed (Dame and Dankers 1988; Prins and Smaal 1990). Edible species of phytoplankton such as diatoms may be consumed at high rates during the spring bloom, and high assimilation provides for an efficient transfer of energy into shellfish biomass, with an upper limit defined by the horizontal and vertical flux of food (Fréchette et al 1989) and the shellfish density distribution (Newell 1991). In less nutrient-rich oligotrophic waters such as those in Maine, U.S.A., shellfish may maximize shell gape during certain periods of the tidal cycle to improve their energy gain when resources are most available (Newell and Gallagher 1992) . In relatively dilute waters (i.e. $<2\ mm^3{\cdot}l^{-1}$), increased filtration rates of algal vs non-algal particles may provide an important mechanism for grazing on energy-rich components of the seston.

An understanding of the factors which control the flux and consumption of particles in shallow waters may be useful in the management of aquacultural operations, as well as in improving our understanding of the role of shellfish populations in coastal ecosystems. Due to the site-specific nature of both biological (phytoplankton) and physical (vertical flux) processes, they are best investigated *in situ* at sites of interest. A commercial lease site (Mud Cove, Stonington, Maine, U.S.A.) provided a shallow, subtidal environment (1-5m depth) with relatively constant bathymetry and bi-directional tidal currents in which to investigate rates of particle supply and demand at a variety of temporal and spatial scales. Following a review of food items of marine bivalves, the physical factors relating to supply (settling, advection, vertical transfer), physiological factors related to shellfish production, and the effects of food quality, we present data which show how a combined oceanographic and biological approach may be used to develop site-specific models of the production of benthic suspension-feeding bivalves.

Food items

While many studies have appeared in the literature regarding the preferential ingestion of organic over inorganic material by filter-feeding bivalve molluscs (e.g. Martin 1925; Newell and Jordan 1983; Bayne et al 1987; Navarro et al 1992; Fegley et al 1992; Stenton-Dozey and Brown 1992), few studies have been made of the specific food items of these shellfish.

In one of the earliest studies of the food of filter-feeding shellfish, Dean (1887) determined that diatoms composed over 88% of the diet of oysters (*Crassostrea virginica*) and Lotsy (1896) believed diatoms to be the sole food source of this species. Moore (1907, 1910), Kellogg (1910) and Grave (1912) all confirmed the dominant role of diatoms in the diet of oysters. Field (1911) examined the digestive tracts of 50 individual mussels (*Mytilus edulis*) and identified 29 species of diatoms and 9 species of protozoa. He also noted the importance of detritus.

Conversely, Petersen (1908) and Petersen and Jensen (1911) concluded that the essential food all bottom-feeding, filter-feeding marine invertebrates was detritus. Blegvad (1914) examined the stomach and intestinal contents of some 45 species of bivalve molluscs representing an array of habitats and feeding mechanisms and concluded that all were "without

Table 1. Organisms identified in gut contents of the mussel, *Mytilus edulis*, from the Gulf of Maine (from Newell et al 1989)

Species	Size (μm)	Habitat[a]	Occurrence
Bacillariophyceae			
Achnanthes longipes	60	B	occasional
Amphipora sp.	80	B	occasional
Amphora spp.	40	B	occasional
Coscinodiscus sp.	85	B/P	occasional
Eucampia zoodiacus	100 (chain)	P	occasional
Leptocylindrus sp.	30-45	P	occasional
Licmophora sp.	20-56	B	common
Melosira sulcata	30-40 (chain)	B	common
Navicula spp.	24-250	B	very common
Nitzschia closterium	70-100	B	common
Nitzschia seriata	100	B	occasional
Nitzschia spp.	10-100	B	very common
Pleurosigma sp.	110	B	common
Skeletonema costatum	15-35 (chain)	P	occasional
Surirella sp.	10-25	B	common
Thalassiosira spp.	15-25	P	very common
Thalassiosira gravida	20	P	occasional
Thalassiosira rotula	60-78 (chain)	P	common
Thalassiothrix nitzschioides	40-70	B	common
unidentified pennates	20-50	B	very common
Dinophyceae			
Dinophysis sp.	30	P	occasional
Dinophysis acuminata	50-55	P	occasional
Dinophysis acuta	50-65	P	occasional
Dinophysis norvegica	50-65	P	occasional
Dinophysis rotundata	35-50	P	occasional
Gonyaulax spinifera	25	P	occasional
Heterocapsa sp.	30	P	occasional
Prorocentrum micans	55	P	very common
Protogonyaulax tamarensis	35	P	common
heterotrophic dinoflagellate	50	P	common
autotrophic *Peridinium*	30-35	P	occasional
dinoflagellate cysts	35-40	B	occasional
Other			
silicoflagellate strew			common
Dictyoca	10-15	P	occasional
Distephanus	30-45	P	occasional
zooplankton strew			common
detritus			very common
bacteria			very common
motile flagellates	3-15		occasional
motile ciliates	75-110		common

[a]B = Benthic; P = pelagic

exception true detritus eaters" and was of the opinion that in Danish waters lamellibranchs depend entirely on detritus as feed (see also Hunt 1925). He indicated that the shellfish were able to "sort through the detritus"; however, specifics of gut analyses were not presented.

Subsequent studies confirm earlier claims regarding the importance of diatoms and other algae as food items. Hunt (1925) examined the stomach contents of over 200 individual shellfish representing over a dozen species and questioned the importance of detritus, noting that in Plymouth waters the proportion of living organisms in the stomachs of bivalves was considerable and seasonal. He further noted that, while the amount of supposed detrital material could be significant at times, the material was undoubtedly the remains of small and delicate organisms taken alive but rendered unrecognizable by the digestive process. Naviculoid diatoms were noted to be the most frequent component and he concluded that living organisms, primarily diatoms, were the most important food organisms. Martin (1925) considered nannoplankton, especially small flagellates and peridines, to constitute the main food source of oysters but acknowledged that diatoms were also important. Nelson (1947) went so far as to call *Skeletonema* "the most valuable of diatoms in New Jersey waters". While pointing out the importance of various particulates as food sources, Galtsoff (1964) noted that "organisms found in the stomach of the oyster reflect the composition of plankton and nannoplankton present in the surrounding water."

Few authors have studied the specific contents of shellfish guts in recent years and our knowledge is still very much limited to the early studies on oysters and mussels. Hummel (1985) noted that during periods of submersion the algal species composition of stomachs of *Macoma balthica* closely resembled that of the overlying water. While there was a shift in species composition during emersion (periods of deposit feeding), the resemblance to water samples remained and Hummel concluded that *Macoma* depends upon food present in the water column for the majority of its energy requirements. Newell et al (1989) identified over 38 items in the guts of mussels

(*Mytilus edulis*) in Boothbay Harbor, Maine, U.S.A. and noted that over two-thirds of the algal species identified were benthic in origin (Table 1). It was also noted that large particles (>110μm) may form a significant portion of the diet of mussels. Studies on gut contents of two species, *Placopecten magellanicus* and *Spisula solidissima* (Shumway et al 1987; unpublished) demonstrated the presence of an array of particles ranging in size from 10-350μm including algae, pollen grains, ciliates, zooplankton tests and considerable detrital material and bacteria. Benthic and pelagic algal species were common and it was concluded that the gut contents generally reflected available organisms in the immediate habitat. In a study of near-bottom seston fluxes, Muschenheim and Newell (1992) found considerable uptake of benthic diatoms and detrital carbon from the bottom 5-10cm of the water column over a Maine mussel bed.

While some species of bivalve molluscs may differentiate between organic and inorganic particles during the ingestion process, and some may preferentially digest specific particle types, the bivalve species studied thus far appear to ingest the particulate matter available to them in the overlying waters.

The presence of poorly edible, noxious, or toxic phytoplankton may reduce filtration rates of shellfish and any subsequent control of the shellfish over algal blooms. Nutrient enrichment has been cited as one of the several factors known to enhance blooms of toxic algal species (Lindahl 1983; Holligan 1985; Smayda 1989, 1990;, Smayda and White 1990; Wyatt 1990). The impacts of these blooms on filter-feeding shellfish range from decreased feeding rates to mass mortalities (Tracey 1988; see Shumway 1990 for review); further, prolonged periods of toxicity result in losses to both wild and cultured shellfisheries operations. These blooms are not limited to any specific algal group, although dinoflagellates are still the most common offenders. Blooms of *Alexandrium* and related species responsible for worldwide outbreaks of paralytic shellfish poisoning also result in mass mortalities of shellfish. Other algal species (e.g. *Gyrodinium, Gymnodinium,*

Aureococcus, Chrysochromulina) are responsible for an array of responses in shellfish including reduced growth and feeding, reproductive and recruitment failure and mass mortalities. Responses recorded to date are species-specific and dependent upon the shellfish and algal species involved (for specific references, see Shumway 1990, 1992; Shumway et al 1990; Shumway and Cembella 1992 and references therein). While many of these blooms are perfectly natural events, there seems little doubt that their increased frequencies are at least partially explained by increased eutrophication. The impact of these blooms on ecosystems is then further exacerbated by reduced feeding/grazing by resident shellfish.

Settling and resuspension of particles.

Sinking and floating of phytoplankton has been considered in great detail by Bienfang (1981), Hutchinson (1967), Smayda (1970) and Walsby and Reynolds (1980). These works all make reference to the number of variables associated with phytoplankton buoyancy, including: hydrodynamics of the water column; size, shape and orientation of cells; ability to form chains and protruberences such as setae; cell density; concentration of silica; and water density. Bienfang (1981) reports sinking rates of heterogeneous, temperate phytoplankton populations to range from 0.3 to $1.7 m \cdot d^{-1}$. Smayda and Boleyn (1965) recorded sinking rates of about $1 m \cdot d^{-1}$ for chain forming centric diatoms such as *Skeletonema costatum*.

The coflocculation of algae and silt has recently been cited as an important mechanism for the vertical transport of seston. Riebesell (1991a, 1991b) observed particles 0.1 - 3.5mm diameter (dominant size class 200-600μm) during and after the spring bloom in the southern North Sea composed primarily of algae, detrital particles and bacteria. Aggregates, defined as agglomerations >100μm diameter as determined with *in situ* photography, have a much higher settling rate (velocity of about $40 m \cdot d^{-1}$, Verhagen, pers. comm.), than their individual constituent particles . This may increase the vertical flux to suspension feeders by 25% or more.

Physical Factors

Food supply is a complex relationship between the volume flow per square meter at a site (e.g. $m^3 \cdot m^{-2} \cdot day^{-1}$), the vertical flow of the water, indicated by U star (U_*), and the size, nature and settling velocity of the particles. Limits to mussel food supply include boundary conditions at the edge of the embayment, as well as modifications to the food supply due to spatial differences in current speed, water depth, bottom roughness, mussel biomass, and mussel density distribution (aggregation). Because physical flow characteristics are site-specific, a rapid method is given (Appendix 1) for estimating volume flux based on measurements made during half a tidal cycle.

Vertical Mixing

Since mussels in bottom culture obtain their food from the overlying water column, their food supply is limited by: transport of food through the site (horizontal advection); mixing of the food to the bottom (vertical diffusion); rate (volume filtered, $m^3 \cdot h^{-1}$ or "filtration velocity" in units of $m^{-1} \cdot h^{-1} \cdot m^{-2}$ of mussel bed); water depth (H); and height above the mussel bed from which the mussels ingest food (Fréchette et al 1989; Sankar 1991). Settling of particles is also important, especially during slack tide (see below). In the previous section, we considered volume flow of seawater over the mussel lease site as a function of the tidal current speed, U, which is controlled by the M2 tide and the local bathymetry. While the volume flow is much greater than the mussels can filter completely, their food supply is limited to the amount available to the bottom. Vertical mixing, or eddy viscosity, A, of the water in unstratified (well mixed) flows depends on the relationship:

$$A = K\, U_* z(1 - z/H) \qquad (1)$$

where K = von Karmen's constant 0.4; z = height above the bottom, and where:

$$U_* = U\,(K)/(\ln(H/Z_0)) \qquad (2)$$

where Z_0 = bottom roughness, or approximately 1/30 x mussel shell length. For a review of relevant boundary-layer physics, see Fréchette et al 1989; Monismith et al 1990, and the paper by Fréchette et al, in this volume.

Mussel filtration

To develop a carrying capacity model for mussel lease sites designed to optimize seed and harvest yields, mussel growth is considered as a mass balance problem bounded by the supply of food particles and mussel demand, or volume of water filtered. Since the mussels live on the bottom, supply is more complicated than in rope culture, since current speed decreases logarithmically near the bottom. Mussel filtration pressure on the total particle spectrum is a function of the volume of water filtered by the mussels, particle concentration, gill retention efficiency, and both horizontal and vertical components of current speed. As mussel density approaches the "carrying capacity" (defined in this case as density-dependent growth due to food depletion), particle densities for mussels downstream may be reduced as much as 50% of the supply to their upstream neighbors. For example, with filtration rates of $3 l \cdot h^{-1}$, particle volumes available to mussels may be reduced from $11.8 mm^3$ to $5.7 mm^3$ inside a patch of mussels with a biomass of $1 kg \cdot m^{-2}$. Above the volume of particles needed for a mussel to maintain a maximum gut volume (i.e. a "full stomach"), mussels will produce pseudofeces, rejecting excess particles. At higher particle densities, they may even close the valves slightly to reduce filtration rate. If pseudofeces are produced, rates of ingestion (I) may then be corrected for total particles cleared by the mussels as:

I = Volume pumped x seston concentration x (1-P)

P = Wt pseudofeces/(Wt pseudofeces + feces)

where I is ingestion, P is percent pseudofeces

Since most areas studied in Maine rarely had particle concentrations above the pseudofeces threshold for mussels, (i.e., 2.0×10^7 particles$\cdot ml^{-1}$, Foster-Smith 1975), the product of clearance rates and food concentration could be used to determine mussel demand in relation to supply. Pseudofeces production was observed on windy days at two sites, and more frequently at the Long Cove sites, where measurements of suspended particulate matter (SPM; GFC filtered seston dried at 85°C) were over 10 mg per liter and particle concentrations were above 2.0×10^7 particles $>3\mu m \cdot l^{-1}$. Most of the occurrences of high particle concentrations in the field were in the vicinity of mudflats, during and after storms, and during the spring bloom. During calm experimental days, the Maine mussels rarely produced pseudofeces under natural particle concentrations of $0.5\text{-}2 \times 10^7$ particles$\cdot l^{-1}$.

Below the pseudofeces threshold, mussels are thought to filter at a relatively constant maximum rate and to retain all particles above 4μm at close to 100% efficiency (Møhlenberg and Riisgård, 1978) and at about 50% for particles at 2μm. Thus, the limit of mussel's filtering ability is at the approximate size of inorganic particles, microflagellates, large cyanobacteria and particles of organic detritus with attached bacteria. Below 2μm, many clay particles and bacteria pass through the gills into the exhalent water. It is noteworthy to mention that the beat frequency of the laterofrontal cirri may also regulate the size at which 100% retention occurs (in the range of 3-14μm, Jørgensen et al, 1986), and these may play a role in feeding selectivity adaptations. Mussels might possibly increase gill "leakage" to numerous inorganic particles in the 3-5μm range when presented with algal particles of slightly higher diameters (Bayne et al 1977), and gill leakage has also been noted during spawning (Newell and Thompson 1984). Variable retention efficiency in the 2-5 μm range may be an important aspect of mussel/particle interactions in the coastal zone.

In very dilute suspensions, e.g. 5×10^6 particles$\cdot l^{-1}$ $>3\mu m$ diameter, (at low tide on the 11/8/90 experiments at Mud Cove, see results) food availability was below the mussels' ability to obtain positive energy gain by active feeding at maximal shell gape, and the mussels may close for several hours, periodically sampling the water to see if conditions have improved. A few authors have mentioned a lower concentration threshold at which mussels pump actively (Thompson and Bayne

1972), but this has not been investigated thoroughly in the field. Thus, during the typical "summer crash" in food concentration in Maine, periods of the tidal cycle may be spent with prolonged shell closure. Mussels fed actively at high tide and through the ebb, until food concentrations fell, after which they closed up for 3-4 hours (see video experiments, below). Low respiration rates were also observed during periods of shell closure.

Most previously published studies on mussel feeding have used Coulter Counters which count particles within a given size range (e.g. >3μm) regardless of particle type. By comparing particle uptake rates of mussels in flow-through feeding chambers to control chambers with no mussels (i.e. Fréchette and Bourget 1985), the difference in particle concentration is multiplied by water flow rate to get rates of particle consumption. Feeding and oxygen consumption of benthic filter feeders are best investigated at the sites of interest using water pumped off the bottom (e.g. 0.5m) through experimental chambers. In a flow-through system, after Fréchette and Bourget (1985), with flow rates >250ml$\cdot$min^{-1} to prevent recirculation (Møhlenberg and Riisgård 1979):

$$\text{Consumption } (mg\cdot h^{-1}) = \text{flow rate } (l\cdot h^{-1}) \times (conc._{control} - conc._{mussel})\ (mg\cdot l^{-1})$$

$$\text{Filtration rate } (l\cdot h^{-1}) = \text{Consumption } (mg\cdot h^{-1})/conc._{control}(mg\cdot l^{-1})$$

In a flow-through efflux chamber with a stir-bar such as used in the Mud Cove field experiments, one considers C_m, the concentration of particles around the mussel, C_i, the concentration in the incoming water, and C_o, the concentration of the outgoing water (Hildreth and Crisp 1976). Then filtration rate is:

$$\text{Filtration rate} = \text{flow rate} \times (C_i - C_o/C_m)$$

If there is a stir bar, C_m is approximately equal to C_o, such that filtration rate is:

$$\text{Filtration rate} = \text{flow rate} \times (C_i - C_o/C_o)$$

Determining pumping rates using the particle tracer method as described above correlates well with more direct measurements using microscopic current probes in the exhalent siphon and laser-beam techniques (Famme et al 1986) above critical flow rates (in the 200-400 $ml \cdot min^{-1}$ range per 45 mm mussel).

Inherent in the use of electronic particle counters ($Coulter_{TM}$) in measuring filtration rates are the assumptions:

- A 100% efficiency in clearance above a given particle size, i.e. 3μm.
- Equal clearance rates on all particles above 3μm, regardless of particle type (e.g. sediment vs algal cell).

In the absence of pseudofeces production, food consumption rates are usually determined by multiplying pumping rate by food concentration (e.g. Chl *a*, algal cells, SPM, POM, PON, POC, energy (joules), particle volume):

Consumption rate = filtration rate x particle concentration

Particle consumption can also be studied in closed systems, where the exponential decline in particle numbers in a well-mixed chamber is followed over time and the filtration rate is calculated according to the Coughlan (1969) method:

$$m = \frac{M \times \log_e (C_o/C_t)}{n \times t}$$

where m = filtration rate, M = volume of suspension, n = number of mussels, C_o = initial concentration, C_t = concentration at time, t.

Williams (1982) cautions that when filtration efficiency is less than 100%, i.e. under 5μm particle diameter for mussels, the use of the above equation will underestimate filtration rate. Particle production by mussels may also bias these types of experiments and should be investigated further. The Coughlan method was used in all the flow cytometry experiments described below.

To obtain near-natural conditions for measurements of physiological rates, techniques have been developed to enclose an area of the shellfish bed in a benthic ecosystem tunnel, BEST (Dame et al 1984; Dame and Dankers 1988; Prins and Smaal 1990), or in a flume with vertical walls (Asmus and Asmus 1991). For the tunnel,

Particle flux ($mg \cdot h^{-1}$) = flow rate x (C_{ti} - C_{to})

where C_{ti} = tunnel inlet; C_{to} = tunnel outlet.

Results are easily converted to a-square meter basis by dividing volume flow ($m^3 \cdot s^{-1}$) by area (m^2) enclosed by the tunnel or flume, and converted into individual rates based on mussel density and biomass within the tunnel. Non-invasive techniques include the time-lapse benthic video monitor (TLBVM, Newell and Gallagher, 1992) which give an indication of pumping rates which are proportional to shell gape (Jørgensen et al 1988; Kramer et al 1989). These are useful for observing tidal variations in feeding in relation to short-time food supply dynamics, but are not as good as direct methods of particle consumption by the mussels.

Food particle (=seston) ingestion rates, forming a size-specific maximum when the mussel gut is full, may be modelled (Hawkins et al 1990) in relation to individual gut volume (indicated by gut content (GC), and gut passage time (GPT):

$$\text{Ingestion rate } (mg \cdot h^{-1}) = GC\ (mg)/GPT\ (h)$$

where $GC = 0.08\ X^{0.68}$; $GPT = 1.30\ X^{0.34}$; X = dry tissue weight (mg) Thus for a bed of given biomass, the vertical flux of particles required for the mussels to maintain a maximum ingestion rate is related to the seston concentration as in Fig 1. Ingestion rate may also be modelled according to Bayne et al, (1989) as a function of particle concentration:

$$\text{Ingestion rate} = IR_{max} * (1-e^{-aC})$$

where IR_{max} = maximum rate based on body size

a = rate at which maximum IR is approached

C = particle concentration

In the Brylinsky model (Brylinsky and Septhon 1991), ingestion rate is modelled as the lesser of the size specific maximum ingestion rate (IR_{max}) and the rate accounted for by somatic body size and temperature. Increases in pumping rates with temperature are due to decreased kinematic viscosity of seawater over the range of 0 - 20°C (Jørgensen et al 1990).

Due to high frequency variability in mussel feeding rates, frequent samples are needed to obtain a reasonable picture of rates at a give time, e.g. stage of the tide (Fréchette and Newell 1988, unpublished). In the Maine field feeding experiments, filtration rate was based on Coulter Counter samples (>100 can be processed per day) or extracts of chlorophyll *a* (>100 can also be processed). In a few cases, direct uptake of particle volume, particle concentration, POC, PON, phytoplankton cells, POM, and SPM were determined. In the 10 flow cytometric experiments in the lab, the Coughlan method was used to determine filtration rates on chlorophyll and non-chlorophyll particles. In the September (1990) drogue experiment, the flow cytometer was also used to analyze surface and bottom water samples upstream and over a commercially seeded mussel farm (Mud Cove).

Scope for growth

Scope for growth is a useful method of obtaining an instantaneous value of a mussel's energy budget with a simplified balance, in energy equivalent terms:

Scope for growth (SFG) (cal h^{-1}) = (Cp x E x AE) - R

where:

Cp = particulate food ingested

E = energy content of food

AE = assimilation efficiency of food

R = respiration.

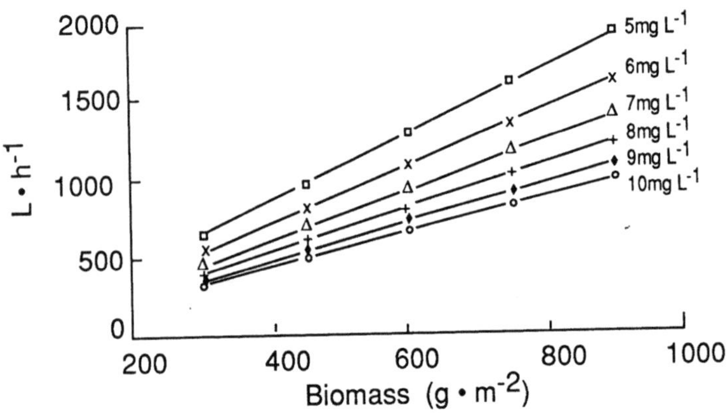

Figure 1. Volume of water required to the mussel feeding zone per square meter for maximum ingestion (Hawkins et al 1990) at ranges of shellfish biomass for varying seston concentrations.

Other terms in the mussel energy budget, such as energy lost due to excretion, are minor and are not considered here. However, since energy available for growth may also be diverted into reproductive products, scope for growth may be different from actual growth, especially in relation to meat size after spawning has taken place.

The scope for growth = zero at the minimum food required to keep the mussel alive but not growing, the so-called "maintenance ration". A reported value for mussels (Bayne et al 1989 and Widdows and Hawkins 1989) has been 4.7 joules$\cdot g^{-1} \cdot d^{-1}$ It is common for mussels to achieve 0 or negative scope for growth and actually lose meat yield during certain times of the year in the absence of spawning. This happened in Mud Cove with low food and no phytoplankton bloom in fall and winter of 1990.

Even during one tidal cycle, conditions for mussel scope for growth may change dramatically:

- At low particle concentrations mussels reduce shell gape and pumping rate causing values of particulate food ingested (Cp) to decline rapidly.
- Reduced shell gape and accompanying reduced water flow decrease oxygen diffusion and reduce mussel respiration to a fraction of the value at maximum pumping rate.

To predict average conditions for growth at any site, daily scope for growth becomes a composite of the values at all tidal cycles. Thus, for annual cycles in mussel scope for growth, the range in food available during a tidal cycle on a single day may equal the range in that parameter over an annual cycle. Similarly, wide ranges in mussel size, age, particle concentration and food quality within a lease site may result in unequal growth rates and reduced yields. Factors important in scope for growth calculations also include mussel size, mussel age, and water temperature.

A measure of a mussel's ability to grow at a given time is called the net growth efficiency, which accounts for the fact that mussels have a higher respiration rate while actively growing: Net growth efficiency = G/G+R; R = mussel respiration which reaches a maximum of about 75% during the productive time of the year (Riisgård and Poulsen 1981) and varies with phytoplankton concentration, i.e. 19% at 3×10^6 cells$\cdot l^{-1}$ vs 61% at 26×10^6 cells$\cdot l^{-1}$ (Riisgård and Randlov 1981).

Assimilation of organic matter ingested by the mussels is most commonly investigated by the ash tracer, or Conover (1966) method. This is based on periodic collection of ambient water samples and fecal pellets during an experiment (typically 1 tidal cycle), obtaining dry weight, ash weight, and weight of organic matter by the difference. Since suspended particles may have weight of structural water in clays, as part of the dry weight (Billen 1978), there is some uncertainty in the organic weight value which is typically determined at 90°C (Bayne and Widdows 1978) and ash weight determined after 3 hours at 450°C. Comparison of lab samples at 80° and 92°C yielded no significant differences in organic weight, and ashing at 450° vs 500°C would account for only about 5% additional weight loss (Mook and Hoskin 1982), but filters become brittle close to 500°C.

Extensive studies on mussel assimilation efficiency in relation to the relative food quality (i.e. % organic matter) of the diet (Bayne and Hawkins, 1990) have shown that gross and net food absorption efficiencies improve with increasing organic content of the diet. Mussels are adapted to long-term changes in the food supply by varying gut passage time and the secretion of digestive enzymes. Bayne et al (1989) reported a regression of absorbed ration (= food ingested x assimilation efficiency) to food quality as:

$$\text{Absorbed ration (joules}\cdot h^{-1}) = 22\ (1 - e^{-2.55}\ \text{POM})$$

A maximum efficiency was obtained at POM levels of 1.9 mg l^{-1}.

Mussel respiration is measured in a similar way to feeding, by comparing chambers with and without mussels at known flow rates, although flow rates are reduced to about 100 ml min^{-1} to see differences:

$$\text{Respiration (ml } O_2\cdot h^{-1}) = \text{flow rate} \times (C_c - C_m)$$

where C_c = concentration O_2 control chamber (ml$\cdot l^{-1}$); C_m = concentration O_2 mussel chamber (ml$\cdot l^{-1}$)

In the tunnel, differences between inlet and outlet are calculated, yielding total mussel bed respiration.

Effects of body size and age

The dry tissue weight, to which most physiological functions are scaled, changes over time. As animals grow larger, rates of essential physiological processes such as feeding, oxygen consumption and assimilation, change in relation to body size, with varying exponents. Since the exponent for respiration is higher than that for feeding, smaller mussels have an energetic advantage during periods of low food supply. Mussel filtration rate per gram tissue weight increases with dry tissue weight (Møhlenberg and Riisgård 1979) as:

$$\text{Filtration rate } (l \cdot h^{-1} \cdot g^{-1}) = 7.45X^{0.66}$$

Mussel oxygen consumption is related to body size W in mg (Riisgård and Randlov 1981) as:

$$\text{Respiration rate: } (\mu l\ O_2 \cdot h^{-1}) = 1.83W^{0.75}$$

Metabolic fecal losses, or the energy required to digest the food, increase with body size W in mg, forming a significant energy drain of as much as 15% of the ingested food (Hawkins et al 1990): Metabolic fecal loss $(\mu g \cdot h^{-1}) = 1.33W^{0.64}$

Metabolic fecal loss per unit ingestion:

$$\mu g \text{ fecal loss per mg ingestion} = 21.6W^{.3}$$

where W = dry tissue weight (mg).

Dependency of the size of particles ingested by mussels on mussel size has not been investigated to this date, however large benthic diatoms and ciliated protozoa of over 100μm long were common in 30 - 60 mm mussel digestive glands (Newell et al, 1989, also Table 4).

Spawning output also increases with mussel size, such that in typical studies (e.g. Griffiths 1981; Bayne 1976), mussel reproductive output rose from about 10% of the net energy gain for 20 mm mussels to over 90% for large 70-80mm mussels. Bayne (1976) reported dry weight of gamete production related to body size as:

$$\text{Fecundity} = 73.5W^{2.79}$$

Mussel net growth efficiency decreases with mussel age due

to reduced efficiency of protein metabolism in the cells as they age (Hawkins 1991). Rates of protein turnover causing higher energy expenditure add to the reduced net energy gain in older, larger spawning adults. Similar differences in protein turnover rates may be found in mussels of different genotype (Koehn and Bayne 1989).

Effects of temperature

Because mussels are poikilotherms, chemical reactions are based on seawater temperatures and metabolic rates increase with temperature such that oxygen consumption may be modelled as (Page and Ricard 1990) as a summary of published works by Thompson (1984); Widdows et al (1979) and Widdows (1978) as:

$$VO_2 = aW^b$$

where: VO_2 = volume of oxygen consumed e.g. ml h^{-1}

$b = 0.782$; $a = 0.117 \times 10^{0.044T}$

W = tissue weight (g)

Similarly, temperature has been modelled in assimilation rate (Bayne 1976) as: $AE = AE(T{=}0) - 0.007 \times T$

AE = percent absorption efficiency

T = temperature (°C)

Shell size can be modelled both as a function of temperature controlling rates of inorganic deposition of the shell,(Almada-Villela et al 1982) and modelled by Brylinksky and Septhon (1991) as Q_{10} shell = 0.1386; and whether or not there is positive scope- for-growth allowing for growth of the organic portion of the shell, e.g. in Brylinsky and Septhon (1991):

Growth Shell inorganic matter = ROI * GRSHO * QSI

where GRSHO = growth of shell organic matter

ROI = ratio organic/inorganic shell input

QSI = Q_{10} shell inorganic deposition rate

Food quality

For a given volume of particulate food ingested, mussel energy gain is due to the food quality, or percent organic matter of the diet. This can be expressed as the weight of organic matter per particle volume (mg POM·mm^{-3}) (Bayne et al

1989), the POM/SPM ratio, the chlorophyll *a* to SPM ratio, the carbon to nitrogen ratio (C:N), or the phytoplankton carbon to total carbon at the site. These indicators of food quality reflect the proximate (protein, carbohydrate, lipid) content of the diet, and the relative dilution by inorganic sediment. In areas of high turbidity and low food quality, mussels reduce filtration rates and increase gut passage time to obtain acceptably high rates of assimilation from the generally nutritionally poor resuspended bottom sediments, and growth is reduced. The proximate composition of phytoplankton has been shown to be closely matched with bivalve tissue, and less so with detrital food sources. The importance of food quality on assimilation is discussed above.

Mussel density

The effects of mussel density, or crowding within a patch, may indirectly reduce scope for growth due to "neighbor interference" beyond the seston depletion, or food competition effects. The ability of mussels to achieve full shell gape, and pump at maximum rates, may be reduced by lateral pressure within a mussel patch, which may overcome the opening moment of the valve ligament. Workers have found a significant correlation between pumping rate and shell gape (Jørgensen et al 1988). Underwater video observations (Newell and Gallagher 1992) indicate active movements of the mussel valves over time scales of 10 minutes which illustrate competition for space within patches. Mussel mantle extension, even at reduced gape, may help to increase filtration rates by increasing the siphon area and elongating the gill axis (Jørgensen 1990). Nonetheless, density- dependent growth of mussels in bottom patches over 1 m diameter is significantly affected by conditions of local mussel density (Newell 1990). The degree to which growth interference is based on local competition for food or crowding, in relation to model results, indicates that food competition may account for most of the reduced growth observed at the lease sites in patches of over 500 mussels$\cdot m^{-2}$. The observation that beds of mussels reaches a constant biomass (i.e. the carrying capacity) at a given site (Hosomi 1985; Newell 1990) suggests that site-specific seston flux limits production in shellfish populations.

Field and laboratory experiments

A five year study (1986-1991) of commercial mussel bottom lease sites in Maine, U.S.A., investigated the sensitivity of mussel growth and seed to harvest yields to conditions controlled by the mussel farmer, namely:

- Seeding density (both mean density and density distribution (aggregation).
- Site selection, including current speed, seston concentration and seston quality.

Also investigated were rates of particle clearance by mussels, respiration, assimilation and growth from commercial lots of 3,000 metric tons of mussels seeded at a variety of sites during the study. A mussel growth model (Newell and Campbell 1992) was produced which predicted growth of Maine mussel populations for each lease section (~3 hectares). Optimal seeding densities, based on meat yield, varied from 300 mussels$\cdot m^{-2}$ to over $1000 \cdot m^{-2}$ at some highly productive sites. Preliminary data from this study (Newell 1991) are presented here, along with a discussion of factors on both spatial (kilometer to centimeter) and temporal (hours to annual cycles) scales.

Over 1500 water samples were taken using a horizonal alpha bottle (Grizzle 1988) of 2.2 l capacity to which a 243μm filter was added to the outlet hose to remove zooplankton. Both bottom (2-12cm off the bed) and surface (1 m below surface) samples were taken. Typical water depths were 2-4 meters during the study. Samples were taken for chlorophyll *a* fluorescence (Phinney and Yentsch 1985), weight of seston (POM, SPM, POC, PON), phytoplankton species and biovolume, particle concentration and particle volume. For phytoplankton and particle counts, only particles above 3μm were counted. Using flow cytometric analysis (see Newell et al 1989), particle sizes and types (chlorophyll vs non-chlorophyll) were determined.

The physical and oceanographic characteristics of the site were determined by surveyor, navigational charts, and at 30m grid intervals in a 2-dimensional flow model (DUCHESS), validated with tide gauges and an electromagnetic S4 current meter moored 0.5m off the bottom or profiling. Tidal range varied from about 3m on neap tides to 4m on spring tides. All sites studied were shallow and subtidal (= 1 - 5m depth at low tide). Bottom roughness was identified with side-scan sonar and direct observations by diver. Boundary layer measurements of current speed above mussels in a flume were used to check theoretical estimates of z_0 and U_*.

Sampling for mussel filtration rate was performed using either flow-through chambers (as in Fréchette and Bourget 1985), in a so-called efflux apparatus (consists of a head tank, a multi-channel peristaltic pump and 10 individual chambers with stir bars), in a benthic ecosystem tunnel, BEST (Dame et al 1984), using a device called the benthic organic seston sampler, BOSS (Muschenheim and Newell 1992), and sampling surface and bottom waters following a drogue over a 20 hectare seeded mussel lease site (Mud Cove).

RESULTS

Physical factors

Boundary layer characteristics above a mussel bed were investigated in 1987 flume experiments (Geyer and Newell 1987, unpubl.), in a 17m long, 20cm deep CRL flume at Woods Hole, MA (Trowbridge et al 1989) to which a 2-meter patch of 60mm mussels was added. Laser-doppler velocimeter (LDV) profiles were made in triplicate at 6 positions along the flume upstream, over and downstream of the mussel bed. Values for U_* and Z_0 were measured for 5 and 15 $cm \cdot s^{-1}$ flows (Table 2). Note that with a mussel shell length of approximately 60mm, a theoretical value for bottom roughness of Z_0 of 6/30 = 0.2cm is reasonably close to the flume run.

U_* is directly proportional to current speed, U, and

decreases slightly with increasing depth. In the flume runs, a 20 cm water depth was used where the boundary layer actually extends close to the surface. At mussel grow-out sites, in 1 - 5m water depths, lower U_* values are expected in relation to U. Since it is difficult to get good U_* measurements in the field, some approximations must be made to get realistic values for equation (1)(see p._). While field roughness is probably higher than that determined on the basis of shell length, and mussel patch form drag would contribute to greater roughness, as a matter of course, U_* can be approximated as U/16 for field mussel grow-out sites in Maine (R. Geyer, pers. commun). See Appendix 2 for a discussion of field profiles (Table I) vs estimated values of U_* (Table II) at subtidal mussel farm sites in Maine.

Once U_* and z_0 have been determined for the site, the food supply to the mussels at the bottom, e.g. a feeding zone of the bottom 2-5cm, is the result of a balance between the rates of horizontal and vertical fluxes over the entire water column. Since the vertical flux is the result of a concentration gradient above the bed, the eddy diffusivity A (equation 1 above) can be solved using a finite difference model (Fréchette et al 1989; Sankar, 1991), where A is determined by a computer and combined with values of U at a series of specified depths (assuming a logarithmic velocity profile) to balance the uptake at the bottom layer by the mussels with a decrease in the food concentration further up in the water column. Food changes on horizontal and vertical scales can be investigated as a function of two major parameters: U_*, as U/16, in $cm \cdot s^{-1}$, and mussel filtration velocity, W_{filt}, $m \cdot h^{-1}$, equivalent to m^3 filtered by the mussels per square meter per hour.

Table 2. Values of U^{*} and Z^{0} obtained from verticle profiles in a flume over a 2.0m mussel bed (Geyer and Newell, unpublished)

RUN ($cm \cdot s^{-1}$)	U_{*} ($cm \cdot s^{-1}$)	Z_{o} (cm)	r^{2}	DISTANCE FROM EDGE (m)
5	0.75	0.01	1.00	0.2
5	0.73	0.18	0.99	0.8
5	0.66	0.17	0.99	1.6
15	1.78	0.16	0.93	0.2
15	1.61	0.105	0.99	0.8
15	1.50	0.134	0.97	2.2

As the water moves downstream, the output of the model is food concentration as a percentage of the initial concentration with a decrease in the downstream direction, and with a vertical gradient in food concentration. Site-specific model runs of food depletion (Fig.2) give direct estimates of flux to the mussel bed in which the balance between ingestion rate required for mussel growth and particle supply can be investigated. The dimensionless parameter, W_{filt}/U_{*} can be examined at varying depths for a given shellfish biomass to find the near-bed food concentration (Figs.3,26). These model runs can then be used to determine different management scenarios on the effects of mussel seeding density on food supply at the given lease site, under a range of environmental conditions. For example, if we determine that a 50% decrease in food availability over a 200m seeded bed is the maximum allowable for a given lease site, the seeding density of mussels may be adjusted to obtain the desired results. For Mud Cove, measured values of W_{filt}/U_{*} at 550 mussels$\cdot m^{-2}$ increased from less than 0.1 at seeding to over 0.2 after 1 growing season, to over 0.4 after two seasons (Newell 1991).

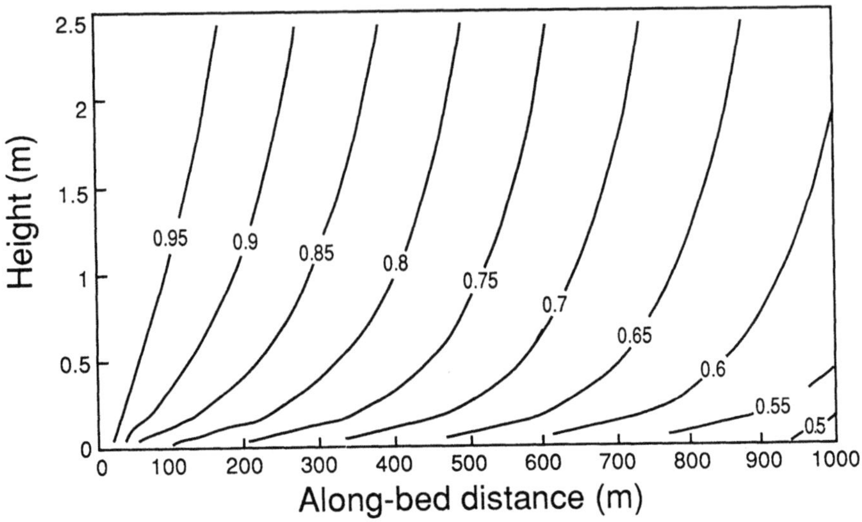

Figure 2. Output of particle contours W_{filt} 1.5 m h^{-1}, bulk current velocity 30 cm s^{-1}, z_0 = .0025m, ingestion height = 4 cm. From Sankar, 1991 for Mud Cove simulation (Newell, 1991).

Characterization of site-specific hydrodynamics and food supply to the boundary layer

To obtain spatial resolution of food supply to a given mussel bed, the results of physical oceanographic models may be useful on length scales of 25-100 meters. By using the DUCHESS 2-dimensional current simulation model (Richardson et al 1992) for Mud Cove (Fig.4) output of volume flow and current speed agree well with the rapid method described in Appendix 1.

Placement of tidal gauges and current meter moorings, along with precision bathymetry provide sufficient data to run models for each site. Because of the strong dependence of eddy diffusivity on U_* as discussed above, current speed at the particular lease section is more critical to mussel growth than volume flux per se. For example, the following comparisons were made at Mud Cove for 960 m^2 grid points within lease sites B, C and D taking into account local water depth and land boundaries:

Table 3. Volume flow, tidal exchange and mean current speed for three stations in Mud Cove, Stonington, Maine

	Spring Tide			Neap Tide		
Station	Flow (m^3 s^{-1})	Exchange	Current (cm s^{-1})	Flow ($m^3 \cdot s^{-1}$)	Exchange	Current ($cm \cdot s^{-1}$)
B	10.8	73%	11.6	7.0	62%	7.8
C	13.8	78%	15.3	8.5	64%	10.2
D	8.0	74%	7.8	4.4	62%	5.2

Tidal exchange, was calculated as (Dyer 1973):

$$\text{Exchange} = \frac{(\text{Volume high tide} - \text{Volume low tide})}{\text{Volume high tide}}$$

Thus at station C, mussel food supply would be almost twice as great as at D, while tidal exchange only varies by 2-4% and volume flow by 25-35%. The vertical flux, proportional to U_* as current divided by 16 varies from 0.95 cm s^{-1} at station C at spring tides to 0.33 $cm \cdot s^{-1}$ at station D during the neap tide. In a study by Carver and Mallet (1990), supply vs demand of POM at a Nova Scotia mussel farm was based on such a tidal exchange. In areas which are not enclosed basins, however, a "flush" type of exchange is more appropriate. Following drogues in Mud Cove revealed flushing rates several times that predicted by tidal exchange. Thus, the use of current speed as an indicator of both horizontal and vertical flux of food is warranted at that site.

Once current speed is determined for the area of interest, a logarithmic profile of speed vs depth (equation 2 above) can be used to predict vertical variations in current speed. Diver observations of z_0 ranged from 0.3cm to 0.6cm in mussel hummocks (Newell, 1991). The effects of increased roughness (ranges z_0 = 0.001 - 0.003cm investigated) or ingestion height of mussels (1.5 to 5cm investigated) on near bed food concentration were, however, small in model simulations relative to differences in water depth, mussel filtration velocity, or current speed.

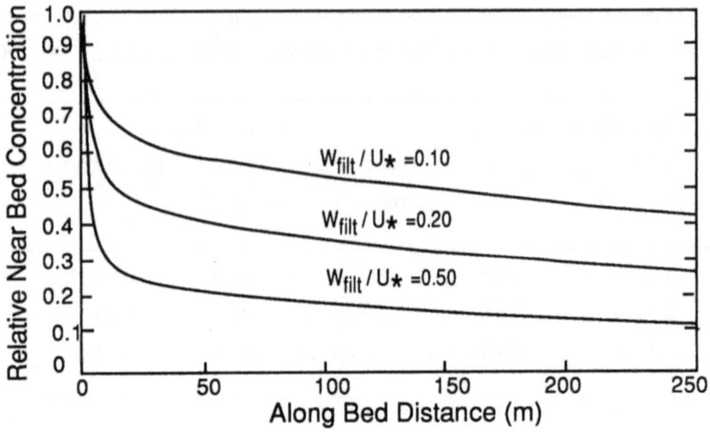

Figure 3. Relative near-bed food concentration in water depth of 2 meters in relation to dimensionless parameter, W_{filt}/U^*. (Sankar 1991).

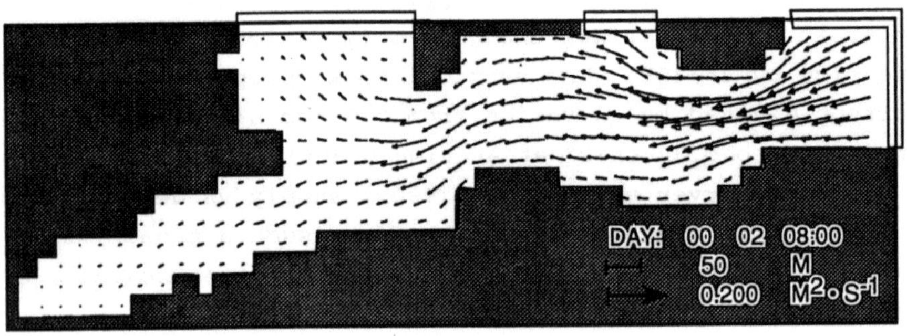

Figure 4. Output of DUCHESS current simulation model for 35 m grid in Mud Cove lease area on the flood tide. Grid mesh size is 31 meters. (Newell and Richardson, in preparation)

Settling and resuspension

Flocs from 0.1 to 2mm diameter were observed frequently at high tide at the Mud Cove site during the video experiments (see below) and near the peak of the spring phytoplankton bloom, but were not present in pre-filtered coulter counter samples. Whether the flocs break up upon entry into the mussel mantle cavity is unknown at this time. Settling rates on the order of 10 - 20 $m \cdot d^{-1}$ are considered realistic, and indicate the importance of flocs in vertical supply of food during certain times of the year. Thus, algae advected into a lease site 3m off the bottom at high tide would be able to settle to the bottom between high water and mid-ebb tide. Settling of particles is more important at low-current sites, such as Webb Cove and Roque Island, Maine with low U_* values of about 0.15 $cm \cdot s^{-1}$.

Due to the high abundance of benthic diatoms and associated high values of chlorophyll *a* in the sediments adjacent to mussel patches (Table 6), any resuspension could result in delivery of nutritionally significant benthic diatom biomass and detrital carbon to the mussel bed. In a study in Nova Scotia (Grant et al 1986) with comparable U_* values to Mud Cove, sediments exported 0.12 mg chl $a \cdot h^{-1} \cdot m^{2}$. It was necessary to include resuspension in a mussel production model (Newell and Campbell 1992) to match observed growth with food supply. In a field investigation of seston consumption by a continuous wild bed of mussels (Muschenheim and Newell 1992), high rates of uptake were noted of benthic diatoms (*Nitzschia* and *Pleurosigma*) from the bottom 5 to 10cm of the water column. Investigations of the current or waves required for the resuspension of phytobenthos might shed some light on the availability of this nearby food resource. The "phytobenthos garden" in the vicinity of shellfish biodeposits may be an important factor in the carrying capacity of embayments for shellfish culture.

Table 3. Characteristics of surface sediment in vicinity of mussel patches at Mud Cove, Stonington. Surface sediment (1-2 cm) sampled by diver.

DATE	N	CHLOROPHYLL ($\mu g \cdot cm^{-2}$)	CARBON ($\mu g \cdot cm^{-2}$)	NITROGEN ($\mu g \cdot cm^{-2}$)	PHYTOPLANKTON ($cells \cdot cm^{-2}$)
3/22/91	6	9.9	2927	399	-
11/9/90	2	-	1917	272	238,840

Kilometer scale studies of seston availability

In a study of the use of aerial and satellite remote sensing for assessing bivalve shellfish aquaculture sites (Campbell et al 1991), a transect was made in September, 1989 in an estuary in Maine, U.S.A. (Frenchman's Bay, Hancock County), Fig.5, while a Landsat photograph, and aerial videos were taken at 3,000, 1,600 and 900m during high and low tide.

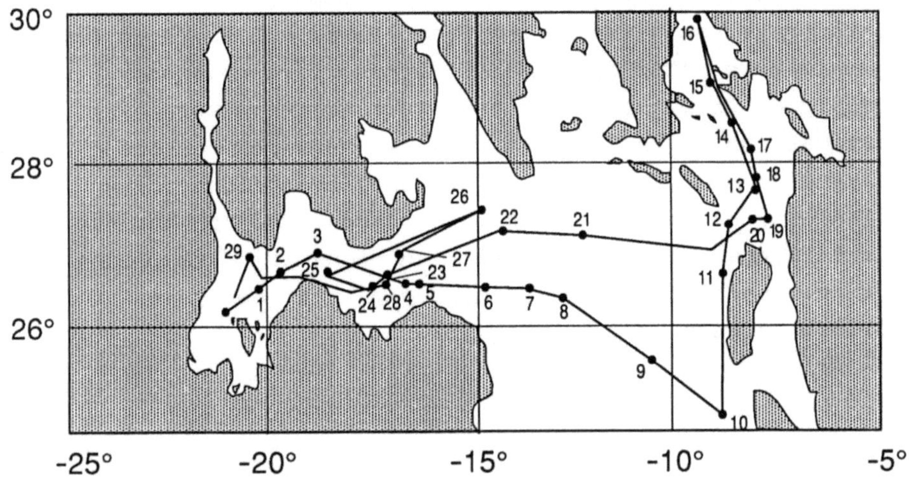

Figure 5. Map of study area, Frenchman's Bay, Maine showing boat's track and location of the sampling stations.

In-vivo fluorescence, temperature, and surface (-1.0m) water samples were taken along an ocean transect (Fig.6).

Results indicate higher chlorophyll in the mouth of the estuary than in the shallow bay at low tide. The phytoplankton are advected into shallow water on the flood tide (compare outgoing transect at low tide, solid line, to incoming transect, dashed line, at high tide, Fig.6). At low tide, phytoplankton cell counts increased dramatically toward the mouth of the estuary (Fig.7), and at the end of the low tide transect into Flander's Bay cell numbers declined in shallow water (Fig.7). The transect data indicate that at about 14-18'N, 68'W a bloom of *Rhizosolenia delicatula* at the mouth of Mt. Desert Narrows and Flander's Bay was a source area of phytoplankton for the nearby embayments, associated with probable upwelling at Googin's Ledge and Half-tide ledge. Refinement of air-borne video sensors holds great promise in tracking these blooms in shallow coastal areas. With the hypothesis of tidal transport of seston into coastal embayments, the effects of shellfish beds on food availability in the boundary layer were studied *in-situ* with the drogue experiments.

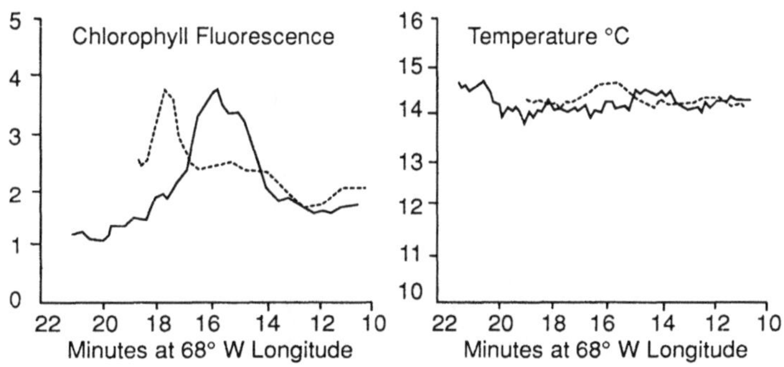

Figure 6. Outgoing (solid) and incoming (dashed) transects of chlorophyll fluorescence and temperature on September 6, 1989. Outgoing was at low tide and incoming was at high tide.

Within embayments: drogue experiments

To examine depletion of seston above mussel beds, it is important to follow the same water mass across an area of known mussel density to reduce error associated with short term (i.e. 15 minute) variability in seston concentration. A weighted drogue at approximately mid-depth with a high cross-sectional area and surface buoy were followed at various positions along the site of interest. At Mud Cove, where 1,000mt of mussels were seeded at an average density of about $500 \cdot m^{-2}$, the drogue was followed over the seeded beds, and water grab samples were taken (Fig. 8) relative to stations along the seeded mussel plots.

Stations were determined as: Upstream (at flood tide, upstream of station B, at ebb tide upstream of station E); stations B to C (350m); stations C to D (200m); stations D to E (275m); surface (1 m below top of water); and bottom (horizontal alpha bottle, effectively the bottom 2-12cm).

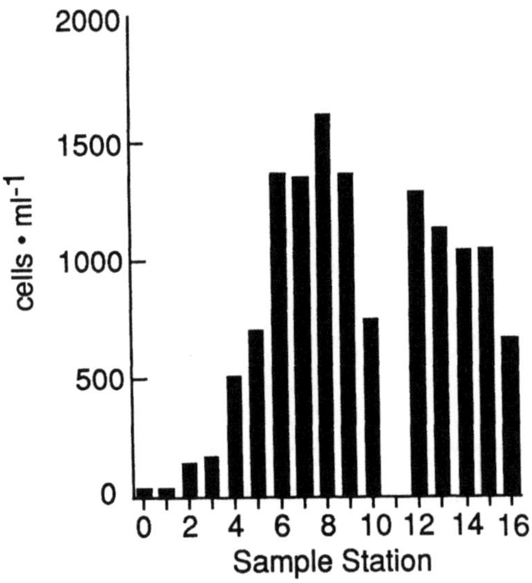

Figure 7. Concentrations of *Rhizosolenia delicatula* (cells ml^{-1}) surface waters along cruise out of Mt. Desert Narrows (stations 1-10) and into Flander's Bay (stations 12-16) on September 6, 1989. Outgoing transect (stations 1-16) was made at low tide and incoming transect was made around high tide.

Drogue experiments were performed during June, July, August and September, 1990, when mean mussel biomass at Mud Cove was 200-500 g dry wt$\cdot$m^{-2}. Drogue positions were compared with current speed and direction data. Since the S4 current meter was 0.5 m off the bottom, it underestimated bulk current speeds. On August 8, 1990, measured speed by the drogue was 15cm$\cdot$s^{-1} vs 10.25 measured by the S4. In most vertical profiles at the site in water depth of 2-4m, current speed at 0.5m depth x 1.2 = bulk current. Drogues could only be used on relatively calm days, for a 2-layered flow was observed with strong NW winds on the floodtide. Since the water was partially stratified during the summer, (in some cases, surface water was 2°C warmer than water at 3m depth) comparisons of surface vs bottom concentrations should be made with caution.

The concentration of seston was measured in surface and bottom waters alongside a moving drogue over tidal cycles (in July, August and September, 1990). The decline in POM concentration across the lease is striking on both the flood and the ebb tides (Figs.9 and 10).

Over a tidal cycle, at station D, POM concentration at surface and bottom waters suggests tidally transported POM imported to the lease in the surface waters at high tide, which settles and reaches the bottom at ebb tide (Fig.11). These results go along with conclusions reached with the video experiments, which show maximal shell gape and feeding activity during high and ebb tides at Mud Cove (Newell and Gallagher 1992).

On September 24, 1990, samples were also taken for phytoplankton carbon from settled biovolumes of phytoplankton, and flow cytometric (FCM) characterization of particle size and type (chlorophyll vs non-chlorophyll). Since the FCM samples had to be prefiltered at 80μm, the results on the larger phytoplankton chains should be viewed with caution. Chlorophyll a on grab samples declined from 2.1 to 1.4μg l^{-1} from Stations B -D on the flood tide. Phytoplankton carbon declined from 79.3μg at station B to 27.9μg at station D, a 283% decline (Fig.12). Diatom cell counts dropped the greatest in bottom water samples, while surface phytoplankton were relatively unchanged (Fig.13).

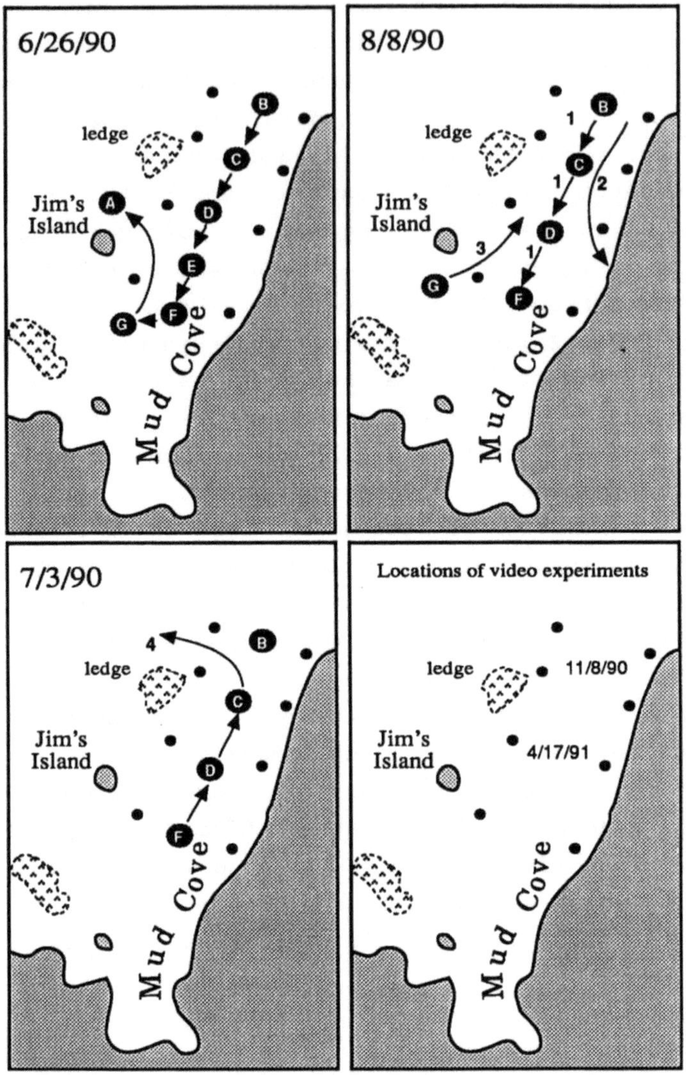

Figure 8. Location of drogue paths at Mud Cove mussel lease site on four sampling dates (see also Fig. 3) and location of underwater video experiments. Black dots are seafarm corner buoys. Water depth is approximately 1m at low tide with a 3m average tide.

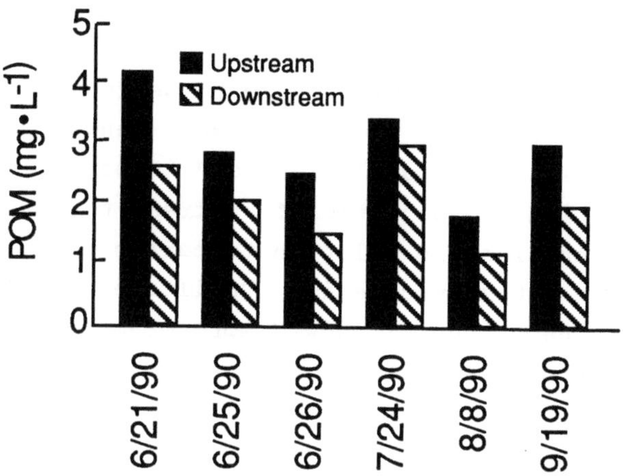

Figure 9. Change in POM across Mud Cove mussel lease on flood tides from grab samples obtained while following a drogue. Bottom water samples.

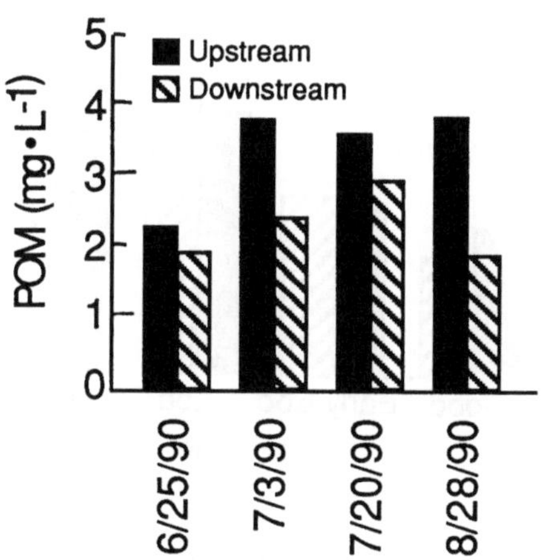

Figure 10. Change in POM across Mud Cove mussel lease on ebb tides from grab samples obtained while following a drogue. Bottom water samples.

The flow cytometer counts of particle concentration and particle type (non-fluorescent, silt particle vs fluorescent algal cell) from surface and bottom samples (Figs.14 and 15) indicate mussels cleared 30-50% of both algal and silt particles in the 5-17μm particle size range from bottom waters, but clearance was lower in the 3-5μm range (under 10%). Other workers have noted mussel gill "leakage" to particles under 5μm (see Review Section). The ratio of upstream to downstream particle concentrations in relation to particle size and type (Fig.15) supports this hypothesis. Unfiltered samples (for phytoplankton direct counts) showed more dramatic declines in total phytoplankton cells than the flow cytometer counts due to prefiltering prior to analysis (removing larger plankton chains, e.g. *Chaetoceros sociale*).

Small-scale (cm) vertical gradients in the food supply

Problems in vertical flux may be investigated by examining the fine vertical gradients of suspended food near the bottom of mussel growing areas. It is difficult to sample natural

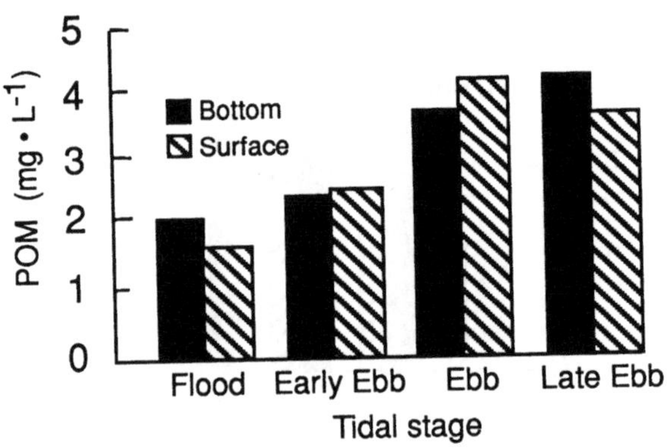

Figure 11. POM at station D, Mud Cove monitored over a tidal cycle 9/24/90.

gradients without a special device, such as the Benthic Organic Seston Sampler (BOSS), which gives duplicate water samples at heights 5-50cm off the bottom at 5 cm intervals (Muschenheim and Newell 1992). A continuous, single-layered mussel bed in Frenchman's Bay, Maine was investigated by BOSS sampling at 2 locations. Upstream of the site, about 20 meters before the mussel bed and over the mussel bed 22 meters from the upstream edge of the bed.

The current was monitored continuously with the S4, and was nearly continuous in speed (9.8 $cm \cdot s^{-1}$) and direction during

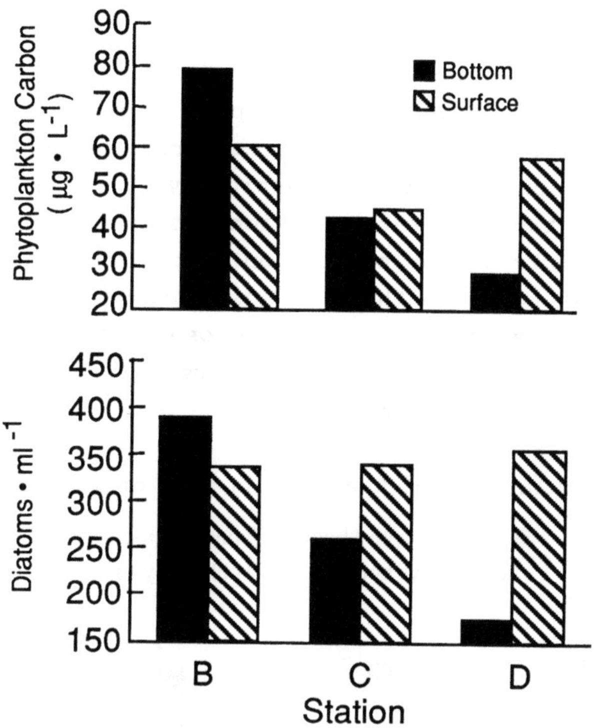

Figure 12. Phytoplankton carbon from grab samples following a drogue in Mud Cove on the flood tide, 9/24/90. Values obtained from direct measurements of phytoplankton cell volume and carbon to biovolume conversions in the literature.

Figure 13. Reduction in numbers of diatoms along drogue path in Mud Cove for bottom and surface samples, flood tide, 9/24/90.

the BOSS deployments. Water density between surface (-0.5m) and bottom (4m) varied less than 0.41%. The upstream samples acted as controls for the effects of mussels on particle concentration in the near-bottom waters. Simultaneous measurements of current speed and mussel pumping rates were used to investigate the feeding zone and filtration capabilities of a productive, high biomass mussel bed. Observations of predicted vs observed organic carbon were also compared with a finite difference model (Fréchette et al 1989; Geyer, pers. commun).

The results show that upstream of the mussel bed, bottom (0-5cm) waters are enriched in benthic diatoms and organic detritus, forming an enhanced food supply to mussels living "on the edge" (Fig.16). Mussels downstream from the edge of the bed rely on surface water for a quantitatively and qualitatively different diet, the flux rate of which limits mussel biomass. Depletion of nearly 50% of the phytoplankton in the mussel feeding zone was noted in the field experiments, indicating a significant refiltering of the water close to the bottom. Calculation of volume flux of water, Q_f relative to the pumping capability of the mussel bed, Q_m:

$$Q_m \cdot m^{-2} \ / \ Q_f \cdot m^{-2} = 1$$

resulted in calculations of the feeding zone of mussels (Fig.17).

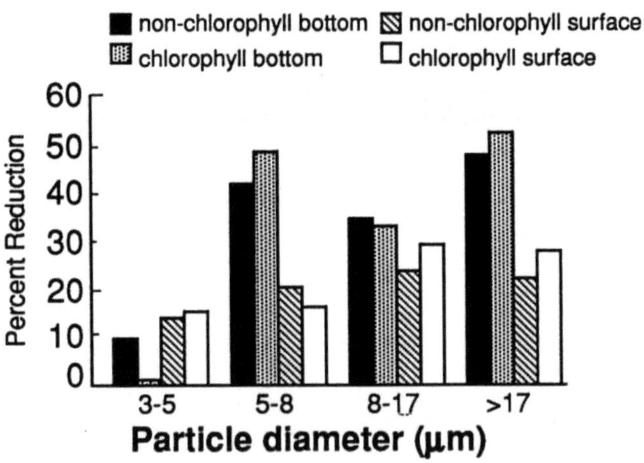

Figure 14. Percent reduction of particles (station B to D) along drogue path 9/24/90 in Mud Cove.

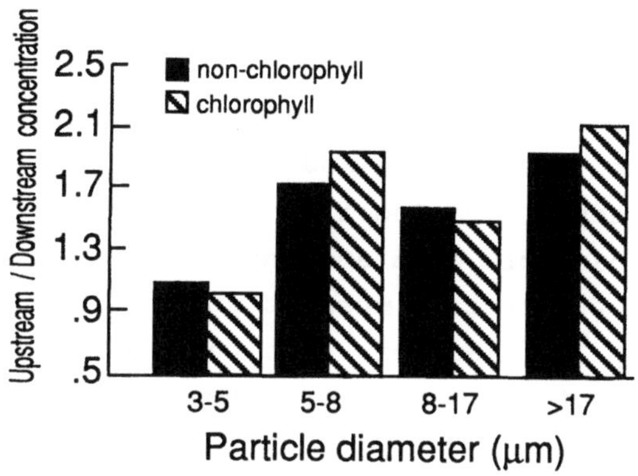

Figure 15. Mud Cove flood tide drogue station B to D 9/24/90, relative concentration of chlorophyll and non-chlorophyll particles by particle size as determined by flow cytometry.

Due to difficulties in obtaining detailed velocity profiles in the field concurrent with measurements in fine vertical gradients in the food supply, sampling with the BOSS has its limitations. Further work is needed in extrapolating fine vertical current gradients obtained from laboratory studies in depth-limited flumes to representative field environments at water depths of several meters.

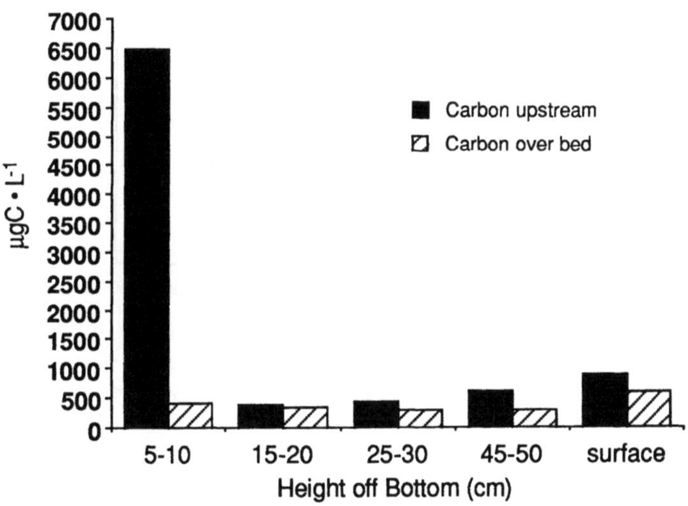

Figure 16. Particulate organic carbon from upstream and above a continuous 22 m mussel bed in 9/89 (Muschenheim and Newell 1992).

The importance of particle selection below the pseudofeces threshold: flow cytometry experiments

In 1987, mussel feeding on natural particulates at lease sites at Ray Point, Camp Island and Webb Cove was investigated using a flow cytometer (see Cucci et al 1989), a device which can measure both particle size and the nature (chlorophyll vs non-chlorophyll) of particles mussels upon which mussels are feeding (e.g. over 3μm diameter). At the study sites in June, 1987, about 15-30% of the particles were fluorescent (chlorophyll-containing). Mussel feeding studies at natural seston levels (1-2 x 10^4 particles$\cdot$$ml^{-1}$) indicated approximately 40% higher clearance rates of phytoplankton cells than silt particles at 5 of the 6 sites (Newell et al 1989). From 1989-

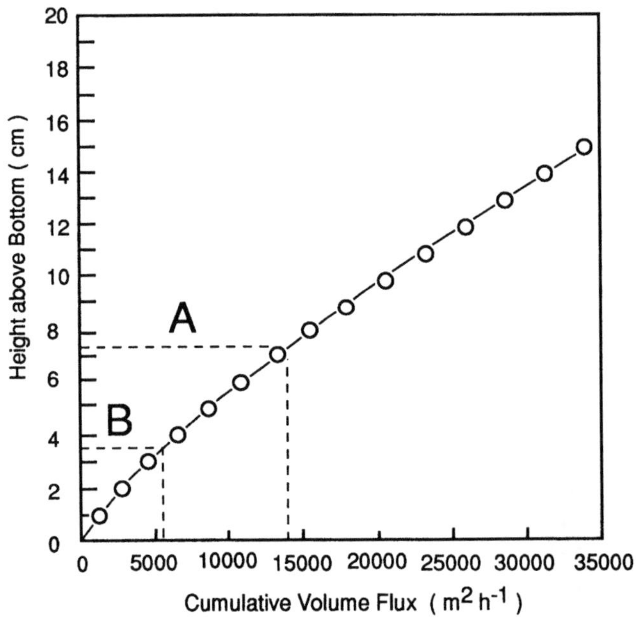

Figure 17. Hourly volume equivalent pumped by the mussels (Region A) and hourly volume equivalent filtered if significant refiltration occurs (Region B) (Muschenheim and Newell 1992).

1991, the effects of particle concentration and food quality (% particles phyto-plankton) were further investigated in two similar experiments using 1 x 10^4 and 2 x 10^4 particles$\cdot$$ml^{-1}$ with varying food quality (Newell and Shumway 1991). Finally, an experiment was performed sampling water from the mussel excurrent siphon to examine possible bias in the data due to particle production by mussels.

The experiments using field water samples are reported in Newell et al. (1989), and data from June 2 and June 9, 1987, are included in the results presented here. Experiments performed in 1989-1991 used Boothbay Harbor water to which ashed mudflat silt, 0.45μm filtered seawater, or cultured algae were added to obtain the following conditions (Fig.18):

Dates	Particle concentration ($number \cdot ml^{-1}$)	% Phytoplankton
11/28 - 12/1/89	8-10,000	18-35%
9/25 - 9/28/90	20,000	10-25%

The 11/28/89 - 12/1/89 experiments at approximately particle concentrations of 1 x 10^4 particles$\cdot ml^{-1}$ and the 9/24/90 - 9/28/90 experiments at concentrations of 2 x 10^4 particles$\cdot ml^{-1}$ (Fig.18)
showed that mussels feed at higher rates on algal particles at

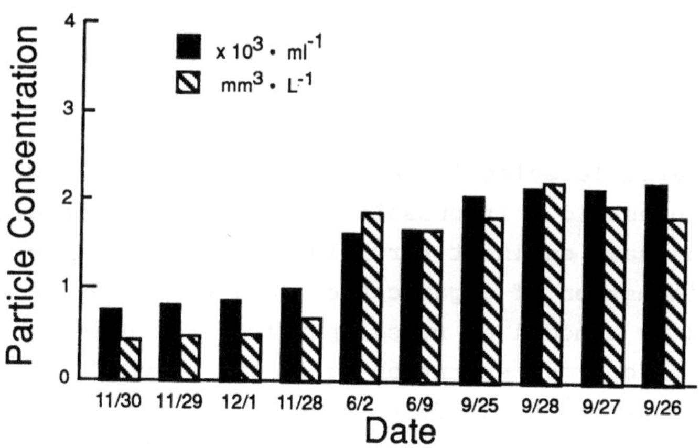

Figure 18. Control particle concentrations (particles ml^{-1} x 10^3 or mm^3 l^{-1}) for flow cytometry experiments.

low concentrations (10,000 particles·ml^{-1}) but lose their ability to remove phytoplankton selectively from silt particles at higher concentrations (20,000 particles·ml^{-1}), at about the pseudofeces threshold where sorting occurs on the labial palps. Differences in filtration rate on chlorophyll and non-chlorophyll particles (Fig.19), along with t-tests of differences in clearance rate due to particle type (Table 5) indicate the importance of selection in relatively clear, non-turbid waters. If selection is occurring, preferential consumption of algae is not indicated using a standard Coulter Counter unless a particular bloom can be identified by its particle size. For example, with 75% silt par-ticles and 25% algal cells at cell concentrations of 1 x 10^8 l^{-1}, mussels may be seen to filter 2 l·h^{-1} while they are actually feeding at a rate of 4 l·h^{-1} on algal cells and 1.5 l·h^{-1} on the silt particles (see Newell et al 1989). In the feeding experiments, 40% higher feeding rates on algal cells have been noted with particle concentrations at 10-20 x 10^6 particles·l^{-1}.Thus, energy gain estimated with the Coulter Counter would be 30-40% less than observed due to selection. These experiments demonstrate a possibly significant factor controlling ingestion of food particles. The results indicate that at normal food levels, 1-2 x 10^7 particles·l^{-1}, the mussels pump at a maximum rate and ingestion rate is determined by food concentration and quality (Newell et al 1989). Higher clearance rates on identically-sized glass coated beads treated with and without algal extracts (Ward and Targett 1989) suggest a chemical cue involved in particle selection prior to ingestion below the pseudofeces threshold. If mussels selectively retain algal particles and leave sediment particles of the same size in suspension, ingestion of organic-rich food particles is greater than expected by a model of mussel particle retention which is independent of food concentration below the pseudofeces threshold. Above the pseudofeces threshold, mussels may also select algal over inorganic particles on the labial palps, rejecting a proportionately higher percent of silt in the pseudofeces (Kiørboe and Møhlenberg, 1981). This results in a model of particle selection for mussels (Table 5) which is maximum between particle concentrations required the initiation

Table 4. Significance of differences in filtration rate (chlorophyll - non-chlorophyll) for each feeding trial.

Date	Significance level (t-test)
11/28/89	0.02
11/29/89	0.01
11/30/89	0.005
12/01/89	0.05
06/02/87	0.05
06/09/87	NS
09/25/90	NS
09/26/90	NS
09/27/90	NS
09/28/90	NS

of feeding to the pseudofeces threshold. The mechanism behind the differential retention of algal particles over silt should be investigated further. This has a significant effect on the ability of mussels to obtain high quality food in relatively clear waters, but also results in greater competition for food quality at the lease site. The results also explain the ability of mussels to sustain high energy gain in clear waters.

Table 5. Conceptual view of algal selection by *Mytilus edulis* in relation to ambient seston levels.

Particle Conc.x $10^6 l^{-1}$	Filtration	Selection of Algae
<5	low	some
8-20	maximum	20-50%
>20	reducing	on palps

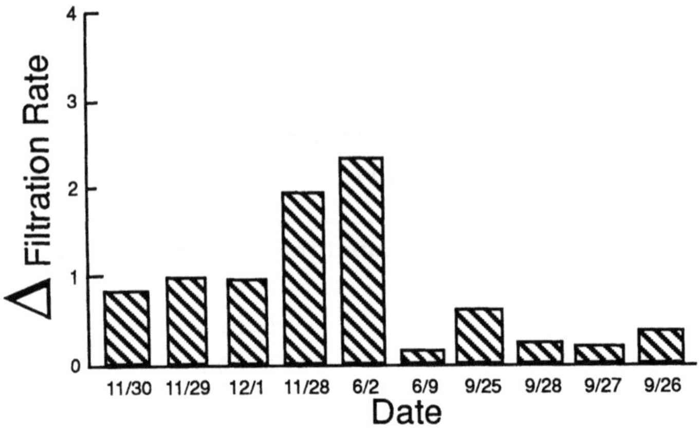

Figure 19. Filtration rate differences: filtration rate on chlorophyll - filtration rate non-chlorophyll particles.

Future models of shellfish control over eutrophication should consider mussel feeding rates not as a simple product of non-selective particle clearance and food concentration, but as modified by enhanced filtration of algal cells in relatively non-turbid waters (0.5-1.5 x 10^7 particles$\cdot l^{-1}$, particle volume 0.5 - 1.5 $mm^3 \cdot l^{-1}$).

Tunnel experiments

The role of filter-feeders in removing suspended particles (seston), and effects on chemical fluxes (especially oxygen and ammonia) is best investigated *in situ*, preferably in an undisturbed bed composed of the shellfish and associated flora and fauna. A benthic ecosystem tunnel (BEST) was developed by Richard Dame et al (1984), which utilizes a 10m long Plexiglass tunnel which is installed over the bed of filter-feeders parallel to tidal flow. The tunnel has been used by workers in the Netherlands to investigate the role of mussel beds in their ecosystem (Prins and Smaal 1990; Dame and Dankers 1988), and is currently being used in France to study uptake rates of bottom cultured oysters. In this study (see Fig.20), a total of seven 1.2 meter sections (8.4 m long) of the original BEST were obtained from Dr. Richard Dame of Coastal Carolina College, South

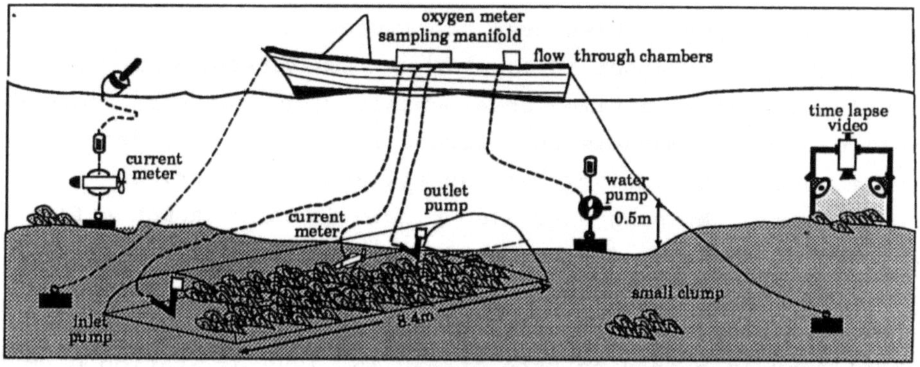

Figure 20. Experimental set-up for April 17, 1991 tunnel experiments in conjunction with underwater video at Mud Cove.

Carolina. A Marsh-McBirney flow meter was attached to the center of the chamber to measure water flow ($cm \cdot s^{-1}$) at mid-depth and two battery driven centrifugal Rule pumps were used to pump water samples from 0.5m in from the inlet, and 0.5m in from the outlet of the tunnel, respectively. Since comparisons between inlet and outlet water can be made only when the tide is running, the tunnel was installed by diver at low tide, samples were made on the flood and ebb tide, and the tunnel was removed at the second low tide. After removal of the tunnel, 8 x 0.61 m^2 cores were made by diver, and mussel biomass, density and mean size were determined. Since a high density mussel bed was needed for tunnel experiments, samples were also taken of lower density mussel patches from the same seed cohort near the tunnel site.

Particle flux was calculated as:

$$\text{Flux } (mg \cdot h^{-1}) = \text{Flow rate } (l \cdot h^{-1}) * (mg \cdot l^{-1} \text{ (inlet-outlet)})$$

Results may be presented on a square meter basis by dividing volume flow ($m^3 \cdot s^{-1}$) by area (m^2) enclosed by the tunnel between the inlet and outlet pumps. Mussel biomass (g dry tissue$\cdot m^{-2}$) and density (number of mussels$\cdot m^{-2}$) can be used to calculate rates of particle removal per individual mussel, and size to weight exponents adjusted accordingly. For the April 17,1991 run at the Mud Cove lease site in Maine (see Fig.20), the specifications listed for the tunnel experiments are given in Appendix 3. At sampling times during the April 17, 1991 experiment, volume fluxes over the mussels in the tunnel (Appendix 3) were about 10-13 $m^3 \cdot h^{-1}$. Simultaneous with the tunnel experiment, a time-lapse benthic video monitor recorded shell gape of 37 mussels at 2 second intervals every ten minutes over 1.5 tidal cycles (Newell and Gallagher, in preparation). Mussels were also placed in flow-though efflux chambers on the moored vessel and were monitored for particle clearance rates using water pumped from 0.5m of the bottom.

Mussels from a cohort of 38 mm (0.2 g dry tissue weight) mussel seed at a seeding date of May 15, 1990 at Station C, Mud Cove were monitored for tissue and shell growth over an eleven month period, to a sample date of April 17, 1991 (Fig.21). Growth (both shell volume and meat size) was over 4.5 times faster in the small clumps (600 m^{-2}) than in the adjacent heavily-seeded patch of mussels (600 per square meter vs 1800) in small clumps. However, the biomass of the low and high density mussel aggregates (patches) was similar (669 vs 595 g dry tissue weight$\cdot$ m^{-2}). Thus, at the time of the experiment the carrying capacity of the lease section studied was about 600-650 g dry tissue weight$\cdot m^{-2}$. Particle clearance per individual mussel (mean dry weight 0.36 g) is presented in Table 7. Water samples were taken during flood and ebb tides, and phytoplankton carbon biomass calculated from measured cell volumes was compared for flood (Fig.22) and ebb (Fig.23) tides. The consumption of about half of the blooming diatom (*Chaetocerus sociale*) by the mussel bed (representing the bottom 0.5 m of the water column) within the tunnel was evident in the water samples. With a particle consumption of 3.88 $mm^3 \cdot h^{-1}$ per mussel, a square meter of bottom would require 7073 $mm^3 \cdot h^{-1}$ or at a seston level of 4 $mm^3 \cdot l^{-1}$, a vertical flux of about 1768 $l \cdot h^{-1}$.

Table 6. Mean particle clearance per mussel as measured in tunnel experiments 4/17/92

POC ($\mu g \cdot h^{-1}$)	376
Particles ($\times 10^6 \cdot h^{-1}$)	10.3
Phytoplankton ($cells \cdot h^{-1}$)	7762
Particle volume ($mm^3 \cdot h^{-1}$)	3.88
Chl *a* ($\mu g \cdot h^{-1}$)	3.93

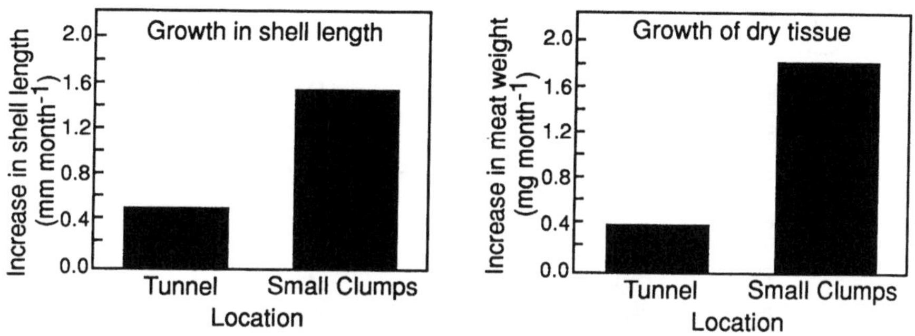

Figure 21. Mussel growth rates over an 11-month period in relation to seeding density for small clumps (595 m^{-2}) and over a continuous patch of mussels in the tunnel (1823 m^{-2}).

A comparison of mean filtration rate and oxygen consumption estimated with the tunnel and the flow-through chambers is presented below:

	Tunnel	Chambers
Filtration rate ($l \cdot h^{-1} \cdot g^{-1}$)	3.24	2.67
Respiration (ml $O_2 \cdot h^{-1} \cdot g^{-1}$)	0.30	0.30

Tidal cycle variations in shell gape: video experiments

An instrument was developed to measure, *in situ*, the effects of particle concentration on feeding activity in undisturbed mussels. The tripod-mounted instrument (Time-Lapse Video Benthic Monitoring, TLBM) uses time-lapse video technology to monitor shell gape of mussels for periods exceeding two weeks. Video tapes, analyzed field by field,

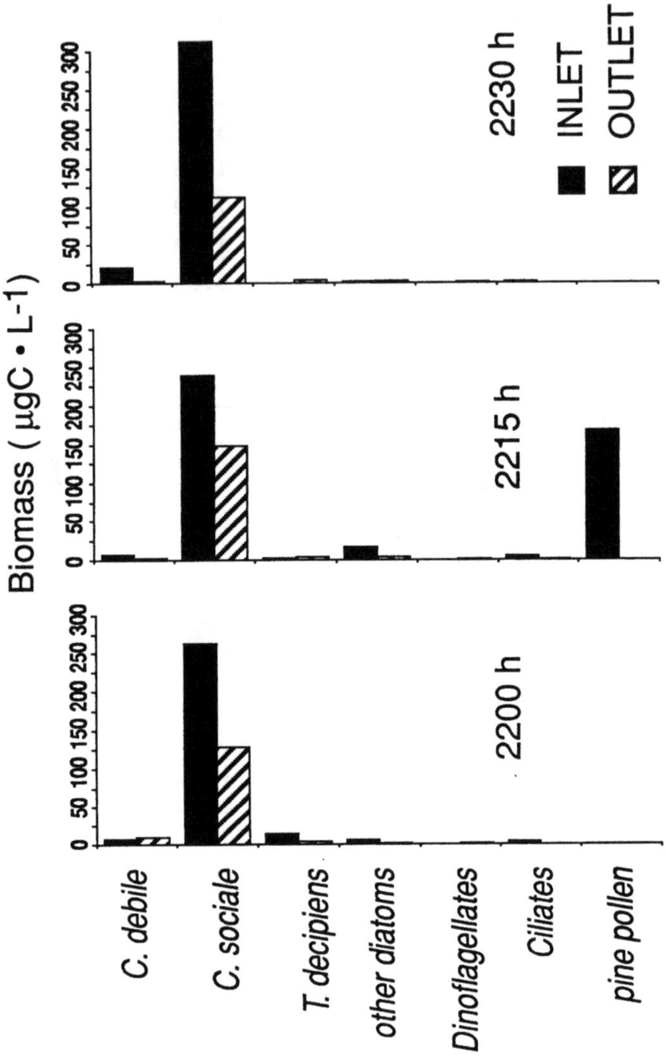

Figure 22. Flood tide differences in phytoplankton biomass taken from inlet and outlet of benthic ecosystem tunnel.

allowed measurements of shell gape of up to 30 mussels at intervals of 1 minute (Newell and Gallagher 1992). Results indicate a positive correlation between mussel shell gape and concentrations of POM and chlorophyll *a*, and a potential particle concentration threshold for the initiation of active feeding by mussels (Newell and Gallagher, in preparation). The results are briefly summarized below.

In the laboratory, mussel shell gape was calibrated to filtration rate by measuring the decrease in chlorophyll with time following pulsed additions of cultured algae to filtered seawater in a running flume.

Clearance rate was measured over a tidal cycle in a flow-through, efflux apparatus using ambient seawater and 10 efflux chambers with stir-bars (Fig.24) in fall (11/8/90), and in a benthic ecosystem tunnel (see section above) while time-lapse video recorded shell gape of over 30 mussels in undisturbed bottom patches from known seeded cohorts of mussels. Mussel filtration rates and shell gape were monitored *in situ* during high food (April, 1991) and low food (November, 1990) conditions. A sine wave pattern of shell gape (Newell and Gallagher, in preparation) indicates a coupling of bivalve pumping rates with mechanisms of particle supply with both tidal (e.g. Stenton-Dozey et al 1992) and seasonal components. During the fall, periods of shell closure were associated with a low seston concentration (under 8×10^6 particles$\cdot l^{-1}$, Fig.24). Low filtration rates and low oxygen consumption (25% of levels during active feeding) were observed around the period of low tide. A time-series of measurements of scope for growth over 4 tidal cycles in October (Newell and Davis, unpublished) in the efflux apparatus resulted in similar results during periods of low seston concentration.

In April, a 20-fold increase in food concentration resulted in higher shell gape during the tidal cycle, with greatest gape on the ebb. While the shapes of the shell gape curves were similar in both cases, the response to flood tide in November

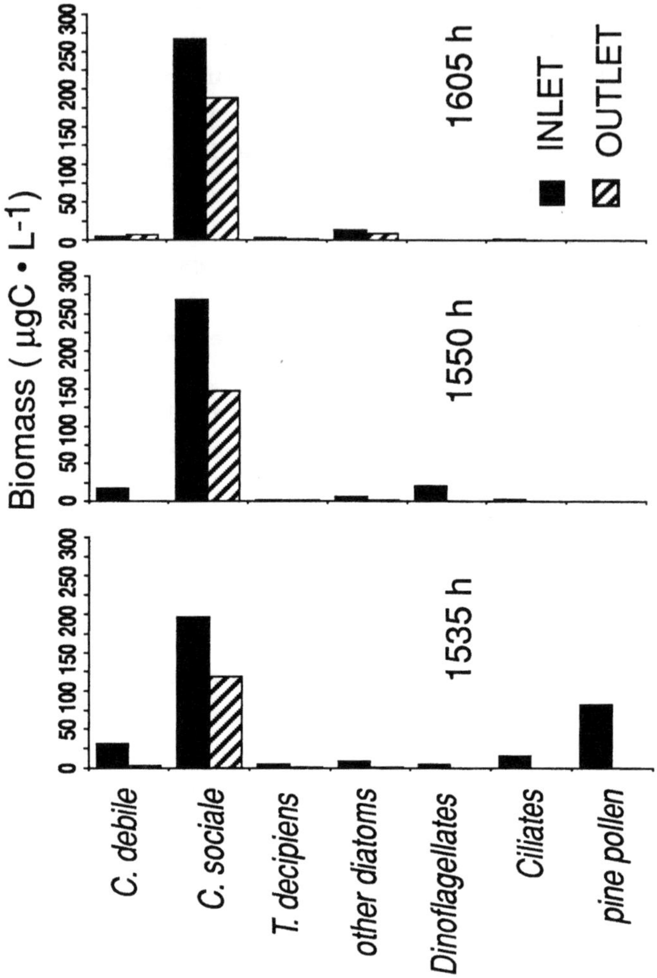

Figure 23. Ebb tide differences in phytoplankton biomass taken from inlet and outlet of benthic ecosystem tunnel 4/17/91.

near station B (edge of lease) and ebb tide in April near station D (about 250m from the edge) was probably due to site-specific effects of mussels on the population's food supply. In November, the lack of competition at station B (see Figs.8 and 9) on the edge of the lease resulted in rapid food supply to the mussels.

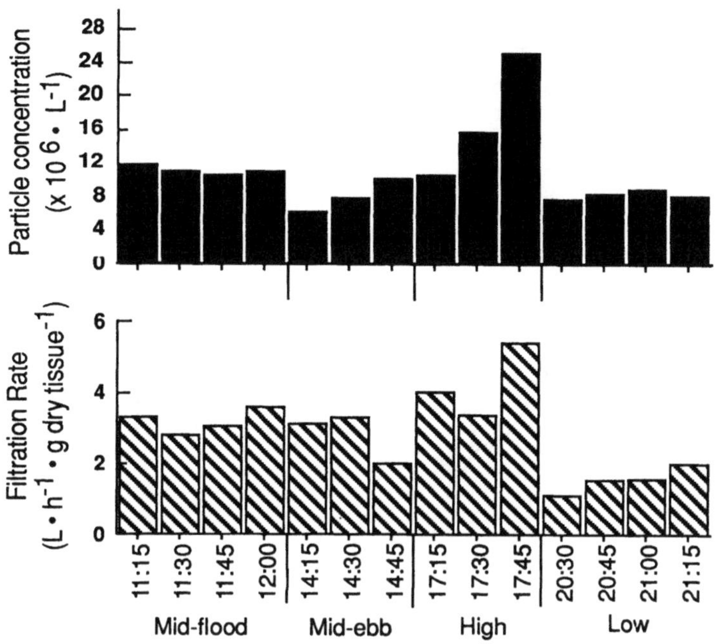

Figure 24. Mussel filtration rates (B) in efflux apparatus in relation to control particle concentration (A) on 11/8/92 at Station B, Mud Cove.

In April, depletion contours would develop in the bottom waters of the middle of the seeded mussel lease on the flood tide, (stations B to C, Fig.8) with settling increasing the supply of food-rich particles by high water/ebb tide (see Fig.11).

Future investigations of the shell gape assay for filtration rate of undisturbed shellfish beds should include more simultaneous measurements of shell gape and filtration rate, and studies of the limitations due to variable gill retention efficiencies.

Seasonal changes in food quality

Seston quantity and quality parameters at the Mud Cove lease site are given in Table 8. Bottom and surface concentrations were similar with the exception of phytoplankton carbon, which was higher in surface waters at all study sites, a combined result of mussel feeding and summer stratification. Both total and detrital C/N ratios were slightly higher in bottom waters, indicating the role of settling of detritus. Carbon to chlorophyll ratio (C:Chl) averaged about 200:1, with low values during the spring bloom in February and March, 1990

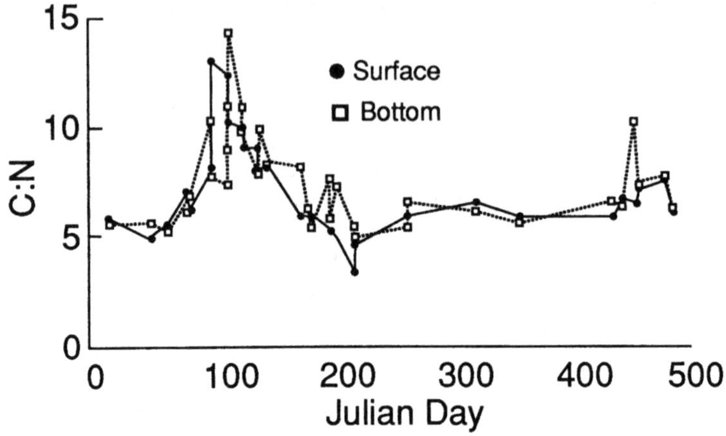

Figure 25. C/N ratio of GFC filtered seston from January, 1990 to June, 1991, bottom and surface samples, at Mud Cove.

and April, 1991 at Mud Cove. A distinct seasonal pattern in seston C/N ratio was observed (Fig. 25) rising from 5 to 10 after the spring bloom, providing mussels with an important source of detrital carbon in late spring and early summer. This food source provides additional energy for the rapid growth obtained by the mussels during this time. Data from all Maine sites show the same trend, with values over 10 during and after the spring bloom, falling to about 5 in the summer. The fraction of POM/SPM changed relatively little during the study, making it a poor indicator of food quality. Phytoplankton nitrogen/ total nitrogen ratio increased during the spring bloom and was lowest during late fall and winter at Mud Cove. Detrital nitrogen lagged behind the phytoplankton bloom by several weeks and was highest during the early summer, following the peak in detrital carbon.

Table 7. Mean values of various indicators of food quality at Mud Cove, Maine from January, 1990 - June, 1991

Seston attribute	units	bottom	surface
SPM (total weight)	$mg \cdot l^{-1}$	8.09	7.52
POM (weight organic matter)	$mg \cdot l^{-1}$	2.84	2.70
Chlorophyll *a*	$\mu g \cdot l^{-1}$	3.09	3.04
Particulate organic carbon	$\mu g \cdot l^{-1}$	396.44	390.31
Phytoplankton carbon	$\mu g \cdot l^{-1}$	88.29	130.40
Detrital carbon	$\mu g \cdot l^{-1}$	368.54	352.04
Particulate organic nitrogen	$\mu g \cdot l^{-1}$	54.38	58.40
Phytoplankton nitrogen	$\mu g \cdot l^{-1}$	9.62	13.07
Detrital nitrogen	$\mu g \cdot l^{-1}$	49.54	53.39
Particle concentration	no. $l^{-1} \times 10^6$	15.15	13.31
Particle volume	$mm^3 \cdot l^{-1}$	0.93	1.66
Organic weight/volume	mg/mm^3	1.72	1.41
POM/SPM	mg/mg	0.36	0.36
Chl *a*/SPM	μg/mg	0.38	0.38
Total POC/PON	$\mu g/\mu g$	7.40	6.90
Detrital C/N	$\mu g/\mu g$	6.67	6.25
Phyto C/Total C	$\mu g/\mu g$	0.15	0.18
Phyto N/Total N	$\mu g/\mu g$	0.15	0.19
C/Chl *a*	$\mu g/\mu g$	211.00	218.30
Detrital N/C	$\mu g/\mu g$	0.15	0.16

Phytoplankton carbon reached a maximum of about 50% of total carbon during the spring bloom and was very low at Mud Cove during the summer period. The detrital N/C ratio increased just after the phytoplankton bloom, and rose again during the summer.

Due to the relative insensitivity of POM measured to changes in food quality, use of POM/particle volume was a relatively poor indicator of food quality over a seasonal cycle in contrast to the laboratory studies of Bayne et al (1989). For different field sites, phytoplankton and total carbon were good indices of food quality during and after the spring bloom, while detrital nitrogen was a better indicator later in the year.

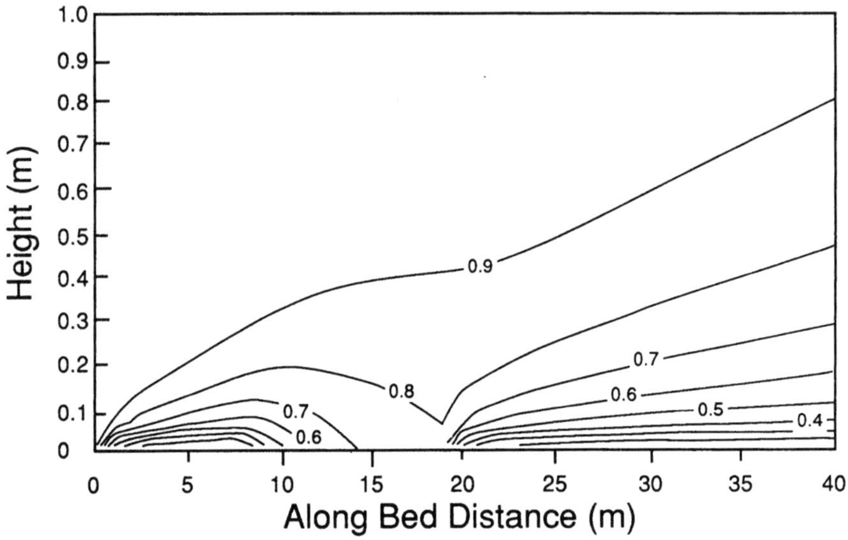

Figure 26. Concentration contours of phytoplankton in areas with no mussels between 9 and 19 meters (from Sankar 1991).

CONCLUSIONS

A review of the literature and recent experiments in Maine have resulted in a dynamic view of mussel/particle interactions from cm to km scales, over periods ranging from tidal cycles to seasonal changes in food quality. Future investigations using natural particulates and relatively undisturbed beds of shellfish of high biomass in the field will elucidate the true role of these organisms in coastal ecosystems. It is recommended that the following approaches be used to investigate the grazing of natural particulates by shellfish:

Problem	Approach	Tool
Current speed and volume flux	Hydrographic flow model	
DUCHESS		
Particle concentration	Seston profiles	BOSS
Scope for growth	Flow-through chambers	EFFLUX
Tidal variation in pumping	Shell gape assay	TLBVM
Meso-scale processes	Drogue experiments	ALPHA
		BOTTLE
Population grazing	*In-situ* tunnels	BEST
Boundary layer processes	Flow profiling	FLUMES
Selective feeding	Feeding experiments	FLOW
CYTOMETER		

Our studies indicate the following attributes on a variety of spatial and temporal scales which exert a major control on shellfish and seston interactions:

Attribute	Spatial Scale	Temporal Scale
Tidal supply of seston	km	hours
Vertical supply of seston	cm	hours
Algal biomass	km	weeks
Seston quality	km	days
Current speed	dm	hours
Shellfish size	cm	months
Temperature	km	weeks
Shellfish density	m	days

Further limits to our knowledge of particle flux include: long-term patterns of shell gape and filtration activity; the role and mechanism of higher algal retention efficiency on the gills; effects of particle flocculation on vertical transfer rates; the relationships among nutrient concentrations and the type of algae present; and resuspension of bottom detritus and benthic algal mats. Combined oceanographic and ecophysiological studies will become more important as we attempt to model shellfish and seston interactions. As we attempt to manage shellfish populations, both for aquaculture production and to maintain control over algal blooms, computer modelling will become an increasingly useful tool for simulations and "what if" scenarios. Studies of current and food gradients in the benthic boundary layer in both field and laboratory flumes will do a great deal to test and validate current models of seston transfer (Fréchette et al 1989; Monismith et al 1990). A computer simulation of depletion contours above a seeded mussel bed (Fig. 26) is an illustration of how mussel farmers may benefit from current computer models by placing spaces between seeded plots to allow greater food supply to the benthic boundary layer.

ACKNOWLEDGMENTS

This research was supported by National Science Foundation SBIR Awards ISI8660201 and ISI8809760 to Great Eastern Mussel Farms. Thanks are due to Deb Murphy and Terri Ainaire (laboratory assistants); Dr. Scott Gallagher (video experiments); Dr. Richard Dame (tunnel experiments); Dr. Rockwell Geyer (flume experiments) and summer support for Sam Sankar; Dr. Cheryl Ann Butman (flume experiments); Dr. Dan Campbell (mussel model); Dr. Charles Gregory (phytoplankton analyses); Joth Davis (efflux apparatus); Dr. Vijay Panchang and John Richardson (hydrographic model); Dr. Janet Campbell and Roger Greene (Remote Sensing Experiments); Dr. Kee Muschenheim (BOSS experiments); Drs. Ray Grizzle, Marcel Fréchette and David Wildish for helpful discussions, and captain Hubba Bradford for help with the commercial seeding experiments. The authors are grateful to Drs. H. and R. Asmus, Dr. Richard Dame and one anonymous reviewer for helpful comments on the manuscript.

Appendix 1. Rapid method for volume flux calcuation at a lease site

If we consider a tide gauge measurement of water depth, H, and a current meter measurement of water velocity, V, to be representative of a region Δ x, Δ y and H, then the flow entering an area with dimensions Δ x times H is q ($m^3 \cdot s^{-1}$).
Assuming velocity varies in time as a sin curve:
$V = V_o \sin(\omega t)$ where V_o = maximum velocity, ω = frequency,
and water depth varies in time as a sin curve:

$H = H_m + \sin(\omega t + a)$ where H_m = mean water depth and
$a = \pi/2 = 90°$ and $\omega = 2\ \pi/12.4$

(i.e. velocity and water depth are out of phase by 90° or 3.1 hours), so $H = Hm + \cos \omega t$.

To find the volume, V, entering the area between times 1 and 2, the integral of qdt from t_1 to t_2, with integration and substitution:

$$V = 2\ H_m\ V_o\ \Delta\ x/\omega$$

Over a period of 6.2 hours, the flow through this vertical plane, $q = (2\ H_m\ V_o\ \Delta\ x)/\pi$ ($m^3 \cdot s^{-1}$)

This average flow goes over a lease area, e.g. 1 square meter, at the rate:

$$2\ H_m\ V_o/\pi \quad (m \cdot s^{-1})$$

Appendix 2. Estimation of U_* at field sites.

One can calculate values of U_* based on velocity profiles made in the field, from the slope of the velocity vs the natural logarithm of depth (U vs ln(z)) curve, by the equation:

$$U_* = K/b \quad \text{where } x = U,\ y = \ln(z) \text{ and } y = ax + b \quad (3)$$

Very high r^2 values are needed for obtaining U_* from velocity profiles, especially if less than five heights are used. Field data of velocity profiles using the S4 current meter and the Marsh-McBirney probe resulted in the following values of U_* using the profile method (Table I). Because field profile r^2 values are less that 0.99, field values of U* values are suspect and must be extrapolated from the flume experiments.

Using the equation above, and representative values of U and H at the Mud Cove lease site, a range in values of U_* may be calculated using a value of 0.16 for z_0, and compared with the simple equation $U_* = U/16$ (Table 4).

Table I U_* calculated from velocity profiles at mussel culture sites.

Location	U ($cm \cdot s^{-1}$)	U_* ($cm \cdot s^{-1}$)	r^2	H (cm)
Mud Cove (Stn B seeded)				
1a	26	3.10	0.93	300
1b	26	2.71	0.88	310
1c	30	1.74	0.68	300
1d	27	2.22	0.82	300
2a	5	1.74	0.29	410
2b	4.5	1.10	0.34	414
Lamoine Lease (not seeded)	15	1.05	0.96	300
Schieffelein Cove (seeded)				
4	5.2	0.80	0.84	500
6	4.5	1.14	0.30	350

Table II. Expected values of U_* from Mud Cove

U ($cm \cdot s^{-1}$)	z_o cm	H (cm)	U_*	U_* as U/16
5	0.16	125	0.30	0.31
5	0.16	250	0.27	
5	0.16	375	0.26	
5	0.16	500	0.248	
10	0.16	125	0.601	0.63
10	0.16	250	0.544	
10	0.16	375	0.515	
10	0.16	500	0.497	
15	0.16	125	0.901	0.94
15	0.16	250	0.816	
15	0.16	375	0.773	
15	0.16	500	0.746	
20	0.16	125	1.201	1.25
20	0.16	250	1.088	
20	0.16	375	1.031	
20	0.16	500	0.994	
25	0.16	125	1.501	1.56
25	0.16	250	1.360	
25	0.16	375	1.289	
25	0.16	500	1.243	
30	0.16	125	1.801	1.87
30	0.16	250	1.632	
30	0.16	375	1.546	
30	0.16	500	1.491	

Appendix 3. Benthic ecosystem tunnel characteristics, April, 1991 field experiment.

- Tunnel length (8.9 m)
- Tunnel width (0.9 m)
- Distance between pumps (6.7 m)
- Effective area enclosed by the tunnel (6 m^2)
- Mussel density within tunnel ($1823 \cdot m^{-2}$)
- Mussel mean individual dry weight (0.357 g, n = 240)
- Mussel mean shell length (43.6 mm, n = 239, s.dev.= 6.5)
- Mussel biomass ($669\ g \cdot m^{-2}$)
- Cross-sectional area tunnel (0.225 m^2)
- Velocity correction factor (0.84, R. Dame pers. commun.)
- Mussel density small clumps (595 m^{-2}, n=4, s. dev. = 164)
- Mussel mean dry meat wt small clumps (0.96 grams, n = 120)
- Mussel mean shell length small clumps (54.5 mm)
- Mussel biomass small clumps ($571\ g \cdot m^{-2}$)

For seven sections of the tunnel, volume flow is calculated as a result of current speed, e.g.:

$3\ cm\ s^{-1} \times 0.225\ m^2 \times 0.84 = .000567\ m^3 \cdot s^{-1}$

Over the 6 square meter mussel bed studied, at 3 cm s^{-1}, volume flux was 0.0034 $m^3 \cdot s^{-1}$ (= instantaneous net flux).

Table III. Current speed and volume flux through a Benthic Ecosystem enclosing 10,938 mussels on flood and ebb tides at Mud Cove on April 17, 1991.

EBB	U ($cm \cdot s^{-1}$)	Flow ($m^3 \cdot h^{-1}$)	normalized flow$\cdot m^{-2}$ ($m \cdot h^{-1}$)
15:35	3.0	12.2	2.0
15:50	3.0	12.2	2.0
16:05	3.0	12.2	2.0
16:20	2.5	10.3	1.7
FLOOD			
22:00	3.2	13.0	2.2
22:15	3.2	13.0	2.2
22:30	2.9	11.9	2.0
22:45	2.4	9.8	1.6

REFERENCES

Almada-Villela PC, Davenport J, and Gruffydd LD (1982) The effects of temperature on the shell growth of young *Mytilus edulis* L. J Exp Mar Biol Ecol 59: 275-288

Bayne BL, Widdows J and Newell RIE (1977) Physiological measurements on estuarine bivalve molluscs in the field. In:Biology of Benthic Organisms eds. Keegan, O'Ceidigh and Boaden, Pergamon Press Oxford p. 57-68

Bayne BL, and Widdows J (1978) The physiological ecology of two populations of *Mytilus edulis* L. Oecologia 37:137-162

Bayne BL, Widdows J, and Thompson RJ (1976) Physiological integrations, p.261-291. In Bayne BL, (ed.), Marine mussels: their ecology and physiology. IBP 10. Cambridge University Press. 506 pp

Bayne BL, Hawkins AJS, Navarro E, and Iglesias IP (1989) Effects of seston concentration on feeding, digestion and growth in the mussel *Mytilus edulis*. Mar Ecol Prog Ser 55:47-54

Bayne BL and Hawkins AJS (1990) Filter-feeding in bivalve molluscs: controls on energy balance. Comparative physiology Animal nutrition and transport processes. Comp Physiol 5:70-93.

Bayne BL, Hawkins AJS, and Navarro E (1987) Feeding and digestion by the mussel *Mytilus edulis* L. (Bivalvia: Mollusca) in mixtures of silt and algal cells at low concentrations. J Exp Mar Biol Ecol 111:1-22

Bienfang PK (1981) Sinking rates of heterogeneous, temperate phytoplankton populations. J Plank Res 3:235-252

Billen G (1978) A budget of nitrogen recycling in North Sea sediments off the Belgian Coast. Estuar Coast Mar Sci 7:127-146

Blegvad H (1914) Food and conditions of nourishment among the communities of invertebrate animals found on the sea bottom in Danish waters. Rep Danish Biol Sta 1914 pp. 45-78

Brylinsky M, and Sephton TW (1991) Development of a computer simulation model of a cultured blue mussel (*Mytilus edulis*) population. Can Tech Rep Fish Aquat Sci 1805.vii+81p

Campbell JW, Garside JC, Greene RH and Newell CR (1991) Use of high resolution aerial and satellite remote sensing for assessing bivalve shellfish aquaculture sites. Maine Aquaculture Innovation Center, Orono, Me. Tech Rep 1991 47 pp

Carver CEA and Mallet AL (1990) Estimating the carrying capacity of a coastal inlet for mussel culture. Aquaculture 88:39-53

Conover RJ (1966) Assimilation of organic matter by zooplankton. Limnol Oceanogr 11:338-354

Coughlan J (1969) The estimation of filtering rate from the clearance of suspensions. Mar Biol 2:356-358

Cucci TL, Shumway SE, Brown WS, and Newell CR (1989) Using phytoplankton and flow cytometry to analyze grazing by marine organisms. Cytometry 10:659-669

Dame RF, Zingmark RG, and Haskin E (1984) Oyster reefs as processors of estuarine materials. J Exp Mar Biol Ecol 83:239-247.

Dame RF and Dankers N (1988) Uptake and release of materials by a Wadden Sea mussel bed. J Exp Mar Biol Ecol 118:207-216
Dyer KR (1973) Estuaries:A Physical Introduction. Wiley and Sons Toronto Ont. pp 109-110
Dean B (1887) The food of the oyster, its conditions and variations. Suppl. to 2nd Rept., Oyster Investigations, State of New York pp. 49-78
Famme P, Riisgard HU, Jørgensen CB (1986) On direct measurements of pumping rates of the mussel *Mytilus edulis*. Mar Biol 92:323-327
Fegley SR, MacDonald BA and Jacobsen TR (1992) Short-term variation in the quantity and quality of seston available to benthic suspension feeders. Estuar Coast Shelf Sci 34: 393-412
Field IA (1911) The food value of sea mussels. Bull US Bur Fish 29:85-128
Foster-Smith RL (1975) The effect of concentration of suspension on the filtering rates and pseudofecal production for *Mytilus edulis* L., *Cerastoderma edule* L. and *Venerupis pullastra* Montagu. J Exp Mar Biol Ecol 17:1-22
Fréchette M and Bourget E (1985) Energy flow between the pelagic and benthic zones: factors controlling particulate organic matter available to an intertidal mussel bed. Can J Fish Aquat Sci 42:1158-1165
Fréchette M, Butman CA, and Geyer WR (1989) The importance of boundary-layer flows in supplying phytoplankton to the benthic suspension feeder *Mytilus edulis* L. Limnol Oceanogr 34:19-36
Galtsoff PS (1964) The American oyster. Fishery Bulletin 64:1-480
Grant J, Mills EL, and Hopper CM (1986) A chlorophyll budget of the sediment-water interface and the effect of stabilizing biofilms on particle fluxes. Ophelia 26:207-219
Grave C (1912) A manual of oyster culture in Maryland. 4th Report of the Board of Shellfish Commissioners of Maryland, Baltimore. 376pp
Griffiths RJ (1981). Production and energy flow in relation to age and shore level in the bivalve *Choromytilus meridionalis* (Kr.). Estuar Coast Shelf Sci 13:477-493
Grizzle RE (1988) The relative effects of tidal currents, seston and bottom sediments on individual growth of *Mercenaria* (Linne): Feeding ecology and aquacultural implications. Ph.D. Thesis, Rutgers University 293 pp
Hawkins AJS (1991) Protein turnover: a functional appraisal. Func Ecol 5:222-233
Hawkins AJS, Navarro E, and Iglesias JIP (1990) Comparative allometries of gut-passage time, gut content and metabolic faecal loss in *Mytilus edulis* and *Cerastoderma edule*. Mar Biol 105:197-204
Herman PMJ, and Scholten H (1990) Can suspension-feeders stabilise estuarine ecosystems? In: M. Barnes and R.N. Gibson (eds.) Trophic relationships in the marine environment. Aberdeen University Press, Aberdeen. pp. 104-116
Hildreth DI and Crisp DJ (1976) A corrected formula for calculation of filtration rate of bivalve molluscs in an

experimental flowing system. J Mar Biol Ass UK 56:111-120
Holligan PM (1985) Marine dinoflagellate blooms - growth strategies and environmental exploitation. pp. 133-139 in: Anderson, White and Baden (eds.) Toxic Dinoflagellates. Elsevier Science Publishers New York/Amsterdam.
Hosomi A (1985) On the persistent trend of constant biomass and the constant total occupation area of the mussel *Mytilus galloprovincialis* (Lamark). Venus Jpn J Malacol 44:33-38
Hutchinson GE (1967) A treatise on Limnology Vol. 2. Introduction to lake biology and the limnoplankton. Wiley New York.
Hummel H (1985) Food intake of *Macoma balthica* (Mollusca) in relation to seasonal changes in its potential food on a tidal flat in the Dutch Wadden Sea. Neth J Sea Res 19:52-76
Hunt OD (1925) The food of the bottom fauna of the Plymouth fishing grounds. J Mar Biol Assoc UK 13:560-599
Jørgensen CB, Larsen PS, Møhlenberg F, and Riisgård HU (1988) The bivalve pump:properties and modelling. Mar Ecol Prog Ser 45:205-216
Jørgensen CB, Larsen PS, and Riisgård HU (1990) Effects of temperature on the mussel pump. Mar Ecol Prog Ser 64:89-97
Jørgensen CB (1990) Bivalve filter feeding: hydrodynamics, bioenergetics, physiology and ecology. Olsen and Olsen, Denmark. 140 p
Jørgensen CB, Famme P, Kristensen HS, Larsen PS, Møhlenberg F, and Riisgård HU (1986) The bivalve pump. Mar Ecol Prog Ser 34:69-77
Kellogg JL (1910) Shellfish industries. Holt and Co. New York 361pp.
Kiørboe T and Møhlenberg F (1987) Particle selection in suspension-feeding bivalves. Mar Ecol Prog Ser 5:291-296
Koehn RK and Bayne BL (1989) Towards a physiological and energetic understanding of the energetics of the stress response. Biol J Linnean Soc 37:157-171
Kramer KJM, Jenner HA, and de Zwart D (1989) The valve movement response of mussels: a tool in biological monitoring. Hydrobiologia 188/189: 433-443. In: Munawar M, Dixon G, Mayfield CI, Reynoldson T, and Sadar MH(eds). Environmental Bioassay Techniques and Their Application. Kluwer Academic Publishers Belgium
Lindahl O (1983) On the development of a *Gyrodinium aureolum* occurrence on the Swedish west coast in 1982. Mar Biol 77: 143-150
Lotsy JP (1896) The food of the oyster, clam and ribbed mussel. Rept U.S. Comm Fish and Fisheries for 1893: 375-386
Martin GW (1925) Food of the oyster. Botanical Gazette 75(2):143-169
Møhlenberg F and Riisgård HU (1979) Filtration rate, using a new indirect technique, in thirteen species of suspension-feeding bivalves. Mar Biol 54:143-147
Møhlenberg F and H.U. Riisgård (1978) Efficiency of particle retention in 13 species of suspension-feeding bivalves. Ophelia 17:239-246
Mook DH and Hoskin CM (1982) Organic determinations by ignition: caution advised. Estaur Coast Shelf Sci 15:697-699

Moore HF (1907) Survey of oyster bottoms in Matagorda Bay, Texas. Document 610, Bureau of Fisheries
Moore HF (1910) Volumetric studies of the food and feeding of oysters. Bulletin of the U.S. Bureau of Fisheries 28:1295-1308
Muschenheim DK and Newell CR (1992) Utilization of seston flux over a mussel bed. Mar Ecol Prog Ser 85:131-136
Navarro E, Iglesias JIP and Ortega MM (1992) Natural sediment as a food source for the cockle *Cerastoderma edule* (L.): effect of variable particle concentration on feeding, digestion and the scope for growth. J Exp Mar Biol Ecol 156:69-87
Nelson TC (1947) Some contributions from the land in determining conditions of life in the sea. Ecological Monographs 17:337-346
Newell CR (1990) The effects of mussel (*Mytilus edulis* L. 1758) position in seeded bottom patches on growth at subtidal lease sites in Maine. J Shellfish Res 9:113-118
Newell CR (1991) Development of a model to seed mussel bottom leases to their carrying capacity. Phase II Report. NSF SBIR ISI8809760. 800 pp
Newell CR, Shumway SE, Cucci TL and Selvin R (1989) The effects of natural seston particle size and type on feeding rates, feeding selectivity and food resource availability for the mussel *Mytilus edulis* L. at bottom culture sites in Maine. J Shellfish Res 8:187-196
Newell CR and Campbell D (1992) Output of a model to seed mussel bottom leases to their carrying capacity: calibration, validation and sensitivity analysis. J Shellfish Res 11:203
Newell CR and Gallagher SM (1992) Short-term variability in seston flux and physiological responses of bottom-cultured mussels: (*Mytilus edulis*) in Maine. J Shellfish Res 11:203
Newell CR and Shumway SE (1991) Mussel feeding selectivity below the pseudofeces threshold: the importance of particle concentration. J Shellfish Res 10.
Newell RIE and Jordan SJ (1983) Preferential ingestion of organic material by the American oyster *Crassostrea virginica*. Mar Ecol Prog Ser 13:47-53
Newell RIE and Thompson RJ (1984) Reduced clearance associated with spawning in the mussel *Mytilus edulis* L.(Bivalvia, Mytilidae). Mar Biol Lett 5:21-33
Officer CB, Smayda TJ and Mann R (1982) Benthic filter feeding: a natural eutrophication control. Mar Ecol Prog Ser 9: 203-210
Page HM and Ricard YO (1990) Food availability as a limiting factor to mussel *Mytilus edulis* growth in California coastal waters. Fish Bull 88:677-686
Petersen CGJ (1908) First report on the oysters and oyster fisheries in the Lim Fjord. Rept Danish Biol Station 15:1-42
Petersen CGJ and Jensen PB (1911) Valuation of the sea. I. Animal life of the sea-bottom, its food and quantity. Rept Danish Biol Station 20:1-78
Phinney DA and Yentsch CS (1985) A novel phytoplankton chlorophyll technique: toward automated analysis. J Plankton

Res 7:633-642
Platt T and B Irwin (1973) Caloric content of phytoplankton. Limnol Oceanogr 18:306-310
Prins TC and Smaal AC (1990) Benthic-pelagic coupling: the release of inorganic nutrients by an intertidal bed of *Mytilus edulis*. Trophic relationships in the Marine Environment pp. 89-103. Proc 24th Europ Mar Biol Symp. Barnes and Gibson (eds.) Aberdeen University Press
Richardson J, Newell CR and Panchang V (1992) Estimation of mussel seeding densities by hydrographic mathematical modeling. J Shellfish Res 11:205
Riebesell U (1991a) Particle aggregation during a diatom bloom.I. Physical aspects. Mar Ecol Prog Ser 69:273-280
Riebesell U (1991b) Particle aggregation during a diatom bloom. II. Biological aspects. Mar Ecol Prog Ser 69:281-291
Riisgård HU and Randlov A (1981) Energy budgets, growth and filtration rates in *Mytilus edulis* at different algal concentrations. Mar Biol 61:227-234
Riisgård HU and Poulsen E (1981) Growth of *Mytilus edulis* in net bags transferred to different localities in a eutrophicated Danish fjord. Mar Pollut Bull 12:272-276
Sankar SN (1991) Modeling plankton depletion over mussel beds: why it's good to live on the edge. Woods Hole Oceanographic Instituion Woods Hole Mass. Summer student fellow under Dr. Rockwell Geyer, final report
Savage RE (1925) The food of the oyster. Fishery Invest. (II)8: 2-50
Shumway SE (1990) A review of the effects of algal blooms on shellfish and aquaculture. J World Aqua Soc 21:65-104
Shumway SE (1992) Mussels and Public Health. Chapter 11 (pp. 511-542) In: The Mussel, *Mytilus* (Gosling EM ed) Elsevier Science Publishers BV Amsterdam
Shumway SE, Selvin R and Schick DF (1987) Food resources related to habitat in the scallop *Placopecten magellanicus* (Gmelin, 1791): A qualitative study. J Shellfish Res 6:89-95
Shumway SE, Barter J and Sherman-Caswell S (1990) Auditing the impact of toxic algal blooms on oysters. Environ Auditor 2:41-56
Shumway SE and Cembella AD (1993) The impact of toxic algae on scallop culture and fisheries. Reviews in Fisheries Science. In press
Small AC (1991) The ecology and cultivation of mussels: new advances. Aquaculture 94:245-262
Smayda TJ (1970) The suspension and sinking of phytoplankton in the sea. Oceanogr Mar Biol Ann Rev 8:353-414
Smayda TJ and Boleyn BJ (1965) Limnol Oceanogr 10:499-509
Smayda TJ (1989) Primary production and the global epidemic of phytoplankton blooms in the sea: a linkage? pp. 449-483 in: Cosper EM, Bricelj VM and Carpenter EJ (eds) Novel Phytoplankton Blooms: Causes and Impacts of Recurrent Brown Tides and Other Unusual Blooms. Springer-Verlag New York
Smayda TJ (1990) Novel and nuisance phytoplankton blooms in the sea: evidence for a global epidemic. pp. 29-40 in: Granéli E *et al* (eds) Toxic Marine Phytoplankton. Elsevier Science Publishers New York/Amsterdam

Smayda TJ and White A (1990) Has there been a global expansion of algal blooms? If so, is there a connection with human activities? pp. 516-517 in: Granéli E *et al* (eds) Toxic Marine Phytoplankton. Elsevier Science Publishers New York/Amsterdam

Stenton-Dozey JM and Brown AC (1992) Clearance and retention efficiency of natural suspended particles by the rock-pool bivalve *Venerupis corrugatus* in relation to tidal availability. Mar Ecol Prog Ser 82:175-186

Thompson RJ and Bayne BL (1972) Active metabolism associated with feeding in the mussel *Mytilus edulis* L. J Exp Mar Biol Ecol 9:111-124

Thompson RJ (1984) The reproductive cycle and physiological ecology of the mussel *Mytilus edulis* in a subartic, non-estuarine environment. Mar Biol 79:277-288

Tracey GA (1988) Feeding reduction, reproductive failure, and mortality in *Mytilus edulis* during the 1985 "brown tide" in Narragansett Bay, Rhode Island. Mar Ecol Prog Ser 50:73-81

Trowbridge JH, Geyer WR, Butman CA, and Chapman RJ (1989) The 17m flume at the Coastal Research Laboratory. Part 2. Flow characteristics. Woods Hole Oceanogr Inst Tech Rep WHOI-89-11

Walsby AE and Reynolds CS (1980) Sinking and floating. In: The physiological ecology of the phytoplankton. I. Morris (ed), U. California Press

Ward JE and Targett NM (1989) Influence of marine microalgal exudates on the feeding behavior of the blue mussel *Mytilus edulis*. Mar Biol 101:313-321

Widdows J (1978) Combined effects of body size, food concentration and season on the physiology of *Mytilus edulis*. J Mar Biol Assoc UK 58:125-142

Widdows J, Fieth P and Worrall CM (1979) Relationships between seston, avaiable food and feeding activity in the common mussel *Mytilus edulis*. Mar Biol 85:217-232

Widdows J and Hawkins AJS (1989) Partitioning of rate of heat dissipation by *Mytilus edulis* into maintenance, feeding and growth components. Physiol Zool 62:764-784

Williams LG (1982) Mathematical analysis of the effects of particle retention efficiency on determination of filtration rate. Mar Biol 66:171-177

Wyatt T (1990) Do algal blooms play homeostatic roles? pp. 249-252 in: Granéli *et al* (eds). Toxic Marine Phytoplankton. Elseviers Science Publishers New York/Amsterdam.

INVERTEBRATE PREDATORS AND THEIR ROLE IN STRUCTURING COASTAL AND ESTUARINE POPULATIONS OF FILTER FEEDING BIVALVES

Raymond Seed
University of Wales Bangor, School of Ocean Sciences, Menai Bridge, Gwynedd LL55 5EY, U.K.

INTRODUCTION

Predators can often play an important role in regulating the distribution and population dynamics of their natural prey. The influence of any specific predator is likely to be most pronounced amongst specialist predators which feed on only a restricted number of prey species, and perhaps least pronounced amongst generalist predators in which the diet includes a much broader spectrum of alternative prey types. Even so, generalist predators can have a significant influence on their prey, especially when the predator occurs in abundance and/or a particular prey species is a preferred component of the diet. Populations of many marine bivalves, however, which are typically either sessile or extremely slow moving, are controlled not by any specific predator but by suites or guilds of predators, often operating at different seasons and/or amongst different size classes of their prey (Kitching and Ebling 1967; Suchanek 1978; Peterson 1979; Menge 1983; Sanchez-Salazar et al 1987a,b; Griffiths 1990).

Many predators concentrate their foraging efforts in areas of high prey density switching to alternative food sources once a particular prey type has become significantly depleted; such foraging behaviour has a regulatory effect and will thus generally prevent local populations of the prey from being driven to extinction. Predators can also alter the structure of prey populations by foraging selectively on particular size classes. Size selective foraging has been documented for several important bivalve predators including gastropod molluscs (Palmer 1984; Hughes and Burrows 1990; Buyanovsky 1992), brachyuran crabs (Jubb *et al* 1983; ap Rheinallt 1986; Seed

1990), lobsters (Elner and Jamieson 1979; Griffiths and Seiderer 1980; Ojeda and Dearborn 1991), echinoderms (O'Neill et al 1983; Penney and Griffiths 1984; Briscoe and Sebens 1988) and shorebirds (Meire and Ervynck 1986; Bustnes and Erikstad 1990; Raffaelli et al 1990). The precise reasons for such foraging behaviour are still not entirely clear though maximizing energy intake (Elner and Hughes 1978; Ojeda and Dearborn 1991; Hughes et al 1992) and/or minimizing handling times (Hughes and Seed 1981; Boulding 1984; McClintock and Robnett 1986) have been suggested as causal factors.

Despite the high levels of predation experienced by many shallow water bivalves, continued coexistence of predator and prey is generally facilitated by virtue of various spatial and temporal (= size) refuges where the impact of the predator is greatly reduced. For those prey which can effectively escape predation by growing out of the size range normally taken by any particular predator, the length of time for which they remain vulnerable will thus depend on growth rate and this in turn will be partly a function of geographic location and tidal elevation. Predation has also been identified as a major factor influencing the structure and diversity of both hard and soft bottom marine communities (Peterson 1979; Paine et al 1985; Dayton 1985; Reise 1985; Thrush 1986; Raffaelli et al 1989).

Bivalve molluscs such as mussels, oysters, cockles and clams are commonly encountered in many coastal and estuarine ecosystems where they provide an abundant source of food for a wide variety of marine and terrestrial predators. The impact of birds and other vertebrate predators is discussed elsewhere in this volume; here the role of invertebrate predators in structuring bivalve populations will be considered. Although three important predatory groups - gastropods, starfish and decapod crustaceans will be examined, the following account is intentionally selective in its coverage, broadly reflecting the author's particular interest in marine mussels and crab-bivalve interactions.

THE PREDATORS

Gastropods: Muricid gastropods belonging to the genus *Nucella (= Thaís)* are widely distributed and important predators in the rocky intertidal zone in many geographical areas and have been the focus of numerous ecological and behavioural investigations (Bayne and Scullard 1978; Palmer 1984; Wickens and Griffiths 1985; Abe 1989; Gosselin and Bourget 1989; Burrows and Hughes 1990; Richardson and Brown 1990). Most research has focused on the dogwhelk, *Nucella lapillus,* which occurs throughout much of northern Europe and along the Atlantic coast of North America. Although occurring on shores ranging from the most sheltered to the most exposed, *N. lapillus* generally achieves its maximum abundance in the mid-low littoral zone on wave exposed coasts

Dogwhelks prefer to feed on mussels and barnacles but will also attack alternative prey, such as cockles (Largen 1967) when their preferred food is scarce or absent. Hughes and Dunkin (1984) showed that *N. lapillus* exhibited no residual preference for either mussels or barnacles though reversible preferences could develop as a result of prolonged diets on one or other of these prey. Ontogenetic shifts in foraging behaviour have also been reported (Hughes *et al* 1992). Thus, whilst *Mytilus edulis* accounted for 75% of the consumed flesh in juvenile *N. lapillus,* this increased to 99% in adult snails. Mussels attacked by *Nucella* can generally be identified by the presence of a small circular hole which the snail drills through the shell by alternate applications of a chemical secretion produced by the accessory boring organ, and radula scraping; small mussels, however, can be opened without resorting to drilling the shell valves. Most mussels are drilled through the thinnest part of the shell (Fig. 1) or above the underlying digestive gland which is rich in glycogen and more easily digested (Hughes and Dunkin 1984); essentially similar results have also been obtained for *N. freycinettii* feeding on *Mytilus trossulus* (Buyanovsky 1992). Dogwhelks take from several hours to well over a day to handle prey and usually move net distances of <0.2 m between meals (Hughes and Drewett 1985). Feeding rates vary from about 0.1-0.6 mussels.whelk^{-1}.d^{-1} depending on shell thickness and temperature (Morgan 1972; Stickle *et al* 1985). Familiarity with the prey, however, shortens handling time and can thus significantly increase prey profitability (i.e. energy intake per unit handling time) by a factor of 1.2-2.0 (Hughes and Burrows 1992).

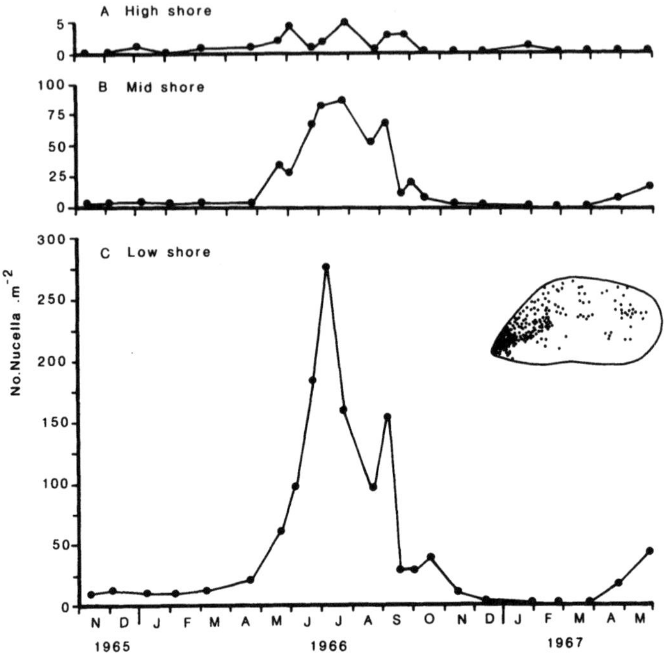

Fig. 1. Seasonal distribution of *Nucella lapillus* at three tidal levels at Robin Hood's Bay, N.E. England: inset, the spatial distribution of drill holes in a sample of *Mytilus edulis* from Filey, N.E. England (After Seed 1969)

The occurrence of dogwhelks on intertidal mussel beds is highly seasonal (Seed 1969; Menge 1978a; Tongiorgi et al 1981). During the colder winter months adult snails aggregate in crevices or pools as part of their annual reproductive cycle (Feare 1972). On the north-east coast of England emergence from these winter aggregations occurs in the spring, and over the summer months large numbers of snails can be observed foraging on mussels in the middle and lower intertidal zone (Fig. 1). Whilst all age groups indulge in some form of winter aggregation, high densities of juvenile snails (up to 1500.m^{-2}) remain dispersed over the mussel beds throughout the winter. On more sheltered shores, fewer, but larger snails emerge from their winter aggregations about a month earlier, and return about a month later than those at the more wave exposed sites, presumably in response to the more benign environmental conditions (Seed 1969).

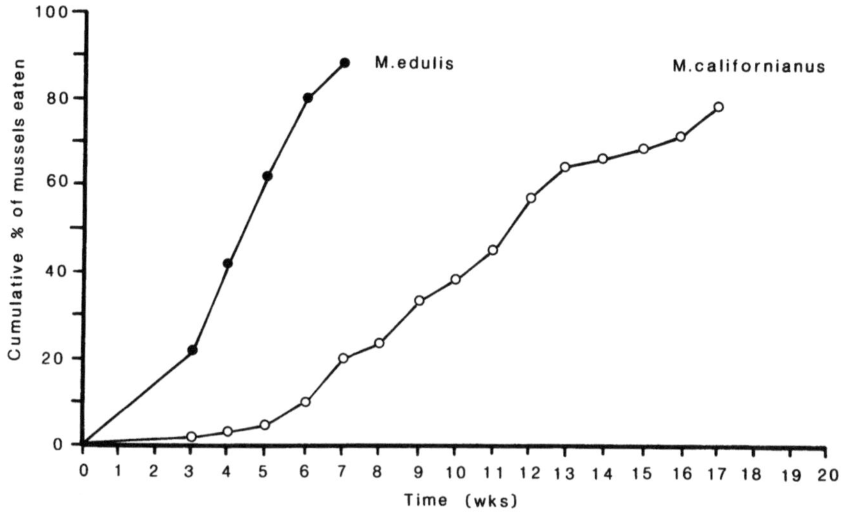

Fig. 2. Rates of predation by twenty *Thais emarginata* feeding on mixed-species clumps of *Mytilus* (max size 4 cm) (After Harger 1972)

Nucella spp. are highly selective both in terms of the species and size range of prey consumed. Selective feeding has been demonstrated unequivocally in the laboratory (Bayne and Scullard 1978; Hughes and Dunkin 1984) and under natural conditions in the field (Palmer 1984). When given a choice, *N. canaliculata* and *N. emarginata* both showed a strong preference for *Mytilus edulis* (but see McDonald *et al* 1991; Gosling 1992, and Seed 1992 for current views on the taxonomy and geographical distribution of *Mytilus)* rather than *M. californianus* (Harger 1972; Suchanek 1978; 1981). The precise mechanism for such selection (Fig. 2) is uncertain but is presumably based on the increased time or energy expended on drilling the thicker shell of *M. californianus* which contains an extra prismatic layer, and/or the potential lower calorific value of *M. californianus* tissue (Suchanek 1981; Palmer 1983). Whatever the reason, laboratory experiments have demonstrated that *N. canaliculata* fed on its preferred prey grew at a significantly faster rate than those fed on *M. californianus*

(Palmer 1983). Indirect evidence of size-selective feeding in *N. lapillus* is provided by comparing the identity and size frequencies of the prey in the diet with those in the natural population. Such comparisons (Fig. 3A) strongly suggest that natural long term diets in *N. lapillus* are biased towards the larger, though less abundant, size classes of mussels (Hughes and Burrows 1992).

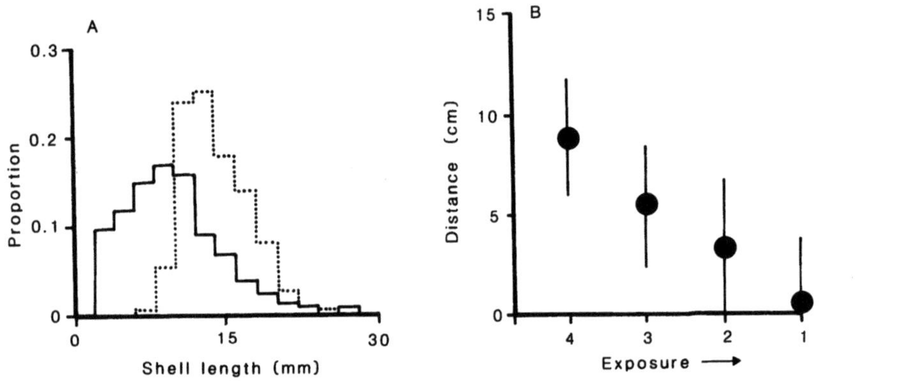

Fig. 3. (A) Size frequency distributions of *Mytilus edulis* available on the shore (continuous line) and those eaten by *Nucella lapillus* (dotted line); (B) The effect of increasing wave exposure on the distance moved from refuges (crevices) by foraging *Nucella lapillus* (after Hughes and Burrows 1992)

The abundance and consequent feeding intensity of *N. lapillus,* like many other marine epibenthic predators, declines sharply with increasing tidal elevation (Fig. 1). Seed (1969) found large numbers of drilled or partially drilled mussels in the lower intertidal zone on wave exposed shores along the north-east coast of England (Table 1) and attributed the absence of *Mytilus edulis* near deep crevices and pools (Plate 1B) to locally intense predation as snails entered and left their dense winter aggregations. Plate 1A shows a typical wave front of N. *lapillus* advancing over the mussel bed in the mid-low intertidal zone. Similar high levels of *Nucella* predation have also been reported amongst low shore populations of *M. edulis* on the New England coast (Menge 1983) and in Alaska where the average percentage of drilled shells ranged from 61% at protected sites to 95% on more wave exposed shores (Suchanek 1978). *Thais (=*

Nucella) canaliculata is known to be a major predator of small *M. californianus* in the low shore along the coast of Washington State in North America (Paine 1974).

Table 1. Frequency of Mytilus edulis shells partially drilled by Nucella lapillus at several locations in N.E. England (after Seed 1969)

Locality	shore type	shore level	n	% partially drilled
Filey Bay	S	low	330	16.4
Filey Bay	S	high	388	4.6
Filey Brigg	WE	low	249	20.5
Filey Brigg	WE	mid	668	6.6
Filey Brigg	WE	high	236	0.4
Robin Hoods Bay	WE	mid	250	13.4
Robin Hoods Bay	WE	high	250	0.5

WE = wave exposed; S = sheltered

Foraging intensity in *N. lapillus* is severely constrained by the prevailing weather conditions (Hughes and Burrows 1990). Strong wave action and desiccation or heat stress exert similar effects by confining snails to the vicinity of refuge sites. Refuges are provided by crevices or algal canopies though even on the open rock surface snails can obtain some mutual protection by forming tightly packed aggregations (Feare 1972). Figure 3B demonstrates how foraging excursions become increasingly confined with increasing wave exposure. As prey become progressively depleted close to refuge sites the scope for selective feeding is greatly reduced and snails will therefore have to forage over greater distances in order to obtain preferred prey items. Consequently, Burrows and Hughes (1990) found that the mean size of prey taken by *N. lapillus* was positively correlated with the proportion of snails foraging on any given occasion, and this in turn was governed by prevailing weather conditions. Using a dynamic programming technique

Hughes and Burrows (1992) were able to predict that at low levels of risk (calm, humid conditions) *N. lapillus* should forage widely over the shore selecting larger, more profitable prey and thereby maximizing energy intake. With increasing risk, however (e.g. dislodgement by waves, desiccation or heat stress) snails should forage close to refuge sites taking whatever prey are available, whilst under extreme conditions foraging should be abandoned altogether unless energy reserves fall below a critical level.

Unlike *N. lapillus,* where feeding on mussels is significantly reduced by wave exposure (e.g. Menge 1983; Burrows and Hughes 1989) *Stramonita (= Thais) haemastoma* actually increased its consumption of oysters, *Crassostrea virginica* and *Ostrea equestris,* at wave exposed sites by switching to smaller prey with shorter handling times (Richardson and Brown 1990). This appears to be an optimal strategy for *Stramonita* since prey profitability does not differ between large or small oysters. For large *N. lapillus,* on the other hand, profitability is greatly reduced when feeding on smaller mussels (Hughes and Dunkin 1984); for these snails, therefore, it is presumably more advantageous to restrict foraging activity and remain within a refuge site when exposed to waves, than to select less profitable prey. These results demonstrate the complexity of predator-prey interactions and suggest that the impact of intertidal predators on prey populations will depend upon the severity of the physical environment, and the relationship between prey size and profitability.

When present at high population density, N. *lapillus* can potentially consume all available prey. This often occurs on wave exposed shores particularly in the vicinity of refuge sites (Seed 1969; Menge 1978a). Close to these refuges snail densities will be high but predation intensity will decline as the snails progressively diverge over the mussel bed. On more sheltered shores, on the other hand, *N. lapillus* is generally less abundant and because these snails have greater access to protected habitats, especially beneath dense canopies of macroalgae, they tend to be far more scattered. Consequently, predation is more diffuse and distinct bare zones within the prey population are rarely observed. Even so, high cumulative mortality and turnover within the mussel population may still occur since foraging activity will be less constrained under these more benign conditions (Menge 1978b; Hughes and Burrows 1992).

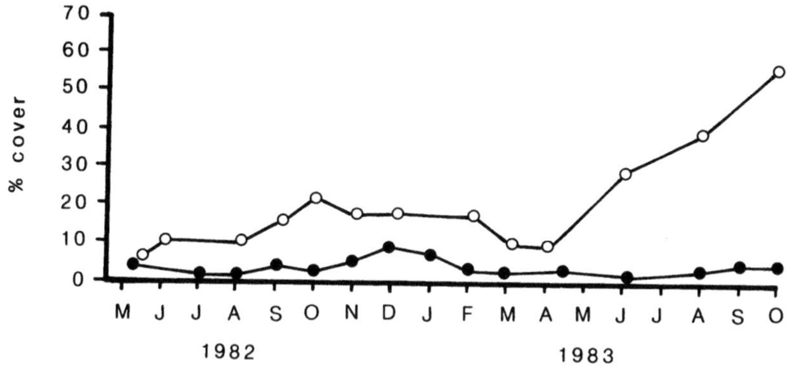

Fig. 4. Changes in the abundance of *Perumytilus purpuratus* in plots from which *Concholepas concholepas* were experimentally excluded (∘) and in control plots (•) (After Moreno *et al* 1986)

The impact of gastropod predation on mussel populations has been elegantly demonstrated by field experiments. Along much of the rocky coastline of central and southern Chile the mid intertidal zone is typically dominated by a monoculture of the mussel *Perumytilus purpuratus.* However, when the commercial collection of *Concholepas concholepas,* a major muricid predator of mussels, was halted by excluding "shell fishermen" from a designated marine reserve, this zone switched dramatically to one dominated by barnacles (Duran and Castilla 1989). Thus, whilst the average width of the mussel belt in 1978 when the reserve was established, measured 2.3 m, by 1981 this mussel had almost disappeared from exposed areas, remaining only as a narrow 0.4 m wide belt in the upper intertidal zone (Moreno et al 1986). No such reduction occurred outside the reserve where *C. concholepas* was still heavily harvested. Moreover, a second mytilid, *Semimytilus algosus* was abundant in the harvested areas but rarely found in the reserve. When *Perumytilus* was experimentally protected from *C. concholepas* predation by cages set out inside the reserve in 1982 the mussel population soon became re-established (Fig. 4). Thus, the mussel dominated community normally observed along much of the Chilean coastline appears to represent an alternative stable pattern (Sutherland 1974) which is strongly shaped and maintained by human factors (Moreno et al 1986).

In South Africa, *Nucella cingulata* is an important predator of the mussels *Choromytilus meridionalis* and *Aulacomya ater* (Penney and Griffiths 1984; Wickens and Griffiths 1985) whilst in New Zealand the littoral mytilids *Perna canaliculus* and *Xenostrobus pulex* are both predated by *Thais orbita (=Dicathais scalaris)* and *Lepsiella scobina* (Luckens 1976). In the Gulf of Mexico *Stramonita (=Thais) haemastoma* prefers to feed on the mussel *Ischadium recurvum* but mainly consumes oysters by virtue of their greater availability (e.g. Garton and Stickle 1980; Richardson and Brown 1990). Other gastropod predators which are known to feed extensively on coastal and estuarine bivalves include the following genera, *Morula* (Fairweather et al 1984; Lively et al 1987; Abe 1989), *Busycon* and *Urosalpinx* (Wood 1968; Peterson 1979, 1982), *Natica* (Broom 1982; Penney and Griffiths 1984; de Angelis *et al* 1985) and *Polinices (Taylor* 1980; Ansell and Morton 1985).

Despite being heavily predated by a variety of gastropods, bivalves are not entirely defenseless against this type of attack. Mussels for example can immobilize predatory snails by means of their byssal threads. This method of defence, in which several individual mussels may actually cooperate in subduing the predator, has been observed in response to several species of predatory snails (Wayne 1980; Carricker 1981; Petraitis 1987; Day et al 1991).

Starfish: Asteroid starfish are perhaps the most destructive predators of natural and cultivated stocks of mussels and oysters in many localities. They feed both extra-and intra-orally and unlike predatory gastropods several prey items can be consumed simultaneously (e.g. Paine *1974)*. *Asterias rubens* is frequently present at low densities on most rocky coasts in northern Europe but periodically its numbers rise dramatically such that it can effectively blanket much of the middle and lower intertidal zone. Such areas may then become virtually denuded of its principal prey species *Mytilus edulis* (Seed 1969). Such intense predation often results in a distinct predation line below which the established prey population is rare or absent. Physiological intolerance to desiccation is thought to be the most likely factor that limits the upshore extension of most epibenthic predators. As such the upper intertidal zone represents an effective refuge for *Mytilus*, though this incurs costs of lowered fitness for these mussels in terms of reduced reproductive output (Suchanek 1981). Dare (1976, 1982) has also reported large invasions of A. *rubens* during certain summers (March-September) in Morecambe Bay in the Irish Sea; starfish densities up to $450.m^{-2}$ were recorded and the swarm, which at one time covered 2.5 hectares of ground and contained at least 2.4×10^6 starfish, may have cleared up to 4000 t of first year *M. edulis* between June and September. The frontal boundary of the swarm sometimes advanced 150-200 $m.month^{-1}$ or 5-7 $m.d^{-1}$. During autumn the starfish retreated into deeper water (see also Menge 1972, 1983; Himmelman and Dutil 1991 for starfish populations elsewhere) and overwintered outside the Bay. The precise causes of these periodic massive aggregations which can effectively decimate molluscan populations are still largely unknown though entrainment of asteroid recruits to recruiting prey cohorts has been demonstrated. Superabundant prey may thus serve as focal points for *A. rubens* outbreaks with successful *Mytilus* recruitment conditions being necessary for successful starfish recruitment (Sloan 1980). Whatever their cause, large aggregations of starfish, which are often very patchy and unpredictable in their occurrence, are clearly a major factor controlling the local distribution and abundance of mussels in the lower shore and subtidally. Asteroid starfish have also been identified as important predators of *Mytilus edulis* on both the Pacific (Suchanek 1978; Paine 1980) and Atlantic (Peterson 1979; Menge 1979; 1983) coasts of North America. In Washington State *M. edulis* settles unpredictably in patches of open space typically just below the intertidal distribution of *M. californianus*. These grow quickly to reproductive size but are soon discovered and consumed by herds of roaming starfish *Pisaster ochraceus* which eliminate large beds of *M. edulis* in a few days (Seed and Suchanek 1992). Interestingly a close association between *M. edulis* and the anemone *Metridium senile* appears to afford the mussel significant protection against its asteroid predator *A. forbesii* (Kaplan 1984).

Each year on the Oregon coast the abundance of *M. trossulus* in the low intertidal zone fluctuates inversely with the intensity of *Pisaster ochraceus* predation (Menge 1992). From late autumn to early spring when starfish are less active and mussel recruitment is high, prey density increases whilst from spring to autumn when starfish are active and mussel recruitment is minimal, prey density decreases. Massive numbers of *Pisaster* feed on mussels and barnacles during low tide exposure and over a period of weeks the lower edge of the *Mytilus* zone retreats upshore until eventually the mussels are restricted to a midshore refuge.

Prey preferences vary with different species of asteroids and some may even vary their prey selection with increase in body size (Sloan 1980; Penney and Griffiths 1984; Himmelman and Dutil 1991). In the low intertidal zone on exposed rocky shores of the Pacific north-west Paine (1976) has documented size-limited predation on *M. californianus* by *Pisaster ochraceus.* From a series of laboratory choice experiments in which *Pisaster* selected medium-sized *N. californianus* and rejected both small and large-sized individuals, McClintock and Robnett (1986) concluded that *P. ochraceus is* maximizing energy intake and minimizing time spent foraging and handling prey. However, Paine's (1976) field observations do not support this evidence since P. *ochraceus* chose mussels of all sizes, though more small individuals were selected. O'Neill et al (1983) similarly found that *Asterias vulgaris* (10-50 mm arm length) fed mainly on small *M. edulis* cultured on nets. Thus, during the six month period of active feeding by starfish over the summer it was estimated that only 1.1% of the total dry weight of mussels >3.S cm was consumed. Loss of medium-sized mussels was 19.6%, whilst almost all small mussels (<2.0 cm) were consumed. These workers concluded, however, that starfish predation presented no serious constraint to the two year mussel production schedule; indeed by culling undersize mussels starfish predation may actually be beneficial since more food and space is then available to the larger mussels. Whilst Feder and Christensen (1967) provide several possible explanations for the existence of size specific predatory behaviour by starfish feeding on molluscs, the energetic costs of foraging and the mechanics of shell opening is clearly an area for further research.

As a result of transplantation experiments Kitching et al (1959) concluded that *Marthasterias glacialis* was at least partially responsible for preventing the establishment of *M. edulis* sublittorally in Lough Ine in south-west Ireland. On the west coast of North America, *Pisaster* effectively controls the distribution of M. *californianus* in the lower shore (Plate 2). Continued removal of this starfish resulted in the encroachment by the mussels into areas not previously occupied, eventually producing a virtual monoculture of mussels occupying the primary substratum (Paine 1974). In

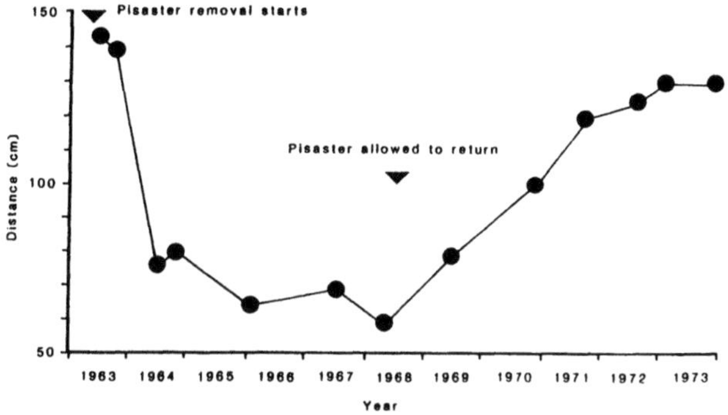

Fig. 5. Changes in the vertical distribution of *Mytilus californianus* following the experimental removal (June 1963) and subsequent return (June 1968) of *Pisaster ochraceus* from a vertical rocky surface in Washington State USA; distance measured from a low shore datum point to the bottom of the mussel zone (After Paine 1974)

other experimental removals, Paine (1971) working with the seastar *Stichaster australis* in New Zealand and Paine et al *(1985)* working with *Heliaster helianthus* in Chile, showed similar trends in altering the lower limits of the indigenous mussels *Perna canaliculus* and *Perumytilus purpuratus* respectively. Typically, when starfish were allowed to re-enter the system the lower limit returned to the previous state (Fig. 5). However, if the mussels are able to grow beyond the size which starfish are capable of consuming, an altered state in which mussels exist below their typical lower limits may persist for up to 10 *(M. californianus)* or 30 *(Perna)* years. This, however, did not occur in Chile because all size ranges of *Perumytilus* were vulnerable to starfish predation.

Marthasterias glacialis is an important predator of mussels along the South African coastline (e.g. Griffiths 1981; Fricke et al 1982). Densities of this starfish are maximal where the

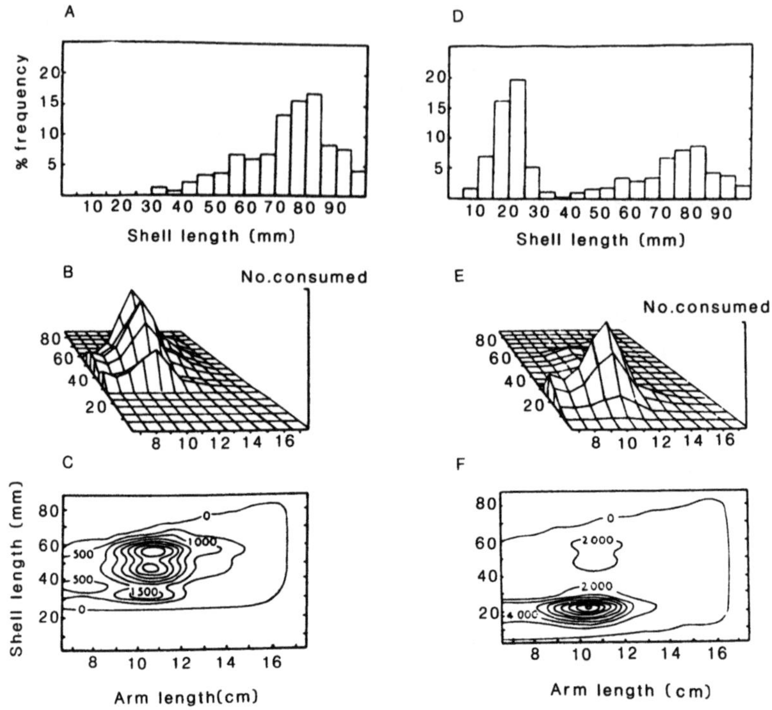

Fig. 6. Size frequency distributions of a single cohort (A) and a two cohort (D) *Choromytilus meridionalis* population. (B, C) and (E, F) predicted surfaces and contour diagrams of the number of mussels consumed.100 $m^{-2}.yr^{-1}$ by a *Marthasterias glacialis* population preying respectively on (A) and (D) (After Penney and Griffiths 1984)

black mussel *Choromytilus meridionalis* is abundant and in such areas mussels form the bulk of the diet (Penney and Griffiths 1984). Using laboratory prey selection and feeding rate data for starfish of different size, predictive three dimensional predation surfaces were developed for a natural starfish population feeding on either one or two cohort mussel populations. The models indicate that predatory effort should be concentrated on the smallest mussels (mainly 40-60 mm) when a single adult cohort is present (Fig. 6A-C) but on recruiting mussels (<30 mm) when two cohorts are present (Fig. 6D-F). Histograms summarizing the predicted total number of mussels consumed.100 $m^{-2}.yr^{-1}$ from the two populations (Fig. 7A,B) show clearly how predation should be concentrated on recruiting cohorts whenever these are available. The extent to which this situation actually occurs under natural conditions can be seen by comparing these predictions with field data collected during similar one or two cohort situations in 1981 and 1982. When only adult mussels were available, starfish were forced to consume larger mussels ~40 mm just within their upper size limit (Fig. 7C). Following the settlement of a second cohort, however, the starfish fed almost exclusively on small individuals (c. 20 mm) just above their lower prey size limit (Fig. 7D). This detailed study by Penney and Griffiths (1984) provides an interesting alternative approach to the study of predator-prey interactions, and convincingly demonstrates how starfish, as well as rock lobsters and naticid whelks, which select a similar size range of prey to the starfish, have the potential to alter significantly the abundance and size frequency structure of *Choromytilus* populations.

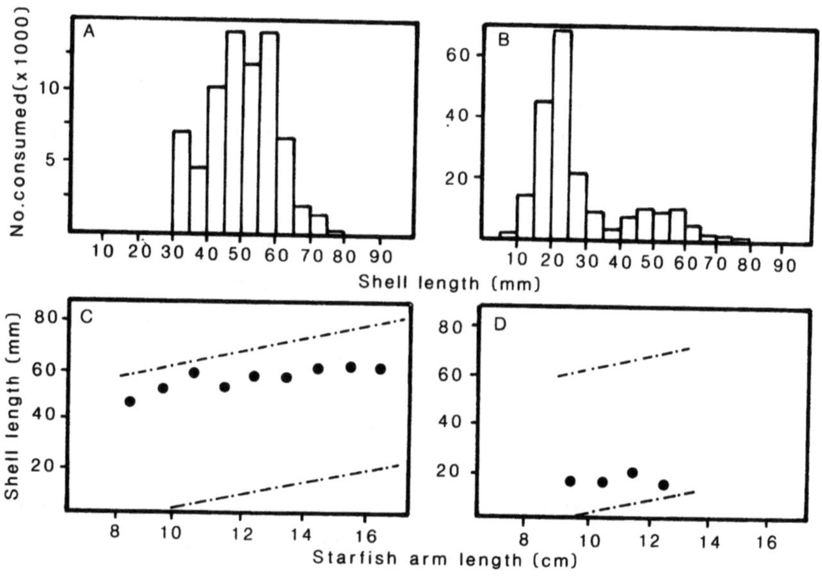

Fig. 7. Predicted size frequency distributions of mussels consumed.100 $m^{-2}.yr^{-1}$ by *Marthasterias glacialis* preying on (A) a single adult cohort of *Choromytilus meridionalis* and (B) a two cohort population. Actual prey selected by field populations of *M. glacialis* during 1981 when a single mussel cohort was present (C) and during 1982 when two mussel cohorts were present (D): pecked lines denote approximate size range of mussels consumed by *M. glacialis* (After Penney and Griffiths 1984)

In the low salinity waters of both the Baltic and Glacier Bay in Alaska the scarcity of mussel predators has effectively allowed *M. edulis* to become the dominant space occupying organism down to depths of 30 m and 3 m respectively (Kautsky 1982; Suchanek 1985). Moreover, in view of the low predation pressures (and presumably the lower Ca^{++} levels) experienced by Baltic mussels, selection has favoured mussels with thinner shells and smaller adductor muscles but a higher fecundity (Kautsky et al 1990). Consequently, when these mussels are transplanted to fully marine sites in the North Sea they are readily attacked and easily pulled open by starfish. Hancock (1965) showed experimentally that mussels with larger adductor muscles were less vulnerable to starfish predation.

Very few studies have investigated the impact of starfish predation subtidally where their effects would be expected to be maximal. The major difficulty here has evidently been the lack of understanding of the dynamics of starfish populations, especially those processes determining recruitment intensity and juvenile survival, which are only poorly understood for even the commonest species. In the northern Gulf of St. Lawrence, however, *Asterias vulgaris (=A. rubens?)* and *Leptasterias polaris* are important mussel predators subtidally (Himmelman and Dutil 1991). Here *M. edulis* forms dense beds extending from the low intertidal zone down to the lower limit of the kelp *Alaria*, where it disappears abruptly. This lower limit appears to be controlled by large aggregations of small (<5 cm) starfish which actively feed on small mussels up to about 2 cm in shell length, since mussels are abundant subtidally on buoys which are inaccessible to these predators (Fradette and Bourget 1980). Larger adult starfish (>15-20 cm) occur mainly at greater water depths on soft substrata where they can feed more efficiently on larger infaunal bivalves such as *Hiatella, Clinocardium* and *Macoma*. Large *Leptasterias* in particular is apparently well adapted to feeding on larger infaunal bivalves which it digs out of the sediment using its tube feet. From analyses of stomach contents, Chiu et al (1985) demonstrated that the semelid bivalve *Theora lata* was the main prey item in the diet of *Luidia* spp in the subtidal waters around Tolo Harbour in Hong Kong. Individual starfish contained up to 59 specimens of *T. lata* though no correlation was found between the number or size of prey items consumed and size of the predator.

Decapod crustaceans: Unlike starfish and predatory gastropods, decapod crustaceans are highly mobile epibenthic predators capable of foraging over large distances even during a single tidal cycle. They are mainly opportunistic generalists which feed on a wide variety of both epifaunal (e.g. Griffiths and Seiderer 1980; Navarrete and Castilla 1988, 1990; Eggleston 1990; Seed 1990) and infaunal (e.g. Hines *et al* 1990; Lin 1990; Sponaugle and Lawton 1990; Mansour and Lipcius 1991) bivalves. Some will also function as detritivores or even scavengers.

Decapod-bivalve predator prey systems have yielded valuable experimental evidence concerning the optimality of prey selection. Studies have embraced numerous families with contrasting chelal morphologies and foraging behaviour. Decapod foraging behaviour and the vulnerability of their prey requires knowledge of the morphological and mechanical characteristics of the predator's chelae. Elner (1978) proposed that chelae serve as a template upon which feeding habits and prey preferences are determined, a hypothesis which has subsequently been verified by several workers (e.g. Blundon and Kennedy 1982a; Boulding 1984; ap Rheinallt and Hughes 1985). Chelae can, however, have functions other than feeding such as defence and reproduction (e.g. Lee and Seed 1992) and these can sometimes confound simple correlations between form and feeding habits. Chelal characteristics and foraging behaviour also vary globally; compared with temperate water species, tropical decapods are considered to be morphologically and behaviourally more specialized for attacking molluscan prey (e.g. Vermeij 1987; Hughes and Elner 1989) probably as a result of the prolonged and intense co-evolution between exploiters and their victims (Hughes 1989).

Most decapods that feed on bivalves use more than one method of attack depending on the prey species, its overall size and relative shell thickness (Table 2). Some techniques, especially those required to open large resistant prey, involve a substantial structural investment in the form of powerful chelae; for other techniques this investment is minimal but the size of prey that can be opened relative to that of the predator is also substantially reduced. Chelate predators generally open bivalves by crushing the shell or, depending on the strength of the chela relative to the prey's shell,

Table 2. Decapod attack methods for opening bivalve molluscs (after Lau 1987)

Tactic	Relative handling time	Relative prey size	Structural Investment
1 Swallowing whole	short	small	low
2 Chipping-biting	moderate	medium	low
3 Wedging	short-moderate	medium	low
4 Prying	moderate	medium	low
5 Crushing	short	medium-large	high
6 Chipping-peeling	long	medium-large	moderate
7 Boring	long	medium	moderate
8 Tubercular peeling	long	medium-large	moderate
9 Shearing	short	medium	moderate

chipping the edge of the valves until the flesh is exposed. Sometimes more specialised claw morphology and shell opening behaviour is used as in the box crab, *Calappa ocellata,* in which a hooked peg and cusp on the outer face of the right chela applies a shearing force across the shell of the mussel *Brachidontes domingensis* (Hughes and Elner 1989).

Elner (1978) described several largely size related mussel opening techniques in *Carcinus maenas.* Crabs sought out and exploited weak spots in the shell by trial and error eventually smashing the mussel through a cumulative process whereby minor fractures in the shell substructure were gradually extended. The ability to open relatively large bivalves by repeatedly loading the shell has also been reported in other predatory decapods. Boulding and Labarbera (1986) showed that *Cancer productus* often squeezed the clam *Protothaca staminea* over two hundred times a three day period before the shell eventually failed. The persistent application of smaller forces enabled crabs to open much larger clams than could be crushed outright, thus greatly extending the size range of prey available to them.

Predatory decapods can be highly selective in their choice of prey. Harger (1972) for example showed that both *Cancer antennarius* and *Pachygrapsus crassipes* preferred *Mytilus edulis* to *M. cali*fornianus. Thus, when these two mytilids co-occurred, the thicker-shelled *M. californianus* was afforded some protection from crab predation by the presence of its more vulnerable congener. Preferences for size classes of prey well below the maximum size than the predator is capable of opening is characteristic of many decapods which feed on shallow water bivalves (ap Rheinallt 1986; Boulding 1984; Penney and Griffiths 1984; Davidson 1986; Navarrete and Castilla 1988; Ojeda and Dearborn 1991) though the degree of selectivity often varies according to hunger levels and food availability. For most bivalves, therefore, vulnerability to decapod predation declines with increasing body size. The crabs *Carcinus maenas* and *Liocarcinus puber* select mussels, *Mytilus edulis,* according to their profitability (Elner and Hughes 1978; ap Rheinallt 1986) thus apparently conforming to predictions of Optimal Foraging Theory. The behavioural basis of such size selection in *C. maenas is* discussed by Jubb et al (1983) and Akumfi and Hughes (1987). For other species, however, minimizing handling time rather than energy maximization may be the basis for size selective predation (Hughes and Seed 1981; Boulding 1984; Cresswell and McClay 1990). Time minimization may be especially important for tidally migrating predators since the time available for feeding will be restricted and during that time the predator itself may be vulnerable to predation. Whilst smaller size categories of prey are generally preferred, most predatory decapods can in fact open larger prey items given sufficient time. In some cases these may be carried to safer habitats and repeatedly loaded until the shell eventually fails. Feeding on larger prey may therefore prove to be a valuable foraging strategy especially if the cost of continuing an attack is low, the risk of leaving a refuge is high and alternative prey are scarce (Boulding and Labarbera 1986).

Size selective predation seems to be less important amongst those decapods which forage on infaunal or semi-infaunal bivalves where a significant investment of time and energy may be required before such prey can be evaluated and excavated from the sediment (Blundon and Kennedy 1982b; West and Williams 1986; Lin 1991). However, even epifaunal bivalves such as *Mytilus* and *Perna* are usually bound firmly together by byssal threads in dense clumps with many of the smaller, more vulnerable, size classes protected between larger mussels. In laboratory experiments Elner and Hughes (1978) found that predation of small mussels by *Carcinus maenas* was lower when these were presented in clumps. Group-living in the marsh mussel, *Geukensia demissa* also appears to offer some protection against predation (Bertness and Grosholz 1985).

Size selective predation can have a potentially significant effect on the structure of the prey population. Thus, preference for smaller size classes of prey is thought to be largely responsible for the distinctly bimodal size distributions reported for several coastal bivalves including *Modiolus modiolus* (Seed and Brown 1975), *Geukensia demissa* (Seed 1980), *Aulacomya ater* (Pollock 1979; Griffiths and Seiderer 1980) and *Perumytilus purpuratus* (Navarrete and Castilla 1988). Estimates from laboratory experiments indicate that natural densities of acanthocyclid crabs along the Chilean coast could potentially remove as many as 210-490 small *Perumytilus (3-6* mm).m^{-2} each day (Navarrete and Castilla 1988).

Ingestion rates of bivalves by predatory decapods can be quite phenomenal. Haddon and Wear (1987) showed that *Ovalipes catharus* could consume over 90cockles.$crab^{-1}.d^{-1}$ (Austrovenus) and at 20°C devoured prey so quickly that the foregut was full in <15 seconds. *Ovalipes punctatus* (du Preez 1984) and *Callinectes sapidus* (Seed 1982) respectively consumed large numbers of Donax (15-20.$crab^{-1}.d^{-1}$) and *Geukensia* (>50.$crab^{-1}.d^{-1}$) whilst individual *Thalamita danae* could ingest up to 20 *Perna viridis* daily (Seed 1990). Temperature has a marked influence on feeding rate. At 15.5°C medium (50-55 mm)and large (55-70 mm) *C. maenas* consumed 37 and 29 cockles, *Cerastoderma edule*.d^{-1} respectively whilst at 6°C predation rates were low and some crabs even stopped feeding for long periods (Sanchez-Salazar et al 1967a).

Mortality from decapod predation is generally most intense in the lower intertidal zone or subtidally where crabs are more abundant and where they are able to forage for longer periods of time. Ebling et al (1964) reported extensive crab predation in Lough Ine in south-west Ireland and tentatively attributed the absence of *Mytilus* sublittorally in many localities to this cause. In the Menai Strait, North Wales, Davies et al (1980) found that when protected by crab-proof fences *M. edulis* survived well whereas unprotected control became completely denuded of live mussels.

Crab predation was identified as a key factor in structuring populations of the cockle, *Cerastoderma edule,* at Traeth Melynog, a small bay (~330 ha) in the Menai Strait, North Wales (Sanchez-Salazar et al 1987b). Figure 8 shows the marked differences in size frequency distributions that exist within this population relative to tidal level. Most

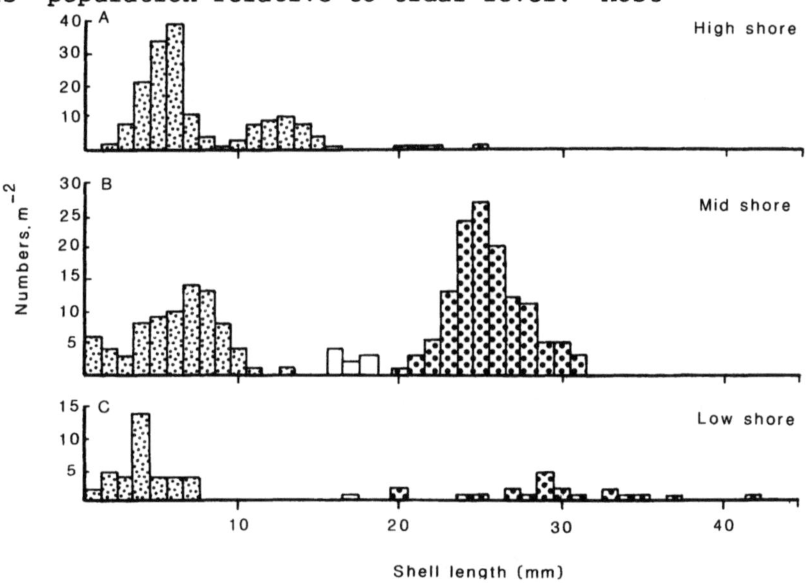

Fig. 8. Size frequency distributions of *Cerastoderma edule* at three tidal levels in the Menai Strait N. Wales in April 1987 before the late summer settlement of cockles: size ranges most vulnerable to predation by *Carcinus maenas* and *Haematopus ostralegus* shown by light and heavy stippling respectively (After Seed 1992b)

of the population is represented by small (<10 mm) first year cockles whilst larger individuals are concentrated in the middle and lower shore. Shore crabs, *Carcinus maenas* move into the intertidal zone to feed with each flood tide and remove large numbers of small cockles from the lower tidal levels, particularly during the warmer months when feeding rates are elevated. The impact of crab predation, however, rapidly declines with tidal elevation. Laboratory experiments show that virtually all (97%) cockles consumed by *C. maenas* are <15 mm (86% are <10 mm) suggesting that the predatory impact of these crabs in natural populations will be restricted to the smaller size classes (Fig. 9B). Consequently, few cockles survive their first summer in the low shore but many more do so at the middle, and particularly the upper tidal levels where crab predation is negligible.

Once beyond their first summer cockles attain a size at which they are seldom taken by crabs and this occurs more rapidly amongst the faster growing individuals in the low shore which attain a mean shell length of ~20 mm by their second winter. With increasing size, however, cockles enter a size range where they become progressively attractive to oystercatchers *Haematopus ostralegus*. The subsequent intensity of oystercatcher predation is a function of cockle size, density and duration of tidal exposure. Accordingly, mortality is highest amongst upper shore cockles with larger (>20 mm) individuals being selectively removed (Fig. 9C). The mortality of larger cockles lower on the shore, conversely, is minimal because they are sparsely distributed as a result of crab

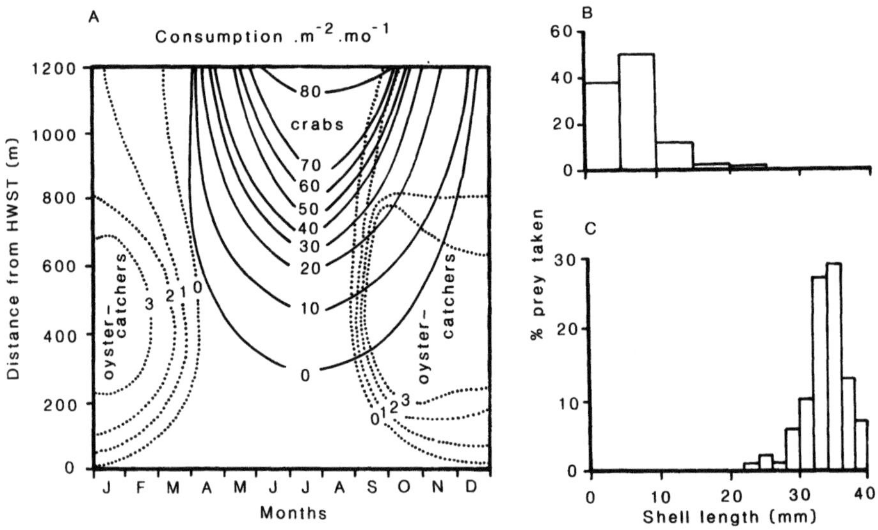

Fig. 9. (A) Predicted consumption rates of *Cerastoderma edule* by *Carcinus maenas* and *Haematopus ostralegus* as functions of tidal level and season. Size frequency distributions of cockles consumed by *Carcinus maenas* (B) and *Haematopus ostralegus* (C) (After Sanchez-Salazar *et al* 1987b)

predation early in life and they are exposed to oystercatchers for only short periods. In the mid-shore, oystercatcher predation is high but the probability of individual cockles surviving is good because cockle density is high. Nevertheless, as these cockles grow they face an ever-increasing risk of predation. The spatial and temporal patterns of mortality inflicted on the cockle population by these two important predators is illustrated in Figure 9A.

Using laboratory and field data, Sanchez-Salazar et al (1987b) estimated that *C. maenas* removed the equivalent of ~950 x 10^6 cockles from Traeth Melynog annually; this is equivalent to 472 x 10^3.linear m^{-1} of shore, i.e. a 1 m wide strip from HWST-LWST. In the Danish Wadden Sea juvenile shore crabs and cockle spat arrive on the tidal flats simultaneously and in July-August consumption by these small crabs (<9 mm) accounted for 26% of the elimination of cockles (Jensen and Jensen 1985). Such observations emphasise still further the impact of predatory crabs in controlling the abundance and population structure of cockles. However, recruitment of juvenile crabs, and their subsequent impact on bivalve populations, is highly variable from year to year depending largely on the severity of weather conditions during winter (Moller 1986; Beukema 1991).

Tidally migrating *Thalamita danae* feed voraciously on small mussels (<15 mm) and appear to be largely responsible for the scarcity of small *Perna viridis* in the low intertidal zone and shallow sublittoral waters of Hong Kong (Seed, 1990). Growth of *P. viridis* at these tidal levels, however, is extremely rapid and those mussels which escape predation quickly enter a size refuge where they become increasingly less vulnerable to crab predation. Such size limited predation has been identified as a key factor permitting *P. viridis* to become successfully established throughout much of southern and eastern Asia (Seed 1990). At higher tidal levels both mussel growth rate and crab foraging time are substantially reduced and consequently the *Perna* population contains proportionately more small size classes. Virtually all size ranges of *Brachidontes variabilis* are vulnerable to *T. danae* but this small mytilid, like *Xenostrobus atrata,* effectively reduces the impact of crab predation by densely occupying higher tidal levels. Of the 14 taxa that could be positively identified in the stomach of *T.*

danae, P. viridis and *B. variabilis* were easily the most important prey items. Only 18% of crabs caught on a falling tide had empty stomachs compared with over 75% of those caught on the rising tide. More significantly, *P. viridis* was present in only 20% and *B. variabilis* in only 4% of crabs caught on the flood tide, whereas both occurred in quantity in over half of all crabs collected on the ebb tide. These observations confirm that *T. danae* undergoes foraging excursions into the intertidal zone in order to feed on small mytilids which are abundant in the middle and upper shore but which, as a result of rapid growth and heavy mortality, are relatively scarce at the lower tidal levels (Seed 1990).

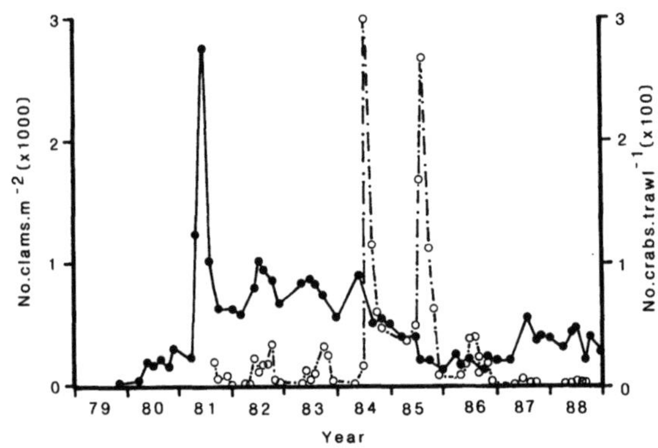

Fig. 10. Changes in the abundance of *Callinectes sapidus* (dashed line) and *Macoma balthica* (solid line) in a subestuary of Chesapeake Bay USA (After Hines *et al* 1990)

Figure 10 shows how, over a ten year period in Chesapeake Bay, natural densities of the Baltic clam, *Macoma balthica,* a major dietary component of the blue crab, *Callinectes sapidus,* responded to annual fluctuations in crab abundance. Moreover, in field experiments clams maintained in caged buckets, and thus inaccessible to predatory crabs, survived well, whilst most of the clams in the uncaged control buckets had been eaten. By cropping clam siphons fish may have an indirect effect on clam survival, because clams are then forced to reside at shallower depths within the sediment where they are more vulnerable to predators (e.g. Reading and McGrorty 1978: Zwarts and Wanink 1989). The mudcrab *Panopeus herbstii* is a major predator of *Geukensia demissa* (Seed 1980; Lin 1990). Juvenile *G. demissa* are especially vulnerable but larger mussels appear to achieve a size refuge from mudcrab predation at about 50-60 mm. Thus, when marked mussels were set out in cages which excluded *Callinectes* and other potential large predators, but which allowed access to mudcrabs, very few of the small size classes (<40 mm) were recovered alive at the end of the experiment, whereas most of the larger mussels (~60 mm) had survived (Fig. 11).

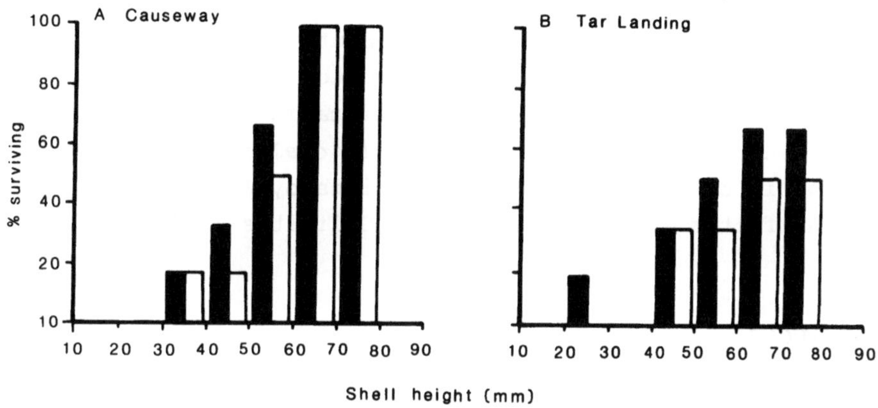

Fig. 11. Survival of *Geukensia demissa* exposed to predation by *Panopeus herbstii* at two sites in N. Carolina USA between April and June (black columns) and April and July (open columns): equal numbers of mussels were available at the start of each experiment (After Lin 1990)

In southern California the spiny lobster *Panulirus interruptus* exerts a significant influence on the population dynamics of intertidal *Mytilus* spp (Robles 1987). Despite continuous recruitment of mussels into these habitats no established mussel beds are present at these sites. Here whelks, fish and lobsters consume *Mytilus* but only the lobsters specialize on mussels. It is not surprising that these lobsters can have such a dramatic impact on mussels since it takes them less than 1.5 min on average to consume a small (<20 mm) *M. californianus*. Whilst intertidal mussel-foraging by lobsters occurs throughout the year, lobster densities and resultant impacts on mussels during the autumn and winter period are less than half those during spring and summer (Robles 1987). American lobsters *(Homarus americanus)* in Nova Scotia also have a dramatic influence on mussel populations (Elner and Campbell 1987). Analysis of stomach contents from both adults and juveniles reveal that the mussels *Mytilus edulis* and *Modiolus modiolus* together comprise 9.6-16.2% of the lobster diet though Elner and Campbell (1987) argue that *M. edulis* is the more important prey species. In South Africa the rock lobster *Jasus lalandii* is a major predator of the mussels *Aulacomya ater* and *Choromytilus meridionalis;* smaller size classes and larger numbers of *C. meridionalis* are consumed, resulting in domination of kelp bed sites by *A. ater* (Pollock 1979; Griffiths and Seiderer 1980; Griffiths 1981; Seiderer et al 1982).

PREY REFUGES

Despite the exceedingly high levels of predation that typify many sessile or slow moving invertebrates in shallow water coastal environments, several structural and ecological characteristics exhibited by bivalves can significantly reduce the risk of predation. These include spatial and temporal (= size) refuges, shell features, and even life history strategies (Table 3). We have already seen how the high intertidal zone often serves as an effective spatial refuge for many epifaunal bivalves; this occurs because the upshore extension of most marine predators is prevented by their physiological intolerance to desiccation. Spatial refuge from predation can similarly occur in regions of extreme (e.g. Baltic Sea) or fluctuating (e.g. estuaries) salinity. Growth to a large body size beyond which the prey is no longer vulnerable to predatory attack may also provide a further channel of escape for some bivalves. Size-limited predation has been reported for several species of epifaunal (Paine 1976; Eggleston 1990: Seed 1990) infaunal or semi-infaunal (Seed and Brown 1978; Whetstone and Eversole 1981; Lin 1990) and even free living bivalves such as scallops (Pohle et al 1991). Moreover, in some species these large bodied relatively predator free individuals, once established, may then provide additional refuge for the smaller more vulnerable size classes (e.g. Pollock 1979; Bertness and Grosholz 1985) thus allowing the prey species to persist despite potentially high levels of predation. Some bivalves even delay the onset of reproduction in order to grow rapidly to a large body size (e.g. Seed and Brown 1977, 1978; Commito 1982). Prey species may have an effective refuge in size well below the critical (maximum) size that a predator can open given sufficient time, if alternative more vulnerable species are available to the predator at the same time.

Wave exposure, mussel growth and crab predation interact to determine the local distribution pattern of *Mytilus edulis* at Lough Ine in south-west Ireland. The absence of mussels from shores of moderate wave exposure is due mainly to intense crab predation (Ebling et al 1964). In more protected locations within the lough, however, mussels grow rapidly into a size refuge, whilst the most waveswept shores provide a spatial refuge where predators are largely absent. When protected by cages, mussels thrived well in those areas from which they were normally excluded by virtue of heavy predation.

The vulnerability of burrowing bivalves is strongly influenced by shell characteristics other than size (Table 3). These include shell thickness, the degree of inflation and the presence or absence of a permanent gape (Blundon and Kennedy 1982a; Boulding 1984). Shell ornamentation, such as spines and lamellae, may also provide antipredator adaptations possibly by effectively increasing body size. Indeed any feature that increases a predators handling time must substantially reduce the probability of the predator persisting long enough to open the prey. Shallow burrowing bivalves, which are more vulnerable to predation, typically have thicker shells and are less likely to have a permanent shell gape than deeper burrowers; tropical species exhibit similar trends which presumably reflect a global pattern in the intensity of predation (e.g. Vermeij and Veil 1978). Thick shells, however, have an energetic cost and some species appear to trade off increased vulnerability against the ability to grow at a faster rate, especially early in life. This life history tactic enables species to persist despite high mortality to predators, because their faster growth and rapid attainment of adult size result in a higher reproductive effort.

Some bivalves escape predation by outdispersing their enemies; continuous recruitment can therefore result in the opportunistic invasion of predator-free habitats or persistence in the face of predation (Blundon and Kennedy 1982b). Inflated shells seem to be less easily opened by crabs. Whilst inflation may thus provide some protection from predation it may also lead to reduced burrowing efficiency,

Table 3. Characteristics evolved by bivalve molluscs which minimize the risk of predation (after Seed 1992b).

	Type	Examples
1	Spatial refuges:	High intertidal zone
		Nestling amongst large conspecifics
		Increased burial depth
		Macrophyte beds
		Substrate heterogeneity
2	Size refuges:	Large body size
3	Structural features:	Thickened shells
		Increased lnflation
		Ornamentatlon (e.g. spines, lamellae)
		Absence of permanent gape
		Strong ligaments and adductors
		Plicate valve margins (prevents shearing)
4	Life history tactics	Continuous or opportunistic recruitment
		Rapid early growth and delayed reproduction
		Increased reproductive effort
		Low population density

indicating how the bivalve shell probably represents an evolutionary compromise between conflicting ecological demands.

In *Protothaca staminea* Boulding and Hay (1984)-showed that predation by *Cancer productus* increased with clam density (up to a maximum of 280 clams.m^{-2}) suggesting that foraging occurs selectively in areas of high prey density. Like *Scylla serrata* (Hill 1979), *C. productus* searches large areas when foraging, returning to or remaining longer within, patches of high prey density. By reducing foraging efficiency, low population densities thus provide an effective refuge for many infaunal bivalves such as clams (Peterson 1982; Lipcius and Hines 1986; Sponaugle and Lawton 1990; Eggleston et al 1992). In the case of *Macoma balthica,* refuge from blue crab predation at low population density is retained even at high crab densities due to mutual interference and intraspecific aggression between crabs. Such results indicate that both predator and prey densities should be examined experimentally if their joint impact on predator-prey dynamics in marine systems is to be fully understood (Mansour and Lipcius 1991). Mortality of *Paphies ventricosa* from *Ovalipes catharus* predation, on the other hand, was negatively density dependent (Haddon et al 1987). Here, however, clams were so closely packed (>2000 m^{-2}) often in multilayers, that the digging ability of crabs was greatly impaired. Increased burial depth can greatly reduce the risk of predation by epibenthic predators such as crabs (Blundon and Kennedy 1982b; Boulding 1984; Haddon et al 1987; Lin 1991) and starfish (Doering 1982) though this can also restrict feeding ability.

Some bivalves obtain refuge from predation in particular types of substrata (Arnold 1984; Lipcius and Hines 1986). The crabs *Ovalipes ocellatus* and *Callinectes sapidus,* for instance, forage more successfully on juvenile *Mercenaria mercenaria* in sand than in more heterogeneous sediments which include shell fragments and gravel (Sponaugle and Lawton 1990). The erect stems of macrophytes such as seagrasses and saltmarsh plants can provide an effective barrier to large mobile epibenthic predators (Blundon and Kennedy 1982b; West and Williams 1986; Pohle et al 1991). This probably explains why blue crabs collected from seagrass meadows had much fuller stomachs than those from adjacent tidal creeks (Ryer 1987). Peterson (1982) found that *Mercenaria mercenaria* and *Chione cancellata* were positively associated with seagrass cover and when this was experimentally removed clams exhibited a higher rate of mortality whilst remaining unaltered in control sites. In this case, however, increased sediment compaction due to the presence of seagrass roots and rhizomes appeared to be the controlling factor since these effectively impaired the digging efficiency of predatory whelks *Busycon* spp. The scallop, *Argopecten irradians,* shows an interesting ontogenetic shift from a spatial refuge to a size refuge as it moves from eelgrass beds to the sediment surface over the course of its post settlement life history (Pohle et al 1991). This shift, however, can result in a critical window of high predation risk between 15-40 mm when scallops are most vulnerable to benthic predators.

In simulated 'saltmarshes' Revelas (1982) found that predation of *Mytilus edulis* by shore crabs, *Carcinus maenas* was 70% lower than in simulated mudflats with no vegetation, as a result of reduced predator-prey encounters. The size frequency structure of natural mussel populations could also be explained on the basis of these laboratory experiments. Mussels within the size range (<3 cm) most frequently consumed by *C. maenas* were significantly more abundant in the saltmarsh than in the mudflat population. This broad agreement between laboratory and field observations suggest that these experimental manipulations reflect natural interactions and that habitat heterogeneity can substantially reduce predator-prey interactions.

CONCLUSIONS

Laboratory and field observations and experiments have unequivocally demonstrated the important role of epibenthic predators in regulating the spatial distributions and population characteristics of many coastal and estuarine bivalves; in turn the availability of these prey species can significantly influence the foraging behaviour and population dynamics of their major invertebrate predators. The precise effects of predation on the prey population, however, will vary from one locality to another depending on the interplay of many factors including the foraging behaviour of the predators, the availability of spatial and temporal refuges, structural features and life history characteristics of the prey, and the nature of the abiotic environment. These factors will interact, often in a complex manner, to produce patterns which are not always predictable either in space or time.

Most bivalves, moreover, have planktonic larvae which are capable of widespread dispersal. Consequently, recruitment success need not necessarily be related to the state of the established resident population. Where such decoupling of juvenile and adult populations occurs, even those populations effectively decimated by predation can be quickly restocked by larvae arising from parents outside the immediate range of the predator. Cycles of abundance and scarcity are characteristic of many populations of sessile marine invertebrates including bivalves, and whilst predation of the established population unquestionably contributes to the determination of ecological pattern, at least on a local scale, other factors such as recruitment success, are also important and these will generally be mediated through climatological and hydrographical processes operating over a much larger scale. However, even recruitment success is likely to be determined partly by predation in the planktonic and immediate post-planktonic stages of life.

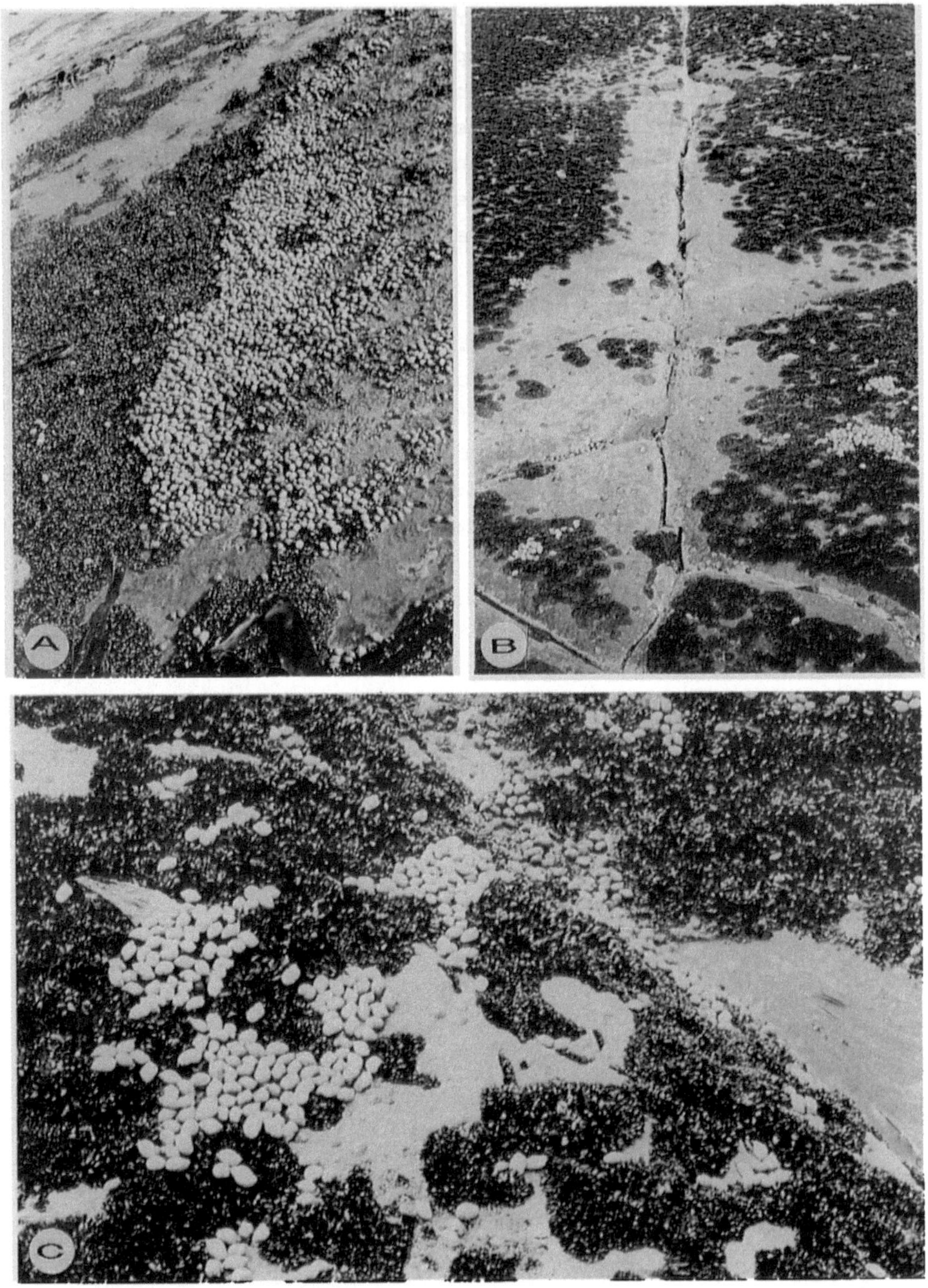

Plate 1 (A) A wavefront of *Nucella lapillus* foraging on low shore *Mytilus edulis* at Ravenscar N.E. England: note the absence of mussels in areas over which the whelks have foraged (right) (B) Absence of *M. edulis* near crevices used as refuges by *N. lapillus* (From Seed 1969) (C) *N. lapillus* feeding amongst patches of juvenile *M. edulis*.

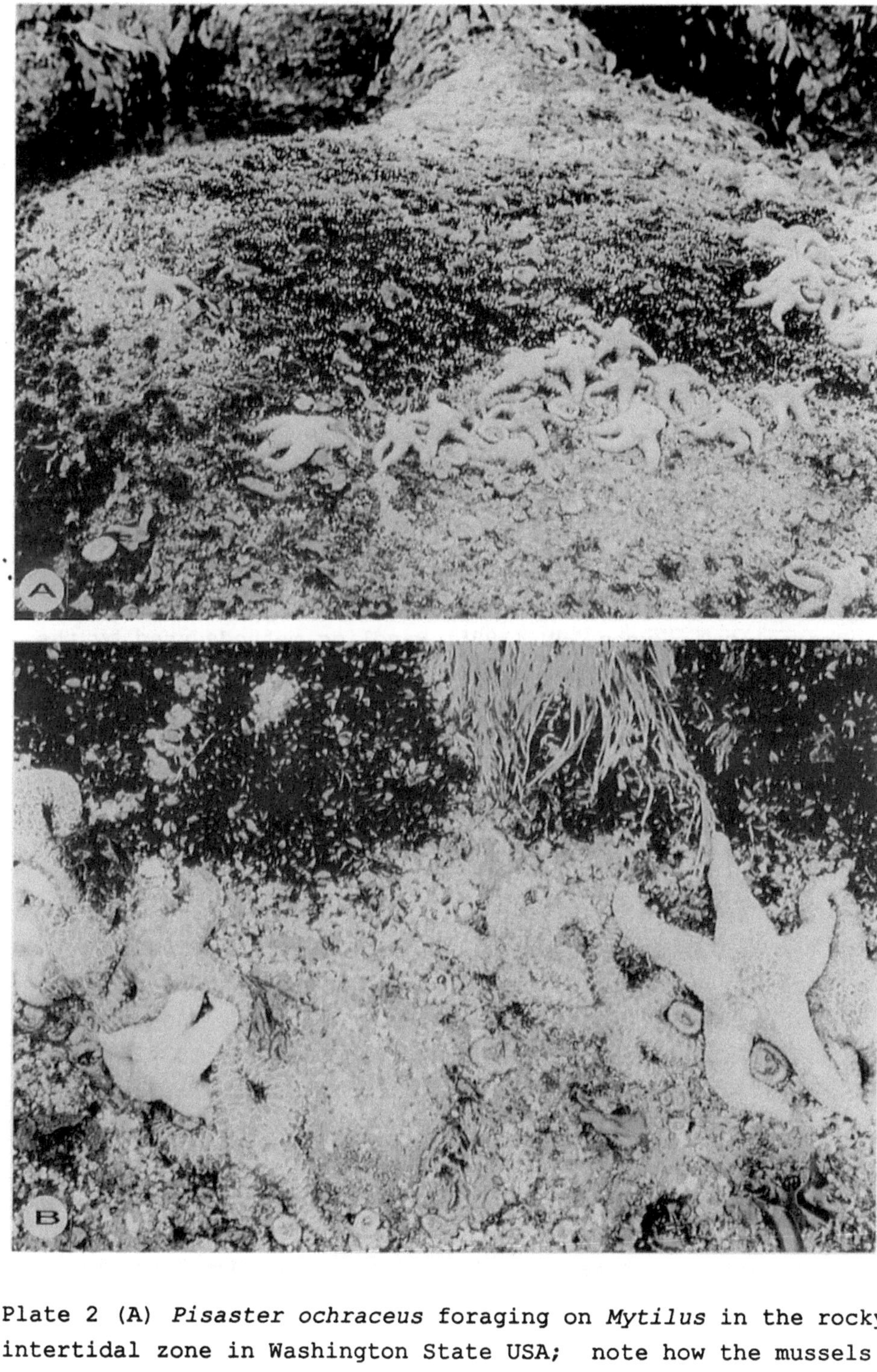

Plate 2 (A) *Pisaster ochraceus* foraging on *Mytilus* in the rocky intertidal zone in Washington State USA; note how the mussels in the central part of the mound have escaped predation by virtue of their higher tidal elevation (B) P *ochraceus* foraging at the lower edge of the *Mytilus* zone.

REFERENCES

Abe N (1989) Interactions between carnivorous gastropods and their sessile prey at a rocky intertidal shore. Physiol Ecol Japan 26:1-38

Akumfi CA Hughes RN (1987) Behaviour of *Carcinus maenas* (L.) feeding on large *Mytilus edulis* L. How do they assess the optimal diet? Mar Ecol Prog Ser 38:213-216

Ansell AD, Morton B (1985) Aspects of naticid predation in Hong Kong with special reference to the defensive adaptations of *Bassina (Callanaitis) calophylla* (Bivalvia). In: Proceedings of the Second International Workshop on the Malacofauna of Hong Kong and Southern China, Hong Xong 1983. (eds) Morton B, Dudgeon D pp. 635-661 Hong Kong University Press, Hong Kong.

ap Rheinallt T (1986) Size selection by the crab *Liocarcinus puber* feeding on mussels *Mytilus edulis* and on shore crabs *Carcinus maenas*. Mar Ecol Prog Ser 29:45-53

ap Rheinallt T Hughes RN (1985) Handling methods used by the velvet swimming crab *Liocarcinus puber* when feeding on molluscs and shore crabs. Mar Ecol Prog Ser 25:63-70

Arnold WS (1984) The effects of prey size, predator size and sediment composition on the rate of predation of the blue crab, *Callinectes sapidus* Rathbun on the hard clam *Mercenaria mercenaria* (Linne). J Exp Mar Biol Ecol 80: 207-219

Bayne BL Scullard C (1978) Rates of feeding by *Thais (=Nucella) lapillus (L.)*. J Exp Mar Biol Ecol 32:97- 111

Bertness MD Grosholz E (1985) Population dynamics of the ribbed mussel, *Geukensia demissa:* the costs and benefits of an aggregated distribution. Oecologia 67:192-204

Beukema JJ (1991) The abundance of shore crabs *Carcinus maenas* (L.) on a tidal flat in the Wadden Sea after cold and mild winters. J Exp Mar Biol Ecol 153:97-113

Blundon JA Kennedy VS (1982a) Mechanical and behavioural aspects of blue crab, *Callinectes sapidus* (Rathbun) predation on Chesapeake bivalves. J Exp Mar Biol Ecol 65: 47-65

Blundon JA Kennedy VS (1982b) Refuges for infaunal bivalves

from blue crab, *Callinectes sapidus* (Rathbun) predation in Chesapeake Bay. J Exp Mar Biol Ecol 65:67-81

Boulding EG (1984) Crab-resistant features of shells of burrowing bivalves: decreasing vulnerability by increasing handling time. J Exp Mar Biol Ecol 76:201- 223

Boulding EG Hay TK (1984) Crab response to prey density can result in density dependent mortality of clams. Can J Fish Aq Sci 41:521-523

Boulding EG Labarbera M (1986) Fatigue damage: repeated loading enables crabs to open larger bivalves. Biol Bull 171:538-547

Briscoe CS Sebens KP (1988) Omnivory in *Strongylocentrotus droebachiensis* (Muller) (Echinodermata: Echinoidea): predation on subtidal mussels. J Exp Mar Biol Ecol 115: 1-24

Broom MJ (1982) Size selection, consumption rates and growth of the gastropod *Natica maculosa (* Lamarck) and *Thais carinifera* (Lamarck) preying on the bivalve, *Anadara granosa*(L.). **J Exp Mar Biol Ecol 56:213-233**

Burrows MT Hughes RN (1989) Natural foraging of the dogwhelk, *Nucella lapillus* (Linnaeus): the weather and whether to feed. J Moll Stud 55:285-295

Burrows MT Hughes RN (1990) Variation in growth and consumption among individuals and populations of dogwhelks, *Nucella lapillus:* a link between foraging behaviour and fitness. J Anim Ecol 59:723-742

Bustnes JO Erikstad KE (1990) Size selection of common mussels, *Mytilus edulis,* by common eider, *Somateria mollissima,* energy maximization or shell weight minimization. Can J Zool 68:2280-2283

Buyanovsky AI (1992) Prey-selection by the dogwhelk *Nucella freycinettii* (Deshayes) feeding on the mussels *Mytilus trossulus* (Gould). J Moll Stud 58:343-345

Carriker MR (1981) Shell penetration and feeding by naticacean and muricacean predatory gastropods: A synthesis. Malacologia 20:403-422

Chiu ST Lam VWW Shin PKS (1985) Mollusc predation by *Luidia* spp (Echinodermata: Asteroidea) in Tolo Harbour and Channel, Hong Kong. In: Proceedings of the Second International Workshop on the Malacofauna of Hong

Xong and Southern China, Hong Kong 1983. (eds) Morton B, Dudgeon D pp. 365-379 Hong Kong University Press Hong Kong
Commito JA (1982) Effects of *Lunatia heros* predation on the population dynamics of *Mya arenaria* and *Macoma balthica* in Maine USA. Mar Biol 69:187-193
Cresswell PD McClay CL (1990) Handling times, prey size and species selection by *Cancer novaezelandiae* (Jacquinot, 1853) feeding on molluscan prey. J Exp Mar Biol Ecol 140: 13-29
Dare PJ (1976) Settlement, growth and production of the mussel, *Mytilus edulis* L., in Morecambe Bay, England. Fish Invest Minist Agric Fish Food Lond Ser II 28:1-25
Dare PJ (1982) Notes on the swarming behaviour and population density of *Asterias rubens* L. (Echinodermata: Asteroidea) feeding on the mussel *Mytilus edulis*. J Cons 40:112-118
Davidson RJ (1986) Mussel selection by the paddle crab, *Ovalipes catharus* (white): evidence of flexible foraging behaviour. J Exp Mar Biol Ecol 102:281-299
Davies G Dare *PJ* Edwards DB (1980) Fenced enclosures for the protection of seed mussels *(Mytilus edulis* L.) from predation by shore crabs *(Carcinus maenas* (L.)). Minist Agric Fish Food Fish Res Tech Rep 56:1-14
Day RW Barkai A Wickens PA (1991) Trapping of three drilling whelks by two species of mussel. J Exp Mar Biol Ecol 149:109-122
Dayton PK (1985) Processes structuring some marine communities: are they general? In: Ecological Communities: conceptual issues and the evidence. (eds) Strong DR, Simberloff D, Lawrence G, Thistle AB pp. 181-197 Princeton University Press Princeton
de Angelis DJ Kitchell JK Post WM (1985) The influence of naticid predation on evolutionary strategies of bivalve prey: conclusions from a model. Am Nat 126:817-842
Doering PH (1982) Reduction of sea star predation by the burrowing response of the hard clam *Mercenaria mercenaria*(Mollusca: Bivalvia). Estuaries 5: 310-315
du Preez HH (1984) Molluscan predation by *Ovalipes punctatus* (de Haan) (Crustacea: Brachyura: Portunidae). J Exp Mar

Biol Ecol 84:55-71

Duran LR Castilla JC (1989) Variation and persistence of the middle rocky intertidal community of central Chile with and without human harvesting. Mar Biol 103:555-562

Ebling FJ Kitching JA, Muntz L, Taylor CM (1964) The ecology of Lough Ine XIII. Experimental observation of the destruction of *Mytilus edulis* and *Nucella lapillus* by crabs. J Anim Ecol 33:73-82

Eggleston DB (1990) Functional responses of blue crabs *Callinectes sapidus* Rathbun feeding on juvenile oyster *Crassostrea virginica* (Gmelin): effects of predator sex and size and prey size. J Exp Mar Biol Ecol 143:73-90

Eggleston DB Lipcius RN Hines AH (1992) Density dependent predation by blue crabs upon infaunal clam species with contrasting distribution and abundance patterns. Mar Ecol Prog Ser 85:55-68

Elner RW (1978) The mechanics of predation by the shore crab *Carcinus maenas* (L.) on the edible mussel, *Mytilus edulis* L. Oecologia 36:333-344

Elner RW Campbell A (1987) Natural diets of lobster *Homarus americanus* from barren ground and macroalgal habitats off southwestern Nova Scotia, Canada. Mar Ecol Prog Ser 37:131-140

Elner RW Hughes RN (1978) Energy maximization in the diet of the shore crab, *Carcinus maenas*. J Anim Ecol 47: 103- 116

Elner RW Jamieson GS (1979) Predation of sea scallops, *Placopecten magellanicus,* by the rock crab, *Cancer irroratus,* and the American lobster *Homarus americanus.* J Fish Res Bd Can 36:531-543

Fairweather PG Underwood AJ Moran MJ (1984) Preliminary investigations of predation by the whelk *Morula marginalba.* Mar Ecol Prog Ser 17:143-156

Feare CJ (1972) The adaptive significance of aggregation behaviour in the dogwhelk *Nucella lapfllus* (L.). Oecologia 7:117-126

Feder HH Christensen AM (1967) Aspects of asteroid biology. In: Physiology of Echinodermata (ed) Boolootian RA, pp. 87-128 Interscience New York

Fradette P Bourget E (1980) Ecology of the epibenthic epifauna of the Estuary and Gulf of St. Lawrence:

factors affecting their distribution and abundance on buoys. Can J Fish Aquat Sci 37:979-999

Fricke AH Koop K Cliff G (1982) Colonization and viability of an artificial steel reef in False Bay, South Africa. Trans R Soc S Afr 44:499-512

Garton D Stickle WB (1980) Effects of salinity and temperature on the predation rate of *Thais haemastoma* on *Crassostrea virginica* spat. Biol Bull 158:49-57

Gosling EM (1992) Systematics and geographic distribution of *Mytilus*. In: The Mussel *Mytilus:* Ecology, Physiology, Genetics and Culture. (ed) Gosling EM, pp. 1-20 Elsevier Press Amsterdam

Gosselin LA Bourget E (1989) The performance of an intertidal predator, *Thais lapillus,* in relation to structural heterogeneity. J Anim Ecol 58:287-303

Griffiths CL (1990) Spatial gradients in predation pressure and their influence on the dynamics of two littoral bivalve populations. In: The Bivalvia-Proceedings of a Memorial Symposium in Honour of Sir Charles Maurice Yonge, Edinburgh, 1986 (ed) Morton B, pp. 321-332 Hong Kong University Press Hong Kong

Griffiths CL Seiderer LJX (1980) Rock-lobsters and mussels limitations and preferences in a predator-prey interaction. J Exp Mar Biol Ecol 44:95-109

Griffiths RJ (1981) Production and energy flow in relation to age and shore level in the bivalve *Choromytilus meridionalis* (Kr.). Estuar Coast Mar Sci 12:101-118

Haddon M Wear RG (1987) Biology of feeding in the New Zealand paddle crab *Ovalipes catharus* (Crustacea, Portunidae). N Z J Mar Freshw Res 21:55-64

Haddon M Wear RG Packer HA (1987) Depth and density of burial by the bivalve *Paphies ventricosa* as refuges from predation by the crab *Ovalipes catharus.* Mar Biol 94: 25-30

Hancock DA (1965) Adductor muscle size in Danish and British mussels in relation to starfish predation. Ophelia 2: 253-267

Harger JRE (1972) Competitive coexistence: maintenance of interacting associations of sea mussels *Mytilus edulis* and *Mytilus californianus.* Veliger 14:387-410

Hill BJ (1979) Aspects of the feeding strategy of the predatory crab *Scylla serrata.* Mar Biol 55:209-214

Himmelman JH Dutil C (1991) Distribution, population structure and feeding of subtidal seastars in the northern Gulf of St. Lawrence. Mar Ecol Prog Ser 76:61-72

Hines AH Haddon AM Wiechart LA (1990) Guild structure and foraging impact of blue crabs and epibenthic fish in a subestuary of Chesapeake Bay. Mar Ecol Prog Ser 67: 105-126

Hughes RN (1989) Foraging behaviour of a tropical crab *Ozius verreauxii.* Proc Roy Soc Lond 237B:201-212

Hughes RN Burrows MT (1990) Energy maximization in the natural foraging behaviour of the dogwhelk *Nucella lapillus* (L.). In: Trophic Relationships in the Marine Environment (eds) Barnes M, Gibson RN, pp. 517-527 Aberdeen University Press Aberdeen

Hughes RN Burrows MT (1992) Predatory behaviour of the intertidal snail, *Nucella lapillus,* and its effect on community structure. In: Mutualism and Community Organisation (eds) Kawanabe H Cohen J Iwasaki K Oxford University Press, Oxford (in press)

Hughes RN Burrows Mt Rogers SEB (1992) Ontogenetic changes in foraging behaviour of the dogwhelk *Nucella lapillus* (L.). J Exp Mar Biol Ecol 155:199-212

Hughes RN Drewett D (1985) A comparison of the foraging behaviour of dogwhelks, *Nucella lapillus* (L.) feeding on barnacles or mussels on the shore. J Moll Stud 51: 73-77

Hughes RN Dunkin SB de (1984) Behavioural components of prey selection by dogwhelks, *Nucella lapillus* (L.) feeding on mussels, *Mytilus edulis* L. in the laboratory. J Exp Mar Biol Ecol 77:45-68

Hughes RN Elner RW (1989) Foraging of a tropical crab: *Calappa ocellata* Holthuis feeding upon the mussel *Brachidontes domingensis* (Lamarck). J Exp Mar Biol Ecol 133:931-1001

Hughes RN Seed R (1981) Size selection of mussels by the blue crab *Callinectes sapidus:* energy maximizer or time minimizer? Mar Ecol Prog Ser 6: 83-89

Jensen KT Jensen JN (1985) The importance of some epibenthic predators on the density of juvenile benthic macrofauna in the Danish Wadden Sea. J Exp Mar Biol Ecol 89: 157- 174

Jubb CA Hughes RN ap Rheinallt T (1983) Behavioural mechanisms of size selection by crabs, *Carcinus maenas (L.)* feeding on mussels, *Mytilus edulis* L. J Exp Mar Biol Ecol 66: 81-87

Kaplan SW (1984) The association between the sea anemone *Metridium senile* (L.) and the mussel *Mytilus edulis (L.)*:reduced predation by the starfish *Asterias forbesii* (Desor). J Exp Mar Biol Ecol 79:1550-1557

Kautsky N (1982) Growth and size structure in a Baltic *Mytilus edulis* population. Mar Biol 68:117-133

Kautsky N Johannesson J Tedengren M (1990) Genotypic and phenotypic differences between Baltic and North Sea populations of *Mytilus edulis* evaluated through reciprocal transplantations. 1. Growth and morphology. Mar Ecol Prog Ser 59:203-210

Kitching JA Ebling FJ (1967) Ecological studies at Lough Ine. Adv Ecol Res 4:198-291

Kitching JA Sloane JF Ebling FJ (1959) The ecology of Lough Ine VIII. Mussels and their predators. J Anim Ecol 28:331-341

Largen MJ (1967) The diet of the dogwhelk, *Nucella lapillus* (Gastropoda: Prosobranchia). J Zool Lond 151:123-127

Lau CJ (1987) Feeding behaviour of the Hawaiian slipper lobster, *Scyllarides squammosus,* with a review of decapod crustacean feeding tactics on molluscan prey. Bull Mar Sci 41:378-391

Lee SY Seed R (1992) Ecological implications of cheliped size in crabs: some data from *Carcinus maenas* and *Liocarcinus holsatus.* Mar Ecol Prog Ser 84:151-160

Lipcius RN Hines AH (1986) Variable functional responses of a marine predator in dissimilar homogeneous microhabitats. Ecology 67:1361-1371

Lin J (1990) Mud crab predation on ribbed mussels in salt marshes. Mar Biol 107:103-109

Lin J (1991) Predator-prey interactions between blue crabs and ribbed mussels living in clumps. Estuar Coastal Shelf Sci 32:61-71

Lively CM Raimondi PT (1987) Desiccation, predation and mussel-barnacle interactions in the northern Gulf of

California. Oecologia 74:304-309

Luckens PA (1976) Settlement and succession on rocky shores at Aukland, North Island, New Zealand. NZ Oceanogr Inst Mem 70:1-64

Mansour RA Lipcius RN (1991) Density-dependent foraging and mutual interference in blue crabs preying upon infaunal clams. Mar Ecol Prog Ser 72:239-246

McClintock JB Robnett Jr TJ (1986) Size selective predation by the asteroid *Pisaster ochraceus* on the bivalve *Mytilus californianus:* a cost-benefit analysis. Pubbl Stn Zool Napoli Mar Ecol 7:321-332

McDonald JH Seed R Koehn RK (1991) Allozymes and morphometric characters of three species of *Mytllus* in the Northern and Southern Hemispheres. Mar Biol 111:323-333

Meire PM Ervynck A (1986) Are oystercatchers *(Haematopus ostralegus)* selecting the most profitable mussels *(Mytilus edulis)?* Anim Behav 34:1427-1435

Menge BA (1972) Competition for food between two intertidal starfish species and its effect on body size and feeding. Ecology 53:635-644

Menge BA (1978a) Predation intensity in a rocky intertidal community. Relation between predator foraging activity and environmental harshness. Oecologia 34:1-16

Menge BA (1978b) Predation intensity in a rocky intertidal community: effect of an algal canopy, wave action and desiccation on predator feeding rates. Oecologia 34:17-35

Menge BA (1979) Coexistence between the seastars *Asterias vulgaris* and *A. forbesi* in a heterogeneous environment: a non-equilibrium explanation. J Exp Mar Biol Ecol 41:245-272

Menge BA (1983) Components of predation intensity in the low zone of the New England rocky intertidal region. Oecologia 58:141-155

Menge BA (1992) Community regulation: under what conditions are bottom-up factors important on rocky shores? Ecology 73:755-765

Moller P (1986) Physical factors and biological interactions regulating infauna in shallow boreal areas.

Mar Ecol Prog Ser 30:33-47
Moreno CA Lunecke KM Lepez MI (1986) The response of an intertidal *Concholepas concholepas* (Gastropoda) population to protection from man in southern Chile and the effects on benthic sessile assemblages. Oikos 46:359-364
Morgan PR (1972) *Nucella lapillus* (L.) as a predator of edible cockles. J Exp Mar Biol Ecol 8:45-52
Navarrete SA Castilla JC (1988) Foraging activities of Chilean intertidal crabs *Acanthocyclus gayi* Milne-Edwards et Lucas and *A. hassleri*. J Exp Mar Biol Ecol 118:115-137
Navarrete SA Castilla JC (1990) Resource partitioning between intertidal predatory crabs: interference and refuge utilization. J Exp Mar Biol Ecol 143:101-131
Ojeda FP Dearborn JH (1991) Feeding ecology of benthic mobile predators: experimental analysis of their influence in rocky subtidal communities of the Gulf of Maine. J Exp Mar Biol Ecol 149:13-44
O'Neill SM Sutterlin AM Aggett D (1983) The effect of size-selective feeding by starfish *(Asterias vulgaris)* on the production of mussels *(Mytilus edulis)* cultured on nets. Aquaculture 35:211-220
Paine RT (1971) A short-term experimental investigation of resource partitioning in a New Zealand rocky intertidal habitat. Ecology 52:1096-1106
Paine RT (1974) Intertidal community structure: experimental studies on the relationship between a dominant competitor and its principal predator. Oecologia 15:93-120
Paine RT (1976) Size-limited predation: an observational and experimental approach with the *Mytilus-Pisaster* interaction. Ecology 57:858-873
Paine RT (1980) Food webs: linkage, interaction strength, and community infrastructure. J Anim Ecol 49:667-685
Paine RT Castilla JC Cancino J (1985) Perturbation and recovery patterns of starfish dominated intertidal assemblages in Chile, New Zealand and Washington State. Am Nat 125:679-691
Palmer AR (1983) Growth rate as a measure of food value in

thaidid gastropods: assumptions and implications from prey morphology and distribution. J Exp Mar Biol Ecol 73:95-124

Palmer AR (1984) Prey selection by thaidid gastropods: some observational and experimental field tests of foraging models. Oecologia 62:162-172

Penney AJ Griffiths CL (1984) Prey selection and the impact of the starfish *Marthasterias glacialis* (L.) and other predators on the mussel *Choromytilus meridionalis* (Krauss). J Exp Mar Biol Ecol 75:19-36

Peterson CH (1979) The importance of predation and competition in organizing the intertidal epifaunal communities of Barnegat Inlet, New Jersey. Oecologia 39:1-24

Peterson CH (1982) Clam predation by whelks *(Busycon* spp): Experimental tests of the importance of prey size, prey density and seagrass cover. Mar Biol 66:159-170

Petraitis PS (1987) Immobilization of the predatory gastropod, *Nucella lapillus* by its prey, *Mytilus edulis*. Biol Bull 172:307-314

Pohle DG Bricelj VM Garcia-Esquivel Z (1991) The eelgrass canopy: an above bottom refuge from benthic predators for juvenile bay scallops *Argopecten irradians*. Mar Ecol Prog Ser 74:47-59

Pollock DE (1979) Predator-prey relationships between the rock lobster *Jasus lalandii* and the mussel *Aulacomya alter* at Robben Island on the Cape West coast of Africa. Mar Biol 52:347-356

Raffaelli D Conacher A McLachlan H Emes C (1989) The role of epibenthic crustacean predators in an estuarine food web. Estuar Coastal Shelf Sci 28:149-161

Raffaelli D Falcy V Galbraith C (1990) Eider predation and the dynamics of mussel bed communities. In: Trophic Relationships in the Marine Environment (eds) Barnes M, Gibson RN pp. 157-169 Aberdeen University Press Aberdeen

Reading CJ McGrorty S (1978) Seasonal variations in the burying depth of *Macoma balthica* (L.) and its accessibility to wading birds. Estuar Coast Mar Sci 6:135-144

Reise K (1985) Tidal Flat Ecology: An Experimental Approach

to Species Interactions. Ecological Studies 54: Springer-Verlag Berlin

Revelas EC (1982) The effect of habitat structure on the predator-prey relationship between the green crab *Carcinus maenas* and the blue mussel *Mytilus edulis*. Biol Bull 163:367-368

Richardson TD Brown KM (1990) Wave exposure and prey size selection in an intertidal predator. J Exp Mar Biol Ecol 142: 105-120

Robles C (1987) Predator foraging characteristics and prey population structure on a sheltered shore. Ecology 68:1502-1514

Ryer CH (1987) Temporal patterns of feeding by blue crabs *(Callinectes sapidus)* in a tidal-marsh creek and adjacent seagrass meadow in the lower Chesapeake Bay. Estuaries 10:136-140

Sanchez-Salazar ME Griffiths CL Seed R (1987a) The effects of size and temperature on the predation of cockles *Cerastoderma edule* (L.) by the shore crab *Carcinus maenas* (L.). J Exp Mar Biol Ecol 111:181-193

Sanchez-Salazar ME Griffiths CL Seed R (1987b) The interactive roles of predation and tidal elevation in structuring populations of the edible cockle, *Cerastoderma edule*. Estuar Coastal Shelf Sci 25:245-260

Seed R (1969) The ecology of *Mytilus edulis* L. (Lamellibranchiata) on exposed rocky shores II. Growth and mortality. Oecologia 3:317-350

Seed R (1980) Predator-prey relationships between the mudcrab, *Panopeus herbstii*, the blue crab, *Callinectes sapidus* and the Atlantic ribbed mussel, *Geukensia (=Modiolus) demissa*. Estuar Coastal Shelf Sci 11:445-458

Seed R (1982) Predation of the ribbed mussel *Geukensia demissa* by the blue crab *Callinectes sapidus*. Neth J Sea Res 16:163-172

Seed R (1990) Behavioural and mechanical aspects of predation by the swimming crab *Thalamita danae* on the green-lipped mussel *Perna viridis*. In: Trophic Relationships in the Marine Environment (eds) Barnes M Gibson RN pp. 528-540 Aberdeen University Press Aberdeen

Seed R (1992a) Systematics evolution and distribution of

mussels belonging to the genus *Mytilus:* an overview. Amer Malac Bull 9:123-137

Seed R (1992b) Crabs as predators of marine bivalve molluscs. Proceedings of the First International Conference on the Marine Biology of Hong Kong and Southern China, Hong Kong 1990. (ed) Morton, B Hong Kong University Press Hong Kong (in press)

Seed R Brown RA (1975) The influence of reproductive cycle, growth and mortality on population structure in *Modiolus modiolus (L.), Cerastoderma edule* (L.) and *Mytilus edulis* L. (Mollusca: Bivalvia). In: Proceedings of the Ninth European Marine Biology Symposium (ed) Barnes H pp. 257-274 Aberdeen University Press Aberdeen

Seed R Brown RA (1977) A comparison of the reproductive cycles of *Modiolus modiolus* (L.), *Cerastoderma (= Cardium) edule (L.)* and *Mytilus edulis* L. in Strangford Lough, Northern Ireland. Oecologia 30:173-188

Seed R Brown RA (1978) Growth as a strategy for survival in two marine bivalves, *Cerastoderma edule* (L.) and *Modiolus modiolus* (L.) J Anim Ecol 47:283-292

Seed R Suchanek TH (1992) Population and community ecology of *Mytilus.* In: The Mussel *Mytilus:* Ecology, Physiology, Genetics and Culture. (ed) Gosling EM pp. 87-169 Elsevier Press Amsterdam

Seiderer LJ Hahn BD Lawrence J (1982) Rock-lobsters, mussels and man: a mathematical model. Ecol Modelling 17:225-241

Sloan NA (1980) Aspects of the feeding biology of asteroids. Oceanogr Mar Biol A Rev 18:57-124

Sponaugle S Lawton P (1990) Portunid crab predation on juvenile hard clams: effects of substrate type and prey density. Mar Ecol Prog Ser 67:43-53

Stickle WB Moore MN Bayne BL (1985) Effects of temperature, salinity, and aerial exposure on predation and lysosomal stability of the dogwhelk *Thais (Nucella) lapillus (L.).* J Exp Mar Biol Ecol 93:235-258

Suchanek TH (1978) The ecology of *Mytilus edulis* L. in exposed rocky intertidal communities. J Exp Mar Biol Ecol 31:105-120

Suchanek TH (1981) The role of disturbance in the evolution

of life history strategies in the intertidal mussels *Mytilus edulis* and *Mytilus californianus*. Oecologia 50:143-152

Suchanek TH (1985) Mussels and their role in structuring rocky shore communities. In: The Ecology of Rocky Coasts. (eds) Moore PG Seed R pp. 70-96 Hodder and Stoughton Sevenoaks

Sutherland JP (1974) Multiple stable points in natural communities. Am Nat 108:859-873

Taylor JD (1980) Diets and habitats of shallow water predatory gastropods around Tolo Channel, Hong Kong. In: Proceedings of the First International Workshop on the Malacofauna of Hong Kong and Southern China, Hong Kong 1977. (ed) Morton B pp. 163-180 Hong Kong University Press Hong Kong

Thrush SF (1986) Community structure on the floor of a sea lough: are large epibenthic predators important? J Exp Mar Biol Ecol 104:171-183

Tongiorgi P Nardi P Galleni L Nigro M Salghetti J (1981) Feeding habits of *Ocinebrina edwardsi* (Mollusca: Prosobranchia) a common mussel drill of the Italian coasts. Pubbl Stn Zool Napoli Mar Ecol 2:169-180

Vermeij GJ (1987) Evolution and Escalation: An Ecological History of Life. Princeton University Press Princeton

Vermeij GJ, Veil JA (1978) A latitudinal pattern in bivalve shell gaping. Malacologia 17:57-61

Wayne TA (1980) Antipredatory behaviour of the mussel *Mytilus edulis*. Am Zool 20:789

West DL Williams AH (1986) Predation by *Callinectes sapidus* (Rathbun) within *Spartina alterniflora* (Loisel) marshes. J Exp Mar Biol Ecol 100:75-95

Whetstone JM Eversole AG (1981) Effects of size and temperature on mudcrab, *Panopeus herbstii* predation on hard clams. Estuaries 4:153-156

Wickens PA Griffiths CL (1985) Predation by *Nucella cingulata* (Linnaeus, 1771) on mussels, particularly *Aulacomya ater* (Molina, 1782). Veliger 27:366-374

Wood L (1968) Physiological aspects of prey selection by the marine gastropod *Urosalpinx cinerea* (Prosobranchia: Muricidae). Malacologia 6:267-320

Zwarts L Wanink J (1989) Siphon size and burying depth in deposit- and suspension-feeding benthic bivalves. Mar Biol 100:227-240

THE IMPACT OF BIRD PREDATION ON MARINE AND ESTUARINE BIVALVE POPULATIONS: A SELECTIVE REVIEW OF PATTERNS AND UNDERLYING CAUSES

Patrick M. Meire
Institute of Nature Conservation
Kiewitdreef 5
B3500 Hasselt
Belgium

INTRODUCTION

Throughout the world, estuaries and coastal seas are important for many species of waders, ducks and geese either as a refueling site on migration or as wintering areas (see Davidson and Pienkowski 1987; Boyd and Pirot 1989 and references therein). For some species they also form important breeding sites. Birds are attracted to these habitats because of the huge amount of food present that can be exploited easily as the tidal flats are exposed at low water or in the shallow parts of the coastal seas. This food source includes plants (seaweeds and seagrasses) but consists mainly of benthic animals, the majority being molluscs, polychaetes and crustaceans.

Estuaries and coastal seas are, however, under increasing pressure from various development schemes and other forms of human influences (Smit et al 1987). This has, at least in Northwest Europe, stimulated considerable research into the ability of these habitats to support waders and waterfowl. Attention has been focused on the birds' responses to their food supply and to other foraging birds (see Goss-Custard 1985 and Goss-Custard and Durell 1990 for reviews). Species feeding on molluscs, especially the Oystercatcher (*Haematopus ostralegus*), have received considerable attention as most parameters of foraging behavior are much easier to measure than in species feeding on softbodied prey species.

NATO ASI Series, Vol. G 33
Bivalve Filter Feeders
in Estuarine and Coastal Ecosystem Processes
Edited by Richard F. Dame

In this paper I shall review the present studies on foraging behavior of bivalve feeding birds. For waders, this will be mainly based on studies carried out along the North-West European coast and estuaries, not only because my main interest is in this area but also because similar studies elsewhere in the world are quite rare or published in the gray literature. For seaducks, more information from North America is included.

Having introduced prey and predators, I will then describe the pattern of predation by the different birds on the different prey species before looking in detail at the factors causing the observed predation pattern; these include the ingestibility of the prey, the availability of the prey and the mechanisms of prey selection. Finally I will try to answer the question what is the impact of birds on bivalve populations.

WHO'S EATING WHAT?

THE PREY SPECIES.

Several intertidal bivalve species occur along the North-West European estuaries. The edible mussel (*Mytilus edulis*), the cockle (*Cerastoderma edule*), and the clams *Macoma balthica, Mya arenaria* and *Scrobicularia plana* are the common prey of several bird species. Recently, Swennen et al (1985) reported that *Ensis directus*, a newcomer on European tidal flats, is also eaten, at least by Oystercatchers. Early this century the edible oyster *Ostrea edulis* was an important food item of the Oystercatcher (Dewar 1922) but this species disappeared and the newcomer *Crassostrea edulis*, which is recently spreading in the Oosterschelde, is apparently not eaten.

In the subtidal part of estuaries and especially in the coastal zone the diversity of bivalves is much larger. Next to cockles and mussels, which remain very important prey species, others, as *Spisula* spec., which can form locally huge banks are also very important. Species or genera, known to be eaten by birds to a reasonable degree are: *Arctica islandica, Astarte, Chlamys, Cardium lamarckii, C. nodosum, Donax, Ensis, Hiatella, Leda, Mactra, Modiolaria, Modiolus, Mya, Nucula, Pecten, Solen, Tellina, Venerupis, Venus,* etc. (Cramp and Simmons 1977).

THE PREDATORS.

Although a few bird species feed on molluscs on rocky shores (e.g., Turnstones, *Arenaria interpres*, Purple Sandpipers, *Calidris maritima*) (see Feare and Summers 1985 for a review) most bivalve feeding birds are found on soft bottom intertidal areas of estuaries and coasts and in shallow coastal seas. The main avian predators here are waders, ducks and gulls. Table 1 summarizes information on the distribution and population estimates for the species occurring in the Palearctic region, based on Monval and Pirot (1989), Laursen (1989) and Cramp and Simmons (1977).

Most seaduck species breed in Iceland, Svalbard, Norway, Sweden, Finland, Russia and Siberia. Some breed more to the south, i.e., the Tufted Duck, *Aythya fuligula* that breeds in the south of France and the Eider Duck, *Somateria mollissima*, which breeds in Scotland, Denmark and the Netherlands. Some species are restricted to the coasts (Eider) others breed more inland. Outside the breeding season large concentrations occur in the Wadden Sea, the Baltic Sea, Kattegat-Skagerak, the North-Sea coasts and some estuaries. Details on the wintering distribution of sea and diving ducks are given by Monval and Pirot (1989) and Laursen (1989). During migration and in winter the species are to a different degree dependent on marine areas. Some species such as Tufted ducks and Goldeneye (*Bucephala clangula*) are found in large numbers on inland waters. Eider, Scoters, etc., occur only in marine areas.

Large numbers of waders breeding from Greenland upto Taymir migrate in winter to the estuaries of NW Europe and the West coast of Africa. Among them two species are to a large extent dependent on bivalves: the Oystercatcher and the Knot (*Calidris canutus*). Oystercatchers breed in, and winter in NW Europe, mainly in the Wadden Sea, the Dutch Delta area and along several estuaries in Great Britain and Ireland. Knots (*C. canutus islandica*), breeding in high Arctic regions of Greenland and Canada, migrate to NW European estuaries, especially the Wadden Sea and a few estuaries in Britain, to winter (Davidson and Wilson 1992). Knots breeding in Siberia (*C. c. canutus*) migrate further south to the West coast of Africa and even as far as South Africa (Piersma et al 1992). Those birds occur along NW European estuaries during their migration. Other waders that feed on bivalves include Curlew (*Numenius arquata*) and to a lesser extend Bar-tailed Godwit (*Limosa lapponica*) and Redshank (*Tringa totanus*).

Gulls breed over the whole Palearctic region. The Iceland Gull (*Larus glaucoides*) and the Glaucous gull (*Larus hyperboreus*) are entirely marine, both with a restricted and very northern distribution. The Great Black-backed Gull (*Larus marinus*) is more widespread along Scandinavian, British and Irish coasts. The Common Gull (*Larus canus*), Lesser Black-backed Gull (*Larus fuscus*) and Herring Gull (*Larus argentatus*) are more widespread, the latter also occurring more inland.

Crows are typical terrestrial birds but some species can feed on bivalves in certain areas.

Diving ducks, especially seaducks, are the most important marine bivalve feeding species in the Palearctic, both in numbers and biomass. However, most information is available on the Oystercatcher.

Table 1. Population estimates of some bivalve feeding species in the Palearctic region. (A: number of subspecies; B: X= occurring also in the Nearctic region; C: population estimate; D: dependence on bivalves: X bivalves are eaten frequently, XX bivalves forms the major part of the diet.

Species		A	B	C	D
Tufted Duck	*Aythya fuligula*	1		750.000	X
Scaup	*Aythya marila*	2	X	150.000	X
Eider	*Somateria mollissima*	6	X	3.000.000	XX
King Eider	*Somateria spectabilis*	1	X	100.000	XX
Spectacled Eider	*Somateria fisheri*	1	X		?
Steller's Eider	*Polysticta stelleri*	1	X	15.000	XX
Harlequin	*Histrionicus hsitrionicus*	1	X	6.500	X
Long-tailed Duck	*Clangula hyemalis*	1	X	2.000.000	XX
Common Scoter	*Melanitta nigra*	2	X	800.000	XX
Velvet Scoter	*Melanitta fusca*	3	X	250.000	XX
Barrow's Goldeneye	*Bucephala islandica*	1	X	2.500	X
Goldeneye	*Bucephala clangula*	2	X	300.000	X
Oystercatcher	*Haematopus ostralegus*	3		874.000	XX
Knot	*Calidris canutus*	2	X	857.000	XX
Bartailed Godwit	*Limosa lapponica*	2	X	822.000	X
Curlew	*Numenius arquata*	2		348.000	X
Common Gull	*Larus canus*	4	X		X
Lesser Black-backed Gull	*Larus fuscus*	5			X
Herring Gull	*Larus argentatus*	10	X		X
Great Black-backed Gull	*Larus marinus*	1	X		X

Birds have essentially three different ways of feeding on bivalves: they may swallow the shell whole, remove the flesh from the shell or rob the prey from another individual. Most birds swallow the whole shell, which is then broken in the gizzard. The shell fragments are either regurgitated as pellets, e.g., Herring Gull and some waders, or defecated (most ducks).

Oystercatchers that are specialized in removing the flesh from the shell use essentially two different methods. The first method, used against mussels that are slightly open, consists of stabbing the bill between the gaping valves. The second method is used against tightly closed mussels and consists of pulling the mussel from the clump and hammering the shell until it cracks (Norton Griffiths 1967). Hammering can be done either on the ventral or the dorsal side of the mussel. Once the bill is in the mussel, either by stabbing or hammering, the adductor muscle is cut and the valves are prized apart by opening the bill inside the shell, pushing one mandible against the left and the other against the right valve. Goss-Custard et al (1982) have shown that individuals that hammer mussels either consistently attack the dorsal or the ventral side. Moreover Hulscher (1982) and Sutherland and Ens (1987) have shown for *M. balthica* and mussels respectively that individuals tend to attack either the right or the left hand valve. The two different methods to open cockles or mussels are also related to the shape of the bill. Wintering European Oystercatchers have essentially three different bill types: pointed, chisel-shaped and blunt (Swennen et al 1983). Birds with chisel-shaped bills are mainly stabbers, those with a blunt bill, hammerers. Birds with pointed bills feed more on non bivalve prey such as ragworms (*Nereis diversicolor*). Swennen et al (1983) showed that the bill shape of individual Oystercatchers varied between the three bill types according to feeding conditions. The same individual could stab or hammer mussels according to its bill shape with equal efficiency. When the food supply is stable individuals keep to one prey and one technique and their bill shape does not alter. When feeding conditions deteriorate the birds are able to switch to other prey, and consequently the shape of the bill will gradually change into the new type associated with the new prey and feeding technique (Hulscher 1985). This adaptation of the bill is possible because the proximal half of the rhamphotheca of the bill grows at a rate of 0.4 mm per day (Hulscher 1985).

Some gulls (Herring Gull; Kelp Gull, *Larus domanianus*; Glaucous-winged Gull, *Larus glaucescens*) and crow species (Rook, *Corvus frugilegus*; Common Crow, *Corvus brachyrhynchus*; Nortwestern Crow, *Corvus caurinus*; Hooded Crow, *Corvus corone*) use a completely different method to remove the flesh from the shell. The prey is taken from the tidal flats and dropped onto a hard surface. This is repeated until the shell breaks and the flesh can be removed from the shell.

Robbing is found in many species. Some gulls, like the Common Gull, obtain most of their bivalve food by robbing other species. It is also seen within one species. Dominant

Oystercatchers could increase their intake rate by robbing mussels from subdominants, whose intake rate declined (Ens and Goss-Custard 1984).

THE PATTERN OF SELECTION

Different bird species feed on different prey species and within one prey species birds may select different size-classes. These two levels of prey selection will be briefly discussed for the most important species.

PREY SELECTION

Different birds, both species and individuals, are dependent on marine bivalves to different degrees. Bivalves, especially the mussel and the cockle are the major food item for many ducks and Oystercatchers.

The diet of single species is, however, highly variable both in space and time. As an example, some of the available data on the diet of Eider are compiled in Table 2. Next to bivalves, depending on the local conditions, echinoderms, polychaetes, fish eggs, crustaceans or even vegetable matter may be important food items. The observed variability in diet is caused by many different factors. Firstly the relative availability of different prey species is reflected in the diet. The diet of Eiders feeding in shallow or deeper parts along the Baltic Sea coast of Schleswig-Holstein differed considerably (Table 2, column a and b). The same differences are found in other studies where the birds are feeding in different habitats (e.g., Nilson 1972; Pehrsson 1976; Guillemette et al 1992). The diet also changes seasonally. Vermeer (1981) found that the principal food of Surf Scoters (*Melanitta perspicillata*) on the coasts of British Columbia are mussels except for about three weeks in spring when Herring (*Clupea harengus*) eggs form the main diet. A similar pattern was found for Eiders in the St. Lawrence estuary (Cantin et al 1974). Nilsson (1969) describes major diet changes in Tufted Ducks during the winter along the South Swedish coast. A more detailed analysis of the factors influencing diet selection is beyond the scope of this paper.

Compared to ducks, the feeding ecology of waders, especially Oystercatchers, has been studied in detail. Data for Oystercatchers have recently been compiled by Hulscher et al (in press) and Zwarts et al (in press). The diet of Oystercatchers is very restricted, most of them feeding either on cockles or mussels. Other prey taken include the bivalves *M. balthica, S. plana, M. arenaria* and the polychaete *N. diversicolor.* Knots feed nearly entirely on molluscs with *Hydrobia ulvae, M. balthica, M. arenaria*, cockles and mussels being the most important.

SIZE SELECTION WITHIN ONE PREY SPECIES

Often birds feeding on bivalves are not faced with the problem of finding their prey, especially on mussel- or cockle beds whereas many as 1000 prey individuals per square meter may be present, but of choosing between individuals. It is

Table 2. Composition of the diet of Eider (*Somateria mollissima*) under different conditions. The percentage wet weight or volume of the different food items in the stomach is given for Eiders feeding in: the Baltic Sea Coast of Shleswig-Holstein in water 6-10 (A) and 18-22 (B) m deep (Meissner and Brager, 1990); C: Sommaroy area, 60 km west of Tromso (Norway) (Bustnes and Erikstad, 1988); D: Dutch Wadden Sea (Swennen, 1976); E: South Sweden (Nilson, 1972); F: Wadden Sea of Schleswig Holstein (Nehls, 1989); G: Firth of Forth (UK) (Player, 1971); H: Central Maine and Long Island, New York (McGilvrey, 1965); this concerns the American Eider, *S. m. dresseri*.

Prey species	A	B	C	D	E	F	G	H
Arctica islandica		86.7	2.8					
Cerastoderma edule				43.2		75		
Macoma balthica				1.1				
Modiolus modiolus			7.3					
Mya arenaria	31.3							
Mya truncata		5.4						
Mytilus edulis	26.5		46.3	41.8	87		70.8	70.1
other molluscs	11.4		5.9	3.5			0.4	5.9
Asterias rubens	21.1							
other echinodermata			6.6					
Other non molluscs	9.7	7.9	2.4	10	13		28.6	24.9
Fish eggs			25.9					

important, therefore, to understand the criteria for prey selection if we are to understand the factors determining the food supply of birds and their potential impact on the prey populations (Sutherland and Ens 1987). The pattern of selection of the most important predators on their dominant prey species is summarized in Fig. 1. Many more species feed on the bivalves mentioned, but either no data on size selection are available or the prey is only taken occasionally.

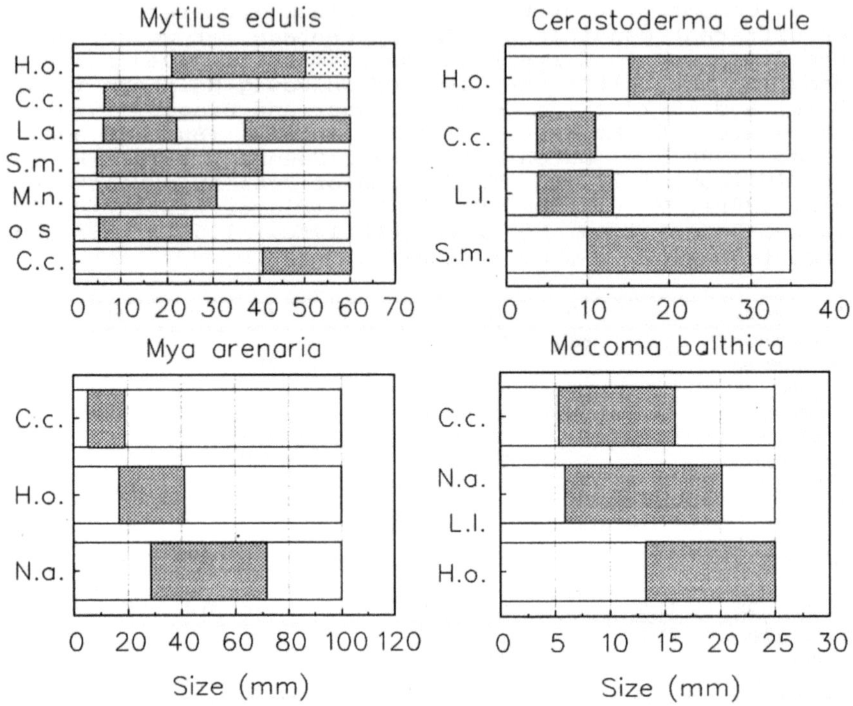

Fig.1 Pattern of selection of the most important predators on their dominant prey species. The stippled area gives the average size range taken by the predator. For the Oystercatcher (*Haematopus ostralegus*) a difference is made for hammering (dark grey) and stabbing birds (dark and pale grey). (H.o. *Haematopus ostralegus*; C.c. *Calidris canutus*; L.a. *Larus argentatus*; S.m. *Somateria mollissima*; M.n. *Melanitta nigra*; o s other seaducks; C.c. *Corvus corone*; N.a. *Numenius arquatus*; L.l. *Limosa lapponica*).

Mytilus edulis

Mussels are consumed by Knots, Oystercatchers, diving ducks and several gulls. This species is beyond doubt the most important bivalve prey for birds. Intertidally the most important mussel feeders are waders and gulls. Knots feed on small mussels between 5 and 23 mm (Zwarts and Blomert 1992; Alerstam et al 1992). They swallow the prey whole. Turnstones feed on small mussels but they form only a small part of their diet (Davidson 1971). An important predator is the Oystercatcher. It is well documented that hammering Oystercatchers select certain size-classes of mussels (Drinnan 1958; Norton Griffiths 1967; Ens 1982; Meire and Ervynck 1986; Sutherland and Ens 1987; Cayford and Goss-Custard 1990). Mainly mussels between 25 and 55 mm are taken. Stabbing Oystercatchers do not take mussels smaller than 20 mm, but accept all larger mussels, often selecting the largest ones (Hartwick and Blaylock 1979; Zwarts and Drent 1981). Herring gulls select a wide range of mussel sizes. The smaller ones (2- 25 mm) are ingested whole, the larger ones (> 35 mm) are dropped onto a firm substrate until they break (Spaans 1971; Milne and Dunnet 1972).

The Eider is a very important predator of the mussel. Like other diving ducks they swallow the mussels whole. Mussels are taken by diving at high tide, when Eiders aggregate in large flocks over the mussel beds, both on intertidal and subtidal areas. The size selection reported for the Eider is very variable. They are able to feed on mussels from 2 to 80 mm but in most studies a clear preference for smaller size classes (10-35) is apparent (Madsen 1954; Dunthorn 1971; Player 1971; Milne and Dunnet 1972; Swennen 1976; Bustnes and Erikstad 1990; Guillemette et al 1992). Some average values are: 18 mm (Milne and Dunnet 1972); 7-8 mm (Guillemette et al 1992); 13.9 mm (Bustnes and Erikstad 1990). The size range for the Goldeneye is between 5-16.3 with a mean of 10.3 mm (Madsen 1954; Pehrsson 1976).

The Hooded Crow takes extremely large mussels, between 45 and 90 mm, size classes rarely or not encountered in the samples from the foraging area (Zwarts and Drent 1981).

Cerastoderma edule

Knots feed on cockles between 5 and 12 mm (Goss-Custard et al 1977; Zwarts et al 1992). The main predator is, however, the Oystercatcher that actively rejects small cockles and feeds mainly on second year and older cockles (15 - 40 mm) (Drinnan 1957; Hancock and Urquhart 1965; Davidson 1967; Brown and O'Connor 1974; Goss-Custard et al 1977; O'Connor and Brown 1977; Sutherland 1982a). In patches where only 2 or more years old cockles occur no apparent size selection was found (O'Connor and Brown 1977; Leopold et al 1989). Only in rare occasions when second year or older cockles were absent did birds feed on smaller individuals (5-10 mm; Davidson 1967; Meire unpublished). Other species of waders may at some time feed on small cockles. Bar-tailed godwits can forage on cockle spat (5-10 mm) which they ingest whole (Meire unpublished).

Eiders take cockles from tidal flats by trampling, this leaves small craters of about 50 cm in diameter and 20 cm deep. (Nehls 1989). However, most cockles are taken by diving. Swennen (1976) found a selection for the smaller sizes when the birds were presented cockles between 15 and 30 mm, a pattern much more pronounced when they were offered cockles between 30 and 45 mm. The maximum length taken was 49, although normally no cockles longer than 40 mm were taken. Common Scoters are also known predators of cockles but no data are available on size selection.

Macoma balthica

M. balthica is eaten by several bird species. Knot are important predators of *M. balthica*. They take shells within the range 5-16 mm (Davidson 1971; Goss-Custard et al 1977; Zwarts and Blomert 1992). Redshanks take a similar size range to Knots (Goss-Custard et al 1977b), Curlew and Bartailed godwit take a wider range: 7-20 mm (Goss-Custard et al 1977b). Oystercatchers feeding by touch on *M. balthica* also select for the larger ones, 10-22 mm (Goss-Custard et al 1977b; Hulscher 1982). Prey smaller than 11 mm were actively rejected.

Other species

Scrobicularia plana is less common than several other bivalve species and predated mainly by Curlew, which take larger clams 15-50 mm, but mainly >30mm (Goss-Custard et al 1977b). Oystercatchers show a similar selection pattern (Hughes 1970). *Mya arenaria* is predated mainly by Curlew, taking larger clams 15-90 mm, but mainly >30 mm, Oystercatchers 15-45 (Zwarts and Wanink 1984). Knots take smaller individuals (7-17 mm)(Zwarts et al 1992).

Arctica islandica is widely taken by diving ducks, but is specially mentioned because the data by Meissner and Brager (1990) clearly show the difference in size selection between Eider and Common Scoters. The first took shells between 7.5 -55 mm with a mean of 23 mm, the second between 3.5 -32 mm with a mean of 15 mm.

FACTORS CAUSING THE OBSERVED PREDATION PATTERN.

Why do not all predators take a wider range of size-classes? This can be due to several reasons (Fig. 2). First of all, a part of the prey population may be inaccessible to birds for several reasons: the prey can be buried too deep so the birds cannot reach it, or the bivalves may be one on top of the other so they are not visible anymore (e.g., on mussel beds). The remaining part of the prey population is the accessible fraction. Of this fraction some individuals may be too large to be ingested by the bird, or else show behavioral or morphological patterns that make them unavailable. Of the available part of the prey population, birds may actively select some individuals. These different factors will be discussed in turn.

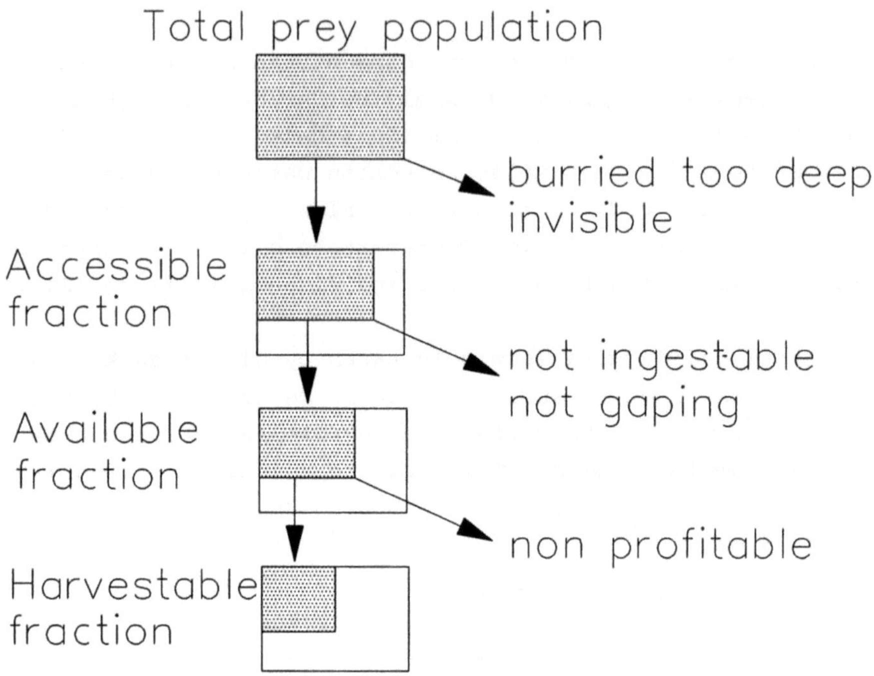

Fig. 2. Different fractions of the total prey population.

ACCESSIBILITY OF THE PREY

Not all individuals of a prey species are actually accessible to foraging birds. Species, like cockles and *Spisula* spec., occur in the upper layer of the sediment and are accessible to most bird predator species, other bivalves live deeper in the sediment and can be out of reach of the bird's bill.

Burying depth of bivalves increases with size of the shell (Reading and McGrorty 1978 for *M. balthica*; Richardson 1985 for *Venerupis japonica* and *Protothaca staminea*; Zwarts and Wanink 1984 for *M. arenaria*; Zwarts 1986 for *S. Plana*). Only Hulscher (1982) found and inverse relation between burying depth and shell size in *M. balthica*. The relationship between burying depth and size was studied in detail for cockles, *M. balthica*, *S. plana* and *M. arenaria* and could be described with an S-curve (Zwarts and Wanink 1989). Burying depth levels off for all four species when they reach a size that equals about 50% of their maximum size. The burying depth of *M. balthica* and *S. plana* is significantly deeper (twice as deep) in winter than in summer, whereas no difference was found in the bivalves (Fig. 3) (Zwarts and Wanink 1989). This difference was explained by a difference in feeding behavior. *M. arenaria* and cockles are always suspension feeders, whereas in summer *M. balthica* and *S. plana* use a part of their siphon to graze the surface around the burrow, whereas deposit feeding does not occur in winter. At that time they are mainly suspension feeding. Burying depth depends to a large degree on siphon weight. This was shown experimentally by Zwarts (1986) who manipulated the burying depth of *S. plana* by removing part of the inhalant siphon.

The importance of burying depth for the predators is shown in Fig. 3. Clams (*M. arenaria*) are eaten by Knots, Oystercatchers and Curlews. It appears that an Oystercatcher cannot find clams of more than 4 cm in length in the upper 7 cm of substrate (the bill length of an Oystercatcher) and that the greater part of the shells above 7 cm in length are out of reach of the female Curlew (bill length 14 cm). Even fewer are accessible to a male bill of 12 cm. (Zwarts and Wanink 1984). Knots cannot reach the deeper living *M. arenaria* and *M. balthica* (Fig. 3).

The accessibility of different sizes of prey is not only a function of its depth distribution but also of the predators search methods. Richardson (1985) calculated the accessibility of Japanese littleneck clams for Crows probing or digging and found that smaller size classes were more available- to digging and larger size classes more available to probing birds.

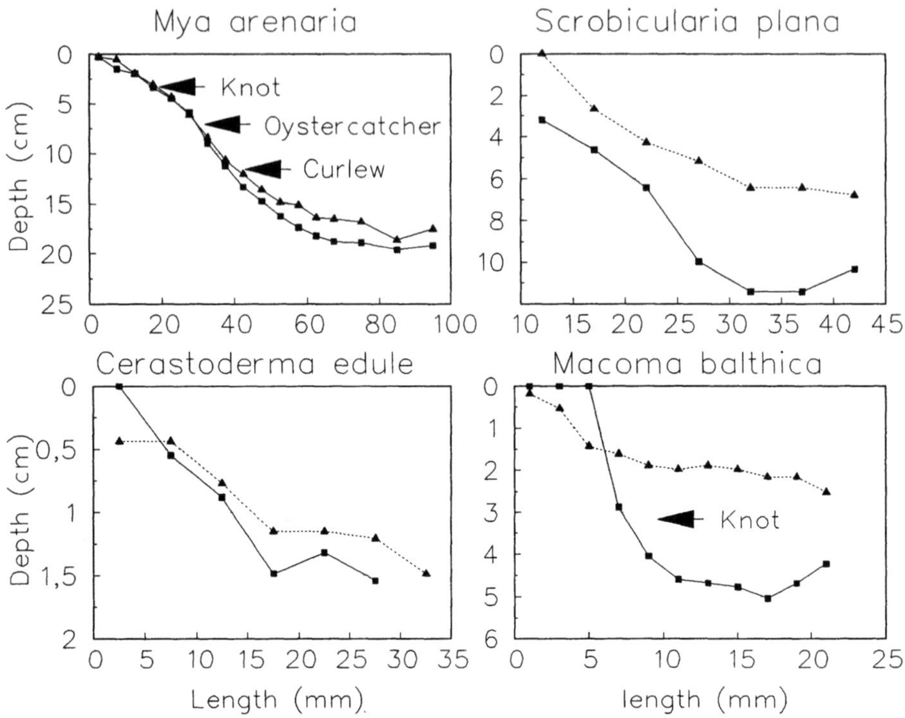

Fig. 3 Depth distribution of 4 bivalves, based on the data of Zwarts & Wanink (1989). The average depth in summer (triangle) and winter (square) is plotted. The bill length of three wader species feeding on these bivalves is indicated with and arrow.

All mussels occurring on mussel beds appear at first sight to be available to predators. Within mussel clumps, however, many mussels may be on top of each other. Smaller ones especially, might be hidden by larger ones. These are, for a visual predator, inaccessable. Meire (1991) showed that there was a clear relation between size and visibility, the smaller mussels being invisible at the surface (Table 3). This relation is likely to be dependent on mussel density but this has not yet been studied.

Table 3. Percentage of mussels (*Mytilus edulis*) which are visible at the surface of the bed (a) (from Meire 1991) and which are not overgrown by barnacles (b) (from Meire and Ervynck 1986). (L length of the mussel in mm).

12.5	17.5	22.5	27.5	32.5	37.5	42.5	47.5	52.5	57.5
100	100	100	100	100	85	42	23	10	5
91.5	94.5	100	72.5	66	51	45.5	28.7	12.6	1

INGESTIBILITY OR INABILITY TO HANDLE THE PREY

For species swallowing the prey individual intact, there might be an upper limit to the size of the shell that can be swallowed. According to Zwarts and Blomert (1992) this limit is probably not determined by length per se, but by the combined effect of shell width and height. Since the gullet has to be stretched widely to allow a large, hard shelled prey to be swallowed, the upper size limit is determined by the circumference of the shell in a plane perpendicular to the length axis. They calculated the upper limit for Knot to be 12 mm for cockles, 17 for *M. arenaria*, 19 for *S. plana* and 21 for mussels. This coincides well with the upper limits of size classes taken as shown in Fig. 1 for this species. For other species like Eiders, Common Scoters and Herring gull, the upper limit must be set by the same restrictions. Pehrsson (1976) found that the maximum length of mussels eaten by 7 species of diving duck wintering in Denmark was proportional to body size (Fig. 4). For all these species larger prey individuals are not suitable prey items, reducing strongly the available prey density. The importance of avoiding large prey was shown by Cootam (1939) who published a photograph of a dead Eider in which a big mussel had stuck between the tongue and palate. Jukema (1979) found a dead Eider with a mussel clenched round the bill. Also Oystercatchers are known to be killed by mussels that have closed their shells onto the bill (Baldwin 1946; Hulscher 1988).

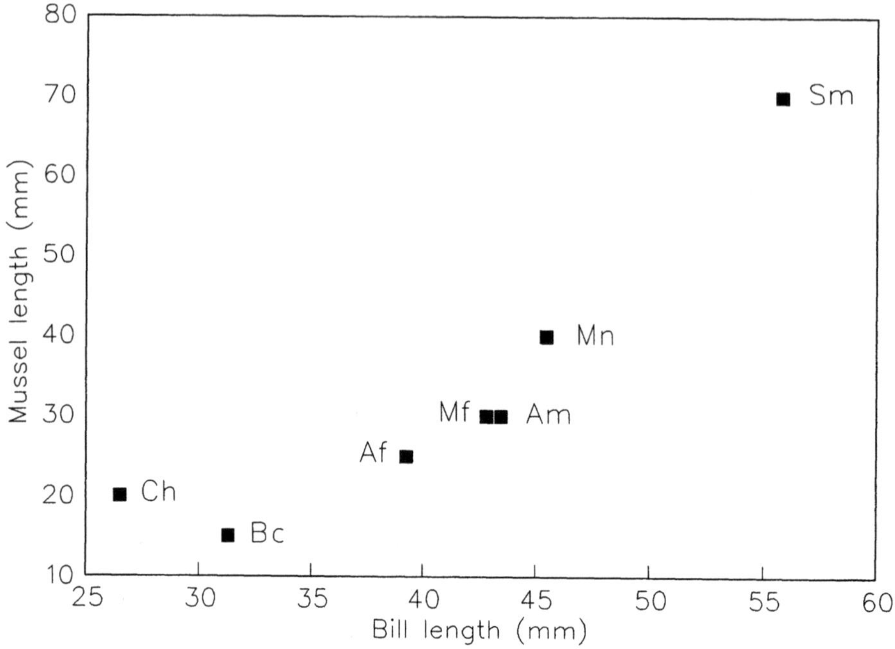

Fig. 4 Maximum mussel (*Mytilus edulis*) length taken by different duck species in relation to bill size (modified after Pehrsson 1976). (Ch *Clangula hyemalis*; Bc *Bucephala clangula*; Af *Aythya fuligula*; Mf *Melanitta fusca*; Am *Aythya marila*; Mn *Melanitta nigra*; Sm *Somateria mollissima*).

OTHER CHARACTERISTICS OF THE PREY INFLUENCING SELECTION.

Oystercatchers also select against mussels that are overgrown by barnacles (*Balanus balanoides*, or other species) (Ens 1982; Durell and Goss-Custard 1984; Meire and Ervynck 1986; Cayford and Goss-Custard 1990). Mussels are either clean of barnacles, have just a few, or are more than half covered. Although this last category was never found in samples of mussels taken by the birds, it could make up an important fraction of the mussels present in the bed (Table 3). We assume that Oystercatchers either cannot open these mussels (as the ventral side is overgrown), or else do not recognize them as mussels because in many cases only the barnacles are visible at the surface. Eiders were also found to reject mussels overgrown by barnacles (Swennen 1976). It was also shown experimentally that Eiders avoid shells with some irregularities on the shell surface. The birds seem to make their decision whether or not to swallow the prey after they manipulated the prey for a short while within the bill (Swennen 1976). The selection for a smooth shell surface is, in birds swallowing whole molluscs, very functional because of the risk of damage of the digestive tract. Swennen (1976) found three dead birds with severe injuries of the stomach wall due to hard remains of a crab and mussels respectively.

Mussels are attached to each other or to dead shells or rock by byssal fibers. Some may be attached too strongly to be taken. Pehrsson (1976) suggests that Goldeneyes cannot tear mussels attached to exposed rocks, contrary to Eiders and Scoters, which have a more effective nail on the upper bill tip and a stronger grasping action allowing a strong pull on a mussel.

Shell thickness may also influence the availability of the prey. As predicted by Ens (1982), Durell and Coss-Custard (1984), Meire and Ervynck (1986), Sutherland and Ens (1987) and Cayford and Goss-Custard (1990) found a strong selection for mussels with a thin shell in hammering Oystercathers (Fig. 5). Clearly, the larger and thicker the mussel, the stronger the the genetic selection for thin shelled individuals. Although no information is available it is very likely that shell thickness is also of importance for the species swallowing their prey whole.

Oystercatchers that stab their prey are dependent on mussels or cockles that slightly gape, since only in these individuals they can push their bill in between the valves. Here, shell thickness and/or barnacle coverage are not important, but the behavior of the prey, whether they are gaping or not. Mussels open their valves to feed when covered

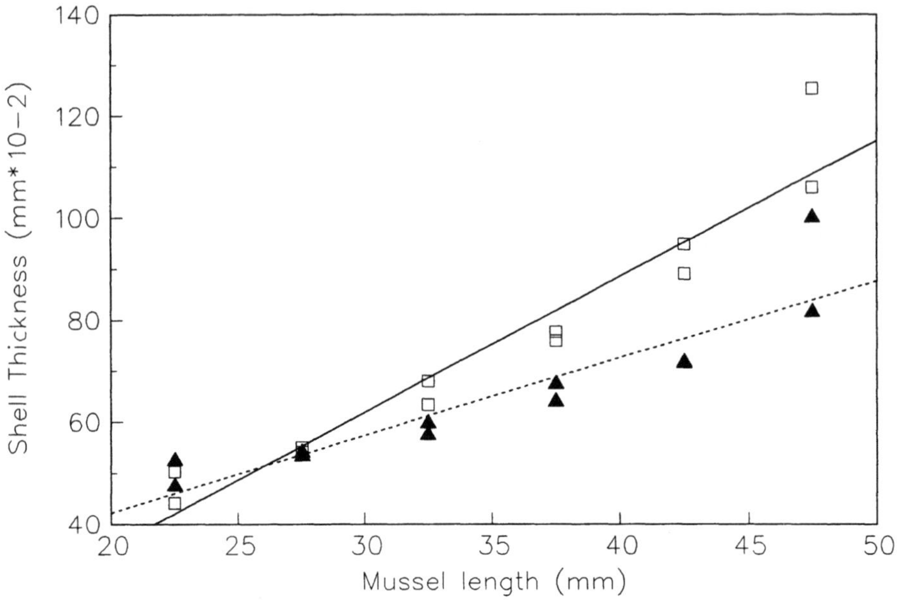

Fig. 5 Average shell thickness (measured at the ventral shell edge near the anterior adductor muscle) of the mussels (*Mytilus edulis*) present on the bed (open squares; solid line) and those taken by ventral hammering Oystercatchers (*Haematopus ostralegus*) (solid squares; broken line).

by a few centimeters of water. Little information on prey behavior in relation to predation is available.

From this discussion it is clear that of all individuals of a prey species present, depending on the predator, only a proportion of the prey population is available to the predator. In some cases the available fraction is only a very small part (see Table 3). Zwarts and Blomert (1992) calculated that for Knot only 14% for *M. balthica*, 12% for cockles and 0.7% for *S. plana* of the total biomass is harvestable by the birds. Calculating the harvestable prey biomass is often very time consuming and needs detailed feeding observations and prey sampling but is a crucial step in understanding the impact of birds on their prey populations and in understanding the distribution of birds over their feeding areas.

PREY SIZE SELECTION

Within the available fraction of the prey population the predators do not take all individuals or do not take them in proportion to their density. The consumption of a greater proportion of one type of prey than another is not itself an indication of selectivity; that type may simply be easier for the bird to detect (Richardson 1985) and the observed pattern results from passive selection. Since large and small shells have different surface areas the changes of encountering a prey of different size are unequal A predator feeding by touch therefore does not encounter different sized prey individuals in proportion to their density. The selection of larger sized *M. balthica* by Oystercatcher could be explained as a result of passive selection with the rejection of smaller clams an active choice (Hulscher 1982). Indeed predators maximizing their intake rate while feeding must ignore prey for which the intake rate while being handled is lower than the average current overall intake, including the time spent both handling and searching for prey. Taking this unprofitable prey would inevitably reduce the overall intake rate. This follows from a simple Optimal Foraging Model, which is a multispecies version of Hollings disc equation,

$$\frac{E}{T}=\frac{\sum E_i P_i}{1+\sum \lambda_i H_i} \tag{1}$$

where T is the total foraging time, λ is the encounter rate and P the probability the prey is taken for class i (Charnov 1976). Prey value is measured by a single dimension, ash free dry weight or energy content (E). Handling time (H) is a fixed constraint. Handling and searching cannot be done at the same time. The prey should be recognized instantaneously and without error. It is assumed that the predator encounters the prey sequentially and randomly and the energetic costs per second of handling are similar for different prey.

The predictions derived from that model are: the most

profitable prey (measured by the ratio E/H) are always taken and new types are added to the diet according to their ranked profitability; a prey type is either always taken upon encounter or never (P is either 1 or 0); the inclusion of a prey type depends only on its profitability and on the characteristics of types of higher rank and not on its own encounter rate. For further details about Optimal foraging models, their assumptions and predictions I refer to Stephenson and Krebs (1986).

As Oysterctachers are conspicuous, feeding on large prey and leaving the empty shells on the surface they are particularly suited for studies on prey selection. Not surprisingly therefore, the predictions of the optimal diet model were tested in several studies, which will be summarized here. A detailed analysis of prey selection in Oystercatcher is given by Zwarts et al (1993).

First I will focus on Oystercatchers, opening mussels by ventral hammering. Ens (1982) concluded that the birds take smaller mussels than predicted by the optimal diet model. The active rejection of the large, most profitable mussels (which frequently occurred) was contrary to the expectation of the model. But as anticipated by Ens (1982), Durell and Goss-Custard (1984) and Meire and Ervynck (1986) have found that mussel selection involves more than size selection. Meire and Ervynck (1986) analyzed in detail the selection of ventral hammerers. From the mussel population present, a limited size range was taken (10-60 mm but mainly between 30 and 45 mm) and there was no preference for the largest mussels present (Fig. 6), although the larger mussels are the most profitable (Fig. 7). However, an important number of large mussels taken were eventually abandoned (wasted handling time WH) without being opened. The probability of opening a mussel once taken decreased significantly with mussel length. Incorporating this into the profitability estimate according to formula (2) shows large mussels are no longer the most profitable (Fig. 7).

$$Profitability = \frac{E_i P_i}{H_i P_i + WH(1-P_i)} \qquad (2)$$

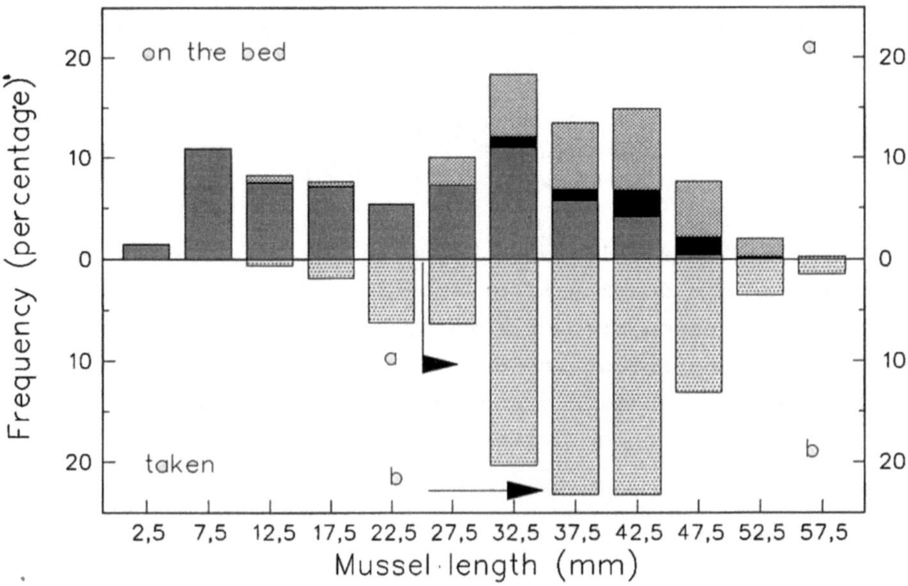

Fig. 6 Length frequency distribution of mussels (*Mytilus edulis*) present on the bed (upper panel) and those taken by ventral hammering Oystercatchers (*Haematopus ostralegus*)(lower panel). Of the mussels present on the bed a fractions is too thick to be eaten ■ , another fraction is covered by barnacles ▨ , leaving the harvestable fraction ▩ . The optimal diet based on the harvestable density includes size classes 27.5 mm and larger (arrow a), based on the total density size classes 37.5 mm and larger (arrow b)(after Meire and Ervynck 1986).

This was also incorporated in the diet model which becomes:

$$\frac{E}{H}=\frac{\sum E_i\lambda_i P_i}{1+\sum\lambda_i\left(H_iP_i+\left(1+P_i\right)WH_i\right)} \tag{3}$$

The optimal diet was calculated based on the total densities of mussels and on the density after subtracting mussels that are too thick to be eaten or overgrown by barnacles (see above). There is a good agreement between the size range taken and that predicted, when we take into account the density of mussels actually available to the Oystercatchers (Fig. 6).

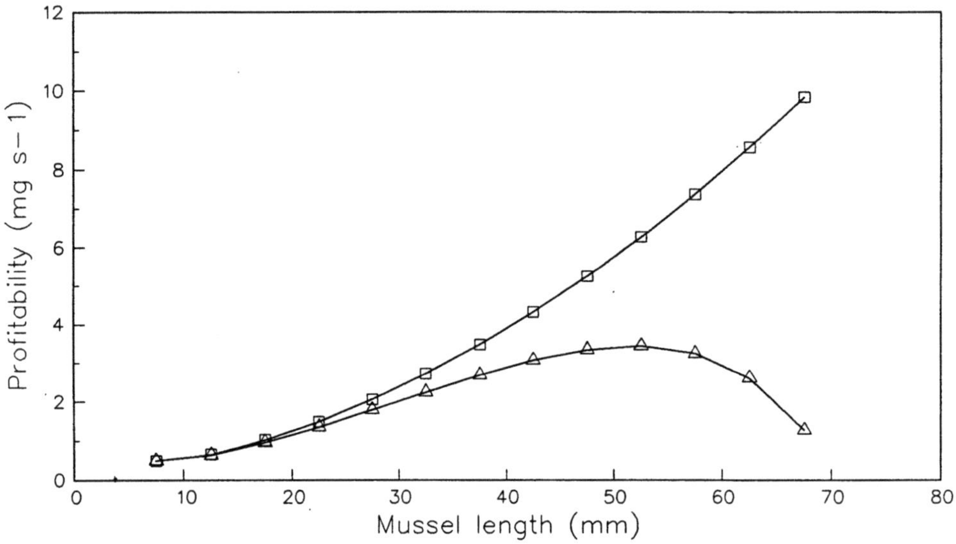

Fig. 7 Profitability of mussels (*Mytilus edulis*) for ventral hammering Oystercatchers (*Haematopus ostralegus*) expressed as E/H (squares) or according to formula 2 (see text) (triangles).

Comparison with the predictions of the model, based on the density of mussels present on the bed, (Fig. 6) clearly indicates the importance of estimating the available prey population. The results of this study were confirmed by Cayford and Goss-Custard (1990) who studied the size selection of Oystercatchers throughout one year. The selection pattern, however, was not constant over the year. In spring the birds switched their diet to much smaller mussels (20-25 mm versus 40-45 in the rest of the year). This coincides with a period in which individual mussel biomass decreases significantly due to spawning. Larger mussels in particular can lose upto 50% of their body weight. This decreased the profitability of the mussels but did not change the relative profitability of the different size classes, hence the selection of smaller size classes was not predicted by the model. This diet change could, however, be explained by several reasons. The biomass of the larger mussels might be very variable depending on their condition after spawning. There is also some evidence that spawning may alter the biochemical composition of mussels and Oystercatchers might select for mussel quality as well as calorific value at this time of the year (Cayford and Goss-Custard 1990). Another possible explanation is that the costs of not selecting the larger mussels was, at that time of the year, very small because adding more size classes to the diet hardly changed the overall intake rate. Indeed April was the only month in which the comparison between the distribution of size-classes taken by ventral hammerers was not significantly different from that predicted by random foraging (Cayford and Goss-Custard 1990). In winter adding smaller mussels to the diet had a significant effect on intake rate. This could explain why Oystercatchers take smaller prey than those which seem to yield the highest E/H. There are, however, many possible explanations in addition to those considered here. Handling time may be more costly for larger prey, the risk of kleptoparasitism may increase with size, and long-handling times may increase the risk of predation of the bird.

Next to selection for shell size, selection for shell thickness has been quite well documented. This selection is not due to the inability of birds to open the thicker shells. Sutherland and Ens (1987) showed that captive Oystercatcher fed with mussels initially select the thinner shelled individuals-; the thicker ones are taken later as the thin shelled ones are depleted. An experiment was carried out to test whether the selection for shell-thickness is in accordance with the energy maximization hypothesis (Meire 1987; in press). In order to estimate the force necessary to open a shell, a copper model Oystercatcher bill was dropped from a standard height on a mussel until the shell broke. An exponential relationship between number of blows required to break the shell and shell thickness was found (Fig. 8). It is obvious that thin shelled mussels taken by Oystercatchers are the easiest to open. Assuming that the number of blows of the copper bill is proportional to handling time, the handling time for thick shelled mussels not eaten by Oystercatchers was estimated. As the profitability of the thicker ones is much lower than the average intake (0.3 mg s^{-1} versus 2 mg s^{-1}) the exclusion of the thick-shelled mussels from the diet is in accordance with the energy maximization principle (Meire 1987; in press).

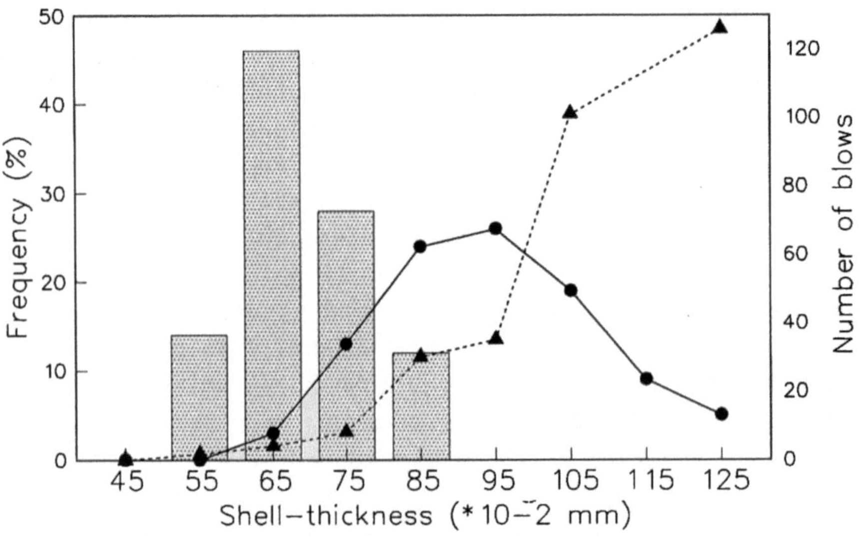

Fig. 8 Frequency distribution of shell thickness of mussels (*Mytilus edulis*) of 40-45 mm present on the bed (line) and those taken by Oystercatchers (*Haematopus ostralegus*) (bars). The broken line gives the number of blows of an artificial bill that were necessary to open a mussel as a function of shell thickness. (after Meire 1987; in press).

For stabbing Oystercatchers it is not possible to compare the observed size selection with predictions, since we cannot yet estimate the proportion of the population which gape (and are thus available to the birds) at any one time. Meire (1987) found no apparent size selection for stabbing Oystercatchers. However, Zwarts and Drent (1981) found that on their mussel beds, with three to four times more mussels, there was strong selection for the larger mussels. The predicted lower acceptance threshold was very close the observed lower limit.

Oystercatchers feeding on cockles take a disproportionate number of large cockles, as a result of active selection for the large more profitable individuals (Sutherland 1982). This selection was in accordance with the prediction of the Optimal Foraging Theory model. When presented with cockles from one year-class, no selection is apparent (Leopold et al 1989). In an extensive review, Zwarts et al (in press) show that the selection of size classes of several bivalves by Oystercatcher accords well to the predictions of the optimal diet model.

Birds dropping hard-shelled prey to gain access to the edible parts must make a number of decisions if they are to realize the maximum rate of net energy intake for their effort. The decisions are: 1) the substrate on which to drop the prey; 2) the height from which to drop the prey; 3) how many times to drop the prey before giving up if the shell does not break; 4) which size of prey to drop (Zach 1979; Richardson and Verbeek 1987). Zach (1979) found that Northwestern Crows feeding on whelks (*Thais lamellosa*) dropped the shells from on average 5.2 m on hard substrate, very close to the predicted optimum of 5 m. At any other than the optimal height, too much energy is expended in repeated trials or in trying to reach a greater height. As the probability of breaking was independent of the number of trials the optimal choice was to drop the whelks until they broke, a behavior that was also observed (Zach 1979). Size selection was studied in detail by Zach (1979)(see above); Kent (1981) and Zwarts and Drent (1981) for Herring Gulls and Hooded Crows feeding on mussels, Richardson and Verbeek (1986; 1987) for Northwestern Crows feeding on littleneck clams (*Venerupis japonica*) and Ward (1991) for Kelp Gulls feeding on *Donax serra*. In all these studies, the birds selected the larger prey individuals as predicted by simple optimal foraging models. The strongest support in favor of the optimization hypothesis is given by Richardson and Verbeek (1987) who compared the size selection of yearling and adult crows. Yearlings took significantly smaller clams despite encountering similar proportions of the different size classes as the adults. They were, however, less efficient in dealing with the prey. They took more time to find and handle clams and required more drops to break them than adults. Within the limits imposed by their inefficiency, the extended diet consumed by the yearlings yielded close to the maximum possible rate of energy intake (Richardson and Verbeek 1987) (Fig. 9).

For diving birds the simple optimal foraging model does not hold. Divers make repeated foraging excursions from a central place (the surface) to which they must return to breathe. Underwater they travel to the bottom, search for, pursue and possibly capture one or more prey, and return to the surface, where the prey is normally consumed. Some time is spent at the surface before the next dive. The time at the surface is related

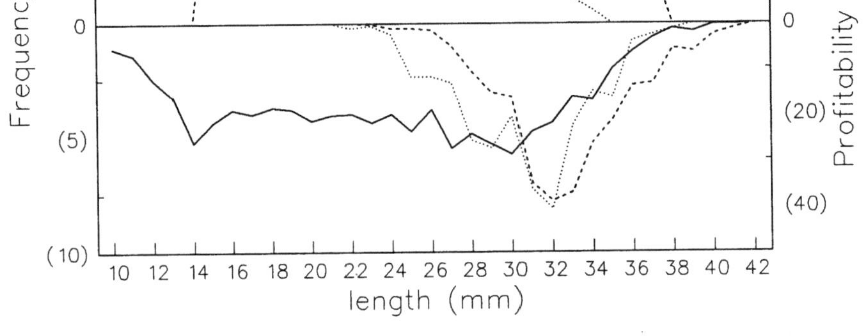

Fig. 9 In the lower panel the frequency distribution of clams (*Venerupis japonica*) present in the mud (solid line) and those taken by adult (dashed line) and juvenile (dotted line) Northwestern Crows (*Corvus caurinus*) is plotted. In the upper panel the profitability of clams for both juveniles (dotted line) and adults (dashed line) is given (adapted from Richardson and Verbeek 1987).

to the length of the preceding dive, which strongly suggests that this is at least partly recovery time during which respiratory gases are exchanged (Butler and Woakes 1979). Birds seem, however, to be able to postpone recovery until a series of dives has been completed (Ydenberg 1988; Ydenberg and Forbes 1988; Ydenberg and Guillemete 1991). The necessity of a pause at the surface was not considered in the studies of Draulans (1982, 1984) or in the "aqualung" model of Dunstone and O'Connor (1979). Houston and McNamara (1985) developed a general theory of central place foraging for single-prey loaders that take account of the fact that longer dives are followed by longer pauses. They showed that the decision policy for accepting and rejecting prey items that maximize the rate of energy gain becomes to be less and less selective as the dive progresses, because rejections become more and more costly. They also showed that it would be optimal eventually to terminate a dive, even if no prey has been captured.

This model was tested by Ydenberg (1988) with captive White-winged Scoters (*Melanitta fusca*) feeding on mussels in the laboratory. The data failed to uphold the predictions of the model, with accepted and rejected prey being equally likely on either side of the accept-reject boundary. When the birds did not have to dive, a consistently increasing tendency to reject larger, more profitable mussels was evident. When the birds had to dive for food, they generally became less selective; these results are similar to those of Draulans (1982, 1984) working with Tufted Ducks.

Many other factors might be involved in prey selection of these diving birds. Bustnes and Erikstad (1990) argue that Eiders select smaller size classes to minimize the daily shell weight intake. Indeed the flesh content was highest (6-8% of the total weight) in the smallest size classes and dropped to 4% in the largest mussels (41-43 mm) as the flesh weight increased 17-fold and the shell weight 26-fold with an increase in mussel length from 14 to 40 mm. Nyström and Pehrsson (1988) hypothesize that the saline sea water that is enclosed within the mussel shells could cause salt-related dehydration problems in diving ducks. As the proportion of sea water increases with increasing mussel size salt stress might be avoided by utilizing smaller mussels. The salt intake per day per unit body-weight decreases from 2.5 to 1 g. Kg^{-1} when feeding exclusively on mussels of 40 or 10 mm respectively (Nyström and Pehrsson 1988). De Leeuw and Van Eerden (1992) found that Tufted Ducks used two distinct feeding methods. Small freshwater mussels (*Dreissena polymorpha*) (<16 mm) are strained from a waterflow generated in the bill (suction feeding), while larger mussels are picked up and handled singly. Suction feeding on small mussels was more profitable than feeding on individual larger mussels. In a non-diving situation, the ducks preferentially took the smallest mussels on offer, smaller than the most profitable ones, probably because the smaller mussels incur lower costs of crushing mussel shells in the gizzard. When diving, a wider range of mussel sizes was taken, although selectivity for the smaller mussels taken by suction feeding increased with diving depth. A functional strategy seemed to be to take up small mussels in a run first and at the end, when bottom time is nearly over, pick up a larger one. As the time spent at the bottom increased with diving depth this explains the increased selectivity. Taking a large prey at the beginning of a dive would prohibit taking any more mussels in a dive, thus not an optimal strategy, whereas not taking a large prey at the end of diving and bringing it to the surface is neither optimal since no further costs are involved to get the extra food item (De Leeuw and Van Eerden 1992). Beauchamp et al (1992) used with reasonable success a dynamic programming model to predict the foraging behavior of Eider.

Another factor that might be very important, especially for birds swallowing the entire bivalve, is the processing rate of food in the stomach. Indeed, Zwarts and Blomert (1992) postulate that a digestive bottleneck might explain prey selection. If the rate of food processing is constrained by the capacity of the gut to digest food, feeding periods in which the intake rate exceeds the processing rate must be followed by digestive pauses. Maximizing the ratio of shell weight/flesh weight becomes a potentially very important point of selection. This ratio varies strongly between species (Zwarts and Blomert 1992) and between size classes of a species (Bustnes and Erikstad 1990). The preference of Knot for *M. balthica* could be explained by its shell weight/flesh weight ratio that is lower than in other bivalves (Zwarts and Blomert 1992) as well as the preference of Eider for small mussels (Bustnes and Erikstad 1990).

From the above discussion we can conclude that the optimal foraging models allow us to better understand the observed patterns of predation as well as to make predictions about the pattern of predation. These models can, however, only be useful when detailed information on both the foraging behavior of the predator and the harvestable fraction of the prey population is available. On both aspects much research is still needed.

IMPACT OF BIRDS ON BIVALVE POPULATIONS

The possible competition between birds feeding on cockles and mussels and commercial fisheries on these bivalves has been the subject of many discussions and stimulated research into the impact of birds on shellfish. In some cases the conflict even resulted in mass killings of Oystercatchers by hunters (Davidson 1968). Some of the available data on the impact of birds on their prey, especially mussels and cockles is summarized here.

The impact of predation on mussel beds was first analyzed by Milne and Dunnet (1972). They found that on mussel beds situated in the Ythan estuary in Scotland the gross production is accounted for by predation and overwintering metabolic requirements. This suggested that the mussel bed was being cropped to a maximum, so that the standing crop at the beginning of each year remained fairly constant and no net changes in size of the mussel population occurred. Birds consume 37% of the production, the Eider taking 21%, gulls 8.6% and Oystercatchers 7.1%. These figures are not based on the fraction of mussels that is actually available to the birds. Baird and Milne (1981) for the same study area estimated that about 73% of mussel production was consumed by birds. Consumption was mainly by Eider (55%), Oystercatcher and Herring Gull (both 21%). Raffaelli et al (1990) estimated the impact of Eiders on the

Inches beds in the Ythan estuary based on exclosure experiments. Eiders removed 36% of the larger mussels (6-30 mm) present on the bed in 60 days (December - January). Galbraith (unpubl. in Raffaelli et al 1990) found that 80% of mussels within this size range disappeared from the Ythan beds between November and April and assumed they were eaten mainly by Eiders. Eiders diving above mussel beds often remove large clumps of mussels along with the focal prey item selected. This leaves bare patches on the mussel bed. Raffaelli et al (1990) found no evidence that this had a major influence on the invertebrate community associated with the mussel bed.

These data are in sharp contrast with the findings of Kautsky (1981) studying the mussel beds in the Askö area in the Baltic Sea about 70 km south of Stockholm. Here only 0.13% of the total standing stock of mussels in July-August is taken by Eider. The main predator is the flounder (*Platicthys flesus*) which does not take more than 0.6 % of the standing stock, leading Kautsky (1981) to the conclusion that the adult mussel population seems to constitute a "dead end" in the food web.

In the Wadden Sea where large numbers of Eider also occur their consumption of benthic invertebrates was estimated by Swennen (1976) and Swennen et al (1989) at 1.19 g ash free dry weight (AFDW) m^{-2} y^{-1}. Between compartments this varies between 0.85 and 1.4 g AFDW m^{-2} y^{-1}. This is only a small proportion of the total biomass (maximum standing stock) which is estimated at 38 g on the intertidal flats (Beukema 1989) and at 43 g AFDW m^{-2} in subtidal areas (Dekker 1989). In the Dutch Wadden Sea, Swennen (1976) estimated the food of the Eider consists of 40% mussels, 40% cockles and 20% other prey. At most 37% of the total food intake consists of cultured mussels (Swennen et al 1989). In Schleswig-Holstein the food of moulting Eiders consists of up to 75% cockles (Nehls 1989). Here consumption is 1.3 g AFDW m^{-2} y^{-1}, or 5% of the macrobenthos biomass or 12.5% of the cockle and mussel biomass (Nehls 1989). This is, however, only including intertidal areas so the overall predation pressure will be lower. The total consumption, however, has tripled since the seventies! In none of these studies the available fraction of the prey population was known. Obert and Michaelis (1991) attributed the significant decline of the mussel population in the catchment area of the Norderey tidal inlet to the co-acting effects of three stock reducing factors: intensified mussel fishery, a series of three severe winters and increased Eider predation.

Although the feeding behavior of Oystercatcher has been studied by many authors few data are available on the predation pressure on mussel beds. Zwarts and Drent (1981) found in three study plots in the Dutch Wadden Sea, that 40% of the mussels larger than 40 mm, the size classes taken by Oystercatchers, were removed by these birds. This decrease was largely due to Oystercatcher predation. The overall decline between 1973 and 1979 in mussel densities, especially of mussels larger than 40 mm, from nearly 2500 to 0 m^{-2} was attributed by Zwarts and Drent (1981) to increasing predation by Herring Gulls. Densities of Herring Gulls increased from 0 in 1967 and 1968 to 47.8 per ha in 1973, increasing even further in the following years. These birds feed on the small first year mussels and hence reduce the mussels remaining for Oystercatchers more than two years later (when the mussels become larger than 40 mm). Meire et al (in press) found the predation pressure of Oystercatchers to vary between less than 10% and over 70% of the standing stock of cockles and mussels, with on average between 30 and 40% on the mussel beds.

On a study plot in the Oosterschelde, Craeymeersch et al (1986) estimated the consumption of Oystercatchers to be about 48 g AFDW m^{-2} y^{-1}. This is 40% of the total production or 60% of the production of the size classes 30-50 mm. As about 40% of the mussels in these size classes are unavailable to the birds (see above) it is highly probable that the Oystercatcher population in this area is food limited. During this 1981/1982 study period, mussel biomass remained constant. Later the mussel population also declined significantly from 300 g AFDW m^{-2} in 1979 to 20 g AFDW m^{-2} in 1988. This is, in my opinion, mainly the result of the severe winters of 1985, 1986 and 1987 and not the result of Oystercatcher predation.

Goss-Custard et al (in press) carried out the most detailed studies on the effect of Oystercatcher predation on mussels. They found up to 30% of the larger mussels (> 40 mm long) present at the start of the season were removed by

Oystercatchers.

Drinnan (1957) estimated that Oystercatchers removed 22% of the cockle population in Morecambe Bay, England, the overall winter mortality was estimated at 74%. Hancock and Urquhart (1965) also found a high mortality of second-winter cockles in the Burry Inlet, South Wales, up to or even more than 90%, but they ascribed this largely to predation by Oystercatchers. O'Connor and Brown (1977) explain the 85% decline in density of second year cockles in Strangford Lough, Northern Ireland, to Oystercatcher predation. Sanchez-Salazar et al (1987) show that mortality of cockles in Traeth Melynog in North Wales, is strongly dependent on tidal elevation. High on the shore mortality is low in the first summer following settlement, but increases during the second winter. This mortality is mainly due to Oystercatchers. In the low shore by contrast, the probability of surviving the first year is very low, due to crab predation, but subsequent life expectancy is excellent. Sutherland (1982b) estimated cockle mortality by Oystercatchers for the same study area at between 28% and almost zero. In this area no cockle fishing occurs. A re-examination of the data of Hancock and Urquhart (1965) by Horwood and Goss-Custard (1977) showed that the high mortality of cockles was not due to Oystercatchers alone. Indeed at high cockle densities the predation capacity of the birds is exceeded and the proportionate effect is small; at low cockle densities the birds could have removed all the second-winter cockles but this did not happen as birds switched prey size below a threshold density of 50-100 second winter cockles m^{-2}. Meire et al (in press) found that in individual study plots, scattered over a tidal flat between 22 and 70 % of the cockles were removed over winter (Fig. 10). The large scatter is to a large extent dependent on the average length of the cockles present. In plots with larger cockles the predation pressure was significantly higher ($r=0.74$, $n=11$, $p<0.01$) (Fig. 11) .

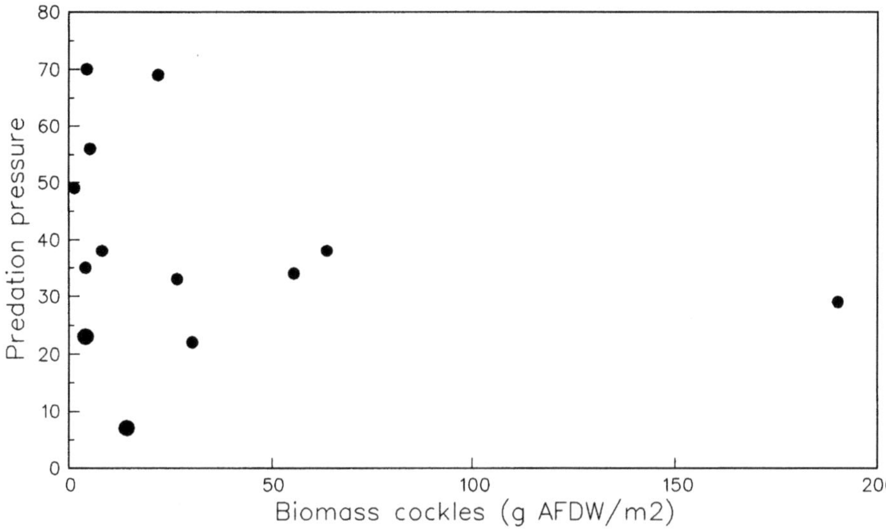

Fig. 10 Predation pressure (% of standing stock) of Oystercatchers (*Haematopus ostralegus*) on cockles (*Cerastoderma edule*) in several study plots in the Oosterschelde as a function of cockle biomass (from Meire et al in press).

What then is the effect of this predation by birds on the prey populations? On the short term there is a decline in density. In the long term the effect can be either positive or negative and this can differ between species. As the Oystercatchers and some other predators remove the larger prey, which are reproductively most productive. These removed pre

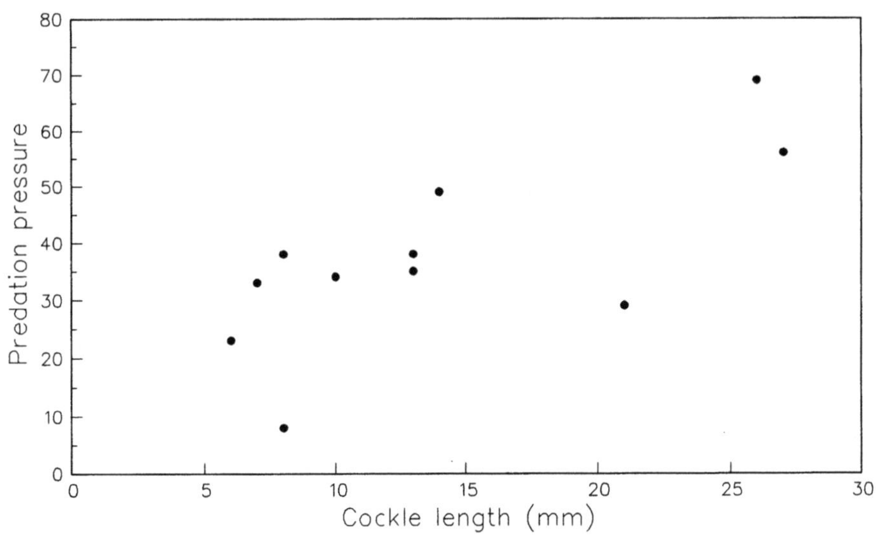

Fig. 11 Predation pressure of Oystercatchers (*Haematopus ostralegus*) on cockles (*Cerastoderma edule*) in several study plots in the Oosterschelde as a function of cockle length (from Meire et al in press).

could have had an influence on reproductive output. Goss-Custard et al (in press) estimated for the Exe estuary that Oystercatchers reduce the potential egg production by mussels up to 10-15% per year. This is, however, unlikely to affect the population in view of the superabundance of larvae in the water.

The survival of mussel spat is strongly density dependent (McGrorty et al 1990). Both the numbers of spat settling during the summer, and the number surviving the first winter, depends on the density of adults already present on the mussel beds. In March a particular density of adults contained a rather similar density of surviving spat in all years. By removing mussels, Oystercatchers reduce the amount of refuge space available to secondary spat that settle amongst the byssus threads of adults, and hence the amount of surviving spat. Hartwick and Baylock (1979) on the other hand, hypothesize that predation by the Black Oystercatcher might be positive because birds did not remove the shells, which provided extra cover for settling spat. In the Oosterschelde also much mussel spat is found on and in dead shells, not only between byssus threads (Meire, unpublished). On the other hand mussels that escape predation might subsequently grow faster due to reduced competition. In this way, however, predation may only reduce the impact of self-thinning, a process of mortality driven by biomass accumulation (Westoby 1984). There is evidence that self-thinning is very strong in mussel populations, with natural mortality occurring as a consequence of assymetric competition (Hughes and Griffiths 1988; Su 1991). In other species, like cockles, large numbers of adults may negatively influence spatfall (André and Rosenberg 1991) so removing adults can have a positive effect.

The impact of birds is also likely to differ significantly between areas. Mussel beds in the Exe estuary are very stable (McGrorty et al 1990) in contrast to those in the Delta area and the Wadden Sea (see, e.g., Obert and Michaelis 1991). Here severe winters and storms can change abruptly and dramatically the whole mussel bed, a phenomenon that is not seen in the Exe estuary or more southern areas in general Also populations of cockles or bivalves as *Spisula* in estuaries or shallow marine areas are known to vary very much in time (see papers in Elliot and Ducrotoy 1991). Therefore the impact of predation is very different between years as shown above for cockles in the Burry Inlet.

CONCLUSIONS

Birds have the potential to deplete severely bivalve populations. These prey species have, however, different possibilities to escape predation. A deep burrow guarantees maximal survival and a burrow below a certain depth reduces predation risk. The potential prey individuals have reached a depth refuge. This depth refuge is determined by the long-billed predators that can probe at maximal depths (Zwarts and Wanink 1985). Cockles remain within reach of their main predators but by growing as soon as possible they can avoid predators feeding only on the smaller size classes. In any case, quick growth is apparently a first priority, to be followed by increase in siphon mass to make possible a greater burying depth. It has been suggested that reproduction in these benthic bivalves is delayed in order to divert their resources into rapid early growth (Seed and Brown 1978). Zwarts and Wanink (1989) indeed found that maximal siphon growth takes place before sexual maturity is reached. The three priorities during the course of the life of a benthic burrowing bivalve therefore seem to be: (1) to grow fast; (2) to increase burying depth; and (3) to produce offspring. Next to burying depth producing a thick shell, or having a low shell/biomass ratio is also a protection against predation. This creates a great variability in the prey populations. Clearly any understanding of the population dynamics requires accepting that prey vary and that different predators may judge prey differently. For example, the composition of the prey population may change over the winter due to depletion of the thin shelled individuals that would change the suitability to the Oystercatchers although having little effect on total prey density. Similarly birds with different feeding techniques may have different requirements.

Ignoring these differences between prey and differences between predators may often hide important relationships in population biology. The ongoing work on waders and their food supply summarized in this paper show that it is possible to disentangle step by step these complex patterns but at present we are still some way from fully understanding the reciprocal interactions between birds and their shellfish prey populations (Goss-Custard et al in press).

ACKNOWLEDGEMENTS

Ray Seed, Aad Smaal, Tom Ysebaert, Eckhart Kuijken and Dinska Amery made many valuable comments on an earlier draft of this paper. Part of the data summarised here were collected while the author hold a grant from the Belgian National Fund for Scientific Research. Much of the research greatly benefitted by the financial support from the Tidal Waters Division of The Ministry of Transport and Public Works (The Netherlands).

REFERENCES

Alerstam T Gudmundsson GA Johannesson K (1992) Resources for long distance migration: intertidal exploitation of *Littorina* and *Mytilus* by Knots *Calidris canutus* in Iceland. Oikos 65:179-189

André C Rosenberg R (1991) Adult-larval interactions in the suspension feeding bivalves *Cerastoderma edule* and *Mya arenaria*. Mar Ecol Prog Ser 71:227-234

Baldwin WP (1946) Clam catches Oystercatcher. Auk 63:589

Baird D Milne H (1981) Energy flow in the Ythan estuary, Aberdeenshire, Scotland. Est Coast Shelf Sci 13:455-472

Beauchamp G Guillemette M Ydenberg RC (1992) Prey selection while diving by common eiders, *Somateria mollissima*. Anim Behav 44:417-426

Beukema J (1989) Long-term changes in macrozoobenthic abundance on the tidal flats of the western Wadden Sea. Helgoländer Meersunters 43:405-415

Bustnes JO Erikstad KE (1990) Size selection of common mussels, *Mytilus edulis*, by common Eiders, *Somateria molissima*: energy maximization or shell weight minimization. Can J Zool 68:2280-2283

Butler PJ Woakes AJ (1979) Changes in heart rate and respiratory frequency during natural behaviour of ducks, with particular reference to diving. J Exp Biol 79:283-300

Boyd H Pirot J-Y (1989) Flyways and reserve networks for water birds. IWRB Special publication No 9, Slimbridge

Brown RA O'Connor RJ (1974) Some observations on the relationships between Oystercatchers *Haematopus ostralegus* L. and cockles *Cardium edule* L. in Strangford Lough. Irish Nat J 18:73-80

Cantin M Bédard J Milne H (1974) The food and feeding of Common Eiders in the St. Lawrence estuary in summer. Can J Zool 52:319-334

Cayford J Goss-Custard JD (1990) Seasonal changes in the size selection of mussels, *Mytilus edulis*, by Oystercatchers, *Haematopus ostralegus*: an optimality approach. Anim Behav 40:609-624

Charnov EL (1976) Optimal foraging: attack strategy of a mantid. Am Nat 110:141-151

Cootam C (1939) Food habits of North American diving ducks. US Dept of Agric Techn Bull 643:1-135

Cramp S Simmons KEL (1977) Handbook of the birds of Europe, the Middle East, and North Africa: the birds of the Western Palearctic. Vol 1: Ostrich-Ducks. Oxford U Press, Oxford

Craeymeersch JA Herman PMJ Meire PM (1986) Secondary production of an intertidal mussel (*Mytilus edulis*) population in the Eastern Scheldt (S.W. Netherlands). Hydrobiologia 133:107-115

Davidson PE (1967) A study of the Oystercatcher (*Haematopus ostralegus* L.) in relation to the fishery for cockles (*Cardium edule* L.) in Burry Inlet, South Wales. Fish Invest London Ser II Vol 25 7:1-28

Davidson PE (1968) The Oystercatcher - a pest of shellfisheries. Pp. 174-180 in Murton RK and Wright EN

(Eds) The problems of birds as pests. Academic Press, London

Davidson PE (1971) Some foods taken by waders in Morecambe Bay, Lancashire. Bird Study 18:177-186

Davidson NC Pienkowski MW (1987) The conservation of international flyway populations of waders. Wader Study Group Bull 49 Supplement

Davidson NC Wilson JR (1992) The migration system of European-wintering Knots *Calidris canutus islandica*. Wader Study Group Bull 64 (Supplement):39-51

Dekker R (1989) The macrozoobenthos of the subtidal western Dutch Wadden Sea. 1. Biomass and species richness. Neth J Sea Res 23:57-68

De Leeuw JJ Van Eerden MR (1992) Size selection in diving Tufted Ducks *Aythya fuligula* explained by differential handling of small and large mussels *Dreissena polymorpha*. Ardea 80:353-362

Dewar JM (1922) Ability of the Oystercatcher to open oysters, and its bearing upon the history of the species. Br Birds 16:118-125

Draulans D (1982) Foraging and size selection of mussels by the Tufted Duck, *Aythya fuligula*. J Anim Ecol 51:943-956

Draulans D (1984) Sub-optimal mussel selection by Tufted Ducks *Aythya fuligula*: test of a hypothesis. Anim Behav 32:1192-1196

Drinnan RE (1957) The winter feeding of the Oystercatcher (*Haematopus ostralegus*) on the edible cockle (*Cardium edule*). J Anim Ecol 26:441-469

Drinnan RE (1958) The winter feeding of the Oystercatcher (*Haematopus ostralegus*) on the edible mussel (*Mytilus edulis*) in the Conway estuary, North Wales. Fish Invest London: Ser II Vol 22:1-15

Dunstone N O'Connor RJ (1979) Optimal foraging in an amphibious mammal I. the aqualung effect. Anim Behav 32:1192-1196

Dunthorn AA (1971) The predation of cultivated mussels by Eiders. Bird Study 18:107-112

Durell SEA le V dit Goss-Custard JD (1984) Prey selection within a size-class of mussels, *Mytilus edulis* by

Oystercatchers, *Haematopus ostralegus*. Anim Behav 32:1197-1203

Elliot M Ducrotoy J-Y (1991) Estuaries and coasts: spatial and temporal intercomparisons. Olson & Olson, Fredensborg, Denmark

Ens B (1982) Size selection in mussel-feeding Oystercatchers. Wader Study Group Bull 34:16-20

Ens B Goss-Custard JD (1984) Interference among Oystercatchers, *Haematopus ostralegus*, feeding on mussels, *Mytilus edulis*, on the Exe estuary. J Anim Ecol 53:217-231

Feare CJ Summers RW (1985) Birds as predators on rocky shores. Pp. 249-164 in Moore PG and R Seed (Eds) The ecology of Rocky Coasts. Hodder & Stonghton

Goss-Custard JD (1985) Foraging behaviour of wading birds and the carrying capacity of estuaries. Pp 169-188 in Sibly RM and Smith RH (eds) Behavioural Ecology. Blackwell Scientific Publications, Oxford

Goss-Custard JD Jones RE Newbery E (1977) The ecology of the Wash I. Distribution and diet of wading birds (Charadrii). J Appl Ecol 14:681-700

Goss-Custard JD Le V Dit Durell SEA Ens BJ (1982) Individual differences in aggressiviness and food stealing among wintering Oystercatchers, *Haematopus ostralegus*. Anim Behav 30:917-928

Goss-Custard JD Le V Dit Durell SEA (1990) Bird behaviour and environmental planning: approaches in the study of wader populations. Ibis 132:273-289

Goss-Custard JD McGrorty S Clarke R (in press) The impact of Oystercatchers on shellfish populations. Ardea

Guillemette M Ydenberg RC Himmelman JH (1992) The role of energy intake in prey and habitat selection of common eiders *Somateria mollissima* in winter: a risk-sensitive interpretation. J Anim Ecol 61:599-610

Hancock DA Urquhart AE (1965) The determination of natural mortality and its causes in an exploited population of cockles (*Cardium edule* L.). Fish Invest London Ser Vol 24 No 2:1-33

Hartwick EB Blaylock W (1979) Winter ecology of a Black Oystercatcher population. Studies in Avian Biology 2:207-

Dan Rev Game Biol 2:159-266

Maron JL (1982) Shell-dropping behaviour of Western Gulls (*Larus occidentalis*). The Auk 99:565-569

McGrorty S Clarke RT Reading CJ Goss-Custard JD (1990) Population dynamics of the mussel *Mytilus edulis*:density changes and regulation of the population in the Exe estuary, Devon. Mar Ecol Prog Ser 67:157-169

Meire PM (1987) Foraging behaviour of some wintering waders: prey selection and habitat distribution. Pp 215-237 in Kamil AC Krebs JR Pulliam HR (eds) Foraging behavior. Plenum Press, New York

Meire PM (1991) Effects of a substantial reduction in intertidal area on numbers and densities of waders. Acta XX Congr Inter Ornit: 2219-2227

Meire PM (in press) Mussel selection in Oystercatchers (*Haematopus ostralegus*): some ecological considerations. Ardea

Meire PM Ervynck A (1986) Are oystercatchers (*Haematopus ostralegus*) selecting the most profitable mussels (*Mytilus edulis*)? Anim Behav 34:1427-1435

Meire PM Schekkerman H Meininger PL (in press) Consumption by waterbirds in the Oosterschelde, SW Netherlands, in relation to carrying capacity and the benthic invertebrate community. Hydrobiologia

Meissner J Brager S (1990) The feeding ecology of wintering Eiders *Somateria mollissima* and Common Scoters *Melanitta nigra* on the Baltic coast of Schleswig-Holstein, FRG. Wader Study Group Bull 58:10-12

Milne H Dunnet GM (1972) Standing crop, productivity and trophic relations of the fauna of the Ythan estuary. Pp 86-106 in Barnes RSK and Green J (Eds) The estuarine environment. Applied Science Publishers, London

Monval J-Y Pirot J-Y (1989) Results of the IWRB International Waterfowl Census 1967-1986. IWRB Special Publication No 8, Slimbridge

Nehls G (1989) Occurrence and food consumption of the Common Eider, *Somateria mollissima*, in the Wadden Sea of Schleswig-Holstein. Helgoländer Meeresunters 43:385-393

Nilson L (1969) Food consumption of diving ducks wintering at

215

Horwood JW Goss-Custard JD (1977) Predation by the oystercatcher, *Haematopus ostralegus* (L.), in relation to the cockle, *Cerastoderma edule* (L.), fishery in the Burry inlet, South Wales. J Appl Ecol 14:139-158

Houston AI McNamara JM (1985) A general theory of central place foraging for single prey loaders. Theor Popul Biol 28:233-262

Hughes RN (1970) Population dynamics of the bivalve *Scrobicularia plana* (Da Costa) on an intertidal mud-flat in North Wales. J Anim Ecol 39:333-356

Hughes RN Griffiths CL (1988) Self-thinning in barnacles and mussels: the geometry of packing. Am Nat 132: 484-491

Hulscher J (1982) The Oystercatcher *Haematopus ostralegus* as a predator of the bivalve *Macoma balthica* in the Dutch Wadden Sea. Ardea 70:89-152

Hulscher J (1985) Growth and abrasion of the Oystercatcher bill in relation to dietary switches. Neth J Zool 35:124-154

Hulscher J (1988) Mossel doodt Scholekster *Haematopus ostralegus*. Limosa 61:42-45

Hulscher J Boates S Johnson R Swennen C (in press) Diet of the Oystercatcher. In Goss-Custard JD (Ed) Behavioral Ecology of the Oystercatcher. Oxford U press, Oxford

Jukema J (1979) Eidereend *Somateria mollissima* als slachtoffer van mossel *Mytilus edulis*. Limosa 52:73-74

Kautsky N (1981) On the trophic role of the blue mussel (*Mytilus edulis* L.) in a Baltic coastal ecosystem and the fate of the organic matter produced by the mussels. Kieler Meeresforsh Sonderh 5:454-461

Kent BW (1981) Prey dropped by Herring gulls (*Larus argentatus*) on soft sediments. Auk 98:350-354

Laursen K (1989) Estimates of Sea Duck winter populations of the Western Palearctic. Dan Rev Game Biol 13 (6):1-22

Leopold MF Swennen C De Bruijn LLM (1989) Experiments on selection of feeding site and food size in Oystercatchers, *Haematopus ostralegus*, of different social status. Neth J Sea Res 23:333-346

Madsen FJ (1954) On the food habits of diving ducks in Denmark.

the coast of South Sweden in relation to food resources. Oikos 20:128-135

Nilson L (1972) Habitat selection, food choice, and feeding habits of diving ducks in coastal waters of South Sweden during the non-breeding season. Ornis Scand 3:55-78

Norton Griffiths M (1967) Some ecological aspects of the feeding behaviour of the Oystercatcher, *Haematopus ostralegus*, on the edible mussel, *Mytilus edulis*? Ibis 109:412-424

Nyström KGK Pehrsson O (1988) Salinity as a constraint affecting food and habitat choice of mussel-feeding diving ducks Anatidae. Ibis 130:94-110

Obert B Michaelis H (1991) History and ecology of the mussel beds (*Mytilus edulis*) in the catchment area of a Wadden Sea tidal inlet. Pp 185-194 in Elliot M and Ducrotoy J-P (Eds) Estuaries and coasts: spatial and temporal intercomparisons. Olson & Olson, Fredensborg

O'Connor RJ Brown RA (1977) Prey depletion and foraging strategy in the oystercatcher *Haematopus ostralegus*. Oecologia 27:75-92

Pehrsson O (1976) Food and feeding grounds of the Goldeneye *clangula* (L.) on the Swedish west coast. Ornis Scand 7:91-112

Piersma T Prokosch P Bredin D (1992) The migration system of Afro-Siberian Knots *Calidris canutus canutus*. Wader Study Group Bull 64 (Supplement):52-63

Player PV (1971) Food and feeding habits of the Common Eider at Seafield, Edinburgh, in winter. Wildfowl 22:100-106

Raffaelli D Falcy V Galbraith C (1990) Eider predation and the dynamics of mussel-bed communities. Pp 157-169 in

Barnes M and Gibson R (Eds) Trophic Relationships in the Marine Environment. Aberdeen U Press, Aberdeen

Reading CJ McGrorty S (1978) Seasonal variations in the burying depth of *Macoma balthica* (L.) and its accessibility to wading birds. Est Coast Mar Sci 6:135-144

Richardson H (1985) Availability of buried littleneck clams (*Venerupis japonica*) to Northwestern crows (*Corvus corone*). J Anim Ecol 54:443-457

Richardson H Verbeek NA (1986) Diet selection and optimization by Northwestern Crows feeding on Japanese littleneck clams. Ecology 67:1219-1226

Richardson H Verbeek NA (1987) Diet selection by yearling Northwestern Crows (*Corvus caurinus*) feeding on littleneck clams (*Venerupis japonica*). The Auk 104:263-269

Sanchez-Salazar ME Griffiths CL Seed R (1987) The interactive roles of predation and tidal elevation in structuring populations of the edible cockle, *Cerastoderma edule*. Est Coast Shelf Sci 25:245-260

Seed R Brown RA (1978) Growth as a strategy for survival in two marine bivalves, *Cerastoderma edule* and *Modiolus modiolus*. J Anim Ecol 47:283-292

Smit CJ Lambeck RHD Wolff WJ (1987) Threats to coastal wintering and staging areas for waders. Wader Study Group Bull 49 (Supplement):105-113

Spaans A (1971) On the feeding ecology of the herring gull (*Larus argentatus* Pont.) in the northern part of the Netherlands. Ardea 59:75-188

Stephenson DW Krebs JR (1986) Foraging theory. Princeton U Press, Princeton, New Jersey

Sutherland WJ (1982a) Do Oystercatchers select the most profitable cockles? Anim Behav 30:857-861

Sutherland WJ (1982b) Spatial variation in the predation of cockles by oystercatchers at Traeth Melynog, Anglesey II. The pattern of mortality. J Anim Ecol 51:491-500

Sutherland WJ Ens BJ (1987) The criteria determining the selection of mussels *Mytilus edulis* by Oystercatchers *Haematopus ostralegus*. Behaviour 103:187-202

Swennen C (1976) Populatie-structuur en voedsel van de Eidereend *Somateria m. mollissima* in de Nederlandse Waddenzee. Ardea 64:311-371

Swennen C De Bruijn LLM Duiven P Leopold MF Marteijn ECL (1983) Differences in bill form of the Oystercatcher *Haematopus ostralegus*; a dynamic adaptation to specific foraging techniques. Neth J Sea Res 17:57-83

Swennen C Leopold MF Stock M (1985) Notes on growth and behaviour of the American razor clam *Ensis directus* in the Wadden Sea and the predation on it by birds. Helgoländer Meeresunters 39:255-261

Swennen C Nehls G Laursen K (1989) Numbers and distribution of Eiders *Somateria mollissima* in the Wadden Sea. Neth J Sea Res 24:83-92

Su M (1991) Self-thinning in some mussel populations (*Mytilus edulis*). Thesis University of Brussels

Vermeer K (1981) Food and populations of Surf Scoters in British Columbia. Wildfowl 32:107-116

Ward D (1991) The size selection of clams by African Black Oystercatchers and Kelp gulls. Ecology 72:513-522

Westoby M (1984) The self-thinning rule. Adv Ecol Res 14:167-225

Ydenberg RC (1988) Foraging by diving birds. Pp 1832-1842 in Ouellet H (Ed) Acta IX Congressus Internationalis Ornithologici. U of Ottawa Press, Ottawa

Ydenberg RC Forbes LS (1988) Diving and foraging in the Western Grebe. Ornis Scand 19:129-133

Ydenberg RC Guillemete M (1991) Diving and foraging in the Common Eider. Ornis Scand 22:349-352

Zach R (1979) Shell dropping:decision-making and optimal foraging in Northwestern Crows. Behaviour 58:106-117

Zwarts L (1986) Burying depth of the benthic bivalve *Scrobicularia plana* (Da Costa) in relation to siphon-cropping. J Exp Mar Biol Ecol 101:25-39

Zwarts L Drent RH (1981) Prey depletion and the regulation of predator density: Oystercatchers (*Haematopus ostralegus*) feeding on mussels (*Mytilus edulis*). Pp 193-216 in Jones NV and Wolff WJ (eds) Feeding and survival strategies of estuarine organisms. Plenum Press, New York

Zwarts L Wanink JH (1984) How Oystercatchers and Curlew successively deplete clams. Pp 69-83 in Evans PR Goss-Custard JD Hale WG (eds) Coastal waders and wildfowl in winter. Cambridge U Press, Cambridge

Zwarts L Wanink JH (1985) Does an optimally foraging Oystercatcher obey the functional response? Oecologia 67:98-106

Zwarts L Wanink JH (1989) Siphon size and burying depth in deposit- and suspension feeding benthic bivalves. Mar Biol 100:227-240

Zwarts L Blomert A-M (1992) Why Knot (*Calidris canutus*) take medium-sized *Macoma balthica* when six prey species are available. Mar Ecol Prog Ser 83:113-128

Zwarts L Blomert A-M Wanink JH (1992 Annual and seasonal variation in the food supply harvestable for Knot *Calidris canutus* staging in the Wadden Sea in late summer. Mar Ecol Prog Ser 83:128-139

Zwarts L Cayford JT Hulscher JB Meire PM Triplet P (in press) Size Selection and intake rate in Oystercatcher. In Goss-Custard JD (Ed) Behavioral Ecology of the Oystercatcher. Oxford U press, Oxford

THE ROLE OF BIVALVE FILTER FEEDER MATERIAL FLUXES IN ESTUARINE ECOSYSTEMS

Richard F. Dame
Coastal Carolina College
Conway, SC 29526 USA

INTRODUCTION

Bivalve reefs and beds are a common feature of many estuarine and coastal environments. Bivalve filter feeders dominate these communities removing food from and releasing waste materials to the overlaying waters. The surrounding shallow tidal waters provide a habitat for growing phytoplankton and an environment for nutrient recycling. In essence, these systems are natural "feed lots" with the herbivorous bivalves playing the same role as large mammals in cultivated terrestrial systems. This feature of bivalves has been recognized and exploited through culture for millennia. With the application of modern technology to bivalve cultivation, many estuarine systems are approaching and even exceeding the capabilities of these ecosystems to maintain their environmental quality. In this paper, I will review modern ecosystem concepts with respect to material processing and nutrient cycling emphasizing the role bivalve filter feeders play in shallow estuarine and coastal ecosystems. In addition, the published evidence of bivalve filter feeder influence on these ecosystems will be examined, synthesized and potential areas of research identified.

MATERIAL PROCESSING AND ECOSYSTEM THEORY

Bivalve filter feeding systems have played a major role in the development of ecosystem theory. Möbius (1877) proposed the term bioceonosis to describe the community of living organisms

NATO ASI Series, Vol. G 33
Bivalve Filter Feeders
in Estuarine and Coastal Ecosystem Processes
Edited by Richard F. Dame

in an oyster-bed on the Schleswig-Holstein flats along the German coast and in mouths of English estuaries. He clearly noted that different species found in these beds had something in common and were a positive influence on each other. Möbius (1877) speculated that not only did the organisms of the oyster bed interact and regulate the oyster community, but they also interacted with the surrounding non-living environment.

The application of energy budgets to whole ecosystems (Juday 1940) led the way for utilizing energy flows and material cycles to describe the functional significance of organisms in ecosystem. Juday's work was the first to emphasize the complexity of studying whole system metabolism and strongly influenced Lindeman's (1942) formulation of the trophic-dynamic concept. The trophic-dynamic concept provided not only a theoretical framework within which to study ecosystems, but also was a stimulus to a profusion of research because of the questions it posed. In this concept, the living components are grouped according to their feeding categories, i.e., trophic level, which are commonly termed primary producers, herbivores, carnivores, decomposers, etc. The energy flow aspects of trophic dynamics were strongly developed by Odum (1953) and led to what is now called the functional approach to ecosystem analysis. In the marine environment, Redfield (1958) argued that in material cycles not only do organisms regulate the environment but they are in turn regulated by the environment.

For the purposes of this paper, I will use Odum's (1983) definition for ecosystems: an ecosystem is any unit that includes all organisms that function together in a given area interacting with the physical environment so that the flow of energy leads to clearly defined biotic structure and cycling of materials between living and non-living parts. This definition clearly belongs to the school of scientist who consider the whole greater than the sum of its parts.

POPULATION/COMMUNITY VERSUS PROCESS/FUNCTIONAL APPROACHES

Recently, the development and application of the ideas of levels of organization, now known as hierarchy theory, have blossomed in ecosystem ecology (O'Neill et al. 1986). Two separate hierarchies, the population/community and the process/functional approaches, have emerged which address ecosystems differently. The former is based on like taxonomic groupings while the latter is predicated on similar ecological processes.

The population/community approach is tangible and envisions a sequence of **ORGANISM - POPULATION - COMMUNITY** (Fig. 1) and this hierarchy generally views the physical environment as a backdrop to the biological processes (O'Neill et al. 1986). The process/functional school places organisms of like functional roles together (Fig. 1) with the functional group at the base and the biosphere at the top (O'Neill et al. 1986). This hierarchy is less tangible than the population/community hierarchy, but the physical environment is interwoven in its understanding.

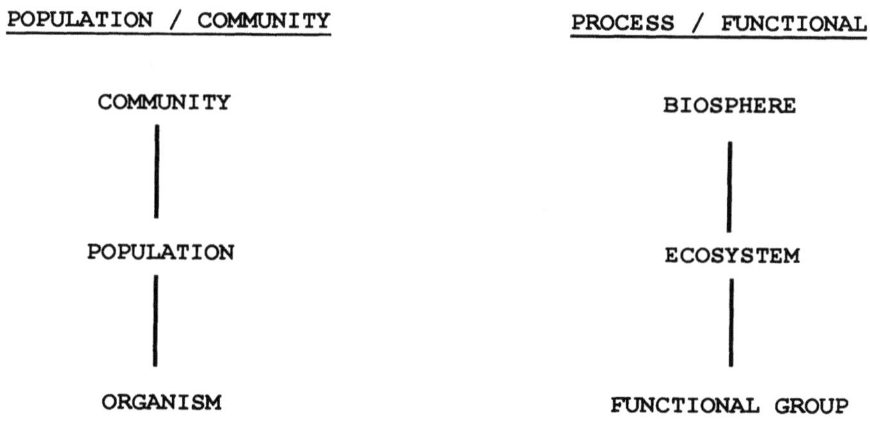

Fig. 1. A comparison of the two common hierarchies of ecosystems.

The two hierarchies do not appear to relate to each other very well, but if, for example, a functional group is dominated by a single species or population there can be a meaningful coherence. Such a relationship exists for bivalve beds and reefs (Fig. 2), where the dominant bivalve is usually a single population and performs a dominant function, i.e., filter feeding, within the estuarine ecosystem. Thus bivalve filter feeders, like bison on the great plains of North America, provide a cross-linkage between the two approaches and the study of ecological aspects in both hierarchies and enrich our understanding of nature.

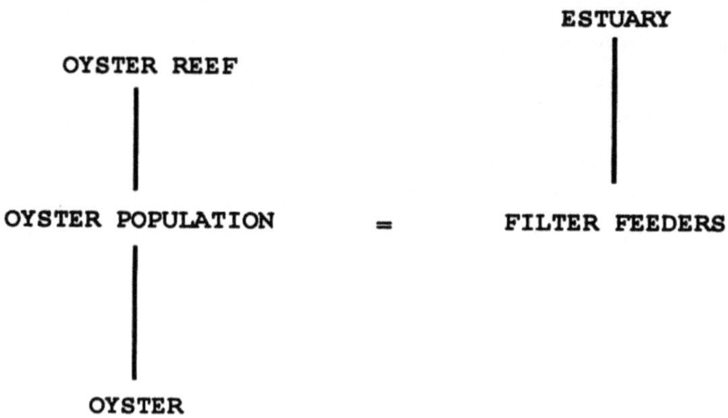

Fig. 2. The relationship between ecosystem hierarchies when bivalve filter feeders dominate.

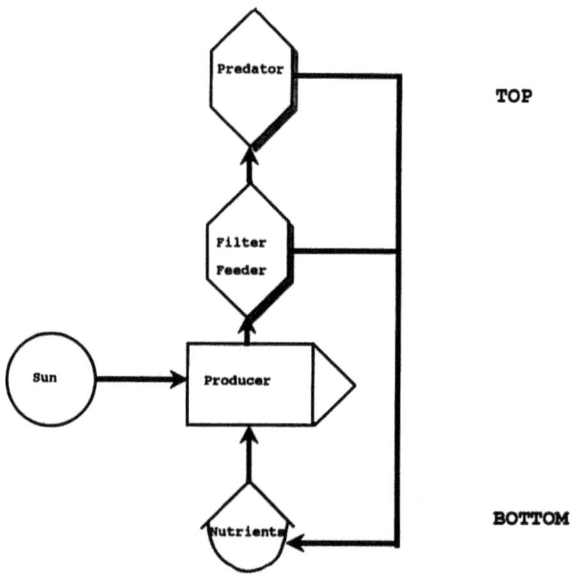

Fig. 3. A generalized food web for bivalve filter feeder dominated systems.

FOOD WEB CONTROLS

The various theories of community structure as explained by food webs (Fig. 3) and the controls on food webs are an area of intense ecological debate (Vadas 1989). The paradigm of bottom-up control of food chains with nutrients and light of major importance to primary producers and the resulting material flows up the food chain affecting the remaining trophic levels is well founded historically (Lindeman 1942).

The top-down (cascading) or food chain dynamics theory was first proposed by Hairston et al. (1960) and expanded by Fretwell (1987) and Oksanen (1988). This theory predicts that the top predator controls the next lower trophic level and results in alternating magnitude (negative correlation) biomass and turnover patterns for adjacent trophic levels. Many of the speculations concerning bivalve filter feeders controlling phytoplankton abundances in the water column subtly fall into this category.

The mutualism-cybernetics theory (Odum and Biever 1984) suggests that plant-herbivore interactions are typically donor-controlled or mutualistic (positive correlation), as herbivores often stimulate the production of the plant species that they feed on. The latter theory is favored by the process/functional group, but attacked by the population/community ecologist because it implies community selection or evolution above the species level. Because of these criticisms, Vadas (1989) has proposed another theory, the donor control-indirect effects theory. In this idea, species interact at ecological time scales with indirect effects caused by nutrient cycling, competition and abiotic fluctuations leading to donor-controlled predator-prey interactions. More recently, Hunter and Price (1992) and Menge (1992) have argued in a similar vein that bottom-up and top-down controls operate simultaneously in most systems and it is the net pattern that we observe in nature.

When bivalve filter feeders occur in abundance, they may functionally and structurally influence their ecosystems. In the following discussions, I will explorer the potential roles of bivalve filter feeders in coastal ecosystems and the evidence supporting their dynamic behavior.

MATERIAL UPTAKE/FILTER FEEDING

Filter feeding by dense groups of bivalves is a major mechanism for moving materials from the water column to the benthos. Numerous studies have attempted to estimate the magnitude of filter feeding at the ecosystem level by applying laboratory derived filtration rate models or estimates to field populations (see Doering and Oviatt 1986). A few of the more prominent studies are reviewed here.

Baird and Milne (1981) developed a steady state carbon model for the Ythan estuary in Scotland. Natural populations of blue mussels are the major benthic macrofauna and they depend on phytoplankton and suspended particulate organic carbon as food. Mussel biomass is thought to be limited mainly by birds. The transfer of material between the mussels and the birds is very high, 71%, and is a major reason this ecosystem is thought to be very efficient in processing carbon to higher trophic levels.

Utilizing the scaled-up method, Rodhouse and Roden (1987) computed a carbon budget for Killary Harbor, Ireland. In this coastal inlet, blue mussels are intensively cultivated and the budget was used to estimate the carrying capacity of the ecosystem for these bivalves. Their calculations suggested that the mussels will extensively crop the resident phytoplankton and the mussels will depend on the tidal input of offshore phytoplankton to sustain their high densities. In addition, sedimentation under the mussels will increase many fold and lead to changes in the benthos. These authors also speculated that the substitution of bivalve filter feeders for filter feeding zooplankton may alter nitrogen cycling by increasing nutrient losses to the sediments. They further contended that changing the rate of nitrogen cycling may enhance primary production and possibly lead to blooms of red tide organisms.

In recent years, there has been a developing argument concerning the role of oysters, *Crassostrea virginica*, in the continuously changing nature of Chesapeake Bay. Before the late 1800's, oysters were not heavily harvested in the Bay and were reported to be a navigation hazard. Newell (1988) calculated from laboratory rate measurements that prior to oysters being overharvested in the 20th century these filter feeders could turnover the shallow (<9m) water in the bay in 1 to 4 days while the doubling times of summer phytoplankton were approximately the same. This coherence suggests the oysters might have exerted a major influence on the phytoplankton in the Bay (see Officer et al. 1982). With the advent of overharvesting and the increasing impact of diseases, the oyster population in the Bay declined from 188 x 10^6 kg db (db=dry body) to the current level of 1.9 x 10^6 kg db. Concurrently nutrient and suspended particulate levels in the Bay increased dramatically, probably due to increased farming and development in the surrounding landscape (Newell 1988). Thus, an increase in suspended sediment was exacerbated by a decrease in a major filtration mechanism, the oysters.

The Bay ecosystem also appears to have changed with the decline of the oysters and consequent reduction in their control of the phytoplankton to a system dominated by microbial and planktonic food webs. The microbial components generate periods of anoxia in the deeper waters of the Bay as they decompose the now unfiltered algae. Without major competition from the benthic filter feeders the planktonic suspension feeding organisms, e.g., copepods and microzooplankton, appear to have expanded their role and now support planktonic predators, e.g., ctenophores and jelly fish, which were previously thought unimportant in the Bay (Newell 1988). A simplistic mass action model of these components of the Chesapeake Bay system with preliminary field data (Ulanowicz and Tuttle 1992) supports this scenario. This Chesapeake Bay story is an interesting example of trying to explain past events with little direct evidence. It is also a persuasive argument for a unified ecosystem approach with laboratory observations, field experiments and systems simulation modeling.

The interest in Chesapeake Bay and other bay systems, i.e., San Francisco Bay, led Officer et al. (1982) to propose the natural control of eutrophication in these systems by bivalve filter feeding. The necessary conditions for such control to take place are shallow water, abundant nutrients, no light, temperature or tubidity limitations, slow hydrodynamic water exchange and abundant populations of benthic filter feeders. Evidence of bivalve filter feeders cropping phytoplankton populations abounds (Doering and Oviatt 1986), but most of these estimates depend on scaled-up filtration rates.

In an extensive mesocosm study, Doering and Oviatt (1986) examined the application of filtration rate models to field populations of bivalves. They pointed out a major problem in using rates determined in the laboratory or under very controlled conditions on individuals or groups of bivalves instead of directly estimating rates in the field. Generally, the application of laboratory determined pumping and filtration rates in order to estimate biodeposition in a particular environment may considerably over estimate observed field biodeposition (Doering and Oviatt 1986). This over estimation occurs because filter feeders in nature are subjected to flowing water of constantly changing velocities during an average tidal cycle. Environmental water velocity and suspended particulate concentrations undoubtedly influence the organism's filtration and pumping rates. In addition, water velocity is a major determining factor in the resuspension of sediments including feces and pseudofeces. Thus many models generate biodeposition rates that could lead to burial of the bivalves. However, recent field studies on mussels and cockles (Smaal et al. 1986) and oysters (Dame 1987) which directly observe the net suspended sediment flux over these systems show a balance. Although large quantities of particulate materials are removed from the water column, equally large quantities are exported. This export must be attributed to resuspension.

BIODEPOSITION

Early estimates of the influence of bivalve filter feeders on sedimentation in coastal ecosystems have been reported by Verwey (1952). His study focused on the relatively shallow Dutch Wadden Sea and its beds of cockles, *Cerastoderma edule*, and blue mussels, *Mytilus edulis*. Because this ecosystem is dominated by pronounced tides and high sediment loads, biodeposition due to bivalve filtration was speculated to be a major factor in the flux of materials into the Wadden Sea.

Verwey thought that the organic material that served as a food source for the cockles and mussels came from the North Sea and that the filter feeders removed this imported material from the water. In addition, Verwey (1952) argued that the filter feeders' biodeposits changed the characteristics of the sediments and the shells of dead animals changed the structure of the substrates. He also observed mussel beds that developed over the bottom of drainage channels changed the geomorphology of the channels. Although Verwey's (1952) estimates of biodeposition are thought to be low because he used underestimates of individual filtration rates (Jørgensen 1990), this study is one of the first which clearly addresses the role of bivalves at the ecosystem level.

Bernard (1974) conducted an extensive study of the biodeposition and energy budget of the Pacific oyster, *Crassostrea gigas*, in the Departure Bay estuary of British Columbia. Like Verwey (1952), Bernard (1974) calculated the cumulative effects of oyster filtration and biodeposition utilizing organismic rates determined in the laboratory. He estimated that the oyster reefs in this area removed as much suspended material as was deposited by gravitational sedimentation and planktonic processes. He also noted that the oyster biodeposits had a higher nutritive potential to browsing organisms, modified the physical and chemical characteristics of the sediments, allowed the establishment of diverse benthic groups, and may also provide a substrate for bacterial development. Bernard's (1974) calculations allowed him to conclude that there was insufficient food in the overlaying water column to support his observed energy budget and materials must be imported to the bivalves via the tides from a much broader source area in order to balance the energy budget. He clearly believed that the concentration of materials by the oyster reef had a major influence on ecosystems far in excess of the actual bivalve biomass.

In a series of studies, Sornin et al (1983), Sornin et al (1986) and Deslous-Paoli et al (1987) used various direct trapping methods to investigate the biodeposition of *M. edulis* and *C. gigas* in a bivalve cultivation zone in France. These authors found that large quantities of suspended sediments were removed from the water column to the benthos. They speculated that the recycling of these organic biodeposits increased the oxygen demand of the sediments and could generate an anaerobic environment fostering ammonification and sulfate reduction. In a more recent study in the same area, Dinet et al (1990) concluded that the bivalve biodeposits appeared to convert the sediments to a reducing environment that lowered meiofaunal abundances.

In a comprehensive study of biodeposition by *M. edulis* in a coastal ecosystem in the Baltic sea, Kautsky and Evans (1986) concluded that mussels increased both the quality and quantity of deposited materials. Their study was important because it was one of the few field studies that used experiments with and without filter feeders and measured sediment deposition directly with sediment traps. Thus using biodeposition rates determined in the field, they found that mussels initiated the sedimentation of smaller particles of high organic content that would otherwise stay in suspension (Fig.4). They also deduced that mussels were an important benthic-pelagic coupling mechanism in their coastal ecosystem because they increased deposition by 10%.

In summary, there is good evidence that bivalve filter feeders remove and increase the deposition of suspended particles. Their biodeposits change the composition of the benthic sediments by enriching them and providing an environment for the development of other organisms. Most of their biodeposits are resuspended and some of these may be refiltered. When the bivalves form dense beds their biodeposition and structure may result in long term geomorphological changes in coastal ecosystems. Although calculated estimates of biodeposition from literature or laboratory rates are better than no estimates at all, rates determined from field measurements and experiments are more compelling.

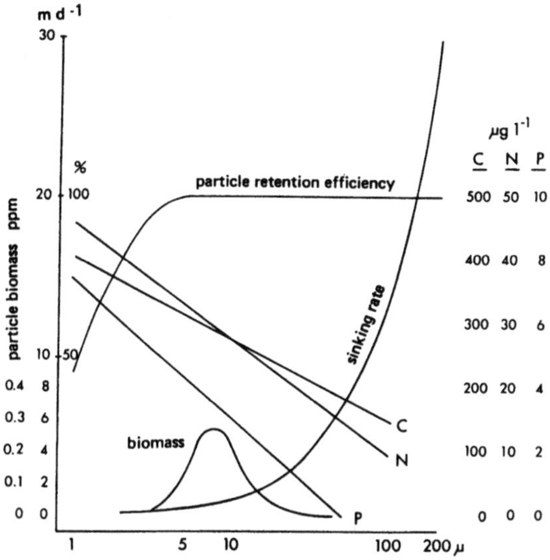

Fig. 4. Biodeposition relationships (after Kautsky and Evans 1986).

NUTRIENT PROCESSING

Many studies of bivalve filter feeding communities have reported high rates of energy flow (Dame 1976; Bahr 1976), but their most important role in estuaries and shallow coastal waters may be as nutrient processors (Kuenzler 1961; Dame et al. 1980; Kautsky 1981; Jordan and Valiela 1982). Bivalve filter feeders remove suspended particulate materials from the water column by both biofiltration and passive sedimentation. In addition to removing phytoplankton and organic materials, they also actively absorb organic compounds, i.e., amino acids, glucose, etc. (Manahan et al 1982; Wright 1982) dissolved in coastal waters. The sediments surrounding the beds are enriched by feces and pseudofeces and play a major role in nutrient processing. The bivalve communities also release materials into the overlaying tidal waters. Most of these nutrients are dissolved and result from the metabolic activities of the bivalves as well as the other organisms living in the sediments.

Nutrient Retention

The salt marsh mussel, *Geukensia demissa* (=*Modiolus demissa*), is a common resident of intertidal marsh-estuarine environments along the Atlantic coast of North America. These mollusks are unusual when compared to other filter feeding bivalves, in that, they live on the marsh mud surrounded by marsh grasses and are exposed to lower current velocities. In an early study, Kuenzler (1961) showed the importance of these filter feeders in the flow of phosphorus and energy in a Georgia salt marsh. Using laboratory filtration and deposition rates applied to field population data, Kuenzler (1961) calculated that the major effect of the marsh mussels on the ecosystem was in the removal of particulate material from the tidal water. This removal amounted to about a third of the suspended particulate phosphorus in the water over the marsh and most was deposited as feces or pseudofeces. Kuenzler (1961) speculated that the activity of the mussels in filtering suspended particulate materials might act as a retention mechanism in the marsh by reducing the export of particulate materials.

Utilizing Kuenzler's (1961)approach, Jordan and Valiela (1982) investigated the nitrogen budget of the marsh mussel in a New England salt marsh. Here again, their calculations indicated that the mussels appeared to act as a retention mechanism for nitrogen filtered from the tidal water. They speculated that if the material filtered by the mussels was exported instead of removed from the water, export nitrogen to the sea would increase several fold. This increased export would greatly alter the nitrogen budget of the salt marsh and the nature nutrient processing in salt marshes. Also, since nitrogen is thought to limit salt marsh productivity, increased retention of nitrogen due to mussel filtration may increase the productivity of salt marshes (Jordan and Valiela 1982).

The phosphorus and nitrogen budget studies of salt marsh mussels are unique in a number of ways. They are the first published efforts to estimate the release of nutrients from a bivalve filter feeding community and to note the nutrient retention potential of these organisms in ecosystems. An examination of these budgets shows that the Kuenzler (1961) budget is balanced because he determined pseudofeces production by difference from other portions of his budget. Similarly, Jordan and Valiela's (1982) budget is balanced because filtration was determined by summing biodeposition, ammonia excretion, byssal thread secretion, gamete production and growth. In the phosphorus budget, Kuenzler (1961) estimated that biodeposition of phosphorus is 94% of filtration suggesting mussels only absorb 6% of the phosphorus they filter. The nitrogen budget of Jordan and Valiela (1982), on the other hand, implies that biodeposition of nitrogen is 50% of filtration indicating mussels absorb half of the nitrogen they filter from suspension. Why is the % adsorption so different for nitrogen as opposed to phosphorus? First, the studies were done in very different systems. The phosphorus study was conducted in warm sub- tropical Georgia with very turbid waters and the nitrogen study was in cool temperate New England with a third the suspended sediment load of Georgia. The Georgia mussels should produce more pseudofeces than the New England mussels as the filtration rates are similar (Jordan and Valiela 1982). Second, both studies assumed steady state conditions to compute their budgets and balanced them by difference for a crucial component, biodeposition or biofiltration. I strongly doubt that either of these mussel systems are in steady state, because marsh-estuarine ecosystems are not mature systems, but systems in constant change. The best rule for nutrient or energy budgets is that all components be measured and no component computed by difference. In addition, direct field estimations of rates are preferable to scaled-up estimates.

Utilizing field experiments, Bertness (1984) directly observed that the presence of marsh mussels facilitates marsh grass growth and overall salt marsh stability. As a filter feeder, *G. demissa* transfers nutrients from the water column to the sediment surface in its fecal material and these additional nutrients increase the growth of the marsh grasses in the surrounding habitat. The mussels also attach to the marsh substrate with byssal threads that bind sediment and prevent the erosion and physical disturbance of the marsh. From these studies, it appears that the salt marsh mussel functions not only as a nutrient retention mechanism, but also as an enrichment mechanism to the grasses in their habitat. They may also have a major roll in the suspended particulate material budgets of nitrogen and phosphorus within the marsh-estuarine ecosystem.

Material Fluxes

The role of bivalve filter feeder beds common to channels or flats on ecosystem nutrient dynamics was first noted by Nixon et al. (1976). In their study of a small embayment, Bissel Cove, Rhode Island, they utilized data from mussels in the laboratory to speculate that abundant blue mussels might import large amounts of suspended particulate nitrogen and export large amounts of ammonia to the adjacent offshore water.

Recently, *in situ* measurements of material uptakes and releases have been conducted on bivalve beds in tidal channels. These studies have used a portable plastic tunnel technique (Dame et al. 1984), flumes (Asmus et al 1990) and domes (Boucher & Boucher-Rondoni 1988). One of the major objectives of each of these studies was to observe material fluxes across bivalve filter feeder beds in as realistic manner as possible.

The initial tunnel work was conducted on oyster reefs in South Carolina and is summarized in Dame et al. (1989). The simultaneous fluxes of various forms of carbon, nitrogen and phosphorus were measured every 10.2 days for 1 year. A regression estimation technique was used to compute annual flux estimates. A striking observation of this study was the considerable variation in material fluxes from one observed tidal cycle to the next. These authors attribute a major portion of this variability to differences in ebb and flood tidal velocity patterns. Because of the design, statistical significance was attached to the *in situ* fluxes as well as the annual fluxes and many were not significantly different from zero. In addition to the general net flux pattern of uptake of particulates and release of dissolved materials, the observed fluxes were almost always an order of magnitude greater than scaled-up estimates from laboratory data. The high material flux rates strongly suggest a major role for oysters in ecosystems where they are abundant. Comparing the ratios of C:N:P fluxes on the oyster reef indicated that proportionally more phosphorus was taken up than would be expected. Since phosphorus does not have a gaseous stage like carbon or nitrogen, the retention of this element by oyster reefs may play a crucial role in estuarine ecosystems. Although this study did provide *in situ* flux estimates, these estimates must still be scaled-up to the level of the ecosystem in order to determine their relative importance.

The portable tunnel technique has also been used to investigate the material fluxes of natural and cultured mussel beds in The Netherlands (Dame and Dankers 1988; Prins and Smaal 1990; Dame et al. 1991). The material fluxes by the mussel beds were similar to those for the oyster reefs in South Carolina. In this case, Prins and Smaal (1990) compared the sources of inorganic nutrients within the estuary and estimated that mussel beds may be important in increasing nutrient availability for phytoplankton and stimulating primary production. Dame et al (1991) determined the specific estuarine material turnover rates assuming the mussel beds were the processors. Shorter material turnover times relative to water mass turnover time implies a greater influence by the mussel beds. Both chlorophyll *a* and ammonia turned over at rates similar to the water mass in the Dutch estuaries implying a major role for mussels in nutrient cycling in these systems.

Flumes are similar to tunnels except they are generally larger, have a larger water flux and are less portable. The most successful flume study to date involved the measurement of material exchange processes over a natural mussel bed near the island of Sylt, Germany in the Eastern Wadden Sea (Asmus et al. 1990). As in the tunnel studies, observed material fluxes were variable from tide to tide. Because several consecutive tidal cycles were observed, some of the material flux variability was relatable to environmental factors. Specifically, phosphate was released on calm days at low water, while phosphate fixation appeared to occur near high water on windy days when seston concentrations were higher which may indicate absorption. Suspended particulate material was taken up on calm days and appeared to be resuspended during turbulent conditions. The C:N ratio of the water leaving the flume showed a distinct increase when mussel filter feeding was dominant. Thus the mussel bed switches between a source and sink for particulate material depending on the concentrations in the water column and wind and tide induced turbulence. These results allow us to further describe mussel beds as dynamic mechanisms within shallow waters. These systems can not only process and aid in recycling nutrients, i.e., carbon, nitrogen, phosphorus and silicon, but also switch their mode of operation depending on local conditions.

As noted earlier, the organic materials that are deposited on and around bivalve beds enrich the sediments (Kaspar et al. 1985; Kautsky and Evans 1987). Organic nitrogen can be mineralized to ammonium and this ammonium may undergo nitrification, the oxidation of ammonium to nitrate. Denitrification, the reduction of nitrate to nitrite and eventually to nitrogen gas, may also occur if the sediments are anaerobic because the increase in organic matter changes the availability of electron acceptors (Tiedje et al. 1982; Kaspar et al. 1985). With anaerobic conditions, organic phosphate may also be decomposed by bacteria into mobile orthophosphate.

Similar events appear to occur in cultivated oyster beds in France. In the large tidal range estuaries of North Brittany, Boucher and Boucher-Rodoni (1988) and Boucher-Rodoni and Boucher (1990) have utilized plastic domes to investigate the role of the oyster, *Crassostrea gigas*, in nutrient fluxes at the sediment water interface. Using relatively low densities of oysters, these authors found that the bivalves contributed only 25% to 40% to the ammonium, urea and oxygen fluxes within their chambers. They noted that the actual contribution of oysters to exchanges at the sediment-water interface rarely fits expected values from potential estimates.

The nutrients from bivalve excretion and biodeposits can have a major influence on the microalgae in the intensive culture facilities (Robert et al. 1982; Maestrini et al. 1986). They found that the microalgae seem to have adapted to the high concentrations of ammonium, nitrate and organic nitrogen. Normally high concentrations of ammonium inhibit the uptake of other nitrogen sources by algae, but the threshold of ammonium inhibition is much higher in these algae than pelagic or neritic species, allowing them to assimilate other sources of nitrogen, such as nitrate and urea, simultaneously with ammonium. Thus the extreme environmental change wrought the cultivation of oysters appears to be within the physiological flexibility of their food source the microalgae.

It is evident from comparative studies using bags (Kautsky and Wallentinus 1980), domes (Boucher and Boucher-Rodoni 1988) and areas of intense cultivation (Kaspar et al. 1985; Maestrini et al. 1986) that both the bivalve filter feeders, the surrounding sediments and the water column are involved in the regulation of material fluxes at the sediment-water interface. These are circular causal networks and bivalves are a major component in their operation.

Bivalve filter feeders probably induce a shortened cycle of nutrients between the benthos and the phytoplankton. The bivalves consume phytoplankton from the water column, release ammonium and orthophosphate as products of metabolism, where these nutrients can be utilized by the phytoplankton. These bivalves are also part of a longer nutrient cycle that includes the nutrient regeneration processes taking place in the sediments surrounding the beds (Fig.5). Thus, bivalve filter feeding beds function as material processors in estuarine ecosystems (Dame et al. 1984) and move large quantities of nutrients from the water column to the sediments and the bodies of bivalves (Kaspar et al. 1985).

The structure of food webs and the control of the food webs themselves have been implicated in the nature of bivalve involvement in nutrient processing (Menge 1992). In comparing two very similar bays along the Oregon coast, Menge (1992), he found one system with an abundant macrophyte population, but with only small numbers of filter feeders, herbivorous fish, carnivores and phytoplankton. The other bay had few benthic macrophytes, but abundant phytoplankton, filter feeders and larger macrofauna. Menge (1992) hypothesized from rather incomplete ecosystem data that bottom-up food web forces should be greater in the system with higher nutrient concentrations. He speculated that top-down and bottom-up factors are tightly linked in the rocky intertidal zone of the Oregon coast where *M. edulis* is dominant and this linkage accounts for much of the macrobenthic variation in these systems. As Hunter and Price (1992), Menge (1992) contended that both top-down and bottom-up forces are at work simultaneously. Future studies concerning food web controls need to be expanded in time and space with an increasing emphasis of factorial approaches and comparative field experiments utilizing different systems should be productive.

The recent comprehensive studies of material fluxes in shallow coastal ecosystems strongly imply that both top-down and bottom-up influences occur. Whether the material flows in these bivalve systems are controlled through a cybernetic mutualism or through the simultaneous operation of top-down and bottom-up processes are not answerable at this time. But it is clear that full-scale ecosystem studies that consider broad temporal and spatial scales will help to clarify our understanding of these questions.

MESOCOSM STUDIES

One approach to simulating large complex estuarine ecosystems is the utilization of mesocosms or actual physical and biological replication of a system on a small scale. The original estuarine mesocosms dealing with bivalve beds were nothing more than large aquaria or ponds (Odum et al. 1963, Bahr and Lanier 1981) which attempted to reproduce specified interesting features of the environment.

The most sophisticated mesocosms used in bivalve research are the Marine Ecosystem Research Laboratory (MERL) tanks that simulate the estuarine ecosystem in Narragansett Bay, Rhode Island (Doering et al. 1986 and Doering et al. 1987). Utilizing a number of these 13-m^3 tanks in an experimental design, the latter authors were able to determine the influence of the clam, *Mercenaria mercenaria*(L.), on system production, sedimentation and benthic remineralization. In general, system production, net and gross sedimentation of particulate carbon and benthic remineralization of dissolved inorganic nitrogen were all greater in mesocosms with clams. In spite of these increases, the clams did not limit phytoplankton biomass in the water column through filtration. The authors argued that production increases were due to the greater flux of nutrients from the benthos. In this same study, Doering and Oviatt (1986) examined a number of prominent bivalve filtration rate models in relation to their ^{14}C filtration data in the mesocosms. They concluded that filtration rates measured under optimal conditions in the laboratory do not apply well to natural systems and may exaggerate the role of filter feeding bivalves to remove suspended particulate materials. Jørgensen (1990) deems this conclusion unwarranted, but does not substantiate his claim or refute Doering and Oviatt (1986). This mesocosm study appears to be another example of the danger scaling-up from individual organism measurements to ecosystem level estimates.

As an alternative to duplicating the complexities of an estuarine ecosystem in mesocosms, a number of studies have utilized enclosures, usually large plastic bags, to determine the role of bivalves in coastal systems (Kautsky and Wallentinus 1980, Kautsky & Evans 1987, Riemann et al. 1988, Horsted et al. 1988). These systems also capture a small portion of the real estuarine environment and utilize replicates in an experimental design. However, as with the mesocosms, they are troubled with circulation and edge effects due to the walls of the chambers. The Danish studies (Riemann et al. 1988, Horsted et al. 1988) are notable in that the food web below and above the bivalve filter feeder, mussels, were examined. They found that the qualitative and quantitative development of the zooplankton community in the enclosures was controlled from the top by fish and from the bottom by mussels. They also found that phytoplankton biomass within the enclosures was a function of nutrient input, phytoplankton size structure, and the physiological state of the mussels. These experimental results appear to support the simultaneous functioning of top-down and bottom up controls within a system dominated by bivalve filter feeders. In summary, the key utility of estuarine enclosures and mesocosms has been the ability to test hypotheses in a replicative and experimental manner rather than depending on inferences from correlative observations of the environment.

ESTUARINE ECOSYSTEM EXPERIMENTS

Because of the complexity of ecosystems in general and estuaries in particular and the drawbacks of mesocosms and enclosures, another approach has recently been utilized which involves the experimental manipulation of estuaries or portions of estuaries. In on-going studies at North Inlet estuary near Georgetown, SC, I have compared the nutrient concentrations in tidal creeks with oysters to creeks with oysters removed and found that significant differences exist between the controls and the experimental creeks. In this experiment, the amount of oyster biomass to water volume in each creek was standardized to that of the North Inlet system of 3.8 g/m^3. The bank-full volume of water in each creek was determined by hypsometric characterization (Eiser and Kjerfve 1986). In a pre-test involving daily sampling for a month, the six creeks were found to be biogeochemically similar in nutrient concentrations. After the oysters were removed from three creeks, nutrient concentrations and chlorophyll *a* were collected daily for two months. Total nitrogen, ammonium, total phosphorus and dissolved phosphorus were found to be significantly different ($P<0.05$) in creeks with oysters as compared to creeks without oysters (Table 1). All other constituents, including chlorophyll *a*, were not significantly different. Although this approach test a specific hypothesis at the estuarine ecosystem level of organization and avoids some of the problems of mesocosms and models, care must be taken to avoid manipulation effects which may confound the interpretation of the observations.

CONCLUSIONS

It is evident from this discussion that bivalve filter feeders play a dynamic role in the structure and function of coastal ecosystems (Fig. 5). While many of their roles are speculative and lack direct evidence, others are well observed. Structurally, they influence creek geomorphology, sediment texture and food web structure. Functionally, they have the potential to influence phytoplankton and seston concentrations in the water column and there by change ecosystem structure from planktonic to benthic dominance. In addition, dense populations of bivalves transfer large quantities of materials from the water column to the sediments where sediment biogeochemistry can

Table 1. Mean (standard error) concentrations of materials in tidal creeks during experimental manipulation from 1 July to 30 August, 1989 for creeks with and without oysters. Number of observation days =60.

Constituent	Oysters	Oysters Removed	Sig.
Chlorophyll *a* (mg/m^3)	8.30(0.27)	8.31(0.29)	
Ammonium (ugh-at N/l)	1.25(0.10)	0.90(0.06)	*
Nitrite+Nitrate (ugh-at N/l)	0.37(0.02)	0.35(0.02)	
Total Dissolved Nitrogen ugh-at N/l)	22.16(0.55)	21.78(0.50)	
Particulate Nitrogen ugh-at N/l)	16.07(0.68)	15.27(0.71)	
Total Nitrogen (ugh-at N/l)	38.22(0.80)	36.94(0.77)	*
Orthophosphate (ugh-at P/l)	0.40(0.01)	0.39(0.01)	*
Total Dissolved Phosphorus (ugh-at P/l)	0.69(0.03)	0.62(0.02)	*
Particulate Phosphorus (ugh-at P/l)	1.08(0.05)	1.05(0.05)	
Total Phosphorus (ugh-at P/l)	1.77(0.06)	1.67(0.05)	*

*Difference between creek type material concentrations significant at $P<0.05$ level.

be dramatically altered. Finally, these organisms seem to play a central role in the processing of nutrients in shallow coastal waters where they speed up nutrient cycling and generate nutrient retention.

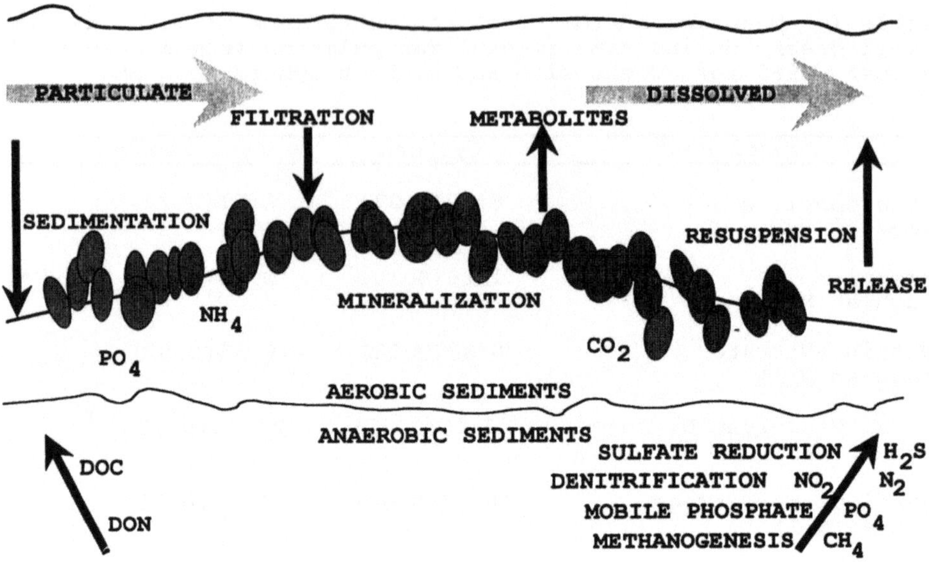

Fig. 5. A conceptual summary of the material processing roles played by bivalve filter feeders in coastal waters.

REFERENCES

Asmus H (1987) Secondary production of an intertidal mussel bed community related to its storage and turnover compartments. Mar Ecol Prog Ser 39:252-266

Asmus H Asmus RM Reise K (1990) Exchange processe in an intertidal mussel bed: a Sylt-flume study in the Wadden Sea. Ber Biol Anst Helgoland 6:1-79

Bahr LM (1976) Energetic aspects of the intertidal oyster reef community at Sapelo Islnad, Gerogia. Ecology 57:121-131

Bahr LM Lanier WP (1981) The ecology of intertidal oyster reefs of the South Atlantic coast: a commuunity profile. US Fish Wldf Ser Washington DC

Baird D Milne H (1981) Energy flow in the Ythan Estuary, Aberdeenshire, Scotland. Est Coast Shelf Sci 13:455-472

Bernard FR (1974) Annual biodeposition and gross energy budget

of mature Pacific oysters, *Crassostrea gigas*. J Fish Res Bd Canada 31:185-190
Bertness MD (1984) Ribbed mussels and the productivity of *Spartina alterniflora* in a New England salt marsh. Ecology 65:1794-1807
Boucher G Boucher-Rodoni R (1988) Insitu measurement of respiratory metabolism and nitrogen fluxes at the interface of oyster beds. Mar Ecol Prog Ser 44:229-238
Boucher-Rodoni R Boucher G (1990) *In situ* study of the effect of oyster biomass on benthic metabolic exchange rates. Hydrobiol 206:115-123
Dame RF (1976) Energy flow in an intertidal oyster population. Est Coast Mar Sci 4:243-253
Dame RF (1987)The net flux of inorganic sediments by an intertidal oyster reef. Cont Shelf Res 7:1421-1424
Dame RF Dankers N (1988) Uptake and release of materials by a Wadden Sea mussel bed. J Exp Mar Biol Ecol 118:207-216
Dame RF Dankers N Prins T Jongsma H Smaal A (1991) The influence of mussel beds on nutrients in the Western Wadden Sea and Eastern Scheldt estuaries. Estuaries 14:130-138
Dame RF Spurrier JD Wolaver TG (1989) Carbon, nitrogen and phosphorus processing by an oyster reef. Mar Ecol Prog Ser 54:249-256
Dame RF Zingmark RG Haskin E (1984) Oyster reefs as processors of estuarine materials. J Exp Mar Biol Ecol 83:239-247
Deslous-Paoli JM Sornin JM Heral M (1987) Variations saisonnieres *in situ* de la production et de la composition des biodepots de trois mollusques estuariens (*Mytilus edulis, Crassostrea gigas, Crepidula fornicata*). Haliotis 16:233-245
Dinet A Sornin J-M Sabliere A Delmas D Feuillet-girard (1990) Influence de la biodeposition de bivalves filtreurs sur les peuplements meiobenthiques d'un mariais maritime. Cah Biol Mar 31:307-322
Doering PH Kelly JR Ovatt CA Sowers T (1987) Effects of the hard clam *Mercenaria mercenaria* on benthic fluxes of inorganic nutrients and gases. Mar Biol 94:377-383
Doering PH Oviatt CA (1986) Application of filtration rate models to field populations of bivalves: an assessment using experimental mesocosms. Mar Ecol Prog Ser 31:265-275
Doering PH Oviatt CA Kelly JR (1986) The effect of the filter feeding clam *Mercenaria mercenaria* on carbon cycling in experimental mesocosms. J Mar Res 44:839-861
Eiser W Kjerfve B (1986) Marsh topograpy and hypsometric characteristics of a South Carolina salt marsh basin. Est Coast Shlef Sci 23:595-605
Fretwell SD (1987) Food chain dynamics: the central theory of ecology? Oikos 50:291-301
Hairston NG Smith FE Slobodkin LB (1960) Community structure, population control, and competition. Am Nat 94:421-425
Hunter MD Price PW (1992) Playing chutes and ladders: Heterogeneity and the relative roles of bottom-up and top-down forces in natural communities. Ecology 73:724-732
Horsted SJ Nielsen TG Riemann B Pock-Stee J Bjornsen PK (1988) Regulation of zooplankton by suspension -feeding bivalves and fish in estuarine enclosures. Mar Ecol Prog Ser 48:217-224

Jordan TE Valiela I (1982) A nitrogen budget of the ribbed mussel *Geukensia demissa* and its significance in nitrogen flow in a New England salt marsh. Limnol Oceanogr 27:75-90

Jørgensen CB (1990) Bivalve filter feeding: Hydrodynamics, bioenergetics, physiology and ecology. Olsen & Olsen, Fredensborg, Denmark

Juday C (1940) The annual energy budget of an inland lake. Ecology 21:438-450

Kaspar HF Gillespie PA Boyer IC MacKenzie AL (1985) Effects of mussel aquaculture on the nitrogen cycle andbenthic communities of Kenepuru Sound, Marlborough Sounds, New Zealand. Mar Biol 85:127-136

Kautsky N (1981) On the trophic role of the blue mussel (*Mytilus edulis*) in a Baltic coastal ecosystem and the fate of the orgnaic matter produced by the mussels. Kieler Meeresforsch 5:454-461

Kautsky N Evans S (1987) Role of biodeposition by *Mytilus edulis* in the circulation of matter and nutrients in a Baltic coastal ecosystem. Mar Ecol Prog Ser 38:201-212

Kautsky N Wallentinus I (1980) Nutrient release form a Baltic *Mytilus*-red algae community and its role in benthic and pelagic productiivty. Ophelia Supp. 1:17-30

Kuenzler EJ (1961) Phosphorus budget of a mussel population. Limnol Oceanogr 6:400-415

Lindeman RL (1942) The trophic-dynamic aspect of ecology. Ecology 23:399-418

Maestrini SY Robert J-M Lefley JW Collos Y (1986) Ammonium thresholds for simultaneous uptake of ammonium and nitrate by oyster-pond algae. J Exp Mar Biol Ecol 102-75-98

Manahan DT Wright SH Stephens GC Rice MA (1982) Trans;ort of dissolved amino acids by the mussel, *Mytilus edulis*: demonstration of net uptake form natural seawater. Science 215:1253-1255

Menge, B (1992) Community regulation: under what conditions are bottom-up factors important on rocky shores? Ecology 73:755-765

Möbius K (1877) Die Auster und die Austerwirthschaft. Wiegundt, Hempel & Parey, Berlin

Newell RIE (1988) Ecological changes in Chesapeake Bay: Are they the result of overharvesting the American oyster, *Crassostrea virginica*? Chesapeake Res Consor Pub 129:536-546

Nixon SW Oviatt CA Garber J Lee V (1976) Diel metabolism and nutrient dynamics of a salt marsh embayment. Ecology 57:740-750

Odum EP (1953) The Fundamentals of Ecology. Saunders, Philadelphia

Odum EP (1983) Basic Ecology. Saunders, Philadelphia

Odum EP Biever LJ (1984) Resource quality, mutualism, and energy partitioning in food chains. Am Nat 124:360-376

Odum HT Siler WL Beyers FJ Armstrong N (1963) Experiments in engineering marine ecosystems. Pub Inst Mar Sci Texas 9:393-404

Officer CB Smyda TJ Mann R (1982) Benthic filter feeding: a natural eutrophication control. Mar Ecol Prog Ser 9:203-210

Oksanen L (1988) Ecosystem organization: mutualism and cybernetics or plain Darwinian struggle for existence? Am

Nat 131:424-444
O'Neill RV DeAngelis DL Waide JB, Allen TFH (1986) A Hierarchical Concept of Ecosystems. Princeton U. Press, Princeton
Prins TC Smaal AC (1990) Benthic-pelagic coupling: the release of inorganic nutrients by an intertidal bed of *Mytilus edulis*. In Barnes M and Gibson RN (eds) Trophic Relationships in the Marine Environment Aberdeen U Press, Aberdeen
Redfield AC (1958) The biological control of chemical factors in the environment. Am. Scientist 46:205-221
Riemann B Nielsen TG Horsted SJ Bjornsen KP Pock-Steen J (1988) Regulation of phytoplankton biomass in estuarine enclosures. Mar Ecol Prog Ser 48:205-215
Rodhouse PG Roden CM (1987) Carbon budget for a coastal inlet in relation to intensive cultivation of suspension-feeding bivalve molluscs. Mar Ecol Prog Ser 36:225-236
Robert JM Maestrini SY Heral M Zanette Y (1982) Production des micro-algues des claires ostreicoles en relation avec l'azote orgnique dissous excrete par les huitres. Oceanol Act 1989:389-395
Smaal AC Verhagen HG Coosen J Haas HA (1986) Interaction between seston quantity and quality and benthic suspension feeders in the Oosterschelde, The Netherlands. Ophelia 26:285-399
Sornin JM Feuillet M Heral M Deslous-Paoli (1983) Effet des biodeposts de l'huitre *Crassostrea gigas* (Thunberg) sur h'accumulation de matires organiques dan les parcs du bassin de Marennes-Oleron. J Moll Stud (Suppt) 12A:185-197
Sornin JM Feuillet M Heral M Fardeau J.-C. (1986) Influence des cultures d'huitres *Crassostrea gigas* sur le cycle du phosphore en zone intertidale: role de la biodeposition. Oceanol Acta 9:313-322
Tiedje JM Sexstone AJ Myrold DD Robinson JA (1982) Denitrification: ecological niches, competition and survival. Antonie van Leeuwenhoek 48:569-583
Ulanowics RE Tuttle JH (1992) The trophic consequences of oyster stock rehabilitation in Chesapeake Bay. Estuaries 15:257-265
Vadas RL (1989) Food web patterns in ecosystems: a reply to Fretwell and Oksanen. Oikos 56:339-343
Verwey J (1952) On the ecology and distribution of cockles and mussels in the Dutch Wadden Sea, their role in sedimentation and the source of their food supply. Arch Neerland Zool 10:172-239
Wright SH (1982) a nutritional role for amino acid transport in filter feeding marine invertebrates. Am Zool 22:621-634

THE UPTAKE OF ORGANIC MATTER AND THE RELEASE OF INORGANIC NUTRIENTS BY BIVALVE SUSPENSION FEEDER BEDS

A.C. Smaal & T.C. Prins[1]
Ministry of Transport and Public Works
Tidal Waters Division
P.O. Box 8039
4330 EA Middelburg, The Netherlands

[1]Netherlands Institute of Ecology
Centre for Estuarine and Coastal Ecology
Vierstraat 28
4401 EA Yerseke, The Netherlands

INTRODUCTION

In many estuarine and coastal ecosystems, bivalve suspension feeders, such as mussels, oysters, cockles and various species of clams, occur in high densities. They are often concentrated in beds and reefs (Wolff 1983). Feeding is performed by pumping and filtering large volumes of water through the gills, and results in a continuous flux of particulate matter from the water column to the bivalve beds. Particles of a size larger than 2 µm in diameter are effectively retained by most species (Jørgensen 1990). In various ecosystems, depletion of organic matter, bacteria and phytoplankton in the overlying water, due to the filtration activity of bivalves, has been observed (Wright et al 1982; Carlson et al 1984; Fréchette and Bourget 1985; Nichols 1985; Peterson and Black 1987; Navarro et al 1991). Observations in San Francisco Bay gave rise to the hypothesis that benthic filter feeding kept the algal biomass at a low level (Cloern 1982; Alpine and Cloern 1992), and could act as eutrophication control (Officer et al 1982). This idea was based on extrapolation of individual clearance rates, taken from the literature, to the total Bay volume.

The potential of bivalve beds to filter considerable quantities of particulate matter from the water column, has been

NATO ASI Series, Vol. G 33
Bivalve Filter Feeders
in Estuarine and Coastal Ecosystem Processes
Edited by Richard F. Dame

confirmed by *in situ* measurements of fluxes on bivalve beds, as developed by Dame et al (1984). Comparison of *in situ* filtration rates of chlorophyll-a by mussels, with laboratory derived data on clearance rates, showed good similarity (Prins and Smaal, 1990, in prep). Moreover, the *in situ* measurements revealed high release rates of inorganic nutrients from the bivalve beds to the water column (Dame and Dankers 1988; Dame et al 1989; Prins and Smaal 1990; Asmus et al 1990, 1992; Zurburg et al 1993). These rates were generally higher than direct excretion rates of nutrients by the bivalves, as measured in the laboratory. The *in situ* measurements are methodologically different from the laboratory measurements performed with individuals, because in the field the fluxes from a bivalve bed as a whole, including sediment and biodeposits, are measured. In addition to direct excretion of nutrients the formation and subsequent mineralization of pseudofeces and feces can explain the higher nutrient release rates. Pseudofeces and feces are biodeposited in bivalve beds and increase sedimentation (Verwey 1952; Haven and Morales-Alamo 1966; Biggs and Howell 1984). Accumulation of biodeposits in mussel beds and oyster reefs is affected by local hydrographic conditions (Fréchette and Bourget 1985; Dame et al 1989), storm events (Asmus et al 1990) and, in case of aquaculture, human activities. Long term accumulation of biodeposits has been observed under suspended culture of bivalves. The biodeposits may influence the benthic community (Tenore et al 1982; Comitto and Boncavage 1989; Dittmann 1990; Dinet et al 1990) and shows high bacterial activity (Dahlbäck and Gunnarson 1981; Kaspar et al 1985; Grenz et al 1990, 1992). The nutrients regenerated by the mineralization of biodeposits may come available again for primary production and the bivalves may thus stimulate the phytoplankton turnover (Asmus and Asmus, 1991).

In conclusion, the *in situ* measurements have shown uptake of particulate organic matter and release of dissolved inorganic nutrients by bivalve beds and consequently enhanced regeneration of inorganic nutrients. This may result in eutrophication control as well as promotion of primary production. The magnitude of bivalve suspension feeders as regulative mechanisms in estuarine ecosystems (Dame 1993), is an important and still open question, which is both of scientific and management interests (Dankers 1993). In this paper data on fluxes of particulate and dissolved material through bivalve suspension feeder beds will be reviewed, with emphasis on the results of *in situ* measurements. These measurements provide a quantitative basis for estimates of the role of bivalves in the turnover of particulate and dissolved material. Some of the consequences for estuarine ecosystems will be discussed.

UPTAKE AND RELEASE OF MATERIAL BY BIVALVE BEDS

Filtration

Biomass, standing stocks and clearance rates of bivalve suspension feeders are shown in Table 1 for a variety of estuarine ecosystems. For comparison of these ecosystems the filtration activity or clearance rate (volume filtered to retain all material) is calculated on the basis of laboratory measurements reported in literature and extrapolated to the field situation and ecosystem scale. To relate filtration activity to the size of various systems, clearance is also calculated as particle clearance time, which is the time needed to filter particles from a volume equivalent to the total system volume. Clearance rate and time are potential values assuming complete mixing of the system. Biomass is expressed as gram per volume (total amount of water at mid-tide) to compare various systems with respect to the impact of the suspension feeders, which by definition depend on the water column for the uptake of material.

In many coastal ecosystems, clearance time is shorter than the residence time of the water body, as shown in Fig. 1. A clearance time in the range of 2-4 days is found in the western Wadden Sea, Oosterschelde, Bay of Brest, Marennes-Oléron Bay and Thau Lagoon. In these systems the bivalve biomass per volume is approximately 2-8 g ash-free dry weight (ADW) m^{-3}. San Francisco Bay, with a biomass of 2.5 g ADW m^{-3}, seems to be an outlier, having a clearance time of 0.7 days and a residence time of 11 days. Indeed, Doering and Oviatt (1986) showed that the filtration rates calculated by Cloern (1982) are probably overestimated. North Inlet and the Sylt area are small and open systems; the residence time is 1 day or less; the clearance time is 0.7 and 2.1 days respectively. This apparently coincides with a bivalve biomass exceeding the level of 2-8 g ADW m^{-3}. In areas with a

Table 1. Mean biomass, standing stock, daily clearance (CL) rate and clearance time (time to filter a volume equivalent to total volume in days) of bivalve suspension feeders in various estuarine ecosystems in comparison with residence time. Biomass per m^2 of total surface; biomass per m^3 of total volume. Biomass and standing stock in ash-free dry weight (ADW) unless otherwise indicated.

area	total vol. $10^6 m^3$	aver. depth m	biomass g m^{-2}	g m^{-3}	stand. stock $10^6 g$	CL rate $l g^{-1} d^{-1}$	CL rate $m^3 m^{-2} d^{-1}$	CL time d	res. time d
SFBay[1]	2500	6	15	2.5	6255	600	9.0	0.7	11.1
NI[2]	22[a]	2.5	10.6[b]	15.4	338	86	3.2	0.7	1.0
Sylt[3]	7.2	1.8	39	21.7	156	22.5	0.9	2.1	0.5
Askö[4]	4000	25	9.4[b]	0.4	1500	26.9	0.3	99	10^4
WSw[5]	4020	2.9	10.4	3.7	14700	48	0.5	5.8	10
OS[6]	2740	7.8	24.3	3.1	8509	87	2.1	3.7	40
Brest[7]	1480	10	74	7.4	10952	48	3.5	2.8	16.7
M-O[8]	675	5	21	4.2	2850	86.4	1.8	2.7	7.1
Thau[9]	265	5	15.9	3.0	800	120	1.8	2.8	-
Arosa[10]	4335	19	30	1.6	6900	51	1.5	12.4	23
Killary[11]	102	14	2.1	0.2	15.5	48	0.1	140	-

a: tidal prism; b: biomass in dry flesh weight.
1: South San Fransisco Bay,CA (USA), various species, after Cloern (1982); 2: North Inlet,SC (USA), oyster reef, after Dame et al (1980); 3: Sylt, Königs-hafen area (G), mussels, after Asmus et al (1990); 4: Askö area, Baltic Sea (S), mussels, after Kautsky & Wallentinus (1980) and Kautsky & Evans (1987); 5: Western Wadden Sea (NL), mussels, after Dame et al (1991); 6: Oosterschelde (NL), mussels (54%) and cockles (46%), after Dame et al (1991) and Smaal et al (1986); 7: Bay of Brest (F), various species, after Hily (1991); 8: Marennes-Oléron Bay (F), mainly oysters, after Héral et al (1988) and Bacher (1989); 9: Thau lagoon (F), oysters, after Deslous-Paoli (pers.comm.); 10: Ria de Arosa (Sp), suspended mussels, after Otto (1975), Tenore et al (1982), Pérez Camacho et al (1991) and Navarro et al (1991); 11: Killary Harbour (IR), suspended mussels, after Rodhouse & Roden (1987).

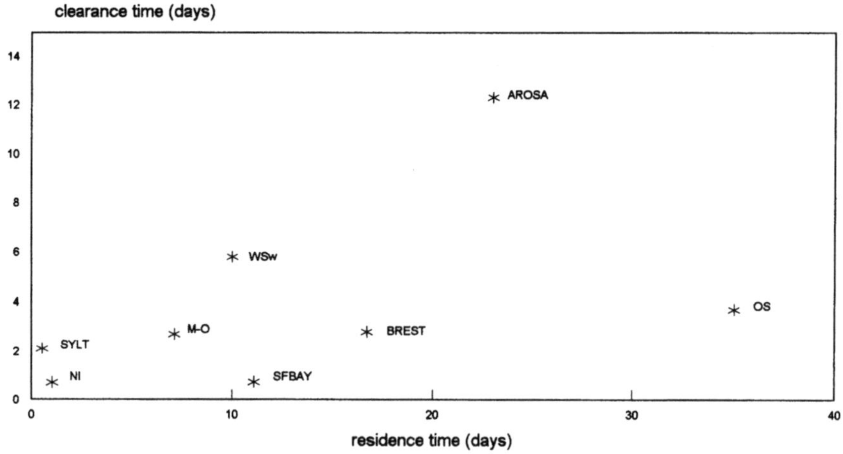

Fig. 1. Residence time of various estuarine ecosystems as compared with clearance time, e.g., the time needed to filter a volume equivalent to the total volume by the bivalve filter feeders (see references in Table 1)

low rate of water renewal (Askö area, Ria de Arosa), lower biomass per volume values are found. The bivalves in Killary harbor are mainly the result of small scale cultivation of mussels on ropes. The clearance rate per m^2 is relatively small and the clearance time is long. Thus, the impact of the mussels on the cycling of material in the bay is limited, as shown in the carbon budget by Rodhouse and Roden (1987) and the main herbivores are zooplankton.

We suggest that, assuming complete mixing and depending on total volume and residence time of the water mass, the impact of the bivalves extends to various scales. These scales are the scale of the bivalve beds, the scale of the defined estuary or bay area, and at the scale of the land-ocean interface. When the biomass per volume is low, the impact of filtration by the bivalves will be limited to the scale of the bivalve communities (Killary Harbor) and the impact on the ecosystem will be relatively small, unless the residence time is very long (Askö area). When the biomass per volume is in the range of ca. 2-8 g ADW m^{-3}, residence times often are longer than clearance times and it seems likely that bivalves have an impact at the scale of the whole ecosystem. A higher biomass per volume is found (North

Inlet, Sylt) when the bivalves occur in areas with short residence times, and the populations seem to be part of an ecosystem with a spatial scale larger than the defined area.

Uptake of organic matter

Net uptake rates of suspended particulate matter (SPM), particulate organic carbon, nitrogen and phosphorus (POC, PON and POP), and phytoplankton-carbon (estimated from chlorophyll-a) by bivalve beds are shown in Table 2, as estimated from *in situ* measurements with benthic ecosystem tunnels (North Inlet, western Wadden Sea, Oosterschelde and Marennes-Oléron), or with the Sylt flume. In the Sylt flume phytoplankton fluxes were measured by cell counts. In the Ria de Arosa samples were taken under mussels, cultivated on ropes. *In situ* measurements with tunnels and flumes reflect the dynamics of seston in the field, as is shown by the large variability of fluxes of suspended particulate

Table 2. Mean net daily uptake rates (and standard error of the mean) of seston components by bivalve suspension feeders, and seston composition (with ranges) from *in situ* measurements in various estuarine ecosystems; values are corrected for control measurements. Phytoplankton-C is calculated from C-Chlorophyll-a ratio = 30, except for Sylt where direct estimates of Phyto-C were made via cell counts. n=number of tidal cycles measured.

area	Seston fluxes				Seston composition			
	SPM $g\ m^{-2}d^{-1}$	POC	PON	Phyto-C	SPM $g\ m^{-3}$	%C	C/N	Phyto-C
NI[1]	-	9.36	0.61	-	-	-	-	-
		(5.2)	(0.54)					
Sylt[2]	81.0	6.06	1.83	3.02	30	3.3	8.3	0.09
	(21.2)	(9.4)	(0.82)	(1.72)	(17/202)	(1.5/3.7)		(0.003/0.2)
WSw[3]	234	6.00	-	2.48	40	3.8	-	0.24
	(174)	(8.1)		(0.9)				
OS[4]	81.8	9.0	2.67	6.0	20	5.0	10	0.15
	(22.3)	(4.6)	(0.96)	(6.6)				
M-O[5]					150	1.3	10	0.15
M.e.	2800	64.0	6.20	3.54				
	(51)	(11)	(1.4)	(2.6)				
C.g.	546	4.6	1.80	3.44				
	(400)	(10)	(0.1)	(1.9)				
Ar[6]	520	186.5	33.6	40.6	1.3	24	7.8	0.05

1: North Inlet (USA), oyster reef 196 $gDFWm^{-2}$ (n=33), Dame et al (1989) ; 2: Sylt, Königshafen (G), mussel bed, 1561 $gADWm^{-2}$ (n=5 except for N: n=10), Asmus et al (1990) ; 3: Western Wadden Sea (NL), mussel bed, 943 $gADWm^{-2}$ (n=4), Dame et al (1991) ; 4: Oosterschelde (NL), mussel bed, 1555 $gADWm^{-2}$, Prins & Smaal (1990, 1993) (n=8); 5: Marennes - Oléron Bay (F), *Mytilus edulis* (M.e.) beds on the bottom or on tables:722 $gADWm^{-2}$ (n=6), *Crassostrea gigas* (C.g.) on tables: 144 $gADWm^{-2}$ (n=4), Zurburg et al (1993); 6: Ria de Arosa (Sp), mussels on rafts, 8130 g $ADWm^{-2}$ (n=?), Pérez Camacho et al (1991).

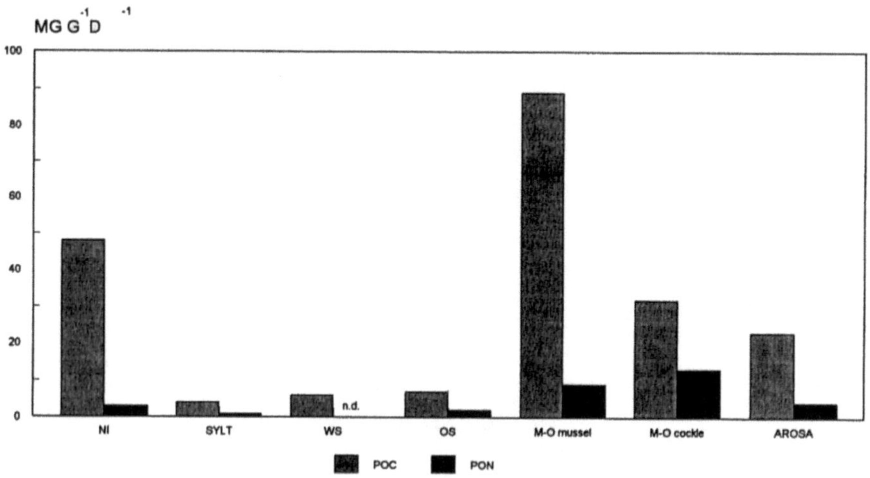

Fig. 2. Daily uptake rates of POC and PON per g ADW in various estuaries, from *in situ* measurements (see references in Table 2).

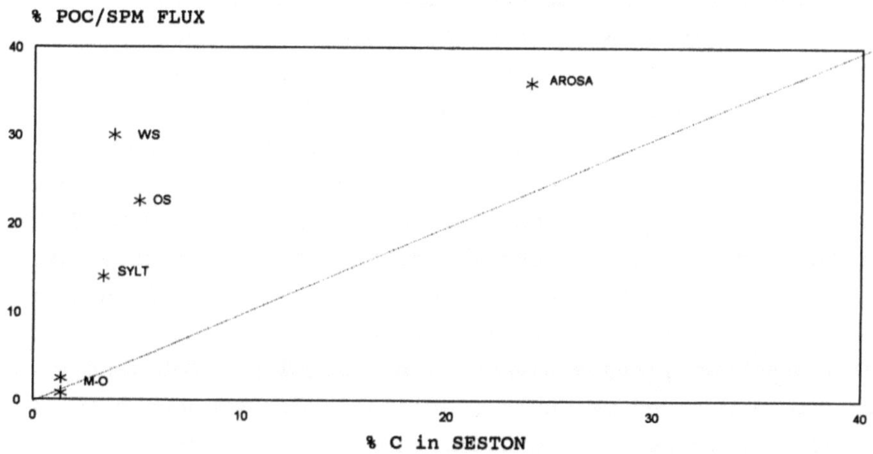

Fig. 3. Particulate organic carbon fluxes as a fraction of total suspended matter fluxes to bivalve filter feeder beds, in comparison with the carbon content of seston in various estuaries.

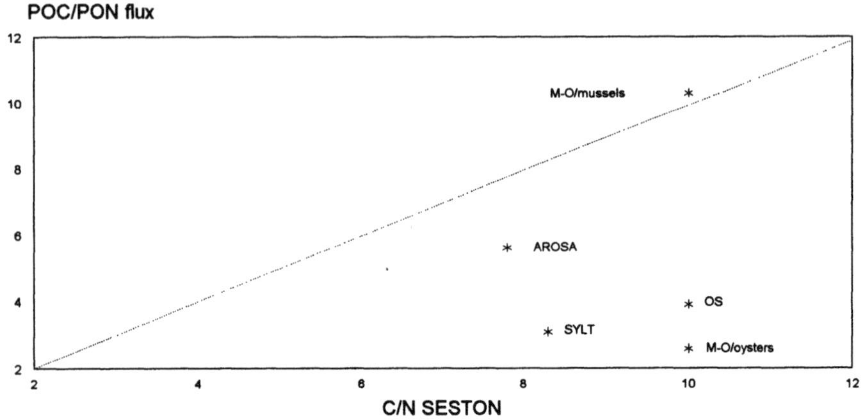

Fig. 4. Ratios of POC and PON fluxes to bivalve filter feeder beds in comparison with C/N ratios (by weight) of seston in various estuaries.

matter (Table 2). They depend largely on the SPM concentration in the water column, and the biomass of the bivalves. The POC uptake rates show high values in the raft culture system, where mussels are concentrated, although the POC concentrations are relatively low. High POC uptake rates are also recorded on mussel populations in Marennes-Oléron bay. In this area bivalves are cultivated on the sediment or on tables, elevated 20-50 cm above the sediment to prevent siltation. The POC fluxes are comparable for the other areas shown in Table 2, ranging from 4.6 to 10.4 g $m^{-2}d^{-1}$ for oysters in Marennes-Oléron and mussels in the Oosterschelde, respectively. In Fig. 2 the POC (and PON) fluxes are expressed per unit of biomass; differences between areas are partly related to differences in concentrations in the water column.

Net uptake rates of POC are higher than might be expected from the carbon content of seston; there is a selective retention of POC over inorganic material on the bivalve beds. This is probably the result of relatively higher resuspension of inorganic material, while POC is retained by the bivalve bed. By comparing the carbon content of seston with the POC flux as a fraction of the total seston flux, it is shown in Fig. 3 that in almost all cases the relative amount of carbon that is retained on the bivalve beds is higher than the carbon content of seston.

The PON uptake rates vary from 0.6 to 33.6 $gm^{-2}d^{-1}$; the variations coincide with variations in POC fluxes, except for North Inlet and the Marennes-Oléron mussels, which show a relatively low PON uptake. In Fig. 2 the PON uptake is expressed per g biomass, ranging from 2 - 12.5 mg per day. Compared with the nitrogen and carbon content of seston, expressed as C/N ratio of seston, the net PON retention on the bivalve beds is higher than the POC retention. In Fig. 4 this is shown by comparing the C/N ratio of seston with the uptake rates of POC and PON, expressed as POC/PON flux.

The phytoplankton uptake rates vary from 2.4 - 6 gC $m^{-2}d^{-1}$ except for the Ria the Arosa, which shows a much higher value that is related to the relatively high mussel density: the flux per gADW is quite similar to the other areas (not shown). As shown in Table 2 the phytoplankton-C flux constitutes in most cases about 50 % of the total net POC flux.

The uptake rates of particulate material show variations related to seston concentrations in the water column, species of bivalves, biomass and other factors such as local climatic conditions. However, in all cases the bivalves create large net particle fluxes from the pelagic to the benthic system. In most cases the net POC fluxes are relatively high, and the PON fluxes are even higher, both in comparison with the seston composition. The relatively high uptake rates of phytoplankton support this observation. The high quality of the net flux is considered to be a consequence of resuspension of biodeposits, predominantly pseudofeces, and selective ingestion of phytoplankton (Asmus and Asmus, this volume).

Filtration pressure

Relative to the primary production, the uptake of phytoplankton exceeds the primary production per m^2. The area required to produce the daily flux of phytoplankton to a m^2 of bivalve bed is presented in Table 3. In most cases 1 unit of bivalve bed takes up phytoplankton produced per day in an area of 11 - 16 x that size. In Ria de Arosa this factor is much higher, which might be due to underestimation of the primary production

(Table 3). In the Wadden Sea the phytoplankton flux per m^2 to the mussel beds is relatively low in comparison with the phytoplankton concentration (Table 2).

Table 3. Primary production and phytoplankton flux to bivalve beds, required production area in m^2 per m^2 bivalve bed, and filtration pressure (phyto-plankton uptake as a fraction of primary production). (Phytoplankton concentration in growing season = 3 * average annual values from Table 2; phytoplankton-C flux from Table 2).

area	primary production gCm^{-2} d^{-1}	10^6gC d^{-1}	surface bivalves 10^6m^2 (% of total)	phytoplankton flux gCm^{-2} d^{-1}	10^6gC d^{-1}	required prod. area	filtr. pressure
Sylt[1]	0.20	0.8	0.48 (12)	3.02	1.5	15.2	1.82
WSw[2]	0.68	994	70(4.8)	2.48	173	3.6	0.17
OS[3]	0.52	182	18(5.1)	6.00	108	11.5	0.59
M-O[4]	0.22	30	32(23.7)	3.49	112	15.9	3.72
Ar[5]	0.27	63	0.85(0.4)	40.6	35	148	0.55

1: after Asmus et al (1990); 2: after Cadée (1986) and Dame et al (1991); 3: after Smaal & van Stralen (1990) and Prins & Smaal (1993); 4: after Bacher (pers. comm.) and Zurburg et al (1993); 5: after Tenore et al (1982) and Pérez Camacho (1991).

The total phytoplankton-C flux to the bivalve beds and the total primary production for various systems are shown in Table 3. The filtration pressure (phytoplankton uptake as a fraction of the primary production), shows values from 0.17 to 0.59 for the Wadden Sea and the Oosterschelde respectively, which means that 17 - 59 % of the primary production are taken up daily by the bivalve beds. It should be noticed that not all consumers are taken into account: calculations are based on the measurements of the phytoplankton flux to the bivalve beds presented in Table 2, so the filtration pressure by all suspension feeders may well be higher. The filtration pressure for Sylt and M-O bay exceeds the daily primary production by a factor 1.8 to 3.7 respectively: the bivalves seem to depend on production from other sources. In the Sylt area this is presumably import from outside as pointed out previously. In the Marennes-Oléron bay, microphytobenthos is considered as an important primary producer, and, due to resuspension, as a food source for the bivalves (Héral, personal comm.).

In conclusion, the time constant of the phytoplankton turnover in the water column rarely exceeds the filtration pressure. In some areas the import of food is a prerequisite to sustain the bivalves. In all cases, the bivalve beds seem to retain a considerable part of the primary production. In this respect, the bivalves show the potential to reduce the biomass of pelagic algae. It depends on the reaction of other primary producers such as macrophytes and microflagellates, and the regeneration rates of dissolved nutrients, whether or not the bivalves can perform eutrophication control.

Biodeposition

Some results of *in situ* measurements of biodeposition rates and biodeposit composition of bivalve suspension feeders, are presented in Table 4. The deposition rates of total seston show considerable variation and depend on variations in local seston concentrations, densities of the bivalves and local hydrographic factors. Results from Tenore et al (1982), Kaspar et al (1985) and Kautsky and Evans (1987) show that the POC content of biodeposits is generally higher than control sediment or material that has settled down by physical or chemical mechanisms. Kautsky and Evans (1987) showed that smaller particles, which contain more organic material, are relatively abundant in biodeposits. Resuspension of low quality elements of biodeposits (such as pseudofeces) as observed by Asmus et al (1990) and Prins and Smaal (1993) also contributes to the relative increase of the sediment quality in bivalve beds.

However, enrichment of the sediment is described in literature relative to reference sediment or sedimentation in a control, and **not** related to the seston composition (Haven and Morales-Alamo 1972; Kautsky and Evans 1987; Grenz et al 1990; Baudinet et al 1990). In contrast to the fluxes towards the bivalve beds, biodeposition does not show a higher organic carbon content in comparison with the seston composition (Table 4). Low organic carbon content of biodeposits compared to both the seston composition and the net flux towards the bivalve beds (Fig. 3) might be an effect off fast decay of biodeposits, as shown y Grenz et al (1990). However, C/N ratios of reference sediment do not show higher values than biodeposits, except in the Kenepuru Sound (see below). In the Queule river (Chile) higher C/N values

Table 4. Biodeposition rates (ranges or average values with standard error) and biodeposit composition from sites under bivalve suspension feeders (ref = reference site)

area	Biodeposition SPM $gm^{-2}d^{-1}$	%C	C/N	N/P	Seston %POC
Askö[1a]					-
M.e.	18.5/23.6	13-15	12/14	-	
ref	12.1	14	14	-	
Askö[1b]					-
M.e. - ref	3.8(0.8)	12.9(2.15)	8.4	9.2	
ref	9.9(2.8)	9.9 (0.76)	8.6	6.4	
M-O[2]	480/6600	1.5/7.7	6.6/12.4	-	5.0/5.6
Fos[3]	702	6.6	15.7	-	11.1
Thau[4]					33.7/43.3
C.g.	-	-	8.5	.7	
M.g.	-	-	8.0	1.0	
Arosa[5]	521	7.4	8.4	-	24
Queule[6]					15
C.c.-ref	271(111)	6.4(1.1)	41(70.8)	-	
M.c.-ref	234(105)	6.0(1.0)	21(13.4)	-	
ref	553(275)	5.5(0.8)	14(5.6)	-	
Kenepuru[7]					-
P.c.	-	0.73-1.41	6.2/7.2	4.3/7.2	
ref.	-	-	7.9/10.0	3.3/6.1	
Dep.Bay[8]	3.9	28.5 (2.5)	-	-	38.5

1:Askö study area in the Baltic Sea(S), *Mytilus edulis* beds, a: (n=3) after Dahlbäck & Gunnarsson (1981) and b: (n=13) after Kautsky & Evans (1987); 2:Marennes-Oléron Bay (F), *Crassostrea gigas* culture, after Sornin et al (1983); 3: Gulf of Fos (F), *Mytilus galloprovincialis* rope culture, after Grenz (1989) and Grenz et al (1990, 1991); 4: Thau lagoon (F), after Deslous-Pauli (pers. comm.); 5: Ria de Arosa (Sp), *Mytilus galloprovincialis* raft culture, after Pérez Camacho et al (1991); 6:Queule River Estuary (Chile), *Choromytilus chorus* and *Mytilus chilensis* (annual mean values with S.D.),after Jaramillo et al (1992); 7: Kenepuru Sound (New Zealand), *Perna canaliculus*, C/N in molar ratio, after Kaspar et al (1985); 8:Departure Bay (Can), *Crassostrea gigas*, after Bernard (1974).

are observed in biodeposits than in reference sediment. According to Jaramillo et al (1992), this might be due to several factors such as low N content of food and resuspended bottom material.

The N/P ratios of biodeposits are low in comparison with composition of phytoplankton (N/P=16, Redfield et al 1963). According to Kaspar et al (1985) and Baudinet et al (1990), denitrification takes place in mussel biodeposits, possibly leading to a low N/P value in the biodeposits. The results of Kaspar et al (1985) indicate that both low C/N and low N/P ratios occur in mussel biodeposits probably as an effect of a fast decay in combination with denitrification. These calculations point out that the biodeposits are of extensive mineralization.

Mineralization of feces and pseudofeces

Preliminary results of experiments on the decomposition rates of feces and pseudofeces are presented in Fig. 5. The experiments were conducted with *Mytilus edulis* in winter and with *Crassostrea gigas* in summer, using natural seston in seawater as food. Because of the different seasons, the food composition differs in both experiments. Directly after 16 hours of feeding, feces and pseudofeces were collected separately, resuspended in filtered seawater and distributed into replicate 1 liter bottles. The bottles were incubated with stirring and gentle reaeration at 20 °C for 30 days. Frequently (every 1,2 or 3 days), subsamples were taken for analysis of particulate and dissolved material The results show higher mineralization rates of feces relative to pseudofeces for both species. The PON content of feces and pseudofeces was quite similar. Per unit PON the mineralization rate of oyster and mussel feces were 1.1 and 4.6 % per day respectively. The values for pseudofeces were 0.5 and 1.6 %.

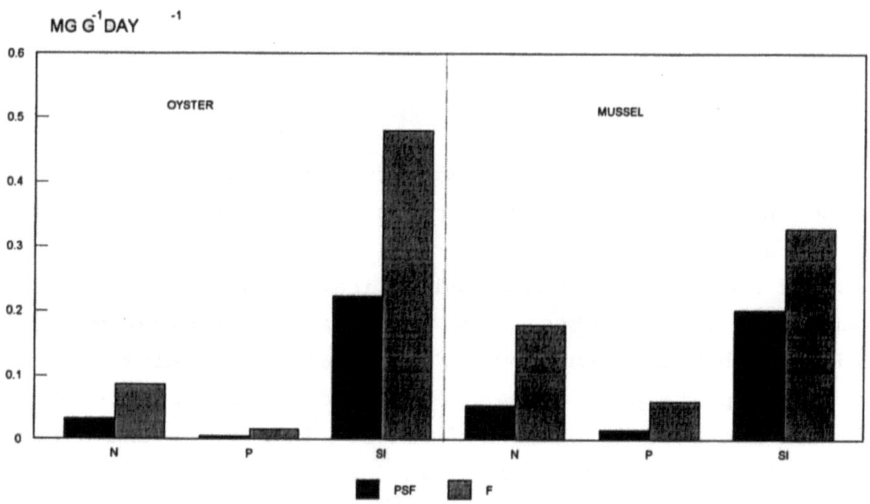

Fig. 5. Mineralization rates of feces and pseudofeces of mussels and oysters during incubation for 30 days. Production of inorganic nutrients per unit of feces and pseudofeces. Data *Crassostrea gigas*, after Zurburg and Armas (1992) (unpublished); mg PON/g biodep = 6.3 for PSF and 7.7 for F. Incubations in August, and *Mytilus edulis*, after Prins and Poppe (1991) (unpubl.); mg PON/g biodeposits = 3.4 for PSF and 3.9 for F. Incubations in Nov

The higher rates of feces breakdown may have been caused by colonization of the feces by bacteria. This is supported by observations of Prieur (1981) who observed the development of high quantities of viable, undamaged Vibrio-like bacterial cells in the hindgut of bivalves that are probably excreted with the feces. The mineralization rates are of the same order of magnitude as found by Garber (1984) for coastal seston. In contrast to Garber (1984), no different phases in mineralization were observed in the these experiments.

The high release of silicate may be an effect of the relatively high incubation temperature (20 °C), because silicate dissolutes relatively fast at higher water temperatures (Helder et al 1983). In the oyster experiment (in summer) the uptake of diatoms by the oysters may play a role as well Moreover, the decomposition kinetics may differ for the various nutrients.

The main conclusion from these preliminary results is that the mineralization rates of feces exceed those of pseudofeces, and are in the same order of magnitude as values reported in literature for coastal seston as substrate.

Release of nutrients

The release rates of nutrients from bivalve beds, mainly based on *in situ* measurements, are shown in Table 5. In all cases there is a net release of ammonia, phosphate and silicate. Nitrate fluxes are in some cases uptake rates. Ammonia fluxes from bivalve beds are considerably higher than from sediment without bivalves. Release rates of phosphate and silicate are in most cases also higher than rates recorded in other sediments. In Fig. 6 the average release rate per g biomass is shown for the various systems, in comparison with excretion rates. The release rates per unit of biomass in the bivalve beds are high in comparison with direct individual excretion rates. The excretion rate of ammonia normally does not exceed about 2 and 3 µmol NH_4 $gADW^{-1}$ h^{-1} for mussels (Bayne and Scullard 1977) and oysters (Boucher and Boucher-Rodoni 1988) respectively. As shown in Table 5 the ammonia release rates are generally higher than the individual excretion rates. The contribution of direct excretion of the bivalves to the net phosphate flux was estimated as 31 - 85 % for the Oosterschelde mussels (Prins and Smaal 1993). A part of dissolved phosphorus will be adsorbed directly to the sediment after remineralization (Balzer et al 1983; Boynton and Kemp 1985). Both Kautsky and Wallentinus (1980) and Asmus et al (1990) reported individual phosphate excretion rates by mussels of 0.01 - 0.7 µmol $gADW^{-1}$ h^{-1}. Direct excretion of silicate is reported by Asmus et al (1990); mussels excreted silicate in various seasons at an average rate of 0.5 µmol $gADW^{-1}$ h^{-1} (min/max: 0.1 - 1.6). This is less than the net release rates from the bivalve beds.

Table 5. Nutrient release rates from bivalve beds to the water column. Measurements throughout the year unless otherwise indicated. Negative fluxes are uptake rates. For comparison nutrient fluxes are given of other types of benthic assemblages. Values are given as min/max or as mean and (s.e.m.)

area	biomass gm^{-2}	NH_4^+ mmol m^{-2} h^{-1}	NO_3^-	PO_4^{3-}	H_2SiO_4
Cal[1]	20/140	0.02/0.40	-0.15/0.03	-0.0/0.05	-
NI[2]	200	1.00	0.00	0.03	-
Sylt[3]					
M.e.	1561	4.95(3.11)	-0.06/1.07	0.87(1.19)	1.14(1.09)
ref	-	-0.82(1.80)	-	0.37(0.29)	0.10(1.05)
WSw[4]	820/940	0.71/15.71	-5.71/-1.43	-1.94/2.90	-0.49/2.66
OS[5]					
M.e	890/2190	-0.02/29.52	-5.46/6.37	-0.43/0.85	-1.66/8.16
C.e.	60	-0.16/0.81	-0.20/0.34	0.06/0.22	0.04/0.98
M-O[6]	899	4.48(2.99)	10.6(12.2)	0.57(0.58)	14.4(15.8)
Morlaix[7]	100/200	0.05/0.38	-0.11/0.05	-	-
Fos[8]	3570[a]	0.00/0.37	-0.02/0.01	-0.00/0.12	0.02/1.21
Formosa[9]	90[b]	0.02/0.04	-0.01/-0.0	0.00/0.01	0.00/0.03
meso-cosm[10]	16[b]	0.23	0.017	0.01	0.10/0.70
Sylt[11]	A.m.	-0.30/0.40	-0.25/0.00	-	-2.50/1.00
Various sediments[12]		-0.13/1.02	-1.15/1.10	-0.02/0.05	-0.46/0.50

a: rope culture; b: numbers*m^{-2}
1: California, *Mercenaria mercenaria* after Murphy & Kremer (1985); 2: North Inlet, SC, *Crassostrea virginica* after Dame et al (1989); 3: Sylt, Königshafen, Eastern Wadden Sea (G), *Mytilus edulis* July-Sept, after Asmus et al (1990); 4: Western Wadden Sea (NL), *Mytilus edulis*, June, after Dame et al (1991); 5: Oosterschelde (NL), *Mytilus edulis* April-Sept, *Cerastorderma edule*, June, after Prins & Smaal (1993); 6: Marennes-Oléron Bay (F), *Mytilus edulis*, October, after Zurburg et al (1993); 7: Bay of Morlaix (F), *Crassostrea gigas* after Boucher & Boucher-Rodoni (1988); 8: Gulf of Fos (F), *Mytilus gallorpovincialis* May-Nov, after Baudinet et al (1990); 9: Ria de Formosa (P), *Ruditapes dessicatus* Nov-March, after Falcao & Vale (1990); 10: MERL mesocosm experiment, *Mercenaria mercenaria* after Doering et al (1987); 11: Sylt, (G), *Arenicola marina* after Asmus (1986); 12: after reviews in Nixon (1981) and Lerat et al (1990).

The measurements of individual excretion rates normally are executed with individual animals in the laboratory. The *in situ* flux measurements are carried out on bivalve beds including biodeposits. The microbial decay of biodeposits seems to contribute significantly to the observed nutrient release fluxes. As concluded from the mineralization experiments reported previously, these fluxes are probably not high as an effect of enhanced microbial activity per unit of organic matter, but due to the larger availability of high quality organic substrates.

Bivalve suspension feeders are an important intermediate in nutrient cycling, through formation and mineralization of biodeposition. The net release fluxes of dissolved nutrients N, P and Si, as shown in Table 5, may not have the same proportion as the net uptake fluxes of PON, POP and, presumably, particle-bound silicon (Table 2). Due to large variation of the fluxes, it is not possible to derive a trend from these data. However, different mineralization rates have been observed for various nutrients, affecting the nutrient ratios in the water column in mesocosms (Doering et al 1987). This may have consequences for the structure of phytoplankton communities in areas with large bivalve beds.

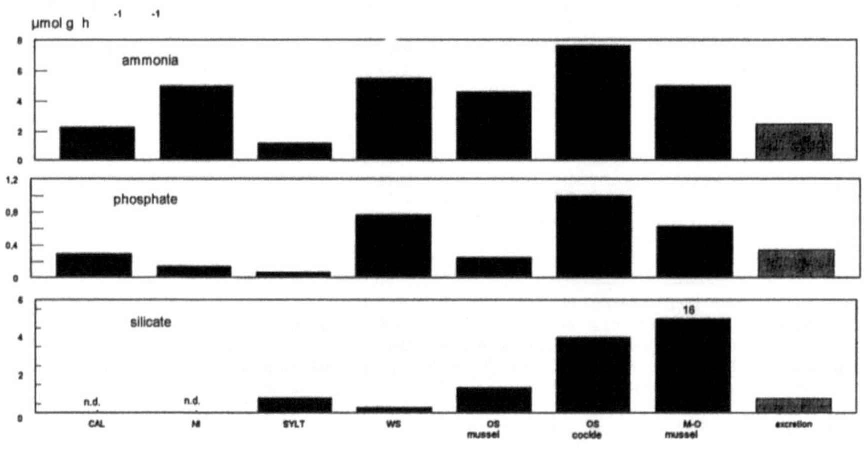

Fig. 6. Average nutrient release rates per g ADW from bivalve beds, based on *in situ* measurements of fluxes in various systems, and individual excretion rates. (see references in Table 5)

Net budgets of nutrient fluxes through bivalve beds

In addition to the uptake and release rates of material by bivalve beds as shown in Tables 2 and 5, net budgets are shown in Fig. 7. This figure summarizes the net uptake rates of particulate matter and the net release rates of dissolved nutrients on an annual basis as reported by Dame et al (1989) for *Crassostrea virginica* in North Inlet (SC-USA), and by Prins and Smaal (1993), for *Mytilus edulis* in the Oosterschelde (NL). In the Oosterschelde, the fluxes are high compared to the North Inlet values, related to differences in biomass (Table 2). Both species show net uptake of phosphorus. Asmus et al (1990), observed phosphate fixation especially during periods of resuspension. As mentioned previously, phosphate can be adsorbed in aerated and mixed sediment (see also Hopkinson 1987). Bivalve beds seem to be a sink for P.

Nitrogen is retained by the oyster bed, and released by the mussel bed. The net uptake of nitrogen by the oyster reef is partly an effect of the uptake of ammonia by macroalgae attached to oyster shells, which flourish during winter, and reduce the net ammonia production (Dame et al 1989). This phenomenon is also observed on natural mussel beds (Asmus et al 1990).

From the net release of dissolved nitrogen from mussel beds in the Oosterschelde it is concluded that denitrification does not play a significant role in this situation. Boucher and Boucher-Rodoni (1988) found high rates of nitrification without direct denitrification in oyster beds, related to limited transport of nitrate to the deeper anoxic sediment

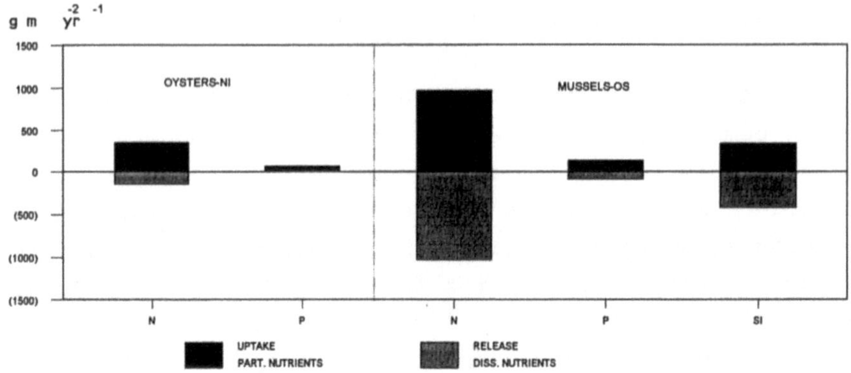

Fig. 7. Annual budgets ($gm^{-2}yr^{-1}$) of net uptake of particulate organic matter and release of dissolved inorganic matter of intertidal bivalve beds from *in situ* estimations, after Dame et al (1989) (*Crassostrea virginica*) and Prins and Smaal (1993) (*Mytilus edulis*).

layers due to limited bioturbation in bivalve beds. Mussel beds probably have low bioturbation as well, as there is an impoverished infauna (Asmus 1987). In contrast, Kaspar et al (1985) and Baudinet et al (1990), observed increased denitrification under rope cultivation sites of mussels. This situation is quite different, however, from the mussel beds presented in Fig. 7, which are intertidal and exposed to strong currents and wave action. Moreover, the nitrogen released consisted of both ammonia and nitrate.

In contrast to many observations that show a flux of nitrate into the sediment (Murphy and Kremer 1985; Dame and Dankers 1988; Dame et al 1989; Asmus et al 1990), nitrate release is observed by Feuillet-Girard et al (1988) and Lerat et al (1990) in oyster cultivation areas. According to Boucher and Boucher-Rodoni (1988) this indicates that both ammonification and nitrification are important at the sediment-water interface.

The net release of nitrogen from mussel beds is not compatible with the uptake and absorption of food on a mussel bed; the net absorption of nitrogen by the mussels is estimated for the summer period to be about 200 g PON m^{-2} yr^{-1} (Prins and Smaal 1993). Due to mortality a large part of the absorbed particulate nitrogen will be mineralized again, and released as dissolved inorganic nitrogen, but there is no net increase of biomass on mussel beds in the Oosterschelde. The amount of mussels harvested by fishermen is in the same order of magnitude as the amount of mussel seed that is imported from other areas, e.g., the Wadden Sea (Van Stralen and Dijkema 1993).

A silicon budget is estimated for the mussels only. The uptake of particulate silicon is estimated on the basis of chlorophyll-a uptake and diatom cell concentrations in the phytoplankton, converted into Si. There is a net release of dissolved Si, which took place mainly in summer and autumn, probably related to higher dissolution of silicate at higher water temperature as mentioned previously.

In conclusion, the uptake of particulate material by bivalve beds occurs in combination with the release of dissolved nutrients. The bivalve beds provide a strong benthic-pelagic coupling, and the processing of material from the water column exceeds by far the retention.

CONSEQUENCES FOR THE ECOSYSTEM

In situ measurements provide a quantitative basis for the hypothesis that bivalve suspension feeder beds play a significant role in coupling the pelagic and the benthic compartments of estuarine and coastal ecosystems, by:

- filtration of large quantities of material from the water column
- reduction and possibly local depletion of the phytoplankton concentration
- formation of biodeposition, consisting of high quality organic material
- remineralization of biodeposits
- release of inorganic nutrients to the water column
- increasing the availability of dissolved nutrients
- affecting the nutrient ratios in the water column.

The uptake of organic matter and the release of nutrients has an impact on processes in the water column. Doering et al (1986) showed that the introduction of 16 specimen.m^{-2} of the bivalve *Mercenaria mercenaria* in a series of mesocosms, resulted in a doubling of phytoplankton production. The impact of mussels on phytoplankton in nutrient-limited periods of the year has been calculated by Asmus and Asmus, 1991; they concluded that the mussels reduce the standing stock of phytoplankton meanwhile promoting the primary production. A stimulation of primary production by grazing was shown in phytoplankton-zooplankton interaction studies by Sterner (1986), and in a model approach by DeAngelis, 1992; the 'herbivore optimization curve' reflects the idea that primary production is stimulated by a certain grazing pressure, related to the biomass of filter feeders. Primary production shows a maximum value at a certain -optimum- filter feeder biomass.

By applying the Simulation Model Oosterschelde ecosystem Herman and Scholten (1991) showed a stabilizing influence of bivalve filter feeders on phytoplankton concentrations, to a large extent independent of nutrient loadings. Increase of phytoplankton biomass, due to a large increase of nutrient loadings was prevented by only a minor increase of bivalve biomass. This was explained by the continuous filtration pressure, as a consequence of the year-round occurrence of the bivalves, in contrast with the seasonal development of zooplankton. By using the same model, Prins and Smaal (1993) have calculated that the contribution of mussel beds to the total nitrogen mineralization in the Oosterschelde is about 30 %. They concluded that the mussel beds increase the turnover of phytoplankton due to regeneration of nitrogen as a limiting nutrient in the Oosterschelde after the spring bloom. In reality, the turnover of phytoplankton in the Oosterschelde increased after the completion of a large scale coastal engineering project. This project decreased nutrient concentrations and increased water residence time and hence increased the benthic-pelagic coupling (Smaal and Nienhuis, 1992).

The bivalve filter feeders also have effects on the structure of the pelagic community. In a set of mesocosm experiments, the impact of the benthos compartment, including suspension feeders, on the structure of the ecosystem was evaluated (Sullivan et al 1991). They stated that biotic (such as predator-prey) interactions between benthic and pelagic compartments should be taken into account when studying effects of benthos on eutrophication and nutrient limitation. Riemann et al (1988, 1990) have shown that mussels in large scale enclosures changed the size distribution and species composition of the planktonic community, resulting in dominance of smaller sized plankton. This is not considered as a benefit for the bivalves.

There is evidence from these experiments that the bivalves can have an impact on the structure of the plankton community (Riemann et al 1988). However, the effects of these structural changes on the bivalves, and the consequences for the ecosystem, need further study.

CONCLUSIONS

Bivalve filter feeders can have a large impact on processes in estuarine and coastal ecosystems. System volumes can be filtered every 0.7 - 13 days. The impact of the bivalves normally extends to areas far outside the bivalve beds. There is a large flux of organic matter to the bivalve beds. In some areas, the local phytoplankton production hardly sustains the phytoplankton flux towards the bivalve beds. However, the regeneration of limiting nutrients may stimulate the primary production and thus provide a positive feed-back on filtration pressure. The filtration pressure may act as a eutrophication control, by limiting the phytoplankton biomass. Nutrients however, will be regenerated and be available again for primary production. The consequences for the ecosystem, e.g., the structure of the phytoplankton community, need further study.

ACKNOWLEDGMENTS

The authors are grateful to Mr I. de Vries for stimulating discussions, to Dr W. Zurburg for critically reading the manuscript and to Dr R. Dame and Dr H. Asmus for review.

REFERENCES

Alpine AE Cloern JE (1992) Trophic interactions and direct physical effects control phytoplankton biomass and production in an estuary. Limnol Oceanogr 37: 946-955

Asmus R (1986) Nutrient flux in short-term enclosures of intertidal sand communities. Ophelia 26: 1-18

Asmus H (1987) Secondary production of an intertidal mussel bed community related to its storage and turnover compartments. Mar Ecol Prog Ser 39: 251-266

Asmus H Asmus R Reise K (1990) Exchange processes in an intertidal mussel bed: a Sylt-flume study in the Wadden Sea. Ber Biol Anst Helgoland 6: 1-79

Asmus R Asmus H (1991) Mussel beds: limiting or promoting phytoplankton? J exp mar Biol Ecol 148: 215-232

Asmus H Asmus RM Prins TC Dankers N Francés G Maasz B Reise K (1992) Benthic-pelagic flux rates on mussel beds: tunnel and tidal flume methodology compared. Helgoländer Meersesunters 46: 341-361

Asmus H Asmus RM (1993) Phytoplankton-mussel bed interactions in intertidal ecosystems. This volume

Bacher C (1989) Capacité trophique du bassin de Marennes-Oléron: couplage d'un modele de transport particulaire et d' un modele de croissance de l' huitre *Crassostrea gigas*. Aquat Living Res 48: 199-214

Balzer W Grasshoff K Dieckmann P Haardt H Petersohn U (1983) Redox-turnover at the sediment/water interface studied in a large bell jar system. Oceanol Acta 6: 337-344

Baudinet D Alliot E Berland B Grenz C Plante-Cuny M-R Plante R Salen-Picard C (1990) Incidence of mussel culture on biogeochemical fluxes at the sediment-water interface. Hydrobiologia 207: 187-196

Bayne BL Scullard C (1977) Rates of nitrogen excretion by species of Mytilus (Bivalvia:Mollusca). J mar biol Ass U K 57: 355-369

Bernard FR (1974) Particle sorting and labial palp function in the pacific oyster *Crassostrea gigas* (Thunberg, 1975). Biol Bull 146: 1-10.

Biggs RB Howell BA (1984) The estuary as a sediment trap:

alternate approaches to estimating its filtering efficiency. In: V Kennedy (ed) The estuary as a filter. Academic Press New York: 107-127

Boucher G Boucher-Rodoni R (1988) In situ measurement of respiratory metabolism and nitrogen fluxes at the interface of oyster beds. Mar Ecol Prog Ser 44: 229-238

Boynton WR Kemp WM (1985) Nutrient regeneration and oxygen consumption by sediments along an estuarine salinity gradient. Mar Ecol Prog Ser 23: 45-55

Cadée GC (1986) Increased phytoplankton primary production in the Marsdiep area (Western Dutch Wadden Sea). Neth J Sea Res 20: 285-290

Carlson DJ Townsend DW Hilyard AL Eaton JF (1984) Effect of an intertidal mudflat on plankton of the overlying water column. Can J Fish aquat Sci 41: 1523-1528.

Cloern JE (1982): Does the benthos control phytoplankton biomass in South San Francisco Bay? Mar Ecol Prog Ser 9: 191-202

Commito JA Boncavage EM (1989) Suspension-feeders and coexisting infauna: an enhancement counterexample. J Exp Mar Biol Ecol 125: 33-42

Dahlbäck B Gunnarson LH (1981) Sedimentation and sulfate reduction under a mussel culture. Mar Biol 63: 269-275.

Dame RF Dankers N (1988) Uptake and release of materials by a Wadden Sea mussel bed. J exp mar Biol Ecol 118: 207-216

Dame RF Spurrier JD Wolaver TG (1989) Carbon, nitrogen and phosphorus processing by an oyster reef. Mar Ecol Prog Ser 54: 249-256

Dame RF Zingmark RG Haskin E (1984) Oyster reefs as processors of estuarine materials. J exp mar Biol Ecol 83: 239-247

Dame RF Zingmark R Stevenson H Nelson D (1980) Filter feeding coupling between the estuarine water column and benthic subsystems. In: Kennedy VS (ed) Estuarine perspectives. Academic Press New York: 521-526.

Dame RF Dankers N Prins TC Jongsma H Smaal AC (1991) The influence of mussel beds on nutrients in the Western Wadden Sea and Eastern Scheldt estuaries. Estuaries 14: 130-138

Dame RF (1993) The role of bivalve filter feeder material fluxes in estuarine ecosystems. This volume

Dankers N (1993) Integrated estuarine management - obtaining a sustainable yield of bivalve resources while maintaining

environmental quality. This volume
DeAngelis DL (1992) Dynamics of nutrient cycling and food webs. Chapman and Hall London
Dinet A Sornin J-M Sabliere A Delmas D Feuillet-Girard M (1990) Influence de la biodéposition de bivalve filtreurs sur les peuplements méiobenthiques d'un marais maritime. Cah Biol Mar 31:307-322
Dittmann S (1990): Mussel beds - amensalism or amelioration for intertidal fauna? Helgoländer Meeresunters 44: 335-352
Doering PH Oviatt CA (1986) Application of filtration rate models to field populations of bivalves: an assessment using experimental mesocosms. Mar Ecol Prog Ser 31: 265-275
Doering PH Oviatt CA Kelly JR (1986) The effects of the filter-feeding clam *Mercenaria mercenaria* on carbon cycling in experimental marine mesocosms. J Mar Res 44: 839-861
Doering PH Kelly JR Oviatt CA Sowers T (1987) Effect of the hard clam *Mercenaria mercenaria* on benthic fluxes of inorganic nutrients and gases. Mar Biol 94: 377-383
Falcao M Vale C (1990) Study of the Ria Formosa ecosystem: benthic nutrient remineralization and tidal variability of nutrients in the water. Hydrobiologia 207: 137-146
Feuillet-Girard M Héral M Sornin JM Deslous-Paoli JM Robert JM Mornet F Razet D (1988) Éléments azotés de la colonne d'eau et de l'interface eau-sédiment du bassin de Marennes-Oléron: influence des cultures d'huîtres. Aquat Living Resourc 1: 251-265
Fréchette M Bourget E (1985) Food-limited growth of *Mytilus edulis* L in relation to the benthic boundary layer. Can J Fish Aquat Sci 42: 1166-1170
Garber JH (1984) Laboratory study of nitrogen and phosphorus remineralization during the decomposition of coastal plankton and seston. Est Coast Shelf Sci 18: 685-702
Grenz C (1989) Quantification et destinée de la biodeposition en zones de production conchylicole intensive en Méditerranée. Thesis Univ Aix-Marseille 144 pp
Grenz C Hermin M-N Baudinet D Daumas R (1990) In situ biochemical and bacterial variation of sediments enriched with mussel biodeposits. Hydrobiologia 207: 153-160
Grenz C Massé H Morchid A Panache A (1991) An estimate of the energy budget between cultivated biomass and the environment

around a mussel-park in northwest Mediterranean Sea. ICES Mar Sci Symp 192: 63-67

Grenz C E Alliot D Baudinet L Helis H Masse, 1992. Influence des opérations de dévasage sur les flux de nutriments a l' interface eau-sédiment (bassin Thau-France). Vie Milieu 42(2):157-164

Haven DS Morales-Alamo R (1966) Aspects of biodeposition by oysters and other invertebrate filter feeders. Limnol Oceanogr 11: 487-498

Haven DS Morales-Alamo R (1972) Biodeposition as a factor in sedimentation of fine suspended solids in estuaries. Geol Soc Am Mem 133: 121-130

Héral M Deslous-Paoli J-M Prou J (1988) Approche de la capacité trophique d'un ecosystème conchylicole. ICES CM 1988/K:22

Helder W De Vries RTP Rutgers van der Loeff MM (1983) Behavior of nitrogen nutrients and dissolved silica in the Ems-Dollard estuary. Can J Fish aquat Sci 40 (Suppl 1): 188-200

Herman PMJ Scholten H (1990) Can suspension-feeders stabilise estuarine ecosystems? In Barnes M Gibson RN (eds). Trophic relationships in the marine environment. Aberdeen University Press Aberdeen: 104-116

Hily C (1991) Is the activity of benthic suspension feeders a factor controlling water quality in the Bay of Brest? Mar Ecol Prog Ser 69: 179-188

Hopkinson CS (1987) Nutrient regeneration in shallow-water sediments of the estuarine plume region of the nearshore Georgia Bight, USA. Mar Biol 94: 127-142

Jaramillo E Bertrán C Bravo A (1992) Mussel biodeposition in an estuary in southern Chile. Mar Ecol Prog Ser 82: 85-94

Jϕrgensen CB (1990) Bivalve filter feeding: hydrodynamics, bioenergetics, physiology and ecology. Olsen and Olsen, Fredensborg Denmark 140 pp.

Kaspar HF Gillespie PA Boyer IC McKenzie AL (1985) Effects of mussel aquaculture on the nitrogen cycle and benthic communities in Kenepuru Sound, Marlborough Sounds, New Zealand. Mar Biol 85: 127-136

Kautsky N Evans S (1987) Role of biodeposition by *Mytilus edulis* in the circulation of matter and nutrients in a Baltic coastal ecosystem. Mar Ecol Prog Ser 38: 201-212

Kautsky N Wallentinus I (1980) Nutrient release from a Baltic

Mytilus-red algal community and its role in benthic and pelagic production. Ophelia suppl 1: 17-30

Lerat Y Laserre P Le Corre P (1990) Seasonal changes in pore water concentrations of nutrients and their diffusive fluxes at the sediment-water interface. J exp mar Biol Ecol 135: 135-160

Murphy RC Kremer JN (1985) Bivalve contribution to benthic metabolism in a California lagoon. Estuaries 8: 330-341

Navarro E Iglesias JIP Perez Camacho A Labarta U Beiras R (1991) The physiological energetics of mussels (*Mytilus galloprovincialis*) from different cultivation rafts in the Ria de Arosa (Galicia, N.W. Spain). Aquaculture 94: 177-196

Nichols F (1985) Increased benthic grazing: an alternative explanation for low phytoplankton biomass in Northern San Francisco Bay during the 1976-1977 drought. Est Coast Shelf Sci 21:379-388

Nixon SW Kelly (1981) Remineralization and nutrient cycling in coastal marine ecosystems. In Neilson BJ Cronin LE (eds) Estuaries and nutrients. Humana Press Clifton New Jersey: 111-138.

Officer CB Smayda T Mann R (1982) Benthic filter feeding: a natural eutrophication control Mar Ecol Prog Ser 9: 203-210

Otto L (1975) Oceanography of the Ria de Arosa (N.W. Spain). KNMI Med en Verh 96 SDU Den Haag

Perez Camacho A González R Fuentes J (1991) Mussel culture in Galicia (N.W. Spain). Aquaculture 92: 263-278

Peterson CH Black R (1987) Resource depletion by active suspension feeders on tidal flats: Influence of local density and tidal elevation. Limnol Oceanogr 32: 143-166

Prieur D (1981) Experimental studies of trophic relationships between marine bacteria and bivalve molluscs. Kieler Meeresforsch Sonderh 5: 376-383

Prins TC Smaal AC (1990) Benthic-pelagic coupling: the release of inorganic nutrients by an intertidal bed of Mytilus edulis. In Barnes M Gibson RN (eds) Trophic relationships in the marine environment. Aberdeen University Press Aberdeen: 89-103

Prins TC Smaal AC (1993) The role of the blue mussel *Mytilus edulis* in the cycling of nutrients in the oosterschelde estuary (The Netherlands). In: Nienhuis PH Smaal AC (eds) The Oosterschelde estuary (The Netherlands): case-study of a changing ecosystem. Kluwer Acad Publ Dordrecht in press

Redfield AC Ketchum BH Richards FA (1963) The influence of organisms on the composition of sea-water. In: Hill HN (ed) The Sea Vol 2 Wiley-Interscience New York:26-77

Riemann B Nielsen TG Horsted SJ Bjrnsen PK Pock-Steen J (1988) Regulation of phytoplankton biomass in estuarine enclosures. Mar Ecol Prog Ser 48: 205-215.

Riemann B Sorensen HM Bjornsen PK Horsted SJ Jensen LM Nielsen TG Sondergaard M (1990) Carbon budgets of the microbial food web in estuarine enclosures. Mar Ecol Prog Ser 65: 159-170.

Rodhouse PG Roden CM (1987) Carbon budget for a coastal inlet in relation to intensive cultivation of suspension-feeding bivalve molluscs. Mar Ecol Prog Ser 36: 225-236

Smaal AC Verhagen JHG Coosen J Haas HA (1986) Interactions between seston quantity and quality and benthic suspension feeders in the Oosterschelde, The Netherlands. Ophelia 26: 385-399

Smaal AC Van Stralen MR (1990) Average annual growth and condition of mussels as a function of food source. In: McClusky DS De Jonge VN Pomfret J (eds) North Sea-Estuaries Interactions Hydrobiologia 195: 179-188

Smaal AC Nienhuis PH (1992) The Oosterschelde (the Netherlands), from an estuary to a tidal bay: a review of responses at the ecosystem level Neth J Sea Res 30: 1-8

Sornin JM Feuillet M Héral M Deslous-Paoli J-M (1983) Effet des biodépôts de l'huître *Crassostrea gigas* (Thunberg) sur l'accumulation de matières organiques dans les parcs du bassin de Marennes-Oléron. J moll Stud suppl 12A: 185-197

Sterner RW (1986) Herbivores' direct and indirect effects on algal populations. Science 231: 605-607

Van Stralen MR Dijkema RD (1993) Mussel culture in a changing environment: the effects of a coastal engineering project on mussel culture (*Mytilus edulis*) in the Oosterschelde (S.W. Netherlands). In Nienhuis PH Smaal AC (eds), The Oosterschelde estuary (The Netherlands): case-study of a changing ecosystem. Kluwer Acad Publ Dordrecht

Stuart V Newell RC Lucas MI (1984) Conversion of kelp debris and faecal material from the mussel *Aulacomya ater* by marine microorganisms. Mar Ecol Prog Ser 7: 47-57

Sullivan BK Doering PH Oviatt CA Keller AA Frithsen JB (1991) Interactions with the benthos alter pelagic food web structure in coastal waters. Can J Fish Aquat Sci 48: 2276-2284.

Tenore KR Boyer LF Cal RM Corall J Garcia-Fernandez C Gonzalez N Gonzalez-Gurriaran E Hanson RB Iglesias J Krom M Lopez-Jamar E McClain J Pamatmat MM Perez A Rhoads DC de Santiago G Tietjen J Westrich J Windom HL (1982) Coastal upwelling in the Rias Bajas, N.W.Spain: contrasting the Rias de Arosa and the Muros. J Mar Res 40: 701-772

Verwey J (1952) On the ecology of distribution of cockle and mussel in the Dutch Waddensea, their role in sedimentation and the source of their food supply. Arch Néerl Zool 10: 171-239

Wolff WJ (1983) Estuarine benthos. In: Ketchum BH (ed) Ecosystems of the world, 26. Estuaries and Enclosed Seas. Elsevier Amsterdam 151-182

Wright RT Coffin RB Ersing CP Pearson D (1982) Field and laboratory measurements of bivalve filtration of natural marine bacterioplankton. Limnol Oceanogr 27: 91-98

Zurburg W Smaal AC Héral M Dankers N (1993) *In situ* estimations of uptake and release of material by bivalve filter feeders in the bay of Marennes-Oléron (France). Proc ECSA/ERF Plymouth 1992 Olsen and Olsen in press

HYDRODYNAMIC CONTROL OF BIVALVE FILTER FEEDERS: A CONCEPTUAL VIEW

D. J. Wildish and D. D. Kristmanson*
Department of Fisheries and Oceans
Biological Station
St. Andrews, New Brunswick E0G 2X0
Canada

INTRODUCTION

Marine bivalve molluscs include a diverse group of mainly benthic forms as adults. Their life histories are complex with a planktonic larval stage. Both larvae and many adult bivalves suspension feed by capturing microscopic particulates, inclusive of detritus, bacteria, microalgae and small animals (=seston). Sestonic particles captured by bivalve suspension feeders typically have diameters of the order of 1 to 100 μm: bacteria to microplankton (Fig. 1) and occur in seasonally variable amounts in seawater. By comparison, the sizes of newly hatched bivalve larvae, such as the giant scallop (*Placopecten magellanicus*) are ~350 μm in length, growing to a maximum valve height (Fig. 1) in excess of 10 cm.

The central theme of this presentation is a brief interdisciplinary review, selecting works that include established hydrodynamic principles (e.g., Schlichting 1968) as a mechanistic explanation for some aspects of the biology of bivalve suspension feeders. We suggest that hydrodynamics is a fundamentally important environmental variable at many levels of benthic biological organization, inclusive of larval transport and settlement, the physiology, ethology and ecology of juvenile and adult filter feeders.

*Department of Chemical Engineering
University of New Brunswick
Fredericton, New Brunswick, Canada

NATO ASI Series, Vol. G 33
Bivalve Filter Feeders
in Estuarine and Coastal Ecosystem Processes
Edited by Richard F. Dame

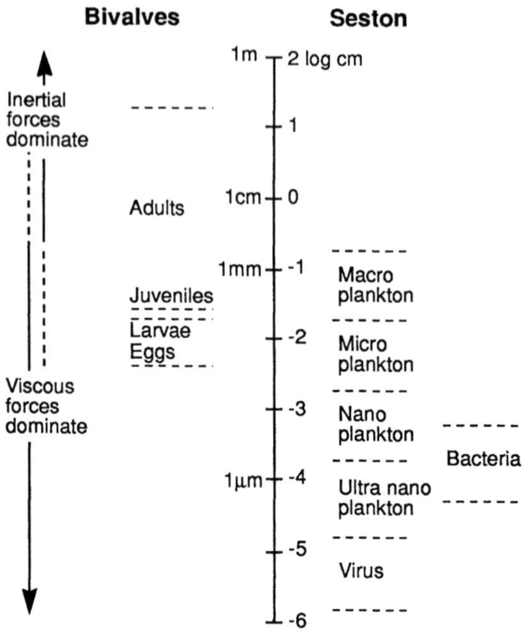

Fig 1. Comparative sizes of suspension feeding bivalves and their potential food (=seston). Based on data in Parsons et al (1977) and Lynch and Poole (1979).

RELEVANT HYDRODYNAMICS AND BIVALVE SIZE

Hydrodynamic processes relevant to bivalves occur at two very different scales, with the relevant hydrodynamics determined by bivalve size (Fig. 1). Large-scale flows include the familiar turbulent transfer and advective processes within a benthic boundary layer, where the flow interacts with the sediment-water interface. Hydrodynamic processes at much smaller size scales involve creeping flows dominated by viscosity and which may behave in surprising ways (Purcell 1977). For instance, under laminar conditions where turbulence and the associated power to mix and "diffuse" is absent, small particles such as seston are transported by much weaker processes and the mechanisms by which they are collected are very different than at a larger scale.

The measure of whether a flow is laminar or turbulent is given by the Reynolds number of the flow. The dynamics of a flow of a real fluid distant from a free surface is primarily dependent on the pressure forces, the inertia of the moving fluid elements and the viscous stresses that are generated by velocity differences. The ratio of the inertial forces to the viscous forces is called the Reynolds number. It is defined as:

$$\mathbf{Re} = \mathbf{L}\ \mathbf{U}/\boldsymbol{\nu}$$

where

L is a characteristic length

U is a characteristic velocity and

$\boldsymbol{\nu}$ is the kinematic viscosity of the fluid that is approximately 0.01 $cm^2 \cdot s^{-1}$.

Flow around a circular cylinder may be used as a typical example. At low **Re**, or creeping flow, streamlines pass smoothly over the obstacle. The drag force (**F**) exerted by the fluid on the cylinder can be easily calculated from:

$$\mathbf{F} = \mathbf{C} * \mathbf{S} * \mathbf{Ps}$$

where

S = frontal area

Ps = stagnation pressure = $0.5 * \mathbf{U}^2/\mathbf{g}$

g = gravitational constant

C = drag coefficient.

At Reynolds number less than about 0.2, the drag coefficient is given by 64/**Re**. At higher Reynolds numbers, a wake starts to develop behind the cylinder and deficiencies in pressure start to appear in the lee of the cylinder. The drag, which is directly proportional to velocity in the creeping flow regime, becomes proportional to a higher power of velocity at higher Re. At Reynolds numbers about 1000, the drag is proportional to the square of velocity (Fig. 2).

In seawater, a **Re** of 0.2 is found with a cylinder of 1 mm diameter moving at 0.02 $cm \cdot s^{-1}$, similar to that of a 100 micron cylinder moving at 0.2 $cm \cdot s^{-1}$. At Reynolds numbers <0.2, we have entered the realm of sinking plankton and of flow by cilia. Low Re flows are not well mixed; in fact, in the absence of molecular diffusion, flows are reversible. As a consequence, the mixing processes with which we are familiar in turbulent flow, and which deliver nutrients to a consumer from far afield, do not exist. Nutrients can only be delivered to sestonic particles by Brownian movement or diffusion by transporting particulate or dissolved materials along a gradient (Lazier and Mann 1990). Somewhat larger seston, e.g. ciliate

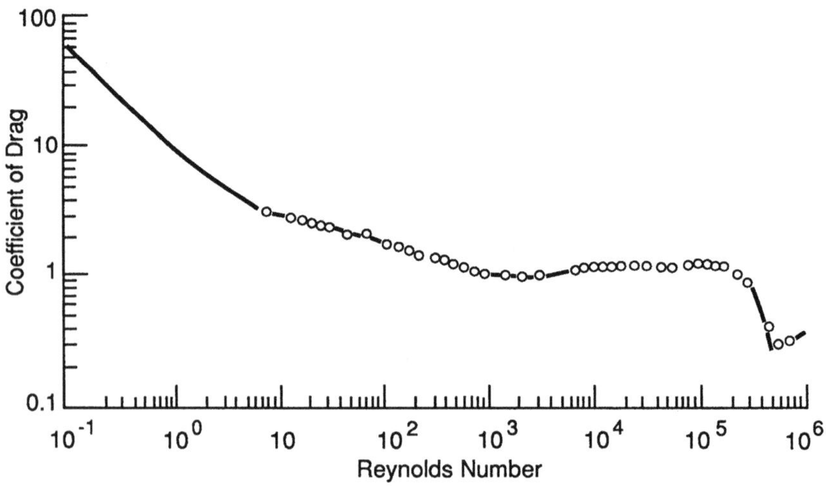

Fig. 2. Logarithmic plot of Reynolds number versus the coefficient of drag (from Schlichting 1968).

Protozoa, or bivalve larvae, must keep moving as a means to ensure that they receive nutrients and avoid localized depletion around their bodies (Purcell 1977).

At a **Re** of 1000, a 1-cm cylinder is in a flow of 10 $cm \cdot s^{-1}$ and is equivalent to the environment external to a small mussel. The actual Re for flow by an isolated mussel is higher than this, but the relationship of drag force to velocity squared is still relevant. In this flow regime, the wake is turbulent and pressure deficiencies of the same order of magnitude as the stagnation pressure will be found on the downstream side of the cylinder. In practice, at 20 $cm \cdot s^{-1}$ velocity, a 2-cm diameter cylinder will have a pressure difference of about 2 mm H_2O between the upstream stagnation point and the center of the aft section of the cylinder. This value is comparable to the pressure that can be exerted by some active suspension-feeding bivalve ciliary pumps (Jϕrgensen et al 1986).

LARVAL TRANSPORT AND SETTLEMENT

The dispersal and settlement of bivalve larvae is a process in which fluid mechanics has an effect on both small and large scales. Bivalve larvae swim by means of metachronally beating cilia on the velum at very low speeds. At sizes >1 mm, the effectiveness of this means of locomotion is severely curtailed (Chia and Buckland-Nicks 1984). During larval life, they are transported passively for great distances, of the order of m to km, by large-scale flows, turbulent and advective processes. The mechanisms by which bivalve larvae are passively transported to the bottom are similar to established boundary layer mechanisms of turbulent transport (Eckman 1990). Evidence that bivalve larvae are widely dispersed in this way is obtained by sampling with appropriately sized plankton samplers (Gallagher and Mann 1986; Tremblay and Sinclair 1990).

On soft sediments, larval metamorphosis, either just before or after attaching to a substrate and settlement, has been viewed as a process that involves active habitat selection by the larva and/or some form of physical deposition process with passive larvae (Butman 1987). Under turbulent conditions, active selection only becomes possible when the larvae are very close to the bottom because turbulent motion is stronger than their ability to sink or swim. In this account we ignore active selection, although we recognize that it may be more important than presently conceived (e.g., Jonsson et al 1991), and concentrate on the hydrodynamics associated with settlement.

During field studies in the Bristol Channel, UK., Tyler and Banner (1977) suggested that metamorphosed (echinoderm) larvae behaved as passive particles and obeyed the same physical laws as equivalent sized sedimentary particles so that they were distributed in the same way by the hydrodynamic regime. Eckman (1979) showed in a field experiment on an intertidal sand flat that real and simulated tubes influenced patterns of flow (mm to cm scale) so that increased larval settling occurred downstream. A one-dimensional, advection-diffusion model was developed by Eckman (1990) to predict the rates of settlement of passively transported larvae onto sediments of differing roughness. The model considers depth-limited boundary layers, variation in boundary shear velocity, larval fall velocity, and the height and aerial density of bottom roughness features such as tubes. Rates of settlement are predicted to increase monotonically with the density of roughness features. Other factors of potential importance in passive settling are the height of the roughness elements and the ratio of larval fall velocity to shear velocity. Flume studies, e.g., by Eckman (1983) and Hannan (1984), have supported the view that hydrodynamic sorting of larvae does occur. Ertman and Jumars (1988) showed that the exhalant flow of infaunal bivalves such as the soft shell clam, *Mya arenaria*, caused enhanced variabilities in recruitment of polychaete larvae within a few cms of the exhalant flow. There was also evidence that local settlement near the incurrent siphon was enhanced. These authors suggested that any negative influence on settling bivalves must be a large-scale phenomenon involving upstream depletion effects.

Larvae of the blue mussel, *Mytilus edulis*, may settle at a valve length of _260 μm but can delay settlement and metamorphosis and grow further to _350 μm in the plankton (McGrath et al 1988). Primary and secondary settlement in the blue mussel was originally described by Bayne (1964). It involves initial settlement on filamentous macroalgae, followed by further growth, and then by a second pelagic phase, during which the mussel migrated to a nearby adult mussel bed. McGrath et al (1988) reported that *Mytilus edulis* larvae on the west coast of Ireland settled directly onto adult blue mussel beds. In laboratory conditions, Eyster and Pechenik (1987) showed that blue mussel settlement onto filamentous red macroalgae was enhanced by culture agitation, suggesting that water movement was important during settlement. After settlement and metamorphosis, byssal threads serve to anchor the juveniles to the algal filaments or adult byssus. In blue mussels, the attachment discs of each byssal thread are polyphenolic, proteinaceous adhesives secreted by the foot, the byssal threads consist of fibrous proteins with ordered crystalline arrays serving to link the attachment site to the foot (Waite 1983). Other byssal products include acid muco-polysaccharides whose function is unclear and the enzyme polyphenoloxidase whose purpose is to induce intermolecular cross-linking of adhesive and fibrous proteins (Waite 1983).

Like blue mussel larvae, the larvae of the bay scallop, *Argopecten irradians*, settle onto the blades of eelgrass, *Zostera marina* and other elevated substrates (Eckman 1987). Byssal threads are used to attach the scallops firmly to the eelgrass. In a field study of bay scallop recruitment, Eckman (1987) showed that it depended on the density of eelgrass shoots (higher where density was lower) and that subsequent growth and survival was better where the settlement density was high, because seawater flux through the eelgrass bed was reduced at high eelgrass densities. Seagrass beds reduce local velocities within the bed because of momentum extraction by the seagrass shoots. This causes decreased shear velocity and results in an increase in the proportion of fine particulates which deposit within the bed. Predator efficiency is also known to decrease as the density of seagrass shoots increase (Eckman 1987).

PHYSIOLOGICAL PARADIGM OF ADULT FILTER FEEDERS IN FLOW

The physiologist studies environmental stimuli as they affect a few cells, a tissue, up to a complete individual animal, usually in a laboratory setting. The aim is to simulate natural physical and chemical conditions around the animal while the selected environmental stimulus is varied. The goal of the physiological experimental approach is to exclude interactions with other individuals of the same or different species.

Because of the size of adult bivalves, the Reynolds number of the ambient flow is large, e.g., for giant scallops, *Placopecten magellanicus*, with a valve height of say 4 cm and at ~10 cm•s^{-1} **Re** = 8000. Thus, the appropriate hydrodynamics involves inertial forces and benthic boundary layer conditions. The filtration responses of bivalves where a sufficiently wide range of ambient velocities has been tested as an environmental variable suggests that it is a ramp function of velocity (Fig. 3), and interactive with seston concentration. Responses shown in Fig. 3 are characterized as follows (Wildish and Saulnier 1993):

a - in motionless seawater and at very low velocities seston becomes locally depleted around the individual bivalve. Velocity increments serve to remove localized seston depletion and enhance bivalve feeding.

b - uptake rates achieve an optimum level for a given concentration and quality of seston and, for a narrow range of increased ambient velocity, no further effects on uptake occur.

c - beyond the optimum velocity range, further reductions of the filtration rate are proportional to the incremental amount of velocity. At high seston concentrations (Fig. 3B), the c part of the response is prolonged to ambient velocities >45 cm•s^{-1}.

d - at high velocities, the filtration system is closed or only intermittently functions for exploration or non-feeding purposes.

Only c in the above scheme has been investigated experimentally (Wildish and Saulnier 1993), with results suggesting that seston uptake rates in the giant scallop are reduced by valve and mantle closure, whose purpose is to reduce external pressure differences between inhalant and exhalant that tends to force seawater through the filter faster than maintained by the ciliary pump. Concomitantly, the volumetric pumping and seston uptake rates are reduced.

Evidence for the ramp function response, either as filtration directly or as

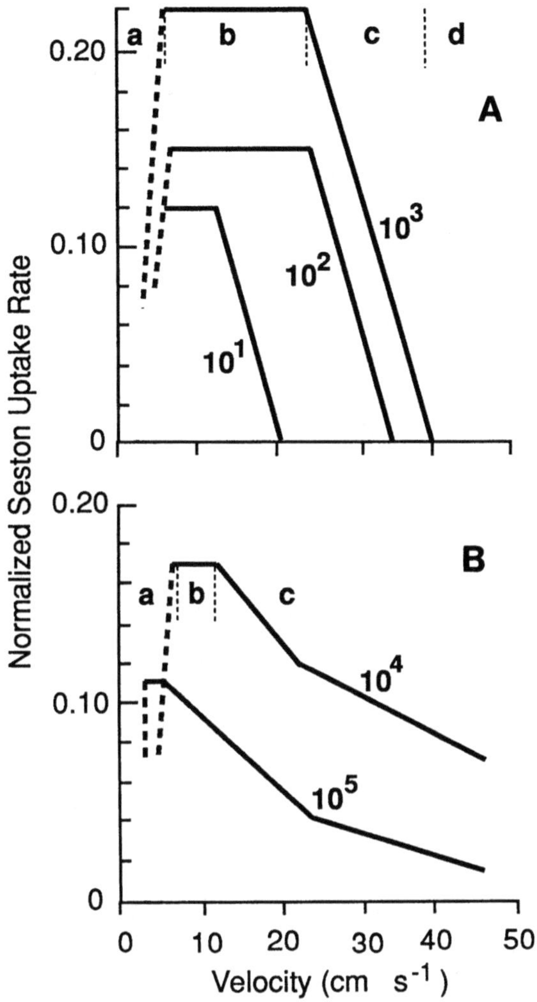

Fig. 3. Diagram showing the effect of velocity, measured near the inhalant opening, on normalized seston uptake by giant scallops at different microalgal concentrations. Data from Wildish et al (1992) each curve represents a concentration (1 x 10^x cells•mL^{-1}) of *Chroomonas salina*.

growth response to a range of velocities, is available only for the bay scallop, *Argopecten irradians concentricus* (Kirby-Smith 1972; Cahalan et al 1989; Eckman et al 1989), and the sea scallop (Wildish et al 1987; Wildish and Saulnier 1993). Preliminary studies with the blue mussel, *Mytilus edulis* (Wildish and Miyares 1990), suggest that it too may show the ramp function filtration response to velocity when its inhalant is opposed to the flow.

Bivalve suspension feeders are *active* filter feeders. This means that seawater within the trophic fluid transport system is primarily moved by a ciliary pump. The ciliary pump has been described in detail by Jørgensen (1981, 1990), Jørgensen et al (1986) and Silvester and Sleigh (1984), and the morphology of the gill filaments is well known. The metachronal beating of lateral cilia are considered as the propulsive force within the trophic fluid transport system. Maximum distances between gill filaments are _40 µm with length of cirri as: frontal - 6, latero-frontal -20-30, and lateral - 15 µm x 0.2 µm diameter (Silvester and Sleigh 1984). These sizes are consistent with very low Reynolds numbers and that the mechanism of seston capture therefore involves viscous flows. The two currents, mutually incompatible, theories regarding the mechanism of capture at the filtering surface are: the shear force theory of Jørgensen (1981) and the filtration theory of Silvester and Sleigh (1984). The shear force theory depends on a complex series of viscous currents produced by each of the three major groups of cilia (or cirri). Sestonic particles are removed from the interfilament through currents created by the lateral cirri at the interface, or shear zone, with the downwardly directed frontal currents. The frontal cirri carry the particles towards the mouth. In the filtration theory, which is compatible with classical aerosol theory (see Rubenstein and Koehl 1977), the side branches of the latero-frontal cirri are considered to form an effective sieve with pore sizes of _1 µm (Silvester and Sleigh 1984). The captured sestonic particles are then passed from latero-frontal cirri to the upstream surface of the gill filament where the frontal cirri help transfer them towards the mouth. As far as we are aware, there is insufficient experimental data to decide between these theories.

ECOLOGICAL PARADIGM OF BIVALVE FILTER FEEDING/GROWTH

The minimum unit of study for the ecologist is the population, e.g., a mussel bed, either maintained in the laboratory or the field and consisting of many individuals of a single or multiple species. More complex units of study include the community or ecosystem. A fundamental difference between the ecological and physiological paradigms is that, whereas ecology considers both temporal and spatial dimensions, physiology need only consider the temporal dimension. In designing field or laboratory ecological experiments, it is important to provide conditions that simulate those occurring in nature as closely as possible. This includes appropriate hydrodynamic conditions (usually at high Reynolds numbers), density and bivalve size, as well as appropriate scaling of the mussel bed path length so that seston depletion occurs at natural seston flux rates.

Field observations on a large horse mussel, *Modiolus modiolus*, bed in the Bay of Fundy showed that mussel production were significantly correlated with the turbulent supply of seston (Wildish and Peer 1983). This result confirmed intuitive observations by earlier bivalve biologists (e.g. Kerswill, 1949) that production or individual growth rate was influenced by water movements. Turbulent mass transfer of seston to the sediment-water interface, N_A, was defined by Wildish and Kristmanson (1979) as:

$$\mathbf{N_A = \gamma\ V(C-C^*)}$$

where γ is a dimensionless coefficient dependent on bottom roughness, **V** is the mean tidal velocity above the benthic boundary layer, **C** is the seston concentration in the bulk layer above the benthic boundary layer and **C*** is the seston concentration at the sediment-water interface near the mussel inhalants. The relationship between $\mathbf{N_A}$ and the $\log_{10}$ of suspension feeder production was shown to be significantly correlated, if some data points with zero or very low production were excluded (Wildish and Peer 1983). For the excluded stations, no or few suspension feeders were present and factors other than seston flux must control the primarily deposit-feeding macrofaunal community present. Using a subset of the data still available from Wildish and Peer (1983). Table 1,shows

Table 1. Velocity (**V**), seston concentration (**C**) and suspension feeder production (**P**) for 35 stations in the Bay of Fundy, from Wildish and Peer (1983). Sediments types: #5-98 - LaHave clay, #9-96 - Scotian Shelf drift, #28-90 - Sambro sand.

Station #	V $cm \cdot s^{-1}$	V_{max} $cm \cdot s^{-1}$	C µg $Atom^{-3}$	P Log_{10} g $wet \cdot m^{-2} \cdot y^{-1}$
5	27	54	27	1.15
7	45	80	27	-1.59
17	44	90	27	0.62
28	44	94	34	0
70	38	75	27	-1.56
73	42	83	27	0.74
79	20	27	50	0.50
81	27	48	34	0.21
82	42	56	27	0
83	35	51	34	1.36
85	44	86	27	0
86	42	96	27	0.99
92	44	87	50	1.12
97	35	65	27	1.06
98	32	60	27	0.68
9	50	99	27	1.75
11	50	97	27	2.10
13	48	93	27	0.72
15	47	92	27	1.40
16	45	93	27	1.35
60	65	135	27	3.18
69	45	87	27	0.98
74	51	99	27	1.81
89	61	120	27	1.99
95	49	90	27	1.98
96	41	76	27	1.56
28	64	128	27	3.28
29	66	137	27	3.07
32	66	136	27	1.87
49	60	125	40	0.47
61	64	130	27	3.10
62	64	135	27	3.23
65	65	133	27	2.66
66	63	127	27	3.12
90	65	102	34	1.99

the suspension feeder production as a function of velocity (based on the grand mean of the modelled bulk layer flow for all tides during 1 year - see Wildish and Peer, 1983). The regression obtained:

$$\log_{10} \mathbf{y} = 5.862\ \mathbf{x} - 1.155,\ n = 29,\ r = 0.85$$

where $\mathbf{y}$ = suspension feeder production, g wet$\cdot m^{-2} \cdot y^{-1}$

$\mathbf{x}$ = mean velocity in $m \cdot s^{-1}$

is significantly correlated at P > 99%. Six stations with 0 or little suspension feeder production are omitted from the regression.

The seston depleted layer over the bed is maintained by mussel pumping (Frechette and Bourget 1985; Frechette and Grant 1991) and supplied by turbulent transfer processes from upper layers of the benthic boundary layer. This turbulent flux is velocity dependent so that increases in velocity cause a corresponding increase in seston supply and, hence, feeding and production rate (Wildish and Kristmanson 1979). Frechette et al (1989) have provided a more complete ecological model of seston supply to a mussel bed, accounting for both turbulent flux and advection from the bulk layer flow. When turbulent mass transfer processes are not limiting, that is when **C>>C'** (where C is the bulk seston concentration and **C'** the seston concentration required for maintenance by a given size and density of mussels) growth is maximal and new recruits may be established. Both these processes, turbulent flux and new recruits, support the logarithmic increase in suspension feeder production due to increased velocity shown in Fig. 4. We do not suppose that production logarithmically increases indefinitely as Fig. 4 seems to suggest. Production must plateau due to limitation by various factors, e.g., competition for space, and eventually decline due to increased mortality linked to spatial competition and/or decreased physiological growth rates due to inhibiting ambient velocities for feeding. Bivalves cannot be maintained at very high velocities due either to difficulties of maintaining their position or to severe growth inhibition.

A BIVALVE SUSPENSION FEEDER-VELOCITY PARADOX?

A paradox is a statement contrary to established or accepted opinion, or a contradictory proposition. The physiological filtration/feeding response of blue mussels (Wildish and Miyares 1990) to increasing ambient velocity seems paradoxical in comparison with the ecological response of a horse mussel bed to increasing velocity in the Bay of Fundy (Fig. 4). The ramp part of the response must occur at a much higher velocity for the population than it does for the individual, e.g., < 6 cm•s^{-1} for the blue mussel feeding on 1 x 10^4 cells•mL^{-1} of *Chroomonas salina* (Wildish and Miyares 1990) versus >65 cm•s^{-1} for the horse mussel bed in natural conditions.

An obvious difficulty in comparing the limited evidence available concerns the very different hydrodynamic conditions of the laboratory flume (uni-directional, constant flow, smooth bottom and short duration experiments) and

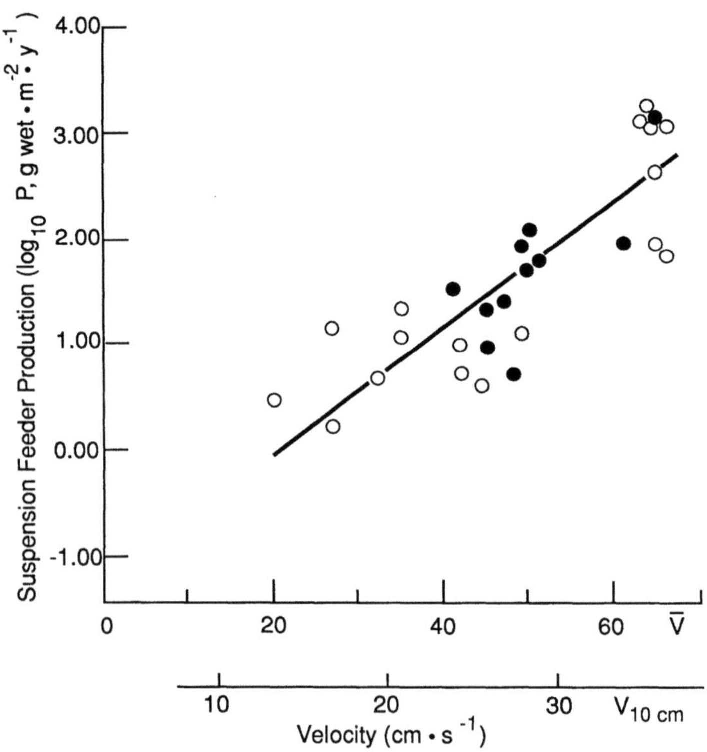

Fig. 4. Relationship between velocity, where is the average in the bulk layer flow, and $V_{10\ cm}$ is the average at 10 cm above bottom, and suspension feeder production in the Bay of Fundy (from Wildish and Peer 1983). Open circles = LaHave clay, closed circles = Scotian Shelf drift.

the field conditions (multi-directional, time-dependent flow, rough bottom and long-term experiments inclusive of tidal and seasonal changes). A limitation of the field data is that velocities are averaged for the bulk layer and so do not represent the inhalant level velocities experienced by the horse mussel. Assuming that this is ~10 cm above the sediment-water interface, which is gravel in particle size, the velocity at the inhalant level can be estimated from Soulsby (1983):

$$\tau_o = \rho C_D V^2$$

where

τ_o = shear stress of the wall, $g \cdot cm^{-1} \cdot s^{-2}$

ρ = density = 1.02 $g \cdot cm^{-3}$

C_D= drag coefficient = 0.0047

V = bulk layer mean velocity, $cm \cdot s^{-1}$

and shear velocity, u^*, is

$$u^* = (\tau_o/\rho)^{1/2}$$

then

$$V_{10} = K \ln Z/Z_o$$

where

V_{10} = the velocity at 10 cm above the bottom, $cm \cdot s^{-1}$

K = von Karamans constant

Z = height above bottom, cm

Z_o = roughness length, in cm, of gravel.

The inhalant level velocity for the horse mussel bed is presented as a separate scale in Fig. 4 and shows that horse mussel production is still increasing at 35 $cm \cdot s^{-1}$. Unfortunately, the laboratory mussel filtration/feeding data are for only one concentration and we do not know how mussels would respond to lower, more natural, seston concentrations at different flows. Assuming that it is reasonable to compare blue and horse mussel feeding/production responses in this way, there is a clear difference in response to velocity. This difference depends on whether individual mussels in short-term tests or populations in natural beds are considered.

ETHOLOGY OF ADULTS IN FLOW

The life form of adult bivalves includes *infauna*, where most of the body is present within the sedimentary substrate, and free-living *epifauna* where the bivalve is present on the substrate surface. For both infauna and epifauna, flow is important as an environmental variable, although it is of greater importance to the latter that must additionally avoid flow-induced dislodgement and, in one superfamily, the Pectinacea, many species have developed swimming capabilities.

Scallops are bilaterally symmetrical animals (Fig. 5) with the inhalant opening over a wide area of the ventral part of the animal. The exhalant produces a more powerful jet over a narrow area near the dorsal/posterior edge of the animal. In natural conditions, scallops tend to adopt a position with the ventral edge and, therefore inhalant opening, facing into the major flow direction (Hartnoll 1967;

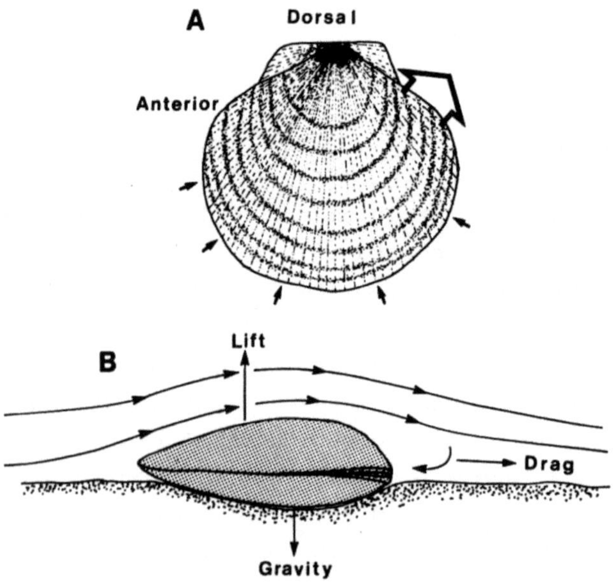

Fig. 5. *Placopecten magellanicus*. A. View from above of the left valve showing orientation and inhalant (small arrows) and exhalant (large arrow) flows; and B. view of the posterior side showing streamlines and forces affecting the instability thresholds.

Caddy 1968). These authors described how the preferred positions were attained by swimming, and by body rotations involving jet thrusting on one side (anterior or posterior) only. Recessing behavior in the giant scallop by directing exhalant flows downward serves to 'blow away' surficial sediments and to allow the animal to settle into a pit in the sediment (Caddy 1968) and hence limit its exposure to energetic flows. Both filtration and growth experiments with giant scallops (Wildish et al 1987; Wildish and Saulnier 1992) have shown that food uptake and growth are limited if the natural position of the scallop with respect to flow direction is reversed. Thus, scallops in which the exhalant opening was directly opposed to unidirectional flume flows suffered reduced seston uptake and growth rates.

In the infaunal, soft-shell clam, *Mya arenaria*, Vincent et al (1988) noticed that the paired inhalant and exhalant openings found at the tip of a mobile siphon were usually oriented normal to the major axis of tidal flow on clam flats and suggested that the preferred position led to optimal feeding because mixing of exhaled, already filtered seawater with the inhalant flow was minimized. How the preferred orientation was achieved on a particular clam flat was not determined by Vincent et al (1988) and might involve behavioral mechanisms or differential mortality. Using model bivalve siphonal currents in flume experiments, Monismith et al (1990) confirmed that an upstream, excurrent flow could involve refiltration depending on the boundary layer conditions, siphon height, orientation and size of the siphons. In their flume experiments, shear velocity and bottom roughness conditions closely simulated those which naturally occurred in the field.

Swimming has been developed by many of the ~7000 extant species of scallops. All use a form of jet propulsion in which the hinged valves are brought rapidly together by action of the large adductor muscle. The hydroelastic walls of the mantle enclose seawater during valve adduction and it is forced out as a propulsive jet. In *Placopecten magellanicus*, swimming is with the ventral edge (Fig. 5) opposed to the current with twin jets on either side of the dorsal hinge, or from the ventral edge during rapid escape movement, downcurrent, in the opposite direction (Dadswell and Weihs 1989). Considerable attention has been given to swimming in scallops, inclusive of studies that have used hydrodynamic principles to explain the complex behaviors involved (Gould 1971; Moore and Trueman 1971; Gruffydd 1976; Thorburn and Gruffydd 1979; Morton 1980; Dadswell and Weihs 1990). Major considerations in the hydrodynamic study of swimming in scallops are lift, which is proportional to valve surface area, gravitational forces resisting lift, which is proportional to buoyant weight, ambient flow velocity and direction that interacts with the scallop body, resulting in drag. This includes form, induced and frictional drag (Gould 1971), all of which affect swimming abilities. Swimming speeds are size-specific and, in juveniles of the giant scallop, up to 3.5 cm in valve height may reach maximum speeds of ~60 and average speeds of ~30 $cm \cdot s^{-1}$ (Manuel and Dadswell 1991).

A common problem for epifaunal bivalves is that they must resist tidal or wave flow forces which tend to dislodge them from their substrate. The problem may be acute because, during settlement, many bivalve larvae appear to be attracted to locations where water movement is energetic, thereby benefiting trophically from the high seston fluxes present. Newly metamorphosed and young juvenile stages of bivalves commonly rely on byssal threads that act like anchor chains and resist flow forces that tend to dislodge the animal. When the byssal threads are under tension, they become very stiff and can support the bivalve against strong dislodgement forces.

These forces include those already diagrammed in Fig. 5 as involved with swimming, with the exception of thrust that is absent. The shape and buoyant weight of the bivalve determine the lift and drag forces the animal will experience at a given current velocity. Frictional drag will be dependent on the sedimentary roughness and the ambient velocity. We are aware of very few studies that consider the ability of flow forces to dislodge bivalves from the sediment, although Ackerman et al (1992) have determined the fluid shear stresses required to detach freshwater zebra mussels from a solid substrate. Arnold and Weihs (1978) developed a hydrodynamic model for benthic organisms in flowing water where **Re** = 10,000-1,000,000 that is applicable to most bivalves because of their characteristic size and velocities experienced in field conditions. Preliminary studies using this model (Wildish and Saulnier 1993) compared live giant scallops with closed valves versus model scallops with the valves propped open. Measurements of slip speed, that is the velocity at which the bivalves first began to slide downstream when exposed to an increased flume flow, is the first indication of instability, occurring at velocities lower than for complete lift-off. Unfortunately, the data provided by Wildish and Saulnier (1993) do not include slip speeds over natural sedimentary bottoms but only on a plexiglass flume floor, so do not adequately simulate natural conditions. Nevertheless, some of the data from Wildish and Saulnier (1993) are of possible interest (Fig. 6) in deciding why older and larger

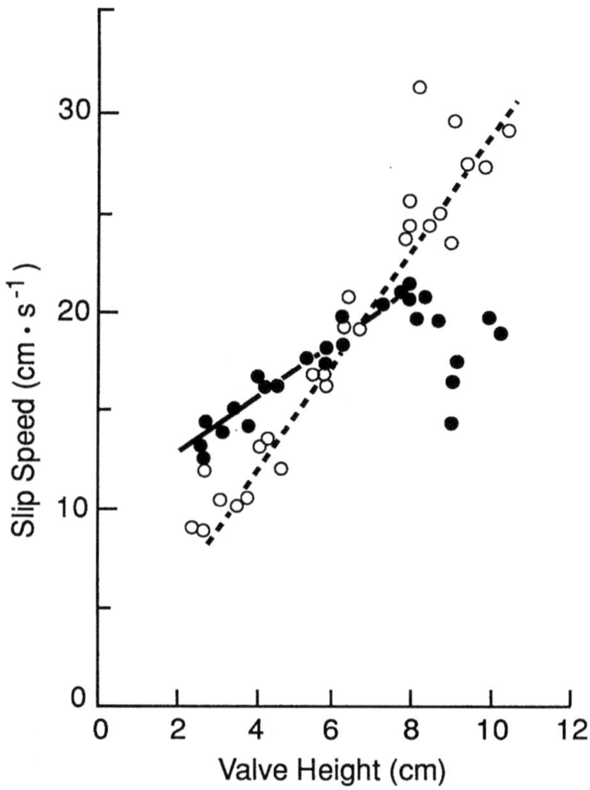

Fig. 6. Relationship between size of *Placopecten magellanicus*, as valve height, and the slip speed determined in a flume (data from Wildish and Saulnier 1993).

scallops (>8 cm valve height) become recessed into the sediment. The data suggest that beyond 8 cm the slip speed does not increase as it does for all sizes of closed valve, and <8 cm valve height open-valved scallops. Presumably, the drag associated with the increased frontal area of the maximally gaped scallop of >8 cm valve height becomes too great to be countered by the frictional resistance exerted by this size of scallop.

DISCUSSION

In this overview, we have shown that hydrodynamic processes are important at many levels of biological organization. The results accomplished to date represent the tip of the iceberg since the bulk of the work to establish hydrodynamic, mechanistic explanations for aspects of the biology of bivalve filter feeders, remains to be done. Although the form and function involved in such mechanisms probably arose by evolutionary processes, we do not feel constrained to inquire how this occurred. This is for practical reasons because the bivalves involved, as well as the environments that help form them are usually poorly known from the geological record, thus precluding direct scientific experimental investigation. It is now recognized that some morphological characteristics may not have constituted Neo-Darwinian adaptations at all but, by chance, have fitted a new function that came to have survival value for the species - the so-called exaptations of Gould and Vbra (1982). In addition, inferential studies of evolution obviously should *follow* studies that link form and function of extant bivalves.

Larval dispersal and settlement processes have yet to be fully understood. Much of the earlier work concerned active larval habitat selection and the role of physical processes in settlement was not begun until the 1970's. Hydrodynamic modelling should continue to be of value in providing mechanistic explanations of larval dispersal and settlement. For example, dispersal patterns of bivalve larvae in relation to large-scale currents and, at a much smaller scale, the possibility that primary settlement of bivalve larvae on elevated substrates, is a means of avoiding the seston depletion near the bottom of the benthic boundary layer, should be tested against appropriate alternative hypotheses.

Although the research effort devoted to bivalve physiology has been the largest body of work reviewed, the effect of flow on feeding/growth had been ignored until the work of Kirby-Smith (1972). A complex interaction involving flow direction, velocity and seston concentration controlling filtration/feeding, and hence growth in filter-feeding bivalves, has been established recently. Some implications from this new work are that the currently used filtration chamber technique (e.g., MacDonald and Thompson 1986; Frechette et al 1989), in which natural seawater is pumped through small chambers containing bivalves, may give results that are not representative of field hydrodynamic conditions. Also, because of the difficulties in obtaining sufficient seasonal input data for each site, it is concluded that predicting growth of bivalves based on filtration/feeding models will not repay the considerable effort involved.

Laboratory flume studies involving the effect of flow at ecological levels of organization are few, although those available are generally supportive of the hydrodynamic models available. Recently, Cole et al (1992) have published preliminary data for a simulated clam bed, *Potamocorbula amurensis*, which suggest that at flume flows <15 $cm \cdot s^{-1}$, population filtration rates were inversely related to velocity. Differences in filtration/ feeding/growth responses of individuals versus populations suggests an apparent bivalve suspension-feeding paradox. Further examination suggests that the paradox results from the divergent paradigms underlying physiology and ecology. Thus, populations respond trophically to seston flux while individuals respond to velocity and seston concentration independently. Populations increase their feeding/growth in response to velocity because it increases the turbulent flux rate. Other factors that may respond to changed flux include changes in density, inclusive of migration, larval settlement and starvation-related mortality. Bivalve populations feeding at suboptimal rates due to velocity inhibition may achieve ecological efficiencies because the seston supply is shared along the downstream path of the bed by a greater number of individuals. The interface between ecology and physiology needs further exploring in view of the very different goals of each discipline.

Suspension-feeding bivalves are sedentary and those which are epifaunal have a common problem in having to resist wave or tidal movements that tend to dislodge them. Among scallops, a swimming ability has been developed and here the two opposing functions of locomotion and resistance to dislodgement must be met by morphological and behavioral adaptations - or exaptations - in the same bivalve. This is a rewarding area for the future application of hydrodynamic principles as they affect sediments and filter-feeding bivalves.

ACKNOWLEDGMENTS

We thank J.D. Ackerman and M. Frechette for critically reviewing the manuscript and Brenda Best and F. Cunningham for preparing it for publication.

REFERENCES

Ackerman JD Ethier CR Allen DG Spelt JK (1992) Investigation of zebra mussel adhesion strength using a rotating disk. J Environ Engin 118: 708-724

Arnold GP Weihs D (1978) The hydrodynamics of rheotaxis in the plaice (*Pleuronectes platessa* L.). J Exp Biol 75: 147-169

Bayne BL (1964) Primary and secondary settlement in *Mytilus edulis* L. (Mollusca). J Anim Ecol 33: 513-523

Butman CA (1987) Larval settlement of soft-sediment invertebrates: the spatial scales of pattern explained by active habitat selection and the emerging role of hydrodynamical processes. Oceanogr Mar Biol Ann Rev 25: 113-165

Caddy JF (1968) Underwater observations on scallop (*Placopecten magellanicus*) behaviour and drag efficiency. J Fish Res Board Can 25: 2132-2141

Cahalan JA Siddall SE Luckenbach MW (1989) Effects of flow velocity, food concentration and particle flux on growth rates of juvenile bay scallops *Argopecten irradians*. J Exp Mar Biol Ecol 129: 45-60

Chia FS Buckland-Nicks J (1984) Locomotion of marine invertebrate larvae: a review. Can J Zool 62: 1205-1222

Cole BE Thompson JK Cloern JE (1992) Measurement of filtration rates by infaunal bivalves in a recirculating flume. Mar Biol 113: 219-226

Dadswell MJ Weihs D (1990) Size-related hydrodynamic characteristics of the giant scallop, *Placopecten magellanicus*, (Bivalvia:Pectinidae). Can J Zool 68: 778-785

Eckman JE (1979) Small-scale patterns and processes in a soft substratum, intertidal community. J Mar Res 37: 437-457

Eckman JE (1987) The role of hydrodynamics in recruitment, growth, and survival of *Argopecten irradians* (L.) and *Anomia simplex* (D'Orbigny) within eelgrass meadows. J Exp Mar Biol Ecol 106: 165-191

Eckman JE (1990) A model of passive settlement by planktonic larvae onto bottoms of differing roughness. Limnol Oceanogr 35: 887-901

Eckman JE Duggins DO Sewell AT (1989) Ecology of understorey kelp environments. I. Effects of kelps on flow and particle transport near the bottom. J Exp Mar Biol Ecol 129: 173-187

Ertman SC Jumars PA (1988) Effects of bivalve siphon currents on the settlement of inert particles and larvae. J Mar Res 46: 797-813

Eyster LS Pechenik JA (1987) Attachment of *Mytilus edulis* L. larvae on algal and byssal filaments is enhanced by water agitation. J Exp Mar Biol Ecol 114: 99-110

Frechette M Bourget E (1985) Food limited growth of *Mytilus edulis* in relation to the benthic boundary layer. Can J Fish Aquat Sci 42: 1166-1170

Frechette M Butman CA Rockwell Geyer W (1989) The importance of boundary layer flows in supplying phytoplankton to the benthic suspension feeder, *Mytilus edulis* L. Limnol Oceanogr 34: 19-36

Frechette M Grant J (1991) An in situ estimation of the effect of wind-driven resuspension on the growth of the mussel *Mytilus edulis* L. J Exp Mar Biol Ecol 148: 201-213

Gallagher SM Mann R (1986) Growth and survival of larvae of *Mercenaria mercenaria* (L.) and *Crassostrea virginia* (Gmelin) relative to broodstock conditioning and lipid content of eggs. Aquaculture 56: 105-121

Gould SJ (1971) Muscular mechanics and the ontogeny of swimming in scallops. Palaeontology 14: 61-94

Gould SJ Vbra ES (1982) Exaptation - a missing term in the science of form. Paleobiology 8: 4-15

Gruffyd LD (1976) Swimming in *Chlamys islandica* in relation to current speed and an investigation of hydrodynamic lift in this and other scallops. Norw J Zool 24: 365-378

Hartnoll RG (1967) An investigation of the movement of the scallop. Helgolander wiss Meeresunters 15: 523-533

Jonsson PR André C Lindegarth M (1991) Swimming behaviour of marine bivalve larvae in a flume boundary layer flow: evidence for near-bottom confinement. Mar Ecol Prog Ser 79: 67-76

Jørgensen CB (1981) A hydromechanical principle for particle retention in *Mytilus edulis* and other ciliary suspension feeders. Mar Biol 61: 277-282

Jørgensen CB (1990) Bivalve filter feeding: hydrodynamics, bioenergetics, physiology and ecology. Olsen and Olsen, Fredensborg, Denmark, 140 p

Jørgensen CB Famme P Kristenson HS Larsen PS Møhlenberg F Riisgard HV (1986) The bivalve pump. Mar Ecol Prog Ser 34: 69-77

Kerswill CJ (1949) Effects of water circulation on the growth of quahogs and oysters. J Fish Res Board Can 7: 545-551

Kirby-Smith WW (1972) Growth of the bay scallop: the influence of experimental water currents. J Exp Mar Biol Ecol 8: 7-18

Lazier JR Mann KH (1989) Turbulence and the diffusive layers around small organisms. Deep Sea Res 36: 1721-1733

Lynch JM Poole NJ (eds) (1979) Microbial ecology: a conceptual approach. Blackwell, Oxford, 266 p

MacDonald BA Thompson RJ (1986) Influence of temperature and food availability on the ecological energetics of the giant scallop *Placopecten magellanicus*. III. Physiological ecology, the gametogenic cycle and scope for growth. Mar Biol 93: 37-48

Manuel JL Dadswell MJ (1991) Swimming behaviour of juvenile giant scallop, *Placopecten magellanicus*, in relation to size and temperature. Can J Zool 69: 2250-2254

McGrath D King PA Gosling EM (1988) Evidence for the direct settlement of *Mytilus edulis* larvae on adult mussel beds. Mar Ecol Prog Ser 47: 103-106

Monismith SG Koseff JR Thompson JK O'Riordan CA Nepf HM (1990) A study of model bivalve siphonal currents. Limnol Oceanogr 35: 680-696

Moore JD Trueman ER (1971) Swimming of the scallop *Chlamys opercularis* (L.). J Exp Mar Biol Ecol 6: 179-185

Morton B (1980) Swimming in *Amusium pleuronectes* (Bivalvia:Pectinidae). J Zool Lond 190: 375-404

Parsons TR Takahashi M Hargrave B (1977) Biological oceanographic processes. Second Edition, Pergamon, Oxford, 332 p

Purcell EM (1977) Life at low Reynolds number. Am J Physics 45: 3-11

Rubenstein DI Koehl MAR (1977) The mechanisms of filter feeding: some theoretical considerations. Amer Nat 111: 981-994

Schlichting H (1968) Boundary-layer theory. Sixth Edition, translated by J. Kestin. McGraw-Hill, New York, 747 p

Silvester NR Sleigh MA (1984) Hydrodynamic aspects of particle capture by *Mytilus*. J Mar Biol Assoc UK 64: 859-880

Soulsby RL (1983) The bottom boundary layer of shelf seas, p 189-266. In B. Johns (ed.) Physical oceanography of coastal and shelf seas. Elsvier Oceanography Series, 35, Amsterdam

Thorburn IW Gruffydd LD (1979) Studies of the behaviour of the scallop *Chlamys opercuralis* (L.) and its shell in flowing seawater. J Mar Biol Assoc UK 59: 1003-1023

Tremblay MJ Sinclair M (1990) Sea scallop larvae *Placopecten magellanicus* on Georges Bank: vertical distribution in relation to water column stratification and food. Mar Ecol Prog Ser 61: 1-15

Tyler PA Banner FT (1977) The effect of coastal hydrodynamics on the echinoderm distribution in the sublittoral of Oxwich Bay, Bristol Channel. Est Coastal Mar Sci 5: 293-308

Vincent R Desrosiers G Gratton Y (1988) Orientation of the infaunal bivalve *Mya arenaria* L. in relation to local current direction on a tidal flat. J Exp Mar Biol Ecol 124: 205-214

Waite JH (1983) Adhesion in bysally attached bivalves. Biol

Rev 58: 209-231

Wildish DJ Kristmanson DD (1979) Tidal energy and sublittoral macrobenthic animals in estuaries. J Fish Res Board Can 36: 197-1206

Wildish DJ Kristmanson DD Hoar RL DeCoste AM McCormick SD White AW(1987) Giant scallop feeding and growth responses to flow. J Exp Mar Biol Ecol 113: 207-220

Wildish DJ Kristmanson DD Saulnier AM (1992) Interactive effect of velocity and seston concentration on giant scallop feeding inhibition. J Exp Mar Biol Ecol 155: 161-168

Wildish DJ Miyares MP (1990) Filtration rate of blue mussels as a function of flow velocity: preliminary experiments. J Exp Mar Biol Ecol 142: 213-220

Wildish DJ Peer D (1983) Tidal current speed and production of benthic macrofauna in the lower Bay of Fundy. Can J Fish Aquat Sci 40 (Suppl. 1): 309-321

Wildish DJ Saulnier AM (1993). Hydrodynamic control of filtration in the giant scallop. Submitted to J Exp Mar Biol Ecol

BIVALVE FEEDING AND THE BENTHIC BOUNDARY LAYER

Marcel Fréchette, Denis Lefaivre and Cheryl Ann Butman[1]
Institut Maurice-Lamontagne
Ministère des Pêches et des Océans
C.P. 1000
Mont-Joli Québec
Canada G5H 3Z4

INTRODUCTION

Benthic suspension feeders particularly some bivalves have strong structuring effects on rocky shore communities which are commonly dominated by the Mytilidae in most oceans (see Paine 1984). This dominance may be related to at least two important aspects of their ecology the sestonic nature of their food and their sessile mode of life. The food of seston feeders is produced in a three-dimensional environment which flows over the animals. Under favourable conditions this food is continuously replenished by currents sinking and reproduction of planktonic organisms. Thus owing to their sedentary nature and feeding behaviour there is strong coupling between the pelagic and benthic environments through bivalves. Energy from the pelagos is channeled directly to the benthos via the filtering activity of the organisms and is eventually used for various metabolic and growth activities (see Dame and Patten 1981). Given the above it can be postulated that the biomass of other feeding guilds (e.g. carnivores which depend directly on primary consumers and perhaps even grazers through the elimination of algal growth by sessile animals Dayton 1973) as well as the overall structure of intertidal communities is dependent on the "success" or "failure" of the suspension feeders. Thus given the central position of suspension feeders

[1]Applied Ocean Physics and Engineering Department, Woods Hole Oceanographic Institution, Woods Hole, Massachusetts, 02543 USA

NATO ASI Series, Vol. G 33
Bivalve Filter Feeders
in Estuarine and Coastal Ecosystem Processes
Edited by Richard F. Dame

as structuring agents of the intertidal communities energy flow from the pelagos to the benthos can be determined by measuring energy flow through this guild. Similar principles hold for subtidal macrofaunal assemblages. Wildish and Peer (1983) for example reported similar relationships between the pelagos and benthos in the Bay of Fundy where suspension feeders contributed over 88% of the total macrofaunal production.

Early studies have suggested current speed as a key factor in promoting secondary production of benthic suspension feeders (Fraser and Smith 1928*a,b*; Kerswill 1949). There is indeed strong evidence that individual growth of active benthic suspension feeders is promoted by increases in current speed (Malouf and Breese 1977; Wildish and Peer 1983; Wilson 1987; Fréchette et al. 1989; Grizzle and Morin 1989; Grizzle and Lutz 1989; Emerson 1990; Newell 1990). At the fastest current speed tested however growth was depressed (Wilson 1987; Grizzle and Morin 1989; Grizzle and Lutz 1989). Thus the relationship between current speed and growth is non-monotonic with negative concavity. This general relationship also has been found at the community level; Wildish and Peer (1983) reported a strong positive relationship between total suspension feeder production and current speed. Low and very high current speeds however resulted in the dominance of deposit feeders and impoverished assemblages respectively (see also Warwick and Uncles 1980).

Negative effects of very high current speed on growth of epifaunal organisms may result from the action of a variety of mechanisms. Wildish et al (1987) and Wildish and Miyares (1990) for example reported direct feeding inhibition attributable to hydrodynamically induced modifications of the internal pressure gradients between the siphonal openings when the exhalent siphon points into the current (see also Eckman et al. 1989 who concluded in addition that turbulence is not involved in the negative response to current speed). Resuspension may also be involved but experiments by Bricelj and Malouf (1984) and Bricelj et al (1984) suggest that resuspension negatively affects growth of *Mya arenaria* only at very high particle concentrations characteristic of storm conditions (see also Turner and Miller 1991 for an account of simulated storm events on the growth of *Mercenaria mercenaria*).

Our goal here is to illustrate how local hydrodynamics may

modulate energy flow in bivalve suspension feeders. We shall limit our analysis to the study of mechanisms that may explain the positive relationship between current speed and growth in field situations. In particular we shall analyse different approaches to field studies and the conceptual framework for these approaches with respect to benthic boundary layer (BBL) processes.

CURRENT SPEED AND GROWTH IN BIVALVE BEDS

(1) Scope for growth

A strategy fundamental to our understanding of reciprocal relationships between benthic suspension feeders and the hydrodynamic environment in which they live has involved the study of scope for growth (SFG) as a function of food availability. Determination of SFG amounts to estimating the contribution of each of the different physiological processes controlling individual growth such as food consumption respiration excretion and so forth and then evaluating the net result. The SFG strategy can be contrasted with growth experiments in that the former decomposes the nature of the growth into its fundamental components whereas the latter considers only net growth.

Potential advantages of the SFG strategy are that (1) it offers the possibility of defining and measuring predictable relationships generally involving allometry (2) the different components controlling growth can be analysed separately (3) results can be integrated into models of energy flow in ecosystems and (4) results represent meaningful correlations between food consumption and growth (Bayne and Worrall 1980; Bayne et al 1989). The SFG approach has been used for example to understand energy flow and individual growth in bivalve populations growing under different thermal regimes (Bayne and Widdows 1978 Bayne and Worrall 1980) or in conjunction with flow analysis to understand the structure of bivalve beds as "ecosystems" (Dame and Patten 1981). SFG or some aspect of it coupled with estimates of food variability and transport has also been used in ecological analyses and for the management of bivalve aquaculture (Rosenberg and Loo 1983; Héral et al 1983; Rodhouse et al. 1985; Bacher 1989; Bacher et al. 1991; Fréchette et al 1991; Smaal et al 1986; Herman and Scholten 1990). Furthermore SFG can be determined at various points within a range of complex temporal variability whereas growth experiments integrate over all conditions during the study period (Fréchette and Grant 1991).

An implicit assumption for using SFG as a tool for understanding the relationship between bivalve growth and the environment is that growth is limited by some metabolic constraint imposed by the environment. If the problem under study is the positive effect of current speed on growth a further point is that growth is biomass/density-dependent (Fréchette and Lefaivre 1990). Chemical competition (accumulation of waste products) oxygen limitation and food limitation have all been proposed as possible explanations. Jørgensen (1980) for example found oxygen depletion in bottom water above a mussel bed in the Limfjord Denmark. Nixon et al (1971) however observed that metabolism in a mussel bed increased with current speed. If the situation in the mussel bed studied by Nixon et al. is analogous to that in the Limfjord it is possible that Nixon et al. observed the effect of enhanced removal of presumably oxygen-depleted water directly above the mussels. Analogous situations but with food as the depleted rate-limiting resource have been observed in intertidal (Fréchette and Bourget 1985a,b; Fréchette et al. 1989; Fréchette and Grant 1991) and subtidal (Newell 1990) mussel beds in eastern North America. It is not clear why the other potentially limiting factors have not received as much attention as food limitation but it could be due to the lack of appropriate instrumentation for such studies.

The number of studies demonstrating or suggesting food depletion above suspension feeder populations continues to increase (see references in Peterson and Black 1991 and Asmus and Asmus 1991). This view of food as the limiting factor to individual growth has replaced the earlier tenet that density-dependence in growth of benthic suspension feeders was caused primarily by space limitation (see references in Buss 1979). Hughes and Griffiths (1988) however have interpreted self-thinning relationships in mussels and barnacles as indicative of space limitation (see also Ardisson and Bourget 1991). Methodologies based on experimental evidence (Peterson and Andre 1980) and analysis of self-thinning relationships (Fréchette and Lefaivre 1990) have been proposed to resolve whether space or food limits a given population. This rather manichaean way of considering food or space competition as mutually exclusive was challenged by Buss (1979) who concluded that in bryozoans food and space should be seen as interdependent limiting factors. In a recent study Fréchette et al (1992) drew similar conclusions for the mussel *Mytilus edulis*. Experiments by both Buss (1979) and Fréchette et al (1992) were done in the laboratory. Whether food limitation acts alone or in conjunction with space limitation SFG is still useful for understanding at least the food limited part of the relationship between growth and current speed.

The versatility and potential of the SFG methodology has prompted a vast research effort on methodology for obtaining unbiased estimates of SFG and for determining environmental factors responsible for its control. Because SFG offers the possibility of developing predictive relationships a general strategy for evaluating energy transfer in sessile suspension feeders has been used where SFG is estimated in the laboratory and relevant environmental variables and bivalve growth or biomass are estimated in the field. This strategy characterized an early phase of research on this topic which we call the "isotropic approach". It paved the way to development of a second research strategy where the emphasis for data acquisition was placed on the resolution of environmental variables that may control growth and food availability rather than on estimates of SFG *per se*. We call this the "boundary-layer approach". The latest trend which we refer to as the "field flume approach" is characterized by attempts to estimate horizontal fluxes of food and nutrients in the field but with virtually no attempt to assess mechanisms acting at the individual level. Each of these approaches and their strengths and weaknesses are discussed below.

(2) The isotropic approach

According to the isotropic approach concentrations of particles and dissolved substances as well as current speed are assumed to be uniformly distributed over small spatial scales (of meters) above bivalve beds. Notwithstanding phytoplankton patchiness which occurs at various spatial scales (Platt et al 1970; Marra et al 1982) it is assumed that sampling the bulk water column is representative of particle concentration available to or hydrodynamic conditions experienced by any given bivalve bed. Many studies have focussed on the seasonal or annual aspects of seston heterogeneity (e.g. Haven and Morales-Alamo 1966) and have sampled at only one height in the water column generally near the bottom but typically at -0.5 m above a bivalve bed.

Whereas spatial resolution of seston concentration in the field has been highly limited in this approach considerable effort has been devoted to determining functional relationships between physiological processes and (bulk) environmental conditions either in closed or in flow-through systems. In a closed system bivalves are placed in a chamber of known water volume with no renewal of water during an experiment and particle concentration or some other variable of interest is monitored as the animals process the water. Sedimentation is accounted for in the calculation of filtration (clearance) rate. Closed systems offer the advantage of being simple to manage because only a single initial batch of seston-laden water is required. The approach assumes however that the animals are being exposed to decreasing particle concentrations as the experiment proceeds. Because food consumption in bivalves is food-dependent (Bayne et al. 1989) flow-through systems may be preferable to closed systems although time-series strategies have been developed to circumvent the problem of variability in filtration rate with decreasing food concentration in closed systems (McClatchie and Lewis 1986; see also Cole et al 1992). Closed systems are also limited in experimental duration.

Flow-through setups were developed to circumvent problems with closed systems. In flow-through systems chambers are arranged in series or in parallel. In the former particle concentration or some other conservative estimator of seston availability (e.g. phytoplankton fluorescence) is sampled at the inflow and outflow of the chamber (see Hildreth and Crisp 1976). In parallel systems particle concentration is sampled at the outflow only of both experimental and control chambers although more sophisticated systems have been used (e.g. see Propp et al. 1982). In the laboratory flow-through systems allow for 1) either the maintenance of experimental steady state conditions (Riisgård and Møhlenberg 1979) or manipulations of experimental conditions without disturbing the animals (Thompson and Bayne 1974) 2) time-series studies of filtration rate (Fréchette and Bourget 1987) 3) pumping of water from the environment to experimental chambers with minimal distortion of bulk environmental conditions.

Selected examples of advances made in studies of bivalve feeding behavior and growth coupled with the estimation of conditions in the water column include the analysis of the effect of seston level on mussels and clams (Widdows et al 1979; Bricelj and Malouf 1984; Bricelj et al 1984) and the effect of seston quality on a scallop (Vahl 1980 1981a,b). Bulk water column measurements coupled with detailed laboratory measurements have provided estimates of energy budgets (Rodhouse 1978) and of the rate of bio-deposit production by oysters (Bernard 1974; Haven and Morales-Alamo 1966) as well as the study of eco-physiological aspects of production and reproduction in mussels (Thompson 1984a,b). Relationships between bivalve feeding growth and seston availability have also been studied from the reverse angle i.e. that of quantifying the potential impact of suspension-feeding communities on seston in the entire water column (Officer 1982; Cloern 1982; Dame et al 1980; Hily 1991). In these studies the extent to which phytoplankton distributions may be controlled by benthic suspension feeders has been ascertained by comparing the time required for the benthos to filter the entire water column relative to time scales associated with other processes such as phytoplankton growth zooplankton grazing and physical transport.

By the end of the late seventies there was growing concern about the spatio-temporal complexity of particle concentration fields in near-bottom waters above benthic suspension feeding populations and the difficulties associated with extrapolating laboratory results to complex field situations. Indeed phytoplankton concentration and other components of seston in coastal areas are heterogeneous both in time and space owing as much to the effect of water-column processes on phytoplankton patchiness (Platt et al 1970) as to the complex interplay between coastal gradients in primary production and resuspension and the associated temporal heterogeneity of these processes (Demers et al 1987; Demers et al 1989; Bacher et al 1991). Furthermore in view of the accumulating evidence for the effect of current speed on energy flow in active suspension feeders (e.g. Nixon et al 1971) and benthic communities in general (Warwick and Uncles 1980; Wildish and Peer 1983) a new paradigm emerged which articulated the effect of current speed on growth via the combined effects of food limitation near the bottom and renewal of food depleted water by the flow regime (Wildish 1977; Wolff 1977).

Despite its usefulness for understanding basic reciprocal relationships between bivalve growth and environmental factors such as seston concentration water temperature and so forth the isotropic approach has been unable to explain effects of current speed on growth other than at the correlative level for the simple reason that mechanistic aspects of current speed and related hydrodynamic characteristics of the environment are not included in the analysis. In particular it overlooks the effect of a major universal feature of flows above the bottom the benthic boundary layer.

(3) The benthic boundary layer approach

Aspects of BBL physics

Water flowing over the sea floor experiences frictional drag at the fluid-surface interface which dissipates energy in the flow and thus slows down the current speed. Because of molecular and turbulent mixing frictional effects at the bottom are transferred within the flow to some distance above the bottom where the flow is no longer influenced by the bed. This conceptually defines the benthic boundary layer region of flow. Descriptions of the hydrodynamical characteristics within the BBL relevant to benthic biologists may be found for example in

Wimbush and Munk (1970) Komar (1976) Nowell and Jumars (1984) Muschenheim et al (1986) Butman (1987) Nowell and Jumars (1987). Thus we briefly summarize here only those characteristics of the BBL that are relevant to active benthic suspension feeders.

In unsteady or gradually varying flows (e.g. as opposed to steady flows over plates) the thickness of the BBL, δ can be approximated by $\delta = \kappa u_* \sigma^{-1}$ where κ is the von Karman constant ($\cong 0.4$) u_* is the friction velocity ($u_* = (\tau \rho^{-1})^{1/2}$ where τ is the boundary shear stress ρ is the fluid density and σ is equal to $2\pi T^{-1}$ where T is the period of the flow). Three regions may be recognized in turbulent boundary layers (Fig. 1). Nearest to the wall is the viscous sub-layer where velocity varies linearly with height above the bed. The viscous sub-layer is largely

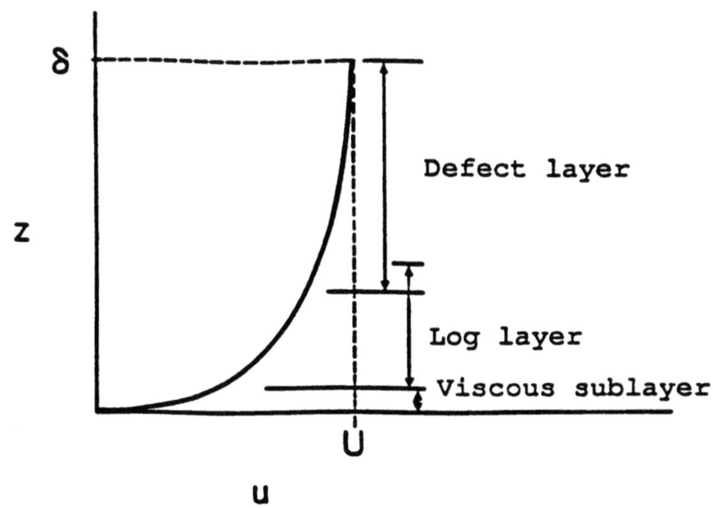

Fig. 1. Diagram of a turbulent bottom boundary layer plotted on a linear scale for both axes showing the relative positions of the viscous sub-layer the log layer and the defect layer. u is water velocity at height z, U is mean-stream velocity and δ is boundary-layer thickness. (After Butman 1987).

continuous and well-defined in flows where roughness elements are sufficiently small such that they are buried within the sub-layer (e.g. sediments of fine sand or smaller) and the roughness Reynolds number Re_* is less than about three ($Re_* = u_* k_b \nu^{-1}$ where k_b is the size of the roughness elements and ν is the kinematic viscosity of the fluid). This kind of flow is referred to as hydraulically smooth-turbulent and viscosity effectively damps turbulent fluctuations within the sub-layer. When Re_* is greater than about 100 roughness elements protrude above the sub-layer and contribute to mixing in the flow so the viscous sub-layer is not continuous. This kind of flow is referred to as hydraulically rough-turbulent and characterizes flow over mussel beds. Under such conditions the transport of momentum within the BBL is determined largely by turbulent mixing processes. Above the viscous sub-layer is a region known as the log layer where velocity varies logarithmically with distance above the bottom according to

$$\frac{u}{u*} = \frac{1}{\kappa} ln\left(\frac{z}{z_0}\right)$$

where u is the mean horizontal flow speed at height z and z_0 is roughness parameter. Finally there is an outer region of flow called the defect layer where the behavior of flow is largely independent of bed roughness.

The discussion above is valid for a steady unstratified flow over a flat bottom of relatively constant roughness. For discussions of the effect of current accelerations waves and moveable bed roughness on the BBL characteristics see for example Nowell (1983) Grant and Madsen (1986) Heathershaw and Langhorne (1988) and Wright et al (1992).

Boundary layers and bivalve growth

Boundary layers are important because they are sites of relatively intense vertical mixing and thus of exchange of particulate and dissolved matter between the pelagic and benthic zones. Because this mixing controls the vertical fluxes near the bed knowledge of BBL processes is essential for understanding the relationships between food available to benthic suspension feeding organisms and the near-bed flow regime. This is the basis for the suggestion by Wildish and Kristmanson (1979) that the mechanistic explanation underlying the relationship between current speed and bivalve growth is that dense populations of benthic suspension feeders deplete the suspended sestonic food source and that food replenishment to the bottom is accomplished by vertical turbulent transport. The potential importance of BBL dynamics was first shown in a flume experiment in which Wildish and Kristmanson (1984) observed that seston depletion was in fact a function of the flow decreasing with mean current speed in the BBL. More recent studies have pointed out however that it is a combination of vertical mixing and horizontal advection that determine ultimately food supply to mussels (e.g. Fréchette et al. 1989; Butman et al. submitted).

Gathering field evidence to test food depletion above the bottom and its effect on energy flow requires sampling designs highly contrasting with those typically used in the isotropic approach. First because flow speed and mixing vary with height above the bed small-scale vertical gradients in current speed and food concentration in the BBL must be resolved. Furthermore it is critical that sampling begin as close to the bed as possible using pumps for example with inlets at fixed heights. Second because phytoplankton are highly patchy in distribution a large number of samples is required to ensure satisfactory resolution of horizontal vertical and temporal gradients. In the field these measurements must occur simultaneously with some assessment of bivalve growth or food consumption over the observed food gradient. Owing to the large sampling effort devoted to resolve spatial gradients it is probably impractical to study the allometry of the physiological processes (e.g. filtration rate). Rather filtration rate measurements can be made on replicate groups of representative individuals in a time-series fashion concurrently with sampling of the seston (see Widdows et al. 1979 for an early example).

This general approach was used by Fréchette and Bourget (1985a,b) to study evidence of phytoplankton depletion and associated growth limitation in an intertidal mussel bed. In a series of experiments sampling effort was devoted to obtaining near-bottom vertical profiles of chlorophyll a and phaeopigments coupled with current speed measurements and estimates of wave activity. Taking advantage of local bottom topography these observations were conducted simultaneously with measurements of the growth of mussels in cages at different heights above the bottom but with same period of tidal immersion. These studies showed that phytoplankton was indeed depleted above the mussel bed for about two thirds of the tidal cycles studied. Depletion was alleviated however by waves and currents. Nevertheless growth was lower near the bottom which was interpreted as an indication of food-limited growth. Despite the mussels being food limited Fréchette and Grant (1991) found that resuspension events did not have a positive effect on growth. Newell (1990) however studied spatial patterns in growth in patches of subtidal mussels and found that growth was faster in relatively small versus large patches. Large patches exhibited edge effects where growth was fastest at the leading edge (relative to the flow direction) of the patches. Furthermore edge zones were wider in zones of high current speed than in zones of low current speed. This is consistent with progressive depletion of food along a mussel bed as well as food-limited growth of mussels.

The discussion above specifically concerns the problem of quantifying food depletion above a bed of suspension feeders and its effect on bivalve growth. The studies cited recognized the potential role of BBL dynamics in determining food availability and sampling designs were adjusted accordingly. The interplay between flow dynamics and bivalve feeding on seston can also modify the qualitative nature of sestonic food. Newell et al (1989) for example sampled seston at two heights above a subtidal mussel bed from the edge of the bed to its inner portion. They found a progressive decrease in seston abundance and quality toward the inner parts of the mussel bed owing to selective feeding. Measurements in the laboratory showed that mussels cleared chlorophyll-bound particles preferentially to particles that did not contain chlorophyll.

The importance of benthic boundary processes on seston quality has also been shown in studies of resuspended sediments and growth of suspension feeders. Turner and Miller (1991) for example showed profiles of resuspended sediments in a laboratory flume which demonstrate the complex nature of suspended particle profiles within the BBL. In short vertical gradients in depleted phytoplankton and resuspended sediments are mirror images of each other. Furthermore owing to the interaction between the dynamics of sheared flows and resuspension/deposition dynamics Muschenheim (1987a,b) found that the food available to benthic suspension feeders was highly structured in the vertical.

Whereas there is general agreement that depletion above a dense assemblage of suspension feeders is related to the BBL flow region the factor(s) that ultimately control bivalve growth has been debated in the recent literature. Studies by Wildish (1977) Wildish and Kristmanson (1979 1984) Fréchette and Bourget (1985a,b) Fréchette et al (1989) Cahalan et al (1989) Eckman et al (1989) Peterson and Black (1991) Turner and Miller (1991) have implicated changes in food concentration (a scalar quantity) near the bottom as the mechanism linking current speed and individual growth. A different view has been offered by Grizzle and Lutz (1989) Grizzle and Morin (1989) (see also Emerson 1990 and Muschenheim and Newell 1992) – that it is food flux rather than food concentration that controls growth. This was based on the fact that growth of *Mercenaria mercenaria* correlated better with seston flux near the bottom than with either seston concentration or current speed. It should be noted however that because food flux incorporates both concentration and velocity it is correlated with both these parameters and thus cannot be viewed as a separate independent factor. Growth experiments with *Argopecten irradians* have failed to show significant effects of seston flux on individual growth (Cahalan et al. 1989). Until further experiments are conducted to test specifically the seston flux hypothesis we view the relationship between flux and mussel growth as an epiphenomenon attributable to the fact that current speed and food concentration are correlated in food-depleted boundary layers.

Studies reporting evidence of seston depletion in the field have generally found depletion in the vertical. Sampling vertical gradients at several points along a bed of benthic suspension feeders is required to adequately assess the spatial pattern of depletion and its effect on bivalve growth. In lieu of this intense sampling effort a two-dimensional numerical model was used to evaluate the magnitude of horizontal and vertical gradients of food availability over an intertidal mussel bed (Fréchette et al. 1989). Based on field data and flow theory model results indicated that strong gradients both in the horizontal and the vertical can occur over a mussel bed (Fig. 2; from Fréchette et al. 1989). Because vertical transport of phytoplankton is determined by turbulence and mixing within the

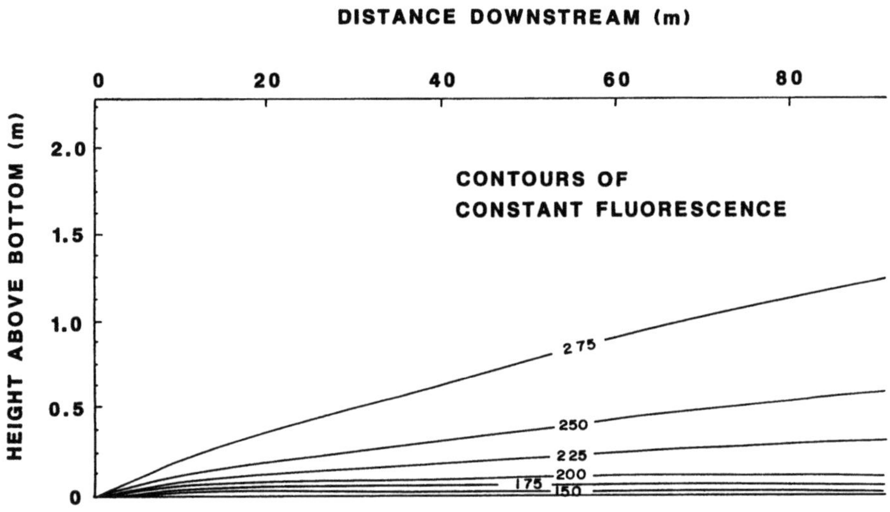

Fig. 2. Horizontal section of model-generated contours of phytoplankton concentration (in fluorescence units liter^{-1}) over the mussel bed for surface current speed u = 15 cm s^{-1} at water depth H = 2.25 m bottom roughness parameter z_0 = 0.1 cm friction velocity u_* = 0.78 cm s^{-1} and mussel uptake height = 3 cm. Distance downstream refers to distance from the leading edge of a mussel bed 100 m long. (From Fréchette et al. 1989)

BBL modeled vertical profiles of phytoplankton concentration closely paralleled logarithmic profiles of mean flow speed. As current speed increases so does vertical transport of phytoplankton to the bed and thus food consumption rate of the mussels. In addition the model showed that roughness due to the mussels themselves increased food availability due to enhanced vertical transport (i.e. turbulent mixing). The results however were sensitive to ingestion height of the mussels.

Monismith et al (1990) in a novel flume study using infaunal bivalve mimics showed that bivalves could influence near-bottom flow not only as roughness elements but also through their pumping jets and siphon behavior. Their results showed that a significant proportion of excurrent water could be refiltered according to complex relationships between the boundary-layer velocity profile the velocity of excurrent water and the siphon height orientation and size. In particular they noticed that in some situations the height at which water is removed from the flow by an incurrent siphon did not match well the height at which water is returned to the flow by excurrent siphon the former being lower than the latter. Because of the effect of pumping on the structure of near-bottom flow Monismith et al. suggested that in models of feeding by bivalve populations phytoplankton loss from the water column should be set at the height of the incurrent siphon whereas for the effect of bivalve feeding on phytoplankton distribution the flow disturbance height should be set by excurrent siphon flow. It seems however that the problem of the height of food intake and distribution very close to mussels which apparently orient more or less haphazardly with respect to flow may well differ from the situation for clams which orient with respect to current direction (e.g. Vincent et al. 1989) in the sediment. In a laboratory flume study of velocity and phytoplankton distributions above a mussel bed Butman et al. (submitted; see Figs. 3 and 4 and "On the necessity of resolving vertical gradients" section for details) found enhanced turbulent mixing due to mussel bed roughness; it was not necessary to invoke direct effects of mussel pumping on flow to explain observed profiles of mean velocity and Reynolds stress. In addition they found reasonably good agreement between measured vertical profiles of phytoplankton concentration and predicted profiles based on the two-dimensional numerical model of Fréchette et al (1989) that neglects near-bottom effects of pumping. This suggests that distinguishing between the heights of incurrent and excurrent siphonal flow is not necessary to model food distribution above mussel beds. The results of Monismith et al (1990) however certainly underline the complexity of small-scale hydrodynamics near infaunal bivalves and the two laboratory flume studies taken together suggest that effects of bivalve suspension feeders on phytoplankton distribution and flow may be somewhat species specific.

Sampling strategy

Despite mounting evidence that hydrodynamic and trophic processes in the BBL create significant small scale gradients studies of food avalability for natural beds of suspension feeders have typically concentrated on temporal heterogeneity and have generally failed to include adequate sampling in the vertical. Nevertheless there is a recent trend toward sampling closer to the bottom (≤ 25 cm above bottom; e.g. Berg and Newell 1986; Grizzle and Lutz 1989; Grizzle and Morin 1989 Fegley et al. 1992) which should at least provide estimates of conditions within the most important region of exchange between the pelagic and benthic zones. Such sampling provides no information on vertical gradients however so vertical and horizontal distributions of phytoplankton abundance and resuspended particles near bivalve beds remain largely unknown. A notable exception is the recent study of Muschenheim and Newell (1992) who sampled at 10 heights within 50 cm of the bottom. Coupled with current speed estimates they estimated horizontal fluxes of particles in two points upstream and above a subtidal mussel bed. It is noteworthy that their single upstream profile showed increased chlorophyll a and diatom concentration near the bottom presumably owing to hydrodynamic sorting (Muschenheim 1987b). Although the amplitude of this effect is probably site-dependent it suggests that if inverted profiles are common upstream of bivalve beds profiling seston above the bivalves only (e.g. Fréchette and Bourget 1985a) could underestimate the actual effect of bivalves on their food resources. It appears that the sampling burden of resolving both spatial and temporal gradients is presently insurmountable without 1) combining modelling efforts with field data (Smaal et al. 1986; Bacher 1989; Fréchette et al. 1989) or 2) developing new techniques that will allow by-passing cumbersome and time-consuming methods such as filtering of water samples. Direct chlorophyll extraction (Phinney and Yentsch 1985) and nephelometry (Fréchette and Bourget 1985a) are two possible methods for decreasing the sampling burden for the chlorophyll and non-chlorophyll components of the seston respectively (these two particular techniques are not however independent). Another approach to reducing the burden of sampling may be flow cytometry (Newell et al. 1989) but the decreased sampling burden may be more-than-compensated for by the time required for processing and analysis (e.g. Demers et al. 1992).

The sampling strategy characteristic of the BBL approach raises some concern about a number of methodological points which are linked to constraints inherent to (1) sampling water in a zone of large vertical gradients in seston (2) the use of filtration chambers with a limited number of organisms and (3) the age-old problems associated with manipulating animals. The first point concerns height of sampling above the bivalve bed. Food consumption is concentration and velocity-dependent (Wildish et al. 1992). In addition the vertical gradient in phytoplankton concentration is steepest close to the bottom (e.g. Fig. 3). Based on the above considerations food ingestion is expected to increase with ingestion height. Indeed in the situation modeled by Fréchette at al (1989) a change in ingestion height from 1 to 5 cm above the bed increased mussel consumption by about 20%. A similar pattern would be expected if sampling height changed accordingly. The problem is to adjust sampling height according to mussel ingestion height which is an unknown. Therefore by setting sampling height lower than 0.05 m it appears that the bias would be of the order of 20% at least for the example given here. It is furthermore critical to sample isokinetically for passive suspension feeders or using actual pumping velocity for active suspension feeders (Monismith et al. 1990; Judge et al. 1992) because concentration gradients are so steep close to the bed.

Regarding the second point Wildish and Miyares (1990) observed direct inhibition of mussel filtration (pumping) rate by current speed. Although this study was done on a limited number of individuals oriented into the flow there is indirect evidence that the average feeding behavior of mussels in beds may be flow-dependent (Butman et al. submitted). Unfortunately direct current speed effects cannot be accounted for in filtration chambers. Furthermore spatial configurations of mussels within chambers are different than in natural situations and chamber volume generally limits the number of individuals that can be tested. Thus direct interference between individuals which is known to occur in nature (Okamura 1986) cannot be accounted for in filtration chambers (but see Fréchette et al. 1992). Moreover physiological rates determined in chambers are for "representative" individuals; considering that size distribution of mussels in the field and in the chambers may differ and that physiological processes generally scale allometrically with body mass linear extrapolation on a weight basis is bound to be biased. Ågren and Axelsson (1980) however found that if rates are measured using individuals of mean mass instead of using individuals representing the whole population structure the (positive) bias error for a physiological rate may be rather small. If the scaling exponent between body mass and a given physiological rate is for example equal to 0.7 the bias is as small as 10% of the true population rate. The calculations of Ågren and Axelsson further suggest that the bias could be even higher because in many cases filtration rate scales with body mass according to exponents lower than 0.7. The figure of 10% however agrees well with comparisons between chamber results and calculations for the natural mussel bed presented in Fréchette et al (1989).

A final point on filtration chamber measurements is that flow-through systems are bound to yield biased estimates of filtration rate (volume of water cleared of particles per unit time) if either filtration rate or input concentration of seston are variable through time. The bias can result in an apparent relationship between animal behavior and environmental characteristics (e.g. phytoplankton concentration) when indeed there actually is none (Northby 1976). In addition Northby (1976) modelled an hypothetical situation which showed that the bias did not affect the mean level of the physiological process measured as much as its apparent functional response to the environment. This bias is attributable to the interplay between mean water residence time in the chamber mixing characteristics of the chamber and the time scale of seston and animal variability. Assuming that mixing time is short with respect to other characteristic time scales in a given situation Northby

(1976) suggested that temporal variability at the output of chambers should be estimated in order to correct for this effect. The procedure outlined by Northby (1976) results in a doubling of sampling effort. To avoid this extra burden of work Fréchette and Bourget (1985a) installed their filtration chambers in parallel. Although this arrangement does not analytically correct all possible time-dependent biases empirical results showed that temporal heterogeneity in this particular situation had a negligible effect on filtration rate. An important point here is that mixing within the chambers is assumed to be complete and to occur on a short time scale. The assumption of complete mixing is an idealized view of the situation. It has been shown that in controlled conditions mixing is generally not complete which can create very complex dynamics in chemical reactors (see Epstein 1990). Because bivalves generally have a very high retention efficiency (Riisgård 1988) experimental animals in chambers egest a more or less steady stream of seawater devoid of particles. Clearly mixing characteristics and water residence time compared to the time scale required by the animals to process the water of the chambers is important if strong food gradients internal to the chambers are to be avoided.

Concerning the third point manipulation of the animals prior to measurements may be a source of stress (Vismann 1990). Fréchette and Bourget (1985a) attempted to minimize such stress by keeping their experimental mussels in plastic mesh cages but it is always difficult to evaluate the extent to which experimental animals behave "naturally" in containers (we note however that Doering and Oviatt 1986 reported that predictions of energy flow in *Mercenaria mercenaria* kept in mesocosms based on filtration rate models obtained using suspensions of natural particles were in reasonable agreement with radio-carbon based estimates of energy flow). In view of such uncertainties other experimental strategies have been considered such as the use of flumes or tunnels installed directly above bivalve beds in the field.

(4) The flume approach

Flumes in the laboratory

Flumes are flow tanks designed to address processes occuring in steady one dimensional open-channel flows. They are particularly useful for studies of processes occuring within the BBL because characteristics of flow within this region of the ocean are independent of total water depth and can be adequately described by U the mean-stream velocity u_* ν and bottom roughness. A discussion of the use of flumes for simulating benthic environments is provided in Nowell and Jumars (1987) and we reiterate here only those points specifically relevant to this paper. In general dynamical similarity between flume and field boundary-layer flows is acheived by matching four dimensionless ratios (Reynolds number Froude number Strouhal number and Euler number) any or all of which may be relevant to a given research problem. For studies involving particle transport other scaling parameters may also be important. The problem of interest here a bed of sedentary bivalves filtering suspended particulate matter in steady field flows over a non-moveable bottom is particularly tricky because dynamical similarity is required in the fluid momentum the suspended particle concentration and the particle flux fields. For subcritical flows (i.e. low Froude number) the relevant scaling parameters for this problem are the roughness Reynolds number and the Rouse number ($w_s/\kappa u_*$ where w_s is the gravitational settling velocity of the particles). For bivalves or siphons that protrude above the bottom the ratio of animal or siphon height relative to the boundary-layer thickness also must be considered. Furthermore water depth must be sufficient relative to flow disturbances resulting from active pumping of inhalent and exhalent siphons. The bivalve bed also must be sufficiently long such that vertical and horizontal gradients have sufficient time to develop.

The use of a laboratory flume to simulate realistic fluid momentum suspended particle concentration and particle flux fields for studies of flow-suspension feeder interactions is much more difficult than many of the research problems for which "flow tanks" have been used or proposed (e.g. Vogel and LaBarbera 1978; Vogel 1981). Determining the drag experienced by an individual mussel attached to a solid surface for example requires dynamical similarity only in the sheared flow approaching the organism. If food capture mechanisms of this organism were of interest then food must be supplied to the animal in the general manner in which it is available in the field (i.e. as suspended or bedload transport) but generally it would not be necessary to simulate the natural field flow because the observation region of interest would be only within the immediate vicinity of the organism. If however the research question involved *growth* of the animal as a function of food supply then dynamical similarity in the upstream flow and food concentration fields would be required. The situation is even more complex when *growth of a bed* of mussels and effects of their feeding on the two-dimensional structure of suspended food concentration are considered because these cases also require dynamical similarity in horizontal and vertical fluxes.

In general very large flumes are required for studies of flow-feeding interactions in dense assemblages of benthic suspension feeders for the following reasons. First to supply a one-dimensional flow (i.e. velocity varying only in the vertical) to the upstream end of a bed of suspension feeders requires a channel width to depth ratio of at least five to minimize secondary flow effects. Thus for a 50-cm-wide flume the maximum allowable water depth would be 10 cm. Second the channel must be adequately long such that flow becomes fully developed (i.e. where the boundary layer has grown to the water surface and no longer changes with distance downstream) at least in terms of the mean velocity profile before reaching the animal bed. Engineering rules of thumb suggest a channel distance of about 50 to 100 water depth for fully developed turbulent flow assuming reasonably adequate entrance conditions. Thus for a 10 cm water depth the minimum channel length *upstream* of the bed of organisms would be 5 m. Third to determine the effects of the bed of suspension feeders on the two-dimensional structure of the phytoplankton concentration field and likewise the effects of the food supply on feeding and growth rates of the animals the bed of organisms must cover a sufficient portion of the flume in the along-channel direction. This distance depends on feeding rates of the animals relative to flow speeds and seston concentration but in general it is reasonable to expect bed lengths of several meters for sufficient two-dimensional structure to develop (e.g. Butman et al submitted). Taken together then these criteria stipulate a flume at least 50 cm wide by 10 cm deep by about 10 m long and this is for animals and/or siphons that do not protude a significant distance above the bottom. Some of these rules of thumb can be violated to a certain extent however depending on the particular research question and the extent to which potential flow artifacts can be evaluated (e.g. a laser Doppler velocimeter can be used to carefully document relevant characteritics of the flume flow field). When detailed measurements of the flume flow regime are not possible then the rules of thumb become very important.

Flumes in the field

Although flumes generally are used in the laboratory to simulate steady boundary-layer flows in the field they have been used directly in the field when it was either not feasible to transport the sediment or organisms to the laboratory for study or when such transport may so disrupt the natural system that a laboratory flume study would be meaningless (e.g. Young and Southard 1978; Nowell et al. 1985; Amos et al. 1992). Field flumes for direct measures of fluxes above and through dense beds of suspension feeders offer a number of potential advantages over the BBL approach: (1) animals or sediments are not disturbed; (2) fluxes are computed using a larger sample of the population studied; (3) direct observations can be made of flow effects on animal behavior or other processes; (4) general environmental conditions (e.g. light water temperature presence of other fauna and flora) are maintained and (5) the flow potentially is simpler and more well-defined than natural field flows. Although the philosophy for field flume studies where flow is not driven by a pump is not to reproduce natural flows but rather to avoid interfering with them several criteria must be met to yield meaningful results. The field flume must (1) not interfere with the natural BBL; (2) be deployed in regions of uni- or bi-directional flow and be oriented parallel to the current otherwise secondary circulation (currents rotating along the longitudinal axis of the flume) will appear and (3) be wide enough (or short enough) to avoid boundary-layer effects from the side walls and top of the flume in the case for example of a tunnel design.

Thus far to our knowledge there are ten published studies involving the use of flumes to quantify particulate and dissolved fluxes above bivalve beds in nature and all are relatively recent studies (Wolaver et al. 1983; Dame et al. 1984; Wolaver et al. 1985; Dame and Dankers 1988; Dame et al. 1989; Prins and Smaal 1990; Asmus et al. 1990; Asmus and Asmus 1991; Dame et al. 1991; Williams et al. 1992). Thus the use of flumes for feeding studies in the field is presently in its infancy. Considering various methodological aspects of these studies we feel that this workshop is an excellent opportunity to discuss the use of flumes for estimating fluxes *in situ*. Our conclusion is that potential advances to our understanding of fluxes of particulate and dissolved substances through and across dense beds of suspension feeders using flumes have been hindered due to various methodological shortcomings. We illustrate these shortcomings in the following by providing estimates of possible biases attributable to inadequate sampling within the BBL using the data from the study of Butman et al. (submitted) which was conducted in a laboratory flume.

On the necessity of resolving vertical gradients: Because of the presence of the BBL where there are vertical gradients in flow and suspended particulate concentrations to avoid biased estimates fluxes must be computed to account for such gradients. In all of the studies that we reviewed however flow is assumed to be uniform in the vertical (Asmus and Asmus 1991 did find variability in current speed with depth which was deemed to be negligible). Closer attention was paid to the seston profile by Wolaver et al (1983) Wolaver et al (1985) Asmus et al (1990) and Asmus and Asmus (1991). In other studies sampling was done at one height only in the middle of the flume. In fact sampling height is generally not specified. Thus there is little or no information on vertical profiles of food and current speed for integral estimates of fluxes over the BBL or enclosed area.

To illustrate potential biases associated with this sampling approach let us consider results from the laboratory flume experiment of Butman et al. (submitted) the purpose of which was testing the model of Fréchette et al (1989). The experiment was done in the 17 m-long channel of a recirculating seawater flume in the Coastal Research Laboratory Woods Hole Oceanographic Institution (flume described in Butman and Chapman 1989; Trowbridge et al. 1989). Mussels were placed in the bottom of the flume across the entire flume width and densely packed (75.5 ±11.4 individuals per 100 cm^2; mean ± SD) to recreate a characteristic mussel bed. The mussel bed was 6.1 m long beginning 7.3 m from the channel entrance. Mussel biomass (dry flesh weight) was 2.45 ± 0.62 g 100 cm^{-2} and mean mussel size was 2.7 ± 0.2 cm. Water depth was 17.4 cm above the mussels (20 cm above flume bottom) and measurements of phytoplankton food concentration (as total fluorescence) and of the flow were made at free-stream velocities of 5 cm s^{-1} and 15 cm s^{-1}. Vertical profiles of current speed were measured using a laser-Doppler velocimeter (LDV) at six positions along the mussel bed and at three cross-stream locations for each along-channel location. Phytoplankton fluorescence was estimated from water samples collected upstream of the mussel bed at 3 cm above the bottom and near the downstream edge of the mussel bed at 1.1 2.2 4.2 and 8.1 cm above mean mussel height. Fluorescence is in units per 1.2 ml (Fl.u.). Because the flume recirculates seawater upstream fluorescence decreased over time due to mussel feeding. To correct for this effect the downstream data were detrended using the upstream fluorescence time series (details in Butman et al. submitted).

Predicted (using the model of Fréchette et al. 1989) fluorescence profiles are shown in Fig. 3 and a fit of the velocity profile data to the universal "law of the wall" is shown in Fig. 4. There is a good fit of logarithmic curves to the flow data indicating that even though mixing is enhanced over a mussel bed the flow is still predictable using classic BBL theory. This suggests that there is only a small direct effect if any of animal pumping on the near-bed flow. Fluorescence profiles are predicted reasonably well by the model particularly for the faster flow. At slow flow there is more fluorescence observed near the bottom than is predicted by the model. This could indicate that the model failed to include some biological aspect(s) of feeding in flow such as flow-dependent feeding rates or regulation of filtration rate by food concentration. Alternatively there might have been some resuspension of sedimented fluorescing particles.

For the purposes of this study fluxes are estimated from the actual fluorescence data not from the model results. To estimate ingoing flux we assume that the fluorescence profile upstream is homogeneous. The flow profile is not however due to the natural BBL that develops upstream of the mussel bed. The flux of fluorescence into the mussel bed is calculated by subtracting the horizontal flux downstream from the flux upstream.

To determine the influence of sampling scheme on fluorescence flux to the mussel bed we made flux calculations for four hypothetical sampling scenarios (Table I Fig. 5). For all scenarios it is assumed that the fluorescence and flow boundary layers extend to 8 and 17.4 cm above the mussel bed respectively. In scenario 1 flux to the mussels is estimated using the actual profiles of fluorescence and flow (Fig. 5A). The results of scenario 1 are assumed to be correct (referred to as the "actual flux") and will be used as a reference for comparison with the other scenarios. In the three other scenarios profiles of fluorescence and flow are not known because only one height is sampled (i.e. the actual thickness of the field BBL is unknown). Thus integration depth (z_{int}) is chosen arbitrarily by the investigator. Scenarios 2 and 3 (Fig. 5B and 5C respectively) depict situations where this one sampling height is used to integrate flux over the whole water column. Scenario 2 differs from scenario 3 in that in the former the BBL is assumed to extend to the surface 17.4 cm above the mussel bed in the present case (e.g. this is probably the case encountered by Wright et al. 1982). In scenario 3 δ is still 17.4 cm thick whereas integration height is arbitrarily fixed at 100 cm by the investigator. Scenario 4 (Fig. 5D) is the situation obtained if the investigator was aware of the problems inherent to extrapolating results over a deeper water depth than is appropriate (e.g. scenario 3) and sampling from within the BBL decided to integrate flux over only that portion of the water column located below the sampling height (z_s). We do not consider the case where the sampling height is above the BBL because by definition there are no bed effects at such heights.

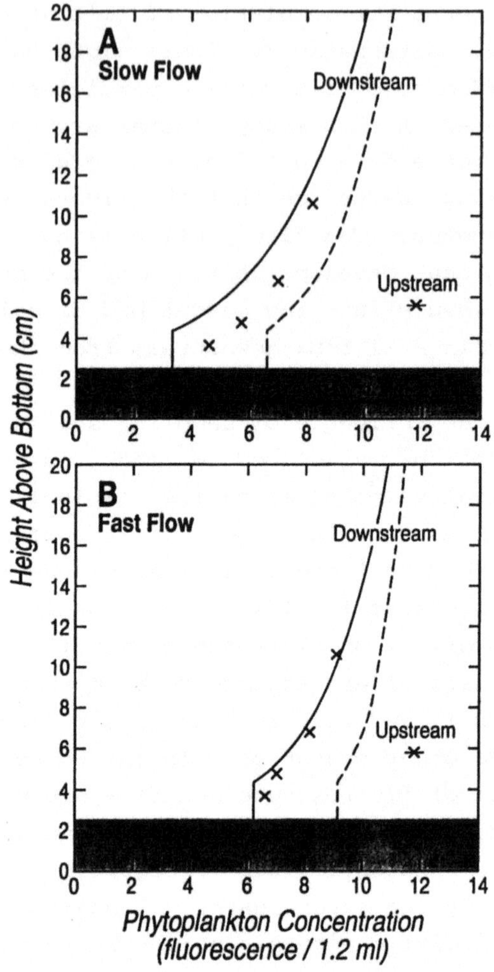

Fig. 3. Measured (crosses) and predicted (solid and dashed lines) vertical profiles of phytoplankton concentration at the end of a flume mussel bed. Solid and dashed lines are for u_{filt} of 5.7 m h^{-1} and 2.1 m h^{-1} respectively; u_{filt} is a model parameter specifying filtration velocity as estimated by mussel bed filtration rate by bed area. Also shown are the mean upstream concentrations (stars). Measured values are the means ±1 S.D for the two slow- and fast-flow runs. Shaded area indicates average mussel bed height. (From Butman et al. submitted)

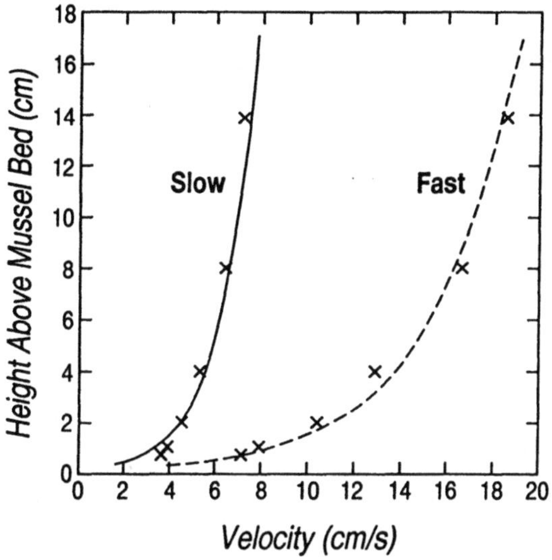

Fig. 4. Measured (crosses) vertical profiles of mean horizontal velocity (average of five centerline profiles along the mussel bed) and the best fit logarithmic regressions with z_0 fixed at 0.1 cm (solid line). (From Butman et al. submitted)

Table 1. Summary of the four hypothetical scenarios for estimating fluorescence flux into a bivalve bed.

Scenario	1	2	3	4
BBL thickness (δ) known	yes	no	no	no
actual water depth (H)	δ	δ	>δ	>δ
no. of sampling heights in the vertical	4	1	1	1
integration height	δ	δ	>δ	<δ

Using the profiles of flow and fluorescence (scenario 1) the estimated fluxes upstream and downstream in slow flow are 1320 and 920 Fl.u. cm s^{-1} respectively. For fast flow the corresponding fluxes are 3350 and 2530 Fl.u. cm s^{-1}. By subtraction we get a flux into the mussels which is 400 and 820 Fl.u. cm s^{-1} for slow and fast flow respectively (Table 2A). The results for scenario 2 are shown in Table 2B. The ratio between observed flux and actual flux varies between 2.1 and 0.6 decreasing with increasing z_s. This suggests that a single sampling height within the BBL is insufficient to estimate a mean flux to the mussel bed without significant bias. This result is specific to the present example where conditions were highly regulated and the upstream phytoplankton concentration profile was assumed to be uniform. Muschenheim and Newell (1992) however reported the presence of an inverted (i.e. with more phytoplankton at the bottom than in surface water) vertical gradient in phytoplankton concentration upstream of a subtidal mussel bed. More complicated upstream conditions than those assumed here obviously would change the quantitative results for scenario 2.

In scenarios 3 and 4 horizontal fluxes are strongly biased negatively or positively depending on whether z_{int} is smaller or larger than δ. In the present case scenario 3 overestimates actual flux to the bed (Table 2C). Scenario 4 on the other hand yields underestimations of actual flux to the bed (Table 2D). In both cases the actual versus estimated flux differ by one order of magnitude if z_s is near the bottom decreasing with increasing z_s. In scenarios 2 3 and 4 the ratio of the observed flux to the bed to the actual flux to the bed is apparently independent of current speed. It is however directly (and strongly) a function of the investigator's assumptions about _ which is unknown relative to z_{int} and z_s. This underscores the necessity for detailed current speed profiles within a field flume in order to adequately define _ and other relevant boundary-layer characteristics.

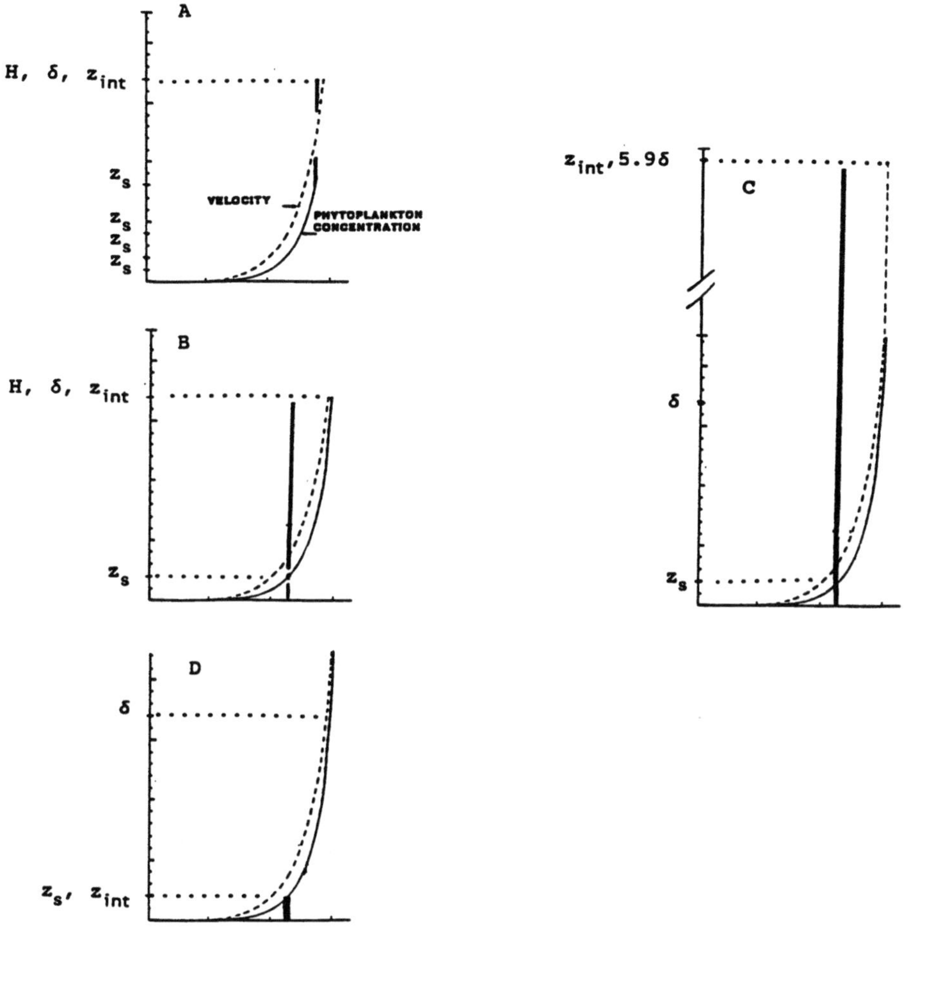

Fig. 5. Graphical representation of sampling schemes of scenarios 1 (A) 2 (B) 3 (C) and 4 (D). Solid line is actual phytoplankton concentration. Broken line is actual velocity. Heavy vertical bars represent the profiles assumed for velocity and fluorescence in each scenario (except in A where velocity is actually measured).

Table 2. Estimates of flux into the mussel bed (Fx_{bed}) as calculated from the difference between flux upstream (Fx_{up}) and flux downstream (Fx_{down}) in a laboratory flume experiment. Four different sampling designs are studied, according to sampling height (z_s) upstream and height of integration (z_{int}), in cm. A, B, C, D: scenarios 1, 2, 3 and 4, respectively. The ratio heading is the ratio of Fx_{bed} observed for a given scenario to Fx_{bed} observed in scenario 1. For further details, see text.

A) Scenario 1.

Slow flow			Fast flow			Ratio	
Fx_{up}	Fx_{down}	Fx_{bed}	Fx_{up}	Fx_{down}	Fx_{bed}	Slow	Fast
1320	920	400	3350	2530	820	–	–

B) Scenario 2.

Height		Slow flow			Fast flow			Ratio	
z_s	z_{int}	Fx_{up}	Fx_{down}	Fx_{bed}	Fx_{up}	Fx_{down}	Fx_{bed}	Slow	Fast
1	20	920	270	650	2420	690	1730	1.6	2.1
2	20	1110	430	680	2660	1140	1520	1.7	1.9
4	20	1180	620	560	3050	1790	1260	1.4	1.5
8	20	1350	1010	340	3310	2750	560	0.9	0.7
17	20	1420	1160	260	3720	3240	480	0.7	0.6

Table 2 (continued)

C) Scenario 3.

Height		Slow flow			Fast flow			Ratio	
z_s	z_{int}	Fx_{up}	Fx_{down}	Fx_{bed}	Fx_{up}	Fx_{down}	Fx_{bed}	Slow	Fast
1	100	4600	1510	3090	12120	3840	8280	7.7	10.1
2	100	5550	2430	3120	13320	6360	6960	7.8	8.5
4	100	5900	3490	2410	15240	10040	5200	6.0	6.3
8	100	6730	5670	1060	16500	15420	1140	2.7	1.4
17	100	7080	6480	600	18600	18120	480	1.5	0.6

D) Scenario 4.

Height		Slow flow			Fast flow			Ratio	
z_s	z_{int}	Fx_{up}	Fx_{down}	Fx_{bed}	Fx_{up}	Fx_{down}	Fx_{bed}	Slow	Fast
1	1	50	20	30	120	40	80	0.1	0.1
2	2	110	50	60	270	130	140	0.2	0.2
4	4	240	140	100	610	410	200	0.3	0.2
8	8	540	360	180	1320	1010	310	0.5	0.4
17	17	1200	960	240	3160	2680	480	0.6	0.6

yields underestimations of actual flux to the bed (Table 2D). In both cases, the actual versus estimated flux differ by one order of magnitude if z_s is near the bottom, decreasing with increasing z_s. In scenarios 2, 3 and 4, the ratio of the observed flux to the bed to the actual flux to the bed is apparently independent of current speed. It is, however, directly (and strongly) a function of the investigator's assumptions about δ, which is unknown, relative to z_{int} and z_s. This underscores the necessity for detailed current speed profiles within a field flume in order to

Wall effects: In contrast to open-channel flow in the laboratory where there is a free surface the flow is bounded on the top bottom and sides of a field flume and thus boundary layers will develop on all four surfaces (not all field flumes have a top however e.g. Asmus et al. 1990 Asmus and Asmus 1991). Therefore the flume height and width must be sufficiently large such that boundary layers on these surfaces do not interfere with the bottom boundary layer (e.g. Table 3 taken from Nowell and Jumars 1987). Even so because of the enhanced mixing and energy dissipation that occurs within boundary layers the mixing of phytoplankton within the water enclosed in the flume may differ from that outside the flume in the unbounded flow (i.e. because there is more "boundary" in the flume). Alternatively the flume could be very short so that boundary layers on the sides and top of the flume have very little distance over which to develop (e.g. Table 3). The disadvantage is that only a small portion of the bed of suspension feeders would then be covered by the flume negating one of the supposed

Table 3. Boundary layer thickness (in cm) for various lenghts of flume channels. From Nowell and Jumars 1987.

Free-stream velocity cm s^{-1}	Downstream distances				
	50	100	200	500	1000
1	3.3	4.6	6.6	10.4	14.7
5	1.5	2.1	2.9	4.6	6.6
10	1.1	1.5	2.1	3.3	23.7
20	0.73	3.3	5.7	11.9	20.6
50	1.6	2.7	4.7	9.9	17.2

advantages of using a field flume--that is obtaining flux estimates for a relatively large bed of suspension feeders.

Because of these problems field flumes built for other purposes such as for estimating the critical erosion velocity of natural sediments usually are designed to achieve fully-developed pipe flow where the boundary layers on all surfaces have grown together to produce a well-defined and characterizable flow and boundary shear stress (e.g. Young and Southard 1978; Nowell et al. 1985; Amos et al. 1992). For these kinds of research questions however only processes occurring at the water-sediment interface are important and so the fact that a boundary layer grows from the flume top down as well as from the flume bottom up is unimportant as long as the boundary shear stress can be characterized. This approach would not be desirable for the purpose of determining food flux to a bed of suspension feeders however because mixing occurring throughout the entire flume volume is important to suspension feeding in flow and will differ substantially from mixing in the unbounded free surface flow.

In summary it seems that although the flume approach appears to be promising for estimating fluxes in the field it may turn out that in practice it is laden with more problems than solutions. If the analysis is borne further it may even be argued that field flumes may be seen from an extreme point of view as unnecessary. Assuming that a given bivalve bed is homogeneous on the spatial axis normal to current it is expected that random fluctuations in horizontal transport normal to current direction will cancel out. It follows that water flow above the bed need not be bounded laterally as horizontal transport does not bias the estimates. Therefore estimates of fluxes along the main stream of the mussel bed without lateral walls can yield the same information as fluxes estimated using field flumes provided that the situation meets our assumption about spatial heterogeneity of the bivalve bed.

This is precisely what Muschenheim and Newell (1992) attempted. Their study is an example of what might be called the "flux approach" in the field. It also highlights what may be the major shortcoming of the flux approach--the enormous sampling burden of estimating flow and particle fluxes repeatedly in at least two locations along the bottom in a variable field environment. The complexity that natural situations can attain is illustrated by Fig. 6 from Fréchette et al (1989). Fluorescence was sampled at three heights (0.05 0.5 and 1.0 m) above an intertidal mussel bed. In this particular case there

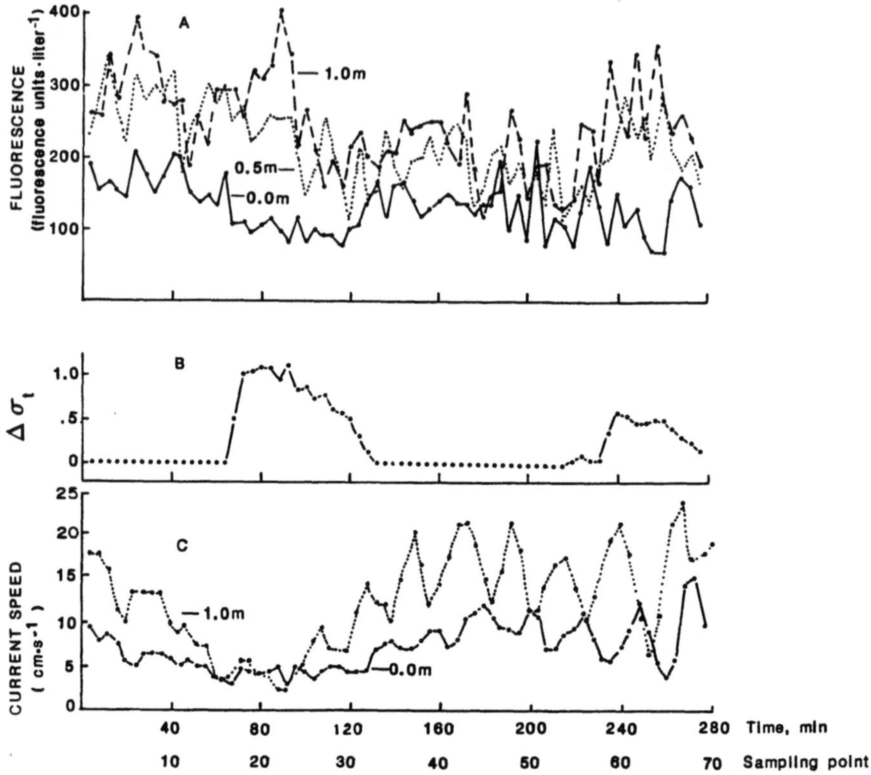

Fig. 6. Time series of field data obtained on 22 August 1982. High tide occurred near sampling point 27. A. Fluorescence. B. Stratification in σ_t units of kg m^{-3} as measured by the difference in density between 0.05 and 1.0 m above the bottom. C. Current speed (From Fréchette et al. 1989)

was evidence of variability in horizontal distributions in phytoplankton in the water approaching the bed as indicated by the fluorescence signal at 1.0 m. Such variability is a fundamental characteristic of phytoplankton spatio-temporal distribution. It is unpredictable and occurs on a variety of scales simultaneously. Its consequences for a flux approach are serious since it requires that sampling schemes be designed to account not only for vertical heterogeneity but also for spatio-temporal heterogeneity in the along-current direction. The situation depicted in Fig. 6 is probably an extreme case because in addition to its features at 1.0 m fluorescence at 0.05 m responded to current speed variations. There were 24-min cyclic variations in current speed at 0.0 m and 1.0 m which were out of phase with respect to each level and episodes of density stratification between sampling heights. Clearly such conditions would make attempts to estimate particle flux to the mussels in the field a nearly insurmountable challenge using either the field flume or the flux approaches.

CONCLUSION

We hope to have demonstrated that boundary-layer dynamics are a capital source of uncertainty in estimating fluxes of particulates and dissolved matter over dense beds of suspension feeders. Three sampling strategies are presently available the BBL approach the field flume approach and the flux approach. The recent trend has been to move toward the field flume and flux approaches which offer the possibility of avoiding shortcomings associated to the measurement of rates in experimental chambers. As the field flume and flux strategies both require a much higher sampling effort than the BBL approach and are strongly dependent on local oceanographic conditions to be useful it is uncertain whether the potential gain in accuracy achieved using them is worth the extra sampling burden involved. Although this issue may be resolved by a rigourous experimental comparison of these three sampling strategies we conclude that each approach is useful under different conditions and that great care should be used in evaluating *a priori* which method can best address the specific research question for specific organisms and habitats.

ACKNOWLEDGEMENTS. CAB was supported by the Office of Naval Research grant N00014-89-J-1637. This is contribution number 8223 from the Woods Hole Oceanographic Institution. We thank E. Bourget for stimulating discussions. L. Pagé provided the technical help for various aspects of draft preparation.

REFERENCES

Ågren GI B Axelsson (1980) Population respiration: a theoretical approach. Ecol Modell 11:39-54

Amos CL GR Daborn HA Christian A Atkinson A Robertson (1992) In situ erosion measurements of fine-grained sediments of the Bay of Fundy. Mar Geol 108:175-196

Ardisson P-L E Bourget (1991) Abundance growth and production estimation of the blue mussel *Mytilus edulis* on moored navigation buoys in the Estuary and northwestern Gulf of St. Lawrence. Can J Fish Aquat Sci 48:2408-2419

Asmus RM H Asmus (1991) Mussels beds: Limiting or promoting phytoplankton? J Exp Mar Biol Ecol 148:215-232

Asmus H RM Asmus K Reise (1990) Exchange processes in an intertidal bed: a Sylt-flume study in the Wadden Sea. Ber Biol Anst Helgoland 6:1-79

Bacher C (1989) Capacité trophique du bassin de Marennes-Oléron: couplage d'un modèle de transport particulaire et d'un modèle de croissance de l'huître *Crassostrea gigas*. Aquat Living Resour 2:199-214

Bacher C M Héral J-M Deslous-Paoli D Razet (1991) Modèle énergétique uniboîte de la croissance des huîtres (*Crassotrea gigas*) dans le bassin de Marennes-Oléron. Can J Fish Aquat Sci 48:391-404

Bayne BL J Widdows (1978) The physiological ecology of two populations of *Mytilus edulis* L.Oecologia (Berl) 37:137-162

Bayne BL CM Worrall (1980) Growth and production of mussels *Mytilus edulis* from two populations. Mar Ecol Prog Ser 3:317-328

Bayne BL AJS Hawkins E Navarro IP Iglesias (1989) Effects of seston concentration on feeding digestion and growth in the mussel *Mytilus edulis*. Mar Ecol Prog Ser 55:47-54

Berg JA RIE Newell (1986) Temporal and spatial variations in the composition of seston available to the suspension feeder *Crassostrea virginica*. Estuarine Coastal Shelf Sci 23:375-386

Bernard FR (1974) Annual biodeposition and gross energy budget of mature Pacific oysters *Crassostrea gigas*. J Fish Res Board Can 31:185-190

Bricelj VM RE Malouf (1984) Influence of algal and suspended sediment concentrations on the feeding physiology of the hard clam *Mercenaria mercenaria*. Mar Biol 84:155-165

Bricelj VM RE Malouf C de Quillfeldt (1984) Growth of juvenile *Mercenaria mercenaria* and the effect of resuspended bottom sediments. Mar Biol 84:167-173

Buss LW (1979) Bryozoan overgrowth interactions -- the interdependence of competition of space and food. Nature (Lond) 281:475-477

Butman CA (1987) Larval settlememt of soft-sediment invertebrates: The spatial scales of pattern explained by active habitat selection and the emerging rôle of hydrodynamical process. Oceanogr Mar Biol Ann Rev 25:113-165

Butman CA RJ Chapman (1989) The 17-meter flume at the Coastal Research Laboratory. Part I: Description and user's manual. Tech Rep Woods Hole Oceanogr Instn 89-10 1-31

Cahalan JA SE Siddall MW Luckenbach (1989) Effects of flow velocity food concentration and particle flux on growth rates of juvenile bay scallops *Argopecten irradians*. J Exp Mar Biol Ecol 129:45-60

Cloern JE (1982) Does the benthos control phytoplankton biomass in South San Francisco Bay? Mar Ecol Prog Ser 9:191-202

Cole BE JK Thompson JE Cloern (1992) Measurement of filtration rates by infaunal bivalves in a recirculating flume. Mar Biol 113:219-225

Dame RF N Dankers (1988) Uptake and release of materials by a Wadden Sea mussel bed. J Exp Mar Biol Ecol 118:207-216

Dame RF BC Patten (1981) Analysis of energy flows in an intertidal oyster reef. Mar Ecol Prog Ser 5:115-124

Dame RF FD Spurrier TG Wolaver (1989) Carbon nitrogen and phosphorus processing by an oyster reef. Mar Ecol Prog Ser 54:249-256

Dame R R Zingmark H Stevenson D Nelson (1980) Filter feeder

coupling between the estuarine water column and benthic subsystems. In Kennedy VS (ed) Estuarine Perspectives Academic Press New York

Dame RF RG Zingmark E Haskin (1984) Oyster reefs as processors of estuarine materials. J Exp Mar Biol Ecol 83:239-247

Dame RF N Dankers T Prins H Jongsma A Smaal (1991) The influence of mussel beds on nutrients in the Western Wadden Sea and Eastern Scheldt Estuaries. Estuaries 14:130-138

Dayton PK (1973) Dispersion dispersal and persistence of the annual intertidal alga *Postelsia palmaeformis* Ruprecht. Ecology 54:433-438

Demers S J-C Therriault E Bourget A Bah (1987) Resuspension in the shallow littoral zone of macrotidal estuarine environment: Wind influence. Limnol Oceanogr 32:327-339

Demers S J-C Therriault E Bourget H Désilets (1989) Small-scale gradients of phytoplankton productivity in the littoral fringe. Mar Biol 100:393-399

Demers S J Kim P Legendre L Legendre (1992) Analyzing multivariate flow cytometric data in aquatic sciences. Cytometry 13:291-298

Doering PH CA Oviatt (1986) Application of filtration rate models to field populations of bivalves: an assessment using experimental mesocosms. Mar Ecol Prog Ser 31:265-275

Eckman JE CH Peterson JA Cahalan (1989) Effects of flow speed turbulence and orientation on growth of juvenile bay scallops *Argopecten irradians concentricus* (Say). J Exp Mar Biol Ecol 132:123-140

Emerson CW (1990) Influence of sediment disturbance and water flow on the growth of the soft-shell clam *Mya arenaria* L. Can J Fish Aquat Sci 47:1655-1663

Epstein IR (1990) Shaken stirred—but not mixed. Nature 346:16-17

Fraser CMcL GM Smith (1928) Notes on the ecology of the little-neck clam *Paphia staminea* Conrad. Trans Roy Soc Can 22:249-269

Fraser CMcL GM Smith (1928) Notes on the ecology of the butter clam *Saxidomus giganteus* Deshayes. Trans Roy Soc Can 22:271-286

Fegley SR BA MacDonald TR Jacobsen (1992) Short-term variation in the quantity and quality of seston available to benthic suspension feeders. Estuarine Coastal Shelf Sci 34:393-412

Fréchette M E Bourget (1985a) Energy flow between the pelagic and benthic zones: factors controlling particulate organic matter available to an intertidal mussel bed. Can J Fish Aquat Sci 42:1158-1165

Fréchette M E Bourget (1985b) Food-limited growth of *Mytilus edulis* in relation to the benthic boundary layer. Can J Fish Aquat Sci 42:1166-1170

Fréchette M E Bourget (1987) Significance of small-scale spatio-temporal heterogeneity in phytoplankton abundance for energy flow in *Mytilus edulis*. Mar Biol 94:231-240

Fréchette M J Grant (1991) An in situ estimation of the effect of wind-driven resuspension on the growth of the mussel *Mytilus edulis* L. J exp mar Biol Ecol 148:201-213

Fréchette M D Lefaivre (1990) Discriminating between food and space limitation in benthic suspension feeders using self-thinning relationships. Mar Ecol Prog Ser 65:15-23

Fréchette M CA Butman WR Geyer (1989) The importance of boundary-layer flows in supplying phytoplankton to the benthic suspension feeder *Mytilus edulis* L. Limnol Oceanogr 34:19-36

Fréchette M AE Aitken L Pagé (1992) Interdependence of food and space limitation of a suspension feeder: consequences for self-thinning relationships. Mar Ecol Prog Ser 83:55-62

Grant WD OS Madsen (1986) The continental-shelf bottom boundary layer. Ann Rev Fluid Mech 18:265-305

Grizzle RE PJ Morin (1989) Effect of tidal currents seston and bottom sediments on growth of *Mercenaria mercenaria*: results of a field experiment. Mar Biol 102:85-93

Grizzle RE RA Lutz (1989) A statistical model relating horizontal seston fluxes and bottom sediment characteristics to growth of *Mercenaria mercenaria*. Mar Biol 102:95-105

Haven DS R Morales-Alamo (1966) Aspects of biodeposition by oysters and other invertebrate filter feeders. Limnol Oceanogr 11:487-498

Heathershaw AD DN Langhorne (1988) Observations of near-bed velocity profiles and seabed roughness in tidal currents flowing over sandy gravels. Estuarine Coastal Shelf Sci 26:459-482

Héral M J-M Deslous-Paoli J-M Sornin (1983) Transferts énergétiques entre l'huître *Crassostrea gigas* et la nourriture potentielle disponible dans un bassin ostréicole: Premières approches. Oceanis 9:169-194

Herman PMJ H Scholten (1990) Can suspension feeders stabilize estuarine ecosystems? In Barnes M Gibson RN (eds) Trophic relationships in the marine environment. Aberdeen University Press Aberdeen

Hildreth DI DJ Crisp (1976) A corrected formula for calculation of the filtration rate of bivalve molluscs in an experimental flowing system. J Mar Biol Ass U K 56:111-120

Hily C (1991) Is the activity of benthic suspension feeders a factor controlling water quality in the Bay of Brest? Mar Ecol Prog Ser 69:179-188

Hughes RN CL Griffiths (1988) Self-thinning in barnacles and mussels. Am Nat 132:484-491

Jørgensen BB (1980) Seasonal oxygen depletion in the bottom waters of a Danish fjord and its effect on the benthic community. Oikos 34:68-76

Judge ML LD Coen KL Heck Jr (1992) The effect of long-term alteration of in situ currents on the growth of *Mercenaria mercenaria* in the northern Gulf of Mexico. Limnol Oceanogr 37:1550-1559

Kerswill CJ (1949) Effects of water circulation on the growth of quahaugs and oysters. J Fish Res Bd Can 7:545-551

Komar PD (1976) Boundary layer flow under steady unidirectional currents. In Stanley DJ Swift DJP (eds) Marine sediment transport and environmental management. John Wiley & Sons New York

Levasseur M J-C Therriault L Legendre (1983) Tidal currents winds and the morphology of phytoplankton spatial structures. J Mar Res 41:655-672

Malouf RE WP Breese (1977) Seasonal changes in the effects of temperature and water flow rate on the growth of juvenile Pacific oysters *Crassostrea gigas* (Thunberg). Aquaculture 12:1-13

Marra J RW Houghton DC Boardman PJ Neale (1982) Variability in surface chlorophyll *a* at a shelf-break front. J Mar Res 40:575-591

McClatchie S MR Lewis (1986) Limitations of grazing rate

equations: the case for time-series measurements. Mar Biol 92:135-140

Monismith SG JR Koseff JK Thompson CA O'Riordan HM Nepf (1990) A study of model bivalve siphonal currents. Limnol Oceanogr 35:680-696

Muschenheim DK (1987a) The dynamics of near-bed seston flux and suspension-feeding benthos. J Mar Res 45:473-496

Muschenheim DK (1987b) The role of hydrodynamic sorting of seston in the nutrition of a benthic suspension feeder *Spio setosa* (Polychaeta: Spionidae). Bio Oceanogr 4:265-287

Muschenheim DK CR Newell (1992) Utilization of seston flux over a mussel bed. Mar Ecol Prog Ser 85:131-136

Muschenheim DK J Grant EL Mills (1986) Flumes for benthic ecologists: theory construction and practice. Mar Ecol Prog Ser 28:185-196

Newell CR (1990) The effects of mussel (*Mytilus edulis* Linnaeus 1758) position in seeded bottom patches on growth at subtidal lease sites in Maine. J Shellfish Res 9:113-118

Newell C R SE Shumway TL Cucci R Selvin (1989) The effects of natural seston particle size and type on feeding rates feeding selectivity and food resource availability for the mussel *Mytilus edulis* Linnaeus 1758 at bottom culture sites in Maine. J Shellfish Res 8:187-196

Nixon SW CA Oviatt C Rogers K Taylor (1971) Mass and metabolism of a mussel bed. Oecologia (Berl) 8:21-30

Northby JA (1976) A comment on rate measurements in open systems. Limnol Oceanogr 21:180-182

Nowell ARM (1983) The benthic boundary layer and sediment transport. Rev Geophys Space Physics 21:1181-1192

Nowell ARM PA Jumars (1984) Flow environments of aquatic benthos. Ann Rev Ecol Syst 15:303-328

Nowell ARM PA Jumars (1987) Flumes: theoretical and experimental considerations for simulation of benthic environments. Oceanogr Mar Biol Ann Rev 25:91-112

Nowell ARM IN McCave CD Hollister (1985) Contributions of HEBBLE to understanding marine sedimentation. Mar Geol 66:397-409

Officer CB TJ Smayda R Mann (1982) Benthic filter feeding: A natural eutrophication control. Mar Ecol Prog Ser 9:203-210

Okamura B (1986) Group living and the effect of spatial position in aggregations of *Mytilus edulis*. Oecologia (Berl) 69:341-347

Paine RT (1984) Ecological determinism in the competition for space. Ecology 65:1339-1348

Peterson CH SV Andre (1980) An experimental analysis of inter specific competition among marine filter feeders in a soft-sediment environment. Ecology 61:129-139

Peterson CH R Black (1991) Preliminary evidence for progressive sestonic food depletion in incoming tide over a broad tidal sand flat. Estuarine Coastal Shelf Sci 32:405-413

Phinney DA CS Yentsch (1985) A novel phytoplankton chlorophyll technique: toward automated analysis. J Plankton Res 7:633-642

Platt T LM Dickie RW Trites (1970) Spatial heterogeneity of phytoplankton in a near-shore environment. J Fish Res Board Can 27:1453-1473

Prins TC AC Smaal (1990) Benthic-pelagic coupling: the release of inorganic nutrients by an intertidal bed of *Mytilus edulis*. In Barnes M Gibson RN (eds) Trophic relationships in the marine environment. Aberdeen University Press Aberdeen

Propp MV MR Garber VI Ryabuscko (1982) Unstable processes in the metabolic rate measurements in flow-through systems. Mar Biol 67:47-49

Riisgård HU (1988) Efficiency of particle retention and filtration rate in 6 species of Northeast American bivalves. Mar Ecol Prog Ser 45:217-223

Riisgård HU F Møhlenberg (1979) An improved automatic recording apparatus for determining the filtration rate of *Mytilus edulis* as a function of size and algal concentration. Mar Biol 52:61-67

Rodhouse PG (1978) Energy transformations by the oyster *Ostrea edulis* L in a temperate estuary. J Exp Mar Biol Ecol 34:1-22

Rodhouse PG CM Roden MP Hensey TH Ryan (1985) Production of mussels *Mytilus edulis* in suspended culture and estimates of carbon and nitrogen flow: Killary Harbour Ireland. J Mar Biol Ass U K 65:55-68

Rosenberg R L-O Loo (1983) Energy flow in a *Mytilus edulis* culture in western Sweden. Aquaculture 53:151-161

Smaal AC JHG Verhagen J Coosen HA Haas (1986) Interaction between seston quantity and quality and benthic suspension feeders in the Oosterschelde The Netherlands. Ophelia 26:385-399

Thompson RJ (1984a) The reproductive cycle and physiological ecology of the mussel *Mytilus edulis* in a subarctic non-estuarine environment. Mar Biol 79:277-288

Thompson RJ (1984b) Production reproductive effort reproductive value and reproductive cost in a population of the blue mussel *Mytilus edulis* from a subarctic environment. Mar Ecol Prog Ser 16:249-257

Thompson RJ BL Bayne (1974) Some relationships between growth metabolism and food in the mussel *Mytilus edulis*. Mar Biol 27:317-326

Trowbridge JH WR Geyer CA Butman RJ Chapman (1989) The 17-meter flume at the Coastal Research Laboratory. Part II: Flow characteristics. Tech Rep Woods Hole Oceanogr Instn 89-11 1-39

Turner EJ DC Miller (1991) Behavior and growth of *Mercenaria mercenaria* during simulated storm events. Mar Biol 111:55-64

Vahl O (1980) Seasonal variations in seston and the growth rate of the Iceland scallop *Chlamys islandica* (OF Müller) from Balsfjord 70° N. J Exp Mar Biol Ecol 48:195-204

Vahl O (1981a) Energy transformations by the Iceland scallop *Chlamys islandica* (OF Müller) from 70°N. I. The age-specific energy budget and net growth efficiency. J Exp Mar Biol Ecol 53:281-296

Vahl O (1981b) Energy transformations by the Iceland scallop *Chlamys islandica* (OF Müller) from 70°N. II. The population energy budget. J Exp Mar Biol Ecol 53:297-303

Vincent B G Desrosiers Y Gratton (1988) Orientation of the infaunal bivalve *Mya arenaria* L in relation to local current direction on a tidal flat. J Exp Mar Biol Ecol 124:205-214

Vismann B (1990) Field measurements of filtration and respiration rates in *Mytilus edulis* L An assesment of methods. Sarsia 75:213-216

Vogel S (1981) Life in moving fluids. The physical biology of flow. Willard Grant Press Boston

Vogel S M LaBarbera (1978) Simple flow tanks for research and teaching. Bioscience 28:638-643

Warwick RM RJ Uncles (1980) Distribution of benthic macrofauna associations in the Bristol Channel in relation to tidal stress. Mar Ecol Prog Ser 3:97-103

Widdows J P Fieth CM Worrall (1979) Relationships between seston available food and feeding activity in the common mussel *Mytilus edulis*. Mar Biol 50:195-207

Wildish DJ (1977) Factors controlling marine and estuarine sublittoral macrofauna. Helgol Wiss Meeresunters 30:445-454

Wildish DJ DD Kristmanson (1979). Tidal energy and sublittoral macrobenthic animals in estuaries. J Fish Res Board Can 36:1197-1206

Wildish DJ DD Kristmanson (1984) Importance to mussels of the benthic boundary layer. Can J Fish Aquat Sci 41:1618-1625

Wildish DJ MP Miyares (1990) Filtration rate of blue mussels as a function of flow velocity: preliminary experiments. J Exp Mar Biol Ecol 142:213-219

Wildish DJ D Peer (1983) Tidal current speed and production of benthic macrofauna in the lower Bay of Fundy. Can J Fish Aquat Sci 40(Suppl 1):309-321

Wildish DJ DD Kristmanson RL Hoar AM DeCoste SD McCormick AW White (1987) Giant scallop feeding and growth responses to flow. J Exp Mar Biol Ecol 113:207-220

Wildish DJ DD Kristmanson AM Saulnier (1992) Interactive effect of velocity and seston concentration on giant scallop feeding inhibition. J Exp Mar Biol Ecol 155:161-168

Williams TM TG Wolaver RF Dame JD Spurrier (1992) The Bly Creek ecosystem study – organic carbon transport within a euhaline salt marsh basin North Inlet South Carolina. J Exp Mar Biol Ecol 163:125-139

Wilson JH (1987) Environmental parameters controlling growth of *Ostrea edulis* L and *Pecten maximus* L in suspension culture. Aquaculture 64:119-131

Wimbush M W Munk (1970) The benthic boundary layer. In Maxwell AE (ed) The Sea vol 4 Interscience New York

Wolaver TG JC Zieman R Wetzel KL Webb (1983) Tidal exchange of nitrogen and phosphorus between a mesohaline vegetated marsh and the surrounding estuary in the Lower Chesapeake Bay. Estuarine Coastal Shelf Sci 16:321-332

Wolaver T G Whiting B Kjerfve J Spurrier H McKellar R Dame T Chrzanowski R Zingmark and T Williams (1985) The flume design – A methodology for evaluating material fluxes between a vegetated salt marsh and the adjacent tidal creek. J Exp Mar Biol Ecol 91:281-291

Wolff WJ (1977) A benthic food budget for the Grevelingen Estuary The Netherlands and a consideration of the mechanics causing high benthic secondary production in estuaries. In Coull BC (ed) Ecology of marine benthos. Belle W Baruch Library in Marine Science no 6 University of South Carolina Press

Wright LD JD Boon JP Xu SC Kim (1992) The bottom boundary of the Bay stem plains environment of the Lower Chesapeake Bay. Estuarine Coastal Shelf Sci 35:17-36

Wright RT RB Coffin CP Ersing D Pearson (1982) Field and laboratory measurements of bivalve filtration of natural marine bacterioplankton. Limnol Oceanogr 27:91-98

Young RA JB Southard (1978) Erosion of fine-grained sediments: seafloor and laboratory experiments. Geol Soc Am Bull 89:663-672

PERSPECTIVES ON FIELD STUDIES AND RELATED BIOLOGICAL MODELS OF BIVALVE GROWTH AND CARRYING CAPACITY

Jon Grant, Michael Dowd, Keith Thompson,
Craig Emerson, and Annamarie Hatcher
Department of Oceanography
Dalhousie University
Halifax, Nova Scotia B3H 4J1
Canada

INTRODUCTION

Marine bivalves are cultured throughout the world, and in many places such as Atlantic Canada and New Zealand, culture of mussels and other groups has grown exponentially in the last decade (Hickman 1989; Mallet 1989). Acceptable culture sites are limited due to habitat suitability, road access, and competing recreational or commercial use such as wild fisheries. As available culture space becomes filled up with stock, there may be a depression of individual bivalve growth rate and an increase in mortality caused by several factors associated with overcrowding. Suspension feeders have a remarkable capacity to filter the water column such that they are food limited at high culture density (Navarro et al. 1991). There are several indications that production maxima have been reached or exceeded as culture is continually expanded to the detriment of food supply (Mallet 1989; Hickman et al. 1991; Pérez Comacho et al. 1991). This phenomenon is also documented for natural populations of both

NATO ASI Series, Vol. G 33
Bivalve Filter Feeders
in Estuarine and Coastal Ecosystem Processes
Edited by Richard F. Dame

infaunal and epifaunal bivalves including mussel beds (Peterson and Black 1987; Fréchette and Grant 1991; Smaal 1991; Bayne and Hawkins 1992). Moreover, culture research has demonstrated major site differences in growth rate (Mallet et al. 1986) confirming that environmental conditions can regulate shellfish production. Overcrowded culture conditions also lead to increased incidence of shellfish diseases (Dijkema and van Stralen 1989). Finally, high culture biomass may produce a negative feedback to the local environment through organic loading and anaerobic conditions beneath culture leases (Dahlbäck and Gunnarsson 1981), potentially leading to degradation of culture environments.

A means to predict the ability of coastal environments to sustain bivalve culture is essential to the continued development of the industry, but there has been little progress along these lines. In this paper, we review developments in this area of research, and present an overview of a field and modelling study of a suspended culture environment representative of coastal inlets and bays. The field study is meant to illustrate a practical approach to assessing carrying capacity. The model is "generic" and portable in its application, and is validated by mussel growth data from coastal Nova Scotia.

In general, one may characterize the approaches to the question of carrying capacity into three categories: empirical studies, calculation of budgets, and simulation modelling. All cases involve determination of bivalve growth based on at least food and temperature, the two most important factors which regulate growth in both cultured and natural populations (MacDonald and Thompson 1986; Hickman et al. 1991; Bayne et al. 1992).

Empirical studies are largely based on correlations between bivalve growth and single or multiple environmental factors. Brown (1988) applied multiple regression techniques to determine that food, temperature, and initial size accounted for up to 62% of the variance in growth of cultured oysters. Smaal and van Stralen (1990) demonstrated that the condition of cultured mussels in the Dutch Eastern Scheldt was correlated with annual primary production. Numerous other examples of this approach abound in the literature (e.g. Wilson 1987; Thompson and Nichols 1988; Grizzle and Lutz 1989; Grant et al. 1990), and serve to identify which environmental variables are important in bivalve growth. Moreover, they verify that prediction of bivalve growth as a function of the environment is feasible, and that more complex models are worthwhile.

A more elaborate extension of this approach involves habitat suitability indices (HSI). Environmental factors known to be important in bivalve habitats are scored between 0 and 1 and used to produce a composite value which then comprises the index. The utility of the index is then verified against field growth trials or population surveys (Brown and Hartwick 1988; Soniat and Brody 1988). Although this a highly effective approach in terms of habitat suitability and aquaculture site selection, it collapses the environment into a single value and does not allow temporal resolution of growth or calculation of carrying capacity.

A second way to estimate the capacity of the environment to support aquaculture production is through a partial ecosystem budget (Rosenberg and Loo 1983; Rodhouse and Roden 1987; Smaal 1991). Phytoplankton standing stock or production may be compared to ingestion requirements of bivalve populations over daily, seasonal or other time scales to determine the biomass of shellfish that can be sustained in a given system. For example, Carver and Mallet (1990) assessed blue mussel *(Mytilus edulis)* carbon requirements based on an energy budget, in relation to tidal fluxes of food in a small Nova Scotia embayment. There are also more complex ecosystem budgets which consider a variety of components such as benthos, zooplankton, detritus, etc. (Rosenberg and Loo 1983; Rodhouse and Roden 1987). These types of calculations are partially the basis for the notion that bivalve populations or communities have the ability to impact or even deplete the water column of seston by suspension feeding (see above). As specified by Smaal (1991), a budgeting approach uses an averaged and long-term time scale (annual), and cannot incorporate feedback from the ecosystem. However, the field studies which comprise this approach are essential in any modelling effort.

The final method, employed as part of the present study, is the use of simulation models in which the culture ecosystem is viewed as distinct compartments or state variables (e.g. shellfish, phytoplankton). Flows of energy or material between compartments are quantified based on internal biological fluxes (e.g. consumption) mediated by external forcing functions (e.g. temperature. The model can be represented as a set of differential equations which correspond to a theoretical perception of which factors are important in bivalve growth, including parameterization of exchange processes. Concurrent field studies are used to verify the importance of these factors. There are numerous examples of simulation models in marine ecology in various forms such as box modelling or network analysis (Kremer and Nixon 1978; Knox 1986; Wulff et al. 1989). These models have also been applied to ecosystems which include natural bivalve populations (Officer et al.

1982; Smaal 1990), but this approach has rarely been applied to the management of shellfish culture.

The prediction of growth and carrying capacity of cultured bivalve populations is important to the continued development of the shellfish industry. In this paper we provide perspectives on a conceptual understanding of mussel growth in relation to environmental factors. We present an overview of the types of field studies and oceanographic instrumentation required to determine carrying capacity at a commercial mussel farm (Upper South Cove), and how they interact with conceptual and simulation models which were developed concurrently. This study was conducted as a large interdisciplinary effort involving an extensive field program and computer simulations conducted over several years.

CULTURE OF *MYTILUS EDULIS* IN ATLANTIC CANADA

Commercial mussel culture in this region has grown rapidly over the past decade with yearly production of several thousand metric tonnes (Mallet 1989). Most culture is in suspension with mussels deployed in mesh sleeves from floating longlines fixed between moored floats. Seed is obtained from collectors as there is abundant spatfall at many sites. Mussels are sleeved in fall or spring at ~20 mm shell length and grown to commercial size (~55 mm) over 1.5-2 years. Harvesting may be carried out year round, but summer product is less desirable due to spawning.

The density of mussel culture is somewhat constrained by logistical considerations of boat access, but approximations may be made of average conditions (Mallet 1989; Mallet and Carver 1989). Longlines are on the order of 100 m in length, with 150 sleeves per line, and 3 lines occupying a distance of 10 m across. Thus an area of 10^3 m^2 contains 450 sleeves. Sleeves are stocked with ~600 mussels m^{-1} of sleeve, and a total vertical length of 3m, but mussels fall off or die, so that harvesting densities are reduced by as much as 50% (Mallet and Carver 1991). Local culture density (i.e. within a lease) is thereby reduced to an estimated 300 mussels m^{-1} or 900 mussels $sleeve^{-1}$ equal to ~400 mussels m^{-2} of culture area. If the volume of the culture system is considered (a well-mixed 7 m water column in our case), then culture density is 57 mussels m^{-3}.

FIELD STUDIES IN SUPPORT OF THE MODEL

Field studies of mussel growth and the culture environment were conducted in Upper South Cove, near Lunenburg, Nova Scotia (Fig. 1). The cove is approximately 3.5 km long by 0.5 km wide, with an average depth of 1-2 m, and a maximum depth of 10 m. Tidal range is about 1.5 m. Much of the cove is ice-covered from late December until late March. The upper end of the cove was closed with a causeway in 1968 so that the uppermost reaches have silted in. Upper South Cove is not exposed to the waves from Lunenburg Bay owing to its protected and narrow single entrance to the sea. Due to tidal choking at the mouth, there is a reduction in tidal amplitude and an increase in tidal harmonics in the cove compared with Lunenburg Bay, resulting in rapid currents (3 m sec^{-1}) through the inlet. The tidal excursion is ~1/2 the length of the cove such that a particle placed midway along the cove will just reach the outside on ebb tide, indicating a boundary between the poorly-flushed upper region which contains most of the commercial mussel culture, and the well flushed lower region.

Freshwater input to Upper South Cove is limited to several small streams, and the water column only contains salinity stratification during ice breakup and heavy rains. The shoreline is largely forested with a rocky or muddy intertidal. Most of the subtidal bottom is muddy, except near the inlet where tidal scour results in a cobble pavement. Eelgrass (*Zostera marina*) grows throughout the cove, and various kelps are abundant in some areas.

Field work was based at the site of Corkum's Island Mussel Farm, where a small field laboratory containing water analysis equipment and a meteorology station were constructed. A variety of small boats were used to sample various stations along the cove and outside into Lunenburg Bay. In winter, an all-terrain vehicle was used to reach sample sites and holes were cut through the ice with a chain saw.

MUSSEL GROWTH

Observed growth of mussels in culture is the most important aspect of the field studies since it allows validation of model output. Growth of suspended mussels was studied with experimental suspended culture at various locations in and out the cove (Fig. 1), corresponding to a horizontal gradient in tidal flushing

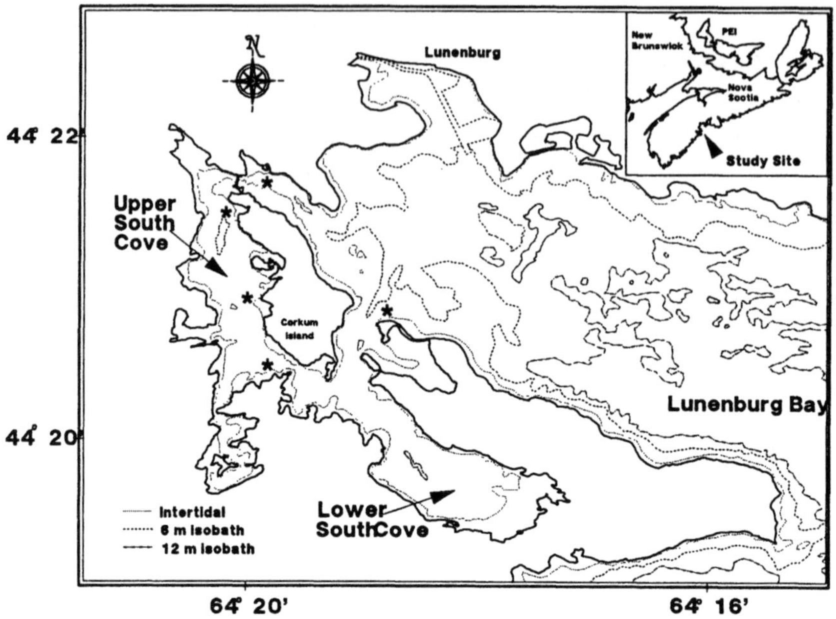

Figure 1. Map of the study site used for mussel field measurements and model simulations. Specific stations for water sampling and/or mussel growth are indicated (*). The causeway to Corkum's Island is at the island connection to the mainland. Bathymetry of the coves and bay is indicated in the lower left corner. The most landward station in the cove (7 m water depth) is poorly flushed; the center station in the cove (7 m) is approximately at the limit of the tidal excursion; the station near the inlet to the cove (5 m) and the stations outside of the cove in Lunenburg Bay (6 m) are well flushed. See text for details.

and temperature. Instrument deployment and/or water sampling (described below) corresponded to these stations. Mussels of 15-25 mm shell length were glued to strips of vexar and suspended from buoys. Shell length was measured biweekly and individuals were sampled monthly for soft tissue mass measurements. On every sampling occasion, a new set of mussels were deployed such that growth trajectories for mussels started at any time could be projected. Tissue weight trajectories were derived from shell length-tissue weight regressions plotted biweekly, and then applied to mean shell length at each date for a given experimental cohort (Mallet and Carver, submitted). The growth data reported here are derived from one such experimental unit of mussels deployed in the upper cove (Fig. 1). The density of these mussels is <1 ind m^{-3}, such that food limitation which may occur at commercial culture densities (see below) is assumed to be unimportant.

INSTRUMENTATION

One of the major questions which arises in the characterization of a culture site is the time scale and spatial coverage of sampling. Coastal ecosystems display a wide variety of spatial gradients and temporal scales in both physical and biological processes, ranging from minutes to seasons (e.g. Grant 1986; MacCaull and Platt 1977). Although a variety of studies have addressed the types of measurements required (e.g. food quality), few have dealt with the frequency and location of sampling schemes relative to the bioenergetic response of organisms. Most studies are constrained by logistic considerations, and monthly sampling is standard.

We felt that high frequency sampling of the environment with a variety of oceanographic sensors was essential to field studies of bivalve carrying capacity. Sensors were chosen to maximize information return on both fluid and particulate properties of the water column. In all cases, internal data loggers allowed remote operation of instruments for two week periods. Fouling and battery life defined the duration of deployment. Calibrations of the instruments described below were derived from both field and laboratory studies, and will be described in separate papers. Electromagnetic current meters (Marsh-McBirney ECM) with conductivity and temperature probes were fixed at several meters below the water surface at two sites along the cove. Tide gauges were placed on the bottom below the ECMs. Curves of tidal height provided input into the one-dimensional model of exchange described below. Current data were used to verify the water flux predicted by the model. Temperature proved to be an invaluable indicator of tidal flushing due to its tracking of internal vs. external water masses (see below). In addition, temperature data has numerous implications for biological processes.

An important criterion for the choice of particle sensors was the ability to distinguish between phytoplankton and non-phytoplankton seston as bivalve food. In our study, a Sea Tech fluorometer and beam transmissometer (25 cm path length) respectively were coupled with the ECM power supply and data logger to perform this task. This combination of particle sensors has been previously used in continental shelf work (Falkowski et al. 1988; Palanques and Biscaye 1992), but less frequently applied to inshore waters. Fluorometers have occasionally been used in assessment of bivalve food resources (Fréchette et al. 1989). The calibration of both instruments is dependent of the origin of suspended material, particle size etc. (Moody et al. 1987). We have been successful in using these instruments to separate phytoplankton from other seston.

Due to the limitations of size and number of instruments, we did not use them to resolve vertical structure of the water column, other than in a manual profiling mode. However, mussel culture is concentrated in the upper water column, and this section is usually well-mixed at our site and adequately characterized by the moored deployments. Our subsequent studies with sea scallop culture have involved vertical resolution of particles using other sensors.

These instruments do not eliminate the need for water sampling, especially as a means of assessing seston quality. In addition, greater spatial coverage is possible with direct sampling since the current meters/particle sensors cover only two sites at any one time. Water sampling using hand bilge pumps and subsequent filtration was conducted every two weeks during 1989-91 at various sites in and out of cove (Fig. 1). Samples were analyzed for total particulate matter, organic content, chlorophyll, and percent carbon and nitrogen. Water sampling was accompanied by CTD-current and irradiance profiles using an Aquamate (Applied Microsystems, Ltd.) and a Li-Corr quantum sensor respectively.

OTHER FIELD STUDIES

Physiological rate measurements on adult mussels (~40 mm shell length) were also conducted biweekly in order to verify scope for growth measurements predicted by the model. Mussel oxygen consumption and nitrogen excretion were measured in respiration chambers at the field lab. Feces were collected for comparison to seston samples in the calculation of absorption efficiency. Clearance rate was not measured on site, but instead the subject of laboratory studies involving the effects of current speed (Shortle 1991) and ingestion of bacteria (Johnson 1990).

Numerous other studies were conducted in support of this field and modelling program including sediment trapping, water column and benthic primary production, sediment chlorophyll and carbon/nitrogen, benthic oxygen demand, nutrient regeneration, and macrofauna biomass, water column nutrients and bacterial biomass, zooplankton biomass, and remote sensing work (synthetic aperture radar). These studies were carried out in order to measure directly the relative importance of certain processes not included in the model, to understand some of the environmental forcing of the ecosystem (e.g. phytoplankton-irradiance-nutrients), and to further validate both model structure and predictions.

CONCEPTUAL MODEL BACKGROUND

In the following sections we provide (a) the conceptual basis on which we have approached a model of culture carrying capacity, (b) a consideration of how field data can be reconciled with modelling, and (c) a proposed structure for a simulation model of mussel growth and carrying capacity, with an example of model output. In our approach, the limiting resource for filter-feeding bivalves is ultimately assumed to be suspended food (Fig. 2). Numerous studies have identified the relationship between food supplies and growth of bivalves (Thompson and Nichols 1988; Beukema and

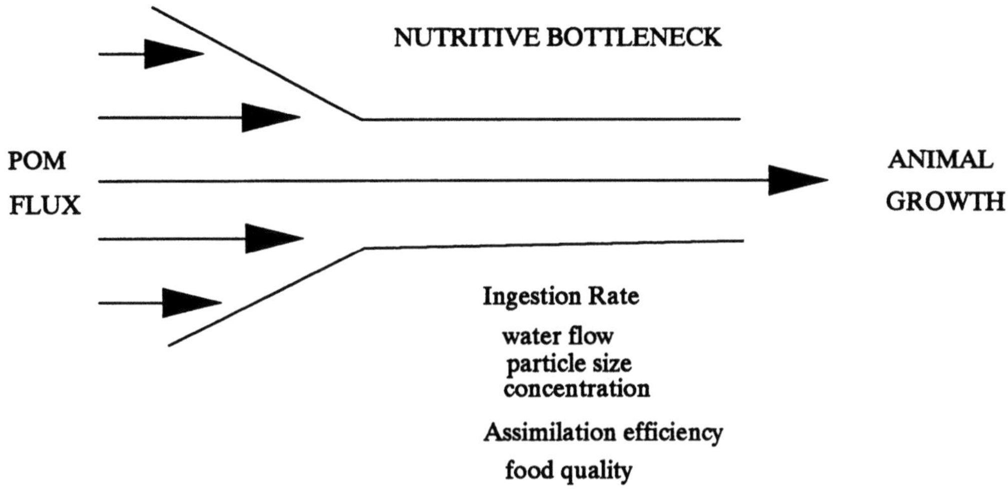

Figure 2. Conceptual diagram of factors responsible for food limitation of suspension feeders. Despite the large flux of particulate organic matter (POM) in the natural environment, several factors discussed in the text create a "nutritive bottleneck" which limit ingestion and absorption of POM, and ultimately suspension feeder growth.

Cadee 1991; see above). Based on organic matter in suspension and the rate of delivery by tidal currents, the bulk flux of organic material in the water column is large, but the ability of bivalves to ingest this material is a function of factors such as particle concentration, size distribution, and ambient fluid flow (Bayne and Hawkins 1992). Moreover, the absorption of suspended material is a function of food quality which encompasses organic/inorganic content and other measures of food value (Grant and Cranford 1991). Selection efficiency during ingestion also alters food quality and subsequent absorption. Together these factors associated with feeding limit the ability of bivalves to utilize organic material in the water column (Fig. 2).

Mussel growth can be defined for a fixed culture area or box of uniform density, rather than for a whole cove or bay. This means that food limitation occurs as a result of local seston depletion as documented for Spanish culture rafts (Navarro et al. 1991) and over natural mussel beds (Smaal et al. 1986; Fréchette et al. 1989; Newell 1990). Carrying capacity is thus applicable first to local culture density (mussels m^{-3} within a lease), and then to a region of a cove whereby an effective density is averaged over an area containing several leases. In practical terms, a box would be defined as a region of the cove with a given tidal flushing rate (see below), and determination of carrying capacity to an entire cove or bay would require more boxes.

If growth of an individual can reliably be predicted by the model, we can determine carrying capacity by increasing mussel numbers until individual growth rate (net biomass at time t) begins to decrease. We thus define carrying capacity as the number of bivalves which can be sustained at a specified growth rate. The model is based in units of organic carbon to standardize all components, except nutrients which limit primary production (see below).

We have not attempted to produce a model of the entire culture ecosystem, with mussels as one of numerous components. Natural mussel beds, other bivalve populations, or cultured mussels may dominate coastal ecosystems in terms of nutrient flux, seston dynamics, or biodeposition (Officer et al 1982; Herman and Scholten 1990; Smaal 1990; Dame 1991; Pérez Comacho et al.

1991), and we have included this feedback in determining carrying capacity. However, for simplicity in the initial efforts, we have produced a limited ecosystem model, and all components of the ecosystem which do not obviously bear on mussel growth have been excluded.

FIELD RESULTS AND CONSIDERATION OF MODEL STRUCTURE

1. WATER AND PARTICLE EXCHANGE

Due to the importance of advection in renewing depleted seston for suspension feeders (Wildish and Kristmanson 1979), considerations of physical exchange are paramount in both field and modelling studies of bivalve aquaculture. Some of these considerations are depicted schematically in Fig. 3. In dense culture, it is important that leases be sited such that seston depleted water is completely exchanged with "outside" water upon leaving a cove, otherwise, the particle-poor water will be recycled back to the bivalves, but with no new input of food (Fig. 3A). In the initial stages of our research we conducted radar-tracked drifter studies to observe patterns of water flow in the cove-bay system (Thompson et al., in prep.). These studies showed a marked asymmetry in the adjacent bay circulation associated with flood and ebb tides. Water was drawn into Upper South Cove from all directions during flood tides, but left as a tight jet into Lunenburg Bay during the ebb. As a first approximation it is reasonable to assume a water parcel leaving the cove does not return on the next incoming tide. These results highlight the importance of studying circulation in the larger bay system, beyond the boundaries of the lease area.

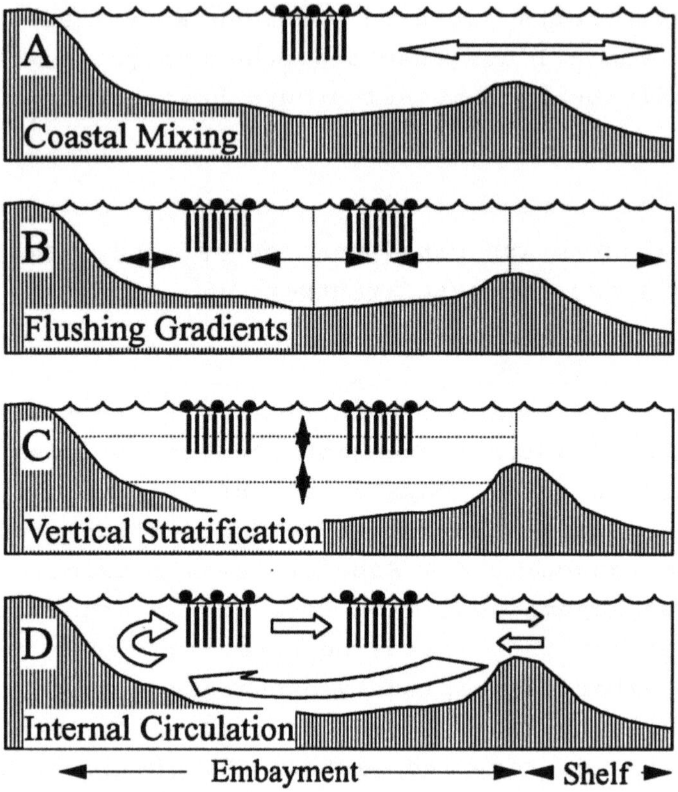

Figure 3. Aspects of tidal exchange and circulation in coastal embayments (cross-section) which affect the delivery of food to bivalves in suspended culture. Culture units are designated by the dark vertical lines with floats (solid circles) on the water surface. The right side of each panel is the seaward side of the bay while the left is landward (land shaded). Arrows depict circulation patterns as described in the text. See text for details of each panel. A. Mixing of cove water with the outer bay. B. Tidal flushing gradient along the cove as a function of the tidal excursion, with decreased exchange (reduced tidal prism) in the upper cove. Dotted lines define regions with different flushing. C. The effects of vertical stratification in the cove. Different regions of the water column are shown as dotted lines. D. Internal circulation between culture leases.

A second consideration, and the one we have emphasized in our approach to flushing, is the tidal prism, or volumetric extent of tidal flux (Fig. 3B). Culture leases in the upper portion of the cove will be less flushed, depending on the streamwise extent of flood tide (tidal excursion), than leases at the mouth of the cove. Evaluation of the importance of this flushing gradient to mussel growth is essential, since it is common to all tidal systems. As mentioned above, a simple conceptualization consists of several boxes (Fig. 3B) each with a characteristic flushing rate. In practice, the spatial dimensions of these boxes can be delineated, as we have done for Upper South Cove.

Two more complex physical situations are also of importance to suspended mussel culture. In many estuarine systems, the occurrence of two layer circulation requires that vertical as well as horizontal mixing must be considered (Fig. 3C). This can be handled by a box model concept as in Fig. 3B, but with exchange values in the vertical dimension. This formulation may be necessary seasonally in systems with freshwater input whereby estuarine circulation causes flushing gradients in two dimensions (Gibbs et al. 1991). In addition, there may be vertical stratification in particle fields due to resuspension, thermoclines, etc. Vertical gradients of mussel growth due to resuspension have previously been demonstrated (Rhoads et al. 1984). Horizontal gradients in mussel growth due to the effect of two-layer circulation on phytoplankton distribution and production have also been suggested (Roden et al. 1987).

Finally, we include the case of interaction between leases as a result of internal circulation within a cove (Fig. 3D). A lease which regularly receives seston depleted water from another lease would obviously be disadvantaged. This type of seston depletion is readily demonstrated in raft culture (Navarro et al. 1991). Incze et al. (1981) formalized this approach by considering flow through successive tiers of a mussel culture raft as a means to model particle depletion and carrying capacity. Leases will also affect the magnitude of currents in the culture area (Gibbs et al. 1991). This situation is a combination of the vertical and horizontal boxes in Figs. 3B and C. The prediction of this type of internal circulation is facilitated by a numerical model which has a relatively fine spatial definition. This class of models is more complex to formulate, and is computation-intensive, but is better able to define physical exchange. Although we have subsequently derived such models for Upper South Cove (D. Sturley, in prep.), the box model approach is suitable for the initial study of carrying capacity.

Based on these considerations we propose a simple formulation of flushing as follows; a detailed explanation of this approach will be provided in another paper. Material such as food particles is exchanged between regions inside of the cove (model boxes) or between a box and the outside as a result of tidal flushing and diffusion. Particle exchange for a box is parameterized based on the spatial differences in particle concentration between the box and the neighboring region, scaled by an exchange coefficient. This flux (F) is represented as

$$(F) = K\ (C_{\infty} - C) \qquad (1)$$

where C represents the particle concentration within the box or region of interest, C_{∞} is the concentration on the outside, and K is the exchange coefficient. For example, a higher concentration on the outside $(C_{\infty} > C)$ would result in a positive flux into the box. More complex models for mixing and exchange could be used (e.g. Fischer et al. 1979), but these would require abandonment of the box model concept.

The exchange coefficient sets the rate for the advective and diffusive fluxes between boxes. The units of K are time^{-1} and the quantity 1/K may be interpreted as the flushing time for the box. K is defined for a regions of a cove (Fig. 3B), and the exchange rate for culture units is dependent on their location within a cove. In our system, K ranges from approximately 0.5 day^{-1} in the upper cove (culture areas) to 2.0 day^{-1} in the outer cove. The determination of a value for K is quite involved since many mixing and exchange mechanisms could be postulated as discussed above. For Upper South Cove, dominated by tidal flushing, a variant of the modified tidal prism method (Dyer 1973) was used to determine K, i.e. the limit of the tidal excursion determines the extent of exchange at various locations along a cove. These values were verified with a summer heat budget calculation based on thermograph records, as well as with drifter studies. Although no mixing or circulation due to winds has been included here, an overall increase in K due to winds could be calculated daily or seasonally.

2. ENVIRONMENTAL VARIABILITY

Long term monitoring of environmental variables with data-logging instruments provides insight into the range of conditions which occur at a culture site, and aids in our conceptualization of the culture ecosystem. These data have been used to provide guidance for initial and boundary conditions, parameters values, model structure, and direction of modelling efforts.

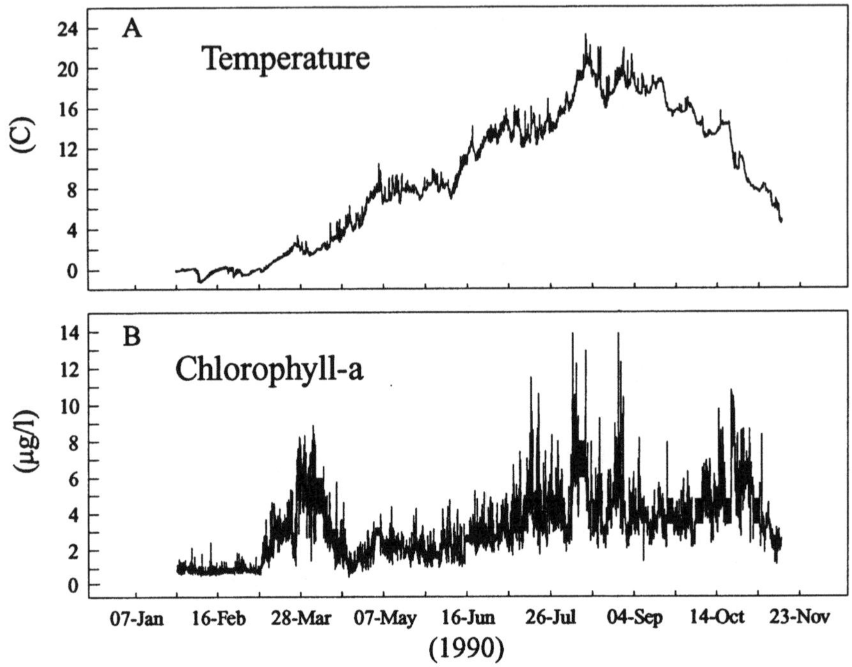

Figure 4. Annual records of temperature and chlorophyll from *in situ* instruments (sampled hourly) at the upper end of Upper South Cove (landward most cove site in Fig. 1). Tidal exchange is typical of quiescent conditions in the upper cove (peak current speed usually 15 cm sec^{-1}). Growth trajectories (Fig. 6B) were measured at this site. A. Water temperature measured by thermistor at 4 m above bottom in a 7 m water column. B. Chlorophyll *a* based on calibrated fluorometer at 4 m above bottom.

Detailed results of these field studies will be the subject of other papers, but we provide a summary here. Examination of temperature data over a year demonstrates a typical seasonal curve, but with numerous high frequency events superimposed on this curve (Fig. 4A). These events signify temperature fluctuation due to tidal exchange. Animals are thus exposed to temperature fluctuations of as much as 5°C over a few hours. In addition, tidal periodicity in temperature provides an indication of flushing rates and a simple means to confirm other physical studies of tidal exchange. There are also low frequency temperature fluctuations which occur as a result of shelf-scale flushing due to episodic frontal systems (Platt et al. 1972). Temperature is thus an easily obtained physical measurement which yields abundant information about the culture environment. Moreover, temperature recorders configured as single instruments are small, inexpensive, and easily deployed.

Some of the implications of rapid temperature change for animal energetics are less apparent. Although mussels and other bivalves show temperature acclimation in feeding and respiration, the significance of tidal or shorter term variation in their bioenergetics is unclear. Studies of acute temperature response rarely incorporate rapid temperature fluctuations, so it is difficult to predict the effects of tidal temperature variability on mussel bioenergetics, especially when filtration and respiration show differing responses (Newell and Branch 1980). Although Newell and Branch (1980) presented metabolic rate-temperature curves based on mollusc body weight and activity level, the curves are polynomial functions for fixed exposure temperatures. It is unclear over what time scale the organism integrates rapid temperature changes in terms of bioenergetics. Widdows (1976) suggested that diel temperature fluctuations produced respiration and clearance rates in *Mytilus edulis* similar to those at constant temperatures corresponding to diel averages. Further work along these lines is required to assess generality and applicability of these results to mussel energy budgets. Our laboratory measurements of respiration were made at non-varying temperatures, and our model uses daily temperature averages with constant Q_{10} values. As discussed below, even seasonal changes in metabolism may be decoupled from temperature. Temperature responses are a clear example of an area where further empirical data would enhance the power of bioenergetic models.

Chlorophyll *a* records from the *in situ* fluorometer also demonstrate high frequency tidal variability superimposed on a seasonal pattern (Fig. 4B). Upper South Cove characteristically has a both a spring and fall bloom. Previous studies of Nova Scotia embayments also indicate significant fall blooms (Platt 1971). Ice cover results in extremely low production during winter. High production also occurs intermittently in summer. Our data and previous studies suggest that these summer peaks are due to the shelf-scale flushing events which bring nutrient-rich shelf water into the cove (Platt et al. 1972; Cloern 1991) as seen in the temperature records. Studies on mussel culture in the Spanish Rias demonstrate that mussel condition is increased in years with enhanced upwelling and water exchange (Blanton et al. 1987). Our data on temperature, chlorophyll, and nutrients are currently being analyzed in this regard. Although we have emphasized physical forcing of primary production, changes in photosynthetic parameters also occur on short time scales (MacCaulay and Platt 1977) and would contribute to variability in phytoplankton biomass. Zooplankton grazing does not seem to limit primary production in the cove (Dowd 1991).

Prediction of the response of bivalves to food fluctuations suffers from a lack of empirical understanding similar to that of temperature response.

Tidal rhythms in both filtration and digestion are documented for various bivalves (Bayne et al. 1988; Newell and Shumway, this volume), but the mechanism for these patterns is not well known. Stenton-Dozey and Brown (1992) demonstrated tidal changes in both clearance rate and particle retention efficiency in rock-pool clams. Experiments measuring short-term changes in mussel filtration rate with changing food concentration show conflicting results (Davenport and Woolmington 1982; Fréchette and Bourget 1987), and it appears that other factors such as particle size, organic/inorganic content, and flow rate interact with concentration to affect feeding (Bayne et al. 1988). Biotic variability such as feeding history and satiation also contribute to non-consistent feeding rates (Riisgård 1991), and bivalve filtration may be erratic despite seemingly constant conditions (Fréchette and Bourget 1987; Bayne, this volume). Because coastal waters show tremendous variability in particle quality and quantity on a variety of time scales (Fegley et al. 1992; Stenton-Dozey and Brown 1992), it would be difficult to incorporate the high frequency fluctuations of food and temperature into a model. Moreover, the time scale over which physiological responses to changing conditions occur is not sufficiently known to be easily modelled, although both experimental and theoretical approaches to these problems have been addressed (see Bayne, this volume). Further empirical studies are necessary to determine appropriate averaging of food and feeding in order to characterize the culture environment. These studies are required to direct continued modelling efforts aimed at simulating environmental variability and its effect on mussel growth. Sensitivity studies with the mussel energy budget (Eq. 2) are a logical starting point.

These types of field studies are clearly a vital complement to the modelling, especially in a coupled physical-biological model. Field results provide insight into the culture ecosystem such that continued development of the model described below can be guided by a realistic appreciation of the factors important in bivalve energetics and growth.

SIMULATION MODEL STRUCTURE

An overview of the model is presented; details of the functional relationships are given elsewhere (Dowd 1991; Dowd et al., in prep). The box model consists of two competing herbivores (mussels and zooplankton), two food sources, phytoplankton and other seston (bacteria, detritus, etc.), and a limiting nutrient. The mussels are fixed within a box (i.e. on culture lines), but zooplankton, phytoplankton and seston are transported in and out of defined boxes by tides (Fig. 5). We have used the box to represent a culture area or lease with a specified density of suspended mussels. Boxes may be anywhere in the cove, but box size is defined by regions of the cove which include limited horizontal gradients (e.g. in flushing), and greater numbers of boxes require calculation of exchange coefficients between them. The goals of the model are prediction of mussel growth rate, elucidation of environmental factors which affect growth, application of this information to optimal culture site selection, and determination of carrying capacity of a site for mussel culture.

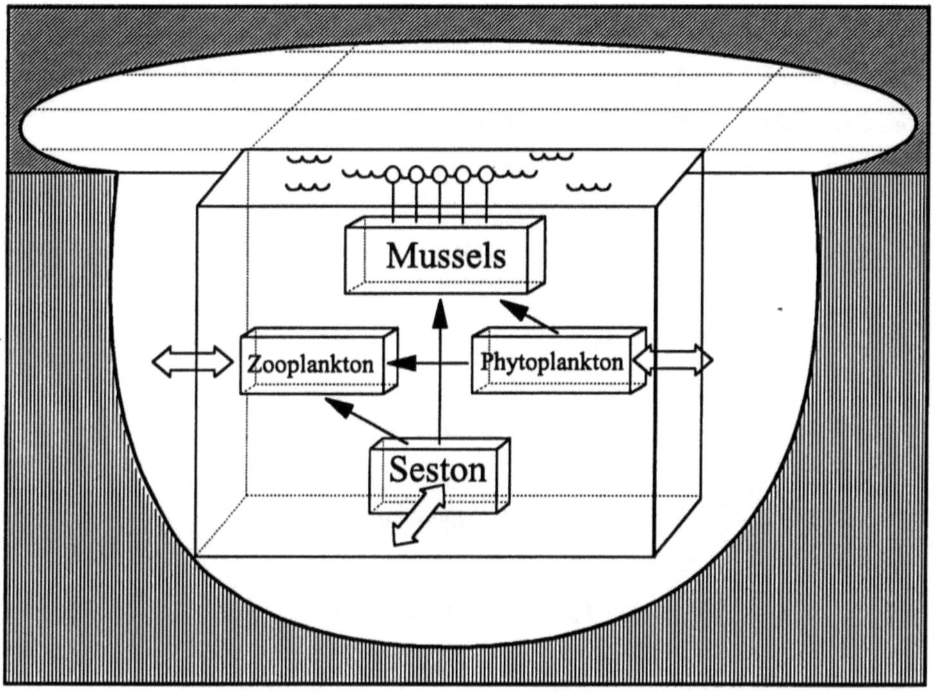

Figure 5. Box diagram for the carbon-based model of suspended mussel culture showing the state variables and their connections. The box represents a defined region of a cove in which the trophic components of the model interact. Mussels and zooplankton feed on both phytoplankton and non-phytoplankton seston (detritus). Mussels are fixed onto suspended longlines, but phytoplankton, zooplankton, and seston are subject to tidal diffusion/advection with the far-field conditions outside of the box.

The box model for mussel growth in an aquaculture ecosystem is described by a system of coupled ordinary differential equations. The limited ecosystem represented by the model includes the following state variables: individual mussel weight (M); mussel numbers (N); phytoplankton (P); and non-phytoplankton seston (S). The carbon flux for a single box and single mussel weight class are

$$\frac{\partial M}{\partial t} = (\varepsilon_M I_M - R_M)M \tag{2}$$

$$\frac{\partial N}{\partial t} = -\lambda_M N \tag{3}$$

$$\frac{\partial Z}{\partial t} = (\varepsilon_Z I_Z - R_Z - \lambda_Z)Z + K(Z_\infty - Z) \tag{4}$$

$$\frac{\partial P}{\partial t} = \gamma P - I_Z^P Z - I_M^P MN + K(P_\infty - P) \tag{5}$$

$$\frac{\partial S}{\partial t} = D_S - I_Z^S Z - I_M^S MN + K(S_\infty - S) \tag{6}$$

A nutrient flux equation is also included to provide a negative feedback on the entire ecosystem via limitation of phytoplankton growth. The nutrient flux equation is measured in the units of the most limiting nutrient (χ) and is represented as

$$\frac{\partial \chi}{\partial t} = E_Z Z + E_M MN + D_\chi - \gamma P + K(\chi_\infty - \chi) \tag{7}$$

The various terms which govern flux rates and their functional dependencies are shown in Table 1. This table shows that the majority of these terms are not simply constants but functionally dependent on the values of the state variables and thetime-dependent forcing functions. This makes the flux equations highly non-linear and sufficiently complex to require numerical solution.

Table 1. Terms used in the box model equations defined with respect to state variables for individual mussel weight (M), zooplankton biomass (Z), phytoplankton biomass (P), seston carbon (S), and a limiting nutrient (χ). Forcing functions refer to sunlight (light) and water temperature (temp).

Term	Description
$\varepsilon_M\ (P,S)$	total absorption efficiency for M
$\varepsilon_Z\ (P,S)$	total absorption efficiency for Z
$I_M\ (P, S, M, temp)$	total ingestion rate for M
$I_Z\ (P, S, temp)$	total ingestion rate for Z
$I_M^P\ (P, S, M, temp)$	ingestion rate for M on P
$I_M^S\ (P, S, M, temp)$	ingestion rate for M on S
$I_Z^P\ (P, S, temp)$	ingestion rate for Z on P
$I_Z^S\ (P, S, temp)$	ingestion rate for Z on S
$R_M\ (P, S, M, temp)$	respiration rate for M
$R_Z\ (temp)$	respiration rate for Z
$\lambda_M\ (temp)$	mortality rate for M
$\lambda_Z\ (temp)$	mortality rate for Z
$E_M\ (P, S, M, temp)$	excretion rate for M
$E_Z\ (P, S, temp)$	excretion rate for Z
$\gamma\ (\chi, light, temp)$	phytoplankton growth rate
K	exchange coefficient
D_S	internal sources and sinks of S
$D\chi$	internal sources and sinks of χ

In order to completely specify the problem (and solve the equations), the following additional information is also required.

1. The initial conditions $M(0)$, $N(0)$, $P(0)$, $Z(0)$, $S(0)$, and $\chi(0)$ all of which must be non-negative to be biologically realistic.
2. The boundary conditions Z_∞, P_∞, S_∞, χ_∞. These values correspond to the values of the state variables in the adjacent box. If there are multiple boxes, then for at least one box (e.g. most seaward), the outside values of the state variables must be prescribed.
3. The forcing functions which here consist of temperature and light.
4. Internal source terms D_S and D_χ. These represent sources or sinks of non-phytoplankton seston (e.g. resuspension, deposition) and nutrients (e.g. benthic regeneration) found within a box.

These required data are obtained either through direct measurement at a specific site, or via approximations from the literature. In the simulations of mussel growth described below, the boundary conditions are derived from time series of environmental variables such as the seasonal cycle of temperature measured in the outer portion of Upper South Cove (e.g. Fig. 4A). Because the model is based on daily averages, smoothed curves were used to respresent temporal changes in the boundary conditions

The general formulation of the flux equations allows a considerable degree of flexibility. The rate functions that have been used in our model are based primarily on well established biological principals such as allometric relationships, Q_{10}-type temperature response, and saturation relationships for ingestion by mussels and zooplankton. The growth curve for phytoplankton was calculated using a photosynthesis-irradiance curve coupled with the underwater light field. Based on these types of generic formulations, it is possible to adapt the model to any site-specific considerations.

The flux equations of the box model describe population energy budgets. For mussels and zooplankton, the growth rate depends on the differences between gains through ingestion and losses through respiration and mortality. Of these two components, the mussels are represented in greater detail by separating total population biomass into individual mussel weight and numbers of mussels. This allows the weight-dependence of the physiological rates to be resolved. The food sources, phytoplankton and non-phytoplankton seston, decline through ingestion by mussels and zooplankton. Gains for these food components occur through growth, in the case of phytoplankton, and through source terms such as resuspension or river input for non-phytoplankton seston. Nutrient inputs occur via excretion by mussels and zooplankton, and nutrient losses take place through uptake by phytoplankton (the feedback mechanism). In addition to these processes which occur within a box, the exchange terms allow phytoplankton, zooplankton, seston, and nutrients to be exchanged between boxes and/or with the outside.

The temporal evolution of the state variables is determined by numerically integrating the differential equations, using daily averages. Although we recognize that daily averages obscure the high frequency variability in environmental factors, finer time scales are not appropriate given the knowledge of bivalve responses as discussed above, and the interest in annual growth rates. Equations are solved using programs written with MATLAB software (Mathworks, Inc., Natick, Mass., USA). Model equations are highly non-linear and contain time-dependent forcing, boundary conditions, and parameters. This results in dynamically complex behaviour characteristic of such models. Consequently, the cause behind every nuance of simulation output is not immediately obvious despite the usefulness of simulations in producing generally interpretable trends. In this regard, sensitivity studies have proved to be informative, especially in suggesting emphases for future field studies.

The key elements of the model are portability and generality. Model specifics concerning bioenergetic terms (e.g. weight-specific respiration) are easily adaptable to other suspension feeding species. Through the application of either local data, robust functional relationships, or literature coefficients, the model can easily be used for other aquaculture sites.

MUSSEL GROWTH PREDICTED AND OBSERVED

Use of the model with a single box in Upper South Cove in a comparison of simulated and observed individual mussel biomass through time indicates reasonable ability of the model to predict mussel growth (Fig. 6). Both simulated growth for low culture density (Fig. 6B) and corresponding observed growth are characterized by low growth in winter and early spring followed by a burst of tissue weight increase during summer. The observed growth tends to begin earlier in the year than the simulated growth, as discussed below. There is a decline in biomass during winter in both observed and predicted trajectories, although it is slightly steeper in the model curve. The absolute value of observed peak annual mussel weight is somewhat less than that of the simulation. Absolute values of peak weight in the model are sensitive to various factors (Dowd 1991) so it is not surprising that there is some spread between the magnitude of predicted and observed biomass. In the second year of growth, there will be spawning events which alter the biomass trajectories (Bayne and Worrall 1980), but for simplicity we have restricted the initial simulations to first year animals.

Examination of environmental data illustrates how food and temperature interact to affect mussel growth. Despite the high phytoplankton abundance during the spring bloom (Fig. 4B), temperatures are low (Fig. 4A) and in the model, mussel ingestion and growth remain low until late spring (Fig. 6A). Growth rate accelerates in July as both temperature and phytoplankton increase. During winter, there is a steep decline in biomass, due to declining food and ingestion coupled with sustained metabolic demands at the relatively high late fall temperatures. Previous studies of growth in *Mytilus edulis* have also noted a winter weight loss (Bayne and Worrall 1980; Rodhouse et al. 1984; Page and Hubbard 1987). Several factors contribute to this decline, the most general in the above studies being a decrease in phytoplankton biomass through the winter. Food quality also decreases during this period because the seston has low organic content (Widdows 1978). In Upper South Cove, there is a decline in chlorophyll and organic content of the seston throughout the winter, and a corresponding decrease in mussel absorption efficiency (Hatcher and Grant, in prep.). This factor, coupled with a decrease in clearance rate at lower winter temperatures, results in a negative scope for growth during this period. Bayne and Widdows (1978) attributed winter weight loss to similar factors in U.K. populations of *Mytilus edulis* .

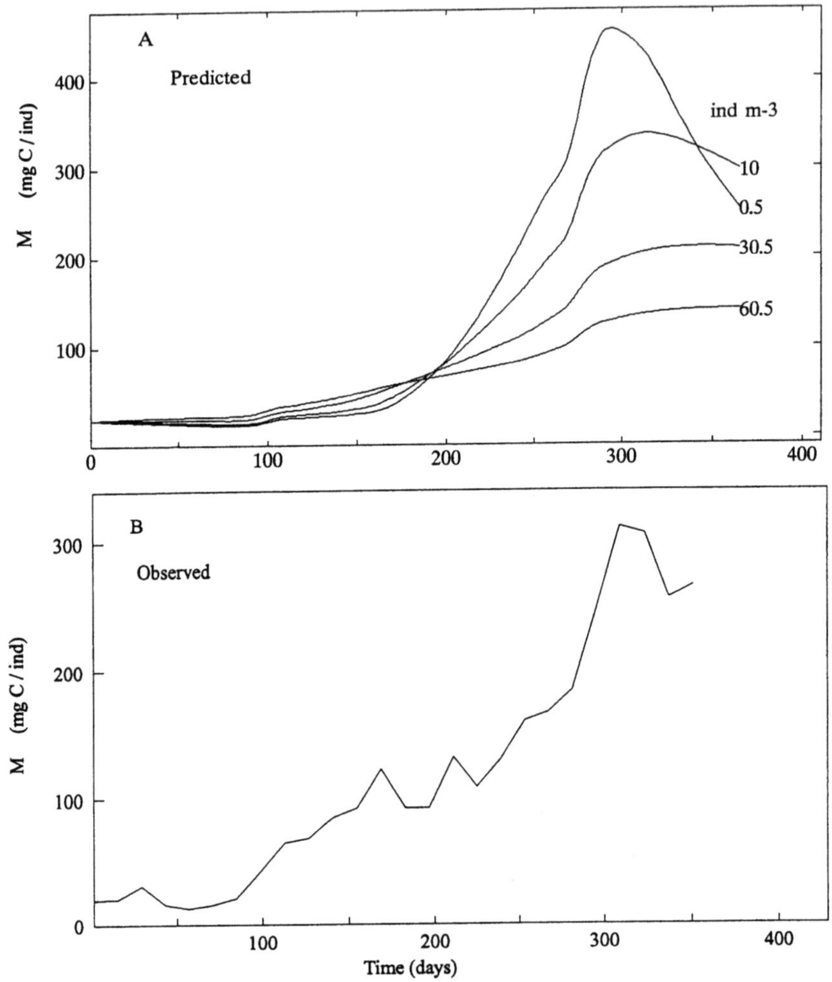

Figure 6. Predicted and observed trajectories of mussel tissue biomass (mg C individual^{-1} over a one year period (January 1 - December 31) in the poorly-flushed upper region of Upper South Cove. The initial mussel weight corresponds to a mussel of ~20 mm shell length. Mussels do not spawn in the first year of growth, so that any buildup of gonad would be accounted for by scope for growth, and reproductive losses can be ignored. Temperature and chlorophyll records for this site are given in Fig. 5. A. Individual mussel biomass as a function of time for different culture stocking densities (ind m^{-3}; indicated at the end of each curve). B. Observed changes in mean (see next pg.)

(Fig. 6 con't.) mussel dry carbon weight over the same time period. Stocking density corresponds to the lowest density in panel A. Carbon weight of tissue was calculated as 40% of dry tissue weight. The error of tissue weight estimated from mean shell length for each measurement period is small considering the generally tight length-weight regressions (Mallet and Carver, submitted).

The relative importance of temperature compared to food as a controlling factor in mussel growth is unclear (Bayne and Widdows 1978; Kautsky 1982; Loo and Rosenberg 1983; Thompson 1984; Page and Hubbard 1987), because high temperatures and increased phytoplankton production are often concurrent. In Upper South Cove, the spring bloom occurs during a period of low temperatures (Fig. 5A) but the observed growth trajectory (Fig. 6B) suggests that mussels do benefit from this pulse of food, presumably due to some temperature acclimation response as suggested above. Thompson (1984) showed that mussel clearance was not depressed by cold Newfoundland water temperatures, and that these animals maintained a positive scope for growth throughout the year. Loo (1992) suggested that *Mytilus edulis* benefits from the spring bloom despite low temperatures, but the ingestion data presented indicates low clearance rates at the studied range of 0 - 4.8°C.

In the predicted growth trajectory (Fig. 6A), the winter decline may be more drastic and the spring tissue increase later than in observed curve (Fig. 6B). The single annual Q_{10} values used for both respiration and ingestion do not allow for seasonal acclimation to temperature change. Previous studies of both respiration and clearance in *Mytilus edulis* have demonstrated that respiration and clearance are adjusted over a matter of weeks as an acclimation response to either colder or warmer temperatures (Widdows and Bayne 1971; Bayne et al. 1973; Jørgensen et al. 1990).

Our own data also suggest that temperature alone does not control the magnitude of mussel respiration (Fig. 7). Although winter respiration is lower than in other seasons, values in spring are higher than in summer, despite a much colder spring water temperature.

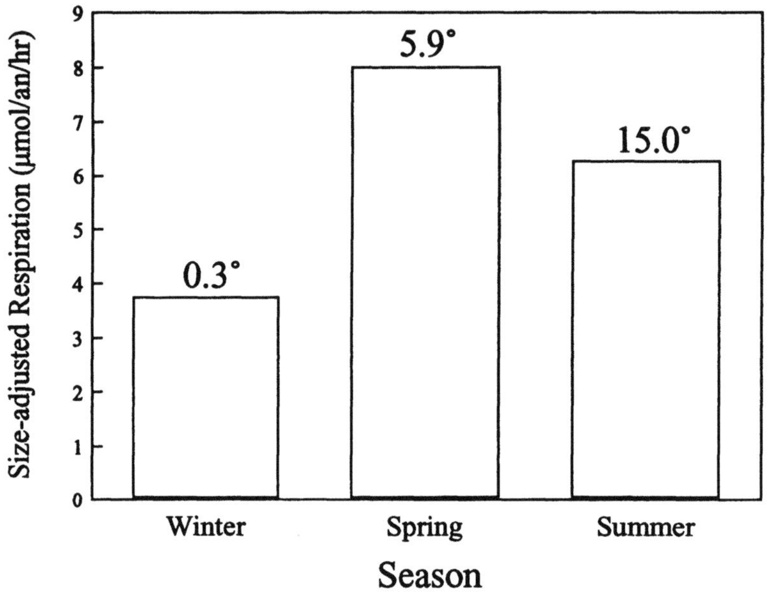

Figure 7. Seasonal means of mussel respiration rate at 3 temperatures (°C) for cultured adult animals in Upper South Cove. Values for individual mussels were standardized to a mussel of 0.7 g dry tissue weight, although only a narrow size range of animals (~40 mm shell length) was used. All three means are significantly different from one another (ANOVA; $p < 0.05$). Coefficients of variation for raw data (non-size adjusted) were 46, 21, and 40% for winter ($n = 46$), spring ($n = 15$), and summer ($n = 24$) respectively.

Because these are adult mussels, there is increased metabolic activity associated with spring gametogenesis (Widdows 1978), an effect that apparently supersedes that of temperature. These results are significant in suggesting improvements in model formulation of mussel bioenergetics and in directing sensitivity analyses. They provide a prime example of the way that empirical studies interact with the model structure.

The question of carrying capacity can be addressed by varying initial density of mussels in culture, and examining the resultant growth trajectories (Fig. 6A). At densities higher than 0.5 mussels m^{-3} the peak weight of mussels during the year declines substantially, although for densities of 10 and ~30 mussels m^{-3} the winter weight is comparable to that of the lowest culture density. At commercial longline densities (~60 mussels m^{-3} within a lease), there is a 40% reduction in winter weight compared to the lowest modelled density. Intraspecific crowding tends to lessen the summer/fall weight gain as well as the winter weight loss (Fig. 6A). This apparently occurs because mussels at higher density are already food limited during the fall bloom and a post-bloom decline in scope for growth is a less abrupt transition in the growth trajectory. These simulations demonstrate the utility of the model in deriving carrying capacity for a coastal inlet. We have presented a simple case with only one box, but other simulations have demonstrated a horizontal gradient in carrying capacity as a result of the tidal flushing gradient in the cove (M. Dowd., in preparation). One consequence of increased exchange is that variables such as phytoplankton biomass are more influenced by boundary conditions than by processes within a box such as photosynthesis (Herman, this volume)

FINAL CONSIDERATIONS OF THE MUSSEL ECOSYSTEM

Empirical studies and modelling of Upper South Cove have played a vital role in understanding of the factors responsible for mussel growth. These studies continue to interact with the modelling to improve the assessment of carrying capacity. The model formulation presented in Fig. 5 is a practical but simplified view of important processes and variables. Below we consider various components of the culture ecosystem and their potential significance to mussel growth.

The food sources available to suspension feeders come from a variety of sources, presenting a complex set of food variables to measure and model. Phytoplankton biomass represented as chlorophyll allows assessment of one portion of suspended carbon which is usually a high quality food source for suspension feeders. Even within this category, there is selection for cell types (Shumway et al. 1985) Previous studies have demonstrated that specific components of the phytoplankton (e.g. flagellates) are correlated with bivalve growth, whereas total chlorophyll is not (Pieters et al. 1980; Beukema and Cadee 1991). In addition, high chlorophyll associated with toxic algal blooms does not always indicate abundant food (Bricelj et al. 1987). Nonetheless, there are many examples where total phytoplankton biomass or chlorophyll is correlated with bivalve growth (e.g. Kautsky 1982; Brown 1988; Thompson and Nichols 1988; see also Bayne and Hawkins 1992) and this premise provides a sound basis for representing mussel food supplies in both field and model studies.

Non-phytoplankton seston includes macrophyte detritus, bacteria, and particulate inorganic matter. Feeding experiments with mussels from Upper South Cove indicate that they do not retain bacterioplankton and derive little nutrition from the low standing stock of particle-bound bacteria (Johnson 1990). A feeding study with *Mytilus edulis* by Lucas et al. (1987) confirms this result, but these authors and others (see Bayne and Hawkins 1992) point out that high bacterial standing stocks (e.g. some upwelling systems) may increase the significance of these cells in bivalve carbon and nitrogen budgets.

There is also evidence that suspension feeders including *Mytilus edulis* benefit from macroalgal detritus such as kelp (see Grant and Cranford 1991). Although this material is abundant in Upper South Cove, our feeding experiments with sea scallops (*Placopecten magellanicus)* indicate that this species is nitrogen limited on a kelp diet (Grant and Cranford 1991). Eelgrass detritus is abundant at the study site, but mussels derived no nutrition from this material in laboratory feeding experiments (J. Grant, unpublished data). Various studies have suggested a positive benefit of resuspended sediment in bivalve nutrition for several species including *Mytilus edulis* (Rhoads et al. 1984; see references in Fréchette and Grant 1991) but the latter authors found no obvious differences in growth of mussels exposed to resuspension. We conclude that the pooling of non-phytoplankton seston as a food source is an appropriate conceptualization given the present understanding of bivalve trophic resources. The importance of this food source may be highly variable at different sites depending on the origin and lability of detritus which makes up this pool. In addition, continued feeding studies with natural seston (e.g. Navarro et al. 1991) are necessary to enhance applicability of this and other studies to a variety of sites

The natural benthos in Upper South Cove might be expected to compete with mussels for primary production (van der Veer 1989), but we have neglected this component in the model. Many of the benthic species beneath the mussel lines are deposit-feeders which utilize mussel biodeposition (e.g. Rosenberg and Loo 1983), a result confirmed by our benthic sampling in the cove. Based on sediment oxygen demand at Upper South Cove (unpublished data), the carbon requirements of this community are significant, but they are utilizing primary production which has already been filtered by mussels, as well as sedimentary detritus. While natural mussel populations can readily dominate benthic energy flow (Smaal 1990; Smaal and Prins, this volume), the situation is less clear for suspended mussels relative to the underlying benthos. Our studies on mussel and benthic carbon/nutrient fluxes will clarify this question for Upper South Cove (Hatcher and Grant, in prep.).

There are a variety of other parameters and variables which might be considered as important to both models and empirical studies of mussel growth. One aspect of mussel bioenergetics which bears discussion is the role of flow velocity in inhibiting mussel feeding at flows >6 cm sec^{-1} as proposed by Wildish in various studies (e.g. Wildish and Miyares 1990; Wildish and Kristmanson, this volume). This velocity range is typical of our upper cove sites and frequently exceeded in the outer cove. Our feeding experiments with *Mytilus edulis* in a laboratory flume showed no feeding inhibition at speeds up to 15 cm sec^{-1} (Shortle 1991). Further experiments are required to incorporate current speed into mussel energy budgets.

CONCLUSIONS

The limiting nature of advected food delivery to bivalves relative to their consumption is well founded in the literature (Wildish and Kristmanson 1979; Incze et al. 1980; Rosenberg and Loo 1983) and well documented in field observations (Blanton et al. 1987; Grizzle and Lutz 1989; Fréchette et al. 1989; Smaal and van Stralen 1990). Moreover, there are sound predictive relationships and empirical approaches whereby bivalve trophic relationships can be quantified and modelled. Nonetheless, further empirical work is necessary to understand the physiological response of bivalves to high frequency changes in environmental conditions, for eventual incorporation into models.

Although we have reported the specifics of a field study and model for coastal Nova Scotia, these efforts are applicable to other aquaculture sites. Temperature, sunlight, and in many places chlorophyll levels are sufficiently known to apply the model to new culture areas. Otherwise, continuous recording instruments, such as those deployed in Upper South Cove can be used to obtain a site-specific data record of environmental variables. Various survey techniques from tide gauges to remote sensing could be used to generate coastal databases suitable for model input. Such databases already exist for many regions and are compiled by government agencies (e.g. Marine Environmental Data Service, Ottawa). The model may then be used to generate maps of potential production for coastal areas.

This research is a significant improvement over many previous approaches to aquaculture carrying capacity because it provides a mechanistic understanding of bivalve growth. The model is useful as a focal point in this work because sensitivity analysis allows assessment of potentially important factors involved in bivalve growth. Modelling interacts with field work in an iterative fashion in that sensitivity analysis can be used to refine the emphasis of field programs. Moreover, field studies allow verification of model structure and output. Field studies additionally identify the scales of spatial and temporal variability in the environment which may be necessary to consider in producing realistic simulations. Other workers have also made progress in empirical/modelling studies of cultured and natural bivalve populations in coastal environments (Smaal 1990; Bacher et al. 1991; Herman, this volume). Physical-biological models and corresponding field programs can thus make a major contribution to the bivalve culture site-slection, management of culture and estimation of carrying capacity, understanding of limiting factors for bivalve growth, and evaluation of the role of suspension feeders in coastal ecosystems.

ACKNOWLEDGEMENTS

We thank Dale Cook for allowing us use of space at Corkums Island Mussel Farms, and for assisting with SCUBA and other field work. Andre Mallet and Claire Carver provided invaluable collaboration, including data on mussel growth. Bryan Schofield supervised instrumentation and the logistics of field work. This research was supported by a Strategic Grant from the Natural Sciences and Engineering Research Council of Canada to JG and KT, and by the Ocean Production

Enhancement Network (OPEN), one of the fifteen Networks of Centres of Excellence supported by the Government of Canada.

REFERENCES

Bacher C, Heral M, Deslous-Paoli JM, Razet D (1991) Modele energetique uniboite de la croissance des huitres (*Crassostrea gigas*) dans le bassin de Marennes-Oleron. Can. J. Fish. Aquat. Sci. 48: 391-404

Bayne BL, Newell RC (1983) Physiological energetics of marine molluscs. In: The mollusca, Vol. 4 (Saleudin ASM, Wilbur KM, eds), pp. 407-515. New York Academic Press

Bayne BL, Thompson RJ, Widdows J (1973) Some effects of temperature and food on the rate of oxygen consumption by *Mytilus edulis* L. In: Effects of temperature on ectothermic organisms, Elsevier Amsterdam (Wieser W, ed), pp. 181-193, Springer-Verlag Berlin

Bayne BL, Worrall CM (1980) Growth and production of mussels *Mytilus edulis* from two populations. Mar. Ecol. Prog. Ser. 3: 317-328

Bayne BL, Hawkins, AJS (1992) Ecological and physiological aspects of herbivory in benthic suspension-feeding molluscs, In: Plant-animal interactions in the marine benthos, (John DM, Hawkins SJ, Price JH, eds), pp. 265-288. Clarendon Press Oxford

Bayne BL, Widdows J (1978) The physiological ecology of two populations of *Mytilus edulis* L. Oecologia 37: 137- 162

Bayne BL, Hawkins AJS, Navarro E (1988) Feeding and digestion in suspension-feding molluscs: the relevance of physiological compensations. Amer. Zool. 28: 147-159

Bayne, B (1993) Feeding physiology of bivalves: time dependence and compensation for changes in food available. This volume

Beukema JJ, Cadee GC (1991) Growth rates of the bivalve *Macoma balthica* in the Wadden Sea during a period of eutrophication - relationships with concentrations of pelagic diatoms and flagellates. Mar. Ecol. Ser. 68: 249-256

Blanton JO, Tenore KR, Castillejo F, Atkinson LP, Schwing FB, Lavin A (1987) The relationship of upwelling to mussel production in the rias on the western coast of Spain. J. Mar. Res. 45: 497-511

Brown JR, Hartwick EB (1988) A habitat suitability index model for suspended tray culture of the Pacific oyster, *Crassostrea gigas* Thunberg. Aquacult. Fish. Manage. 19: 109-126

Bricelj VM, Epp J, Malouf RE (1987) Intraspecific variation in reproductive and somatic growth cycles of bay scallops *Argopecten irradians*. Mar. Ecol. Prog. Ser. 36: 123-137

Brylinsky M, Sephton TW (1991) Development of a computer simulation model of a culture blue mussel (*Mytilus edulis*) population. Can. Tech. Rep. Fish. Aquat. Sci. 1805

Carver CEA, Mallet AL (1990) Estimating the carrying capacity of a coastal inlet for mussel culture. Aquaculture 88: 39-53

Cloern JE (1991) Tidal stirring and phytoplanton bloom dynamics in an estuary. J. Mar. Res. 49: 203-221

Dahlbäck B, Gunnarsson LÅH (1981) Sedimentation and sulfate reduction under a mussel culture. Mar. Biol. 63: 269-275

Davenport J, Woolmington AD (1982) A new method of monitoring ventilatory activity in mussels and its use in a study of the ventilatory patterns of *Mytilus edulis* L. J. Exp. Mar. Biol. Ecol. 62: 55-67

Dijkema R, van Stralen M (1989) Mussel cultivation in the Netherlands. World Aquacult. 20: 56-62

Doering PH, Oviatt CA (1986) Application of filtration rate models to field populations of bivalves: an assessment using experimental mesocosms. Mar. Ecol. Prog. Ser. 31: 265-275

Dowd, M (1991) On the prediction of bivalve growth in an aquaculture site. M.Sc. thesis, Dalhousie University

Dyer KR (1973) Estuaries: a physical introduction. John Wiley London

Falkowski PG, Flagg CN, Rowe GT, Smith SL, Whitledge TE, Wirick D (1988) The fate of a spring phytoplankton bloom: export or oxidation. Cont. Shelf Res. 8: 457-484

Fegley SR, MacDonald BA, Jacobsen TR (1992) Short-term variation in the quantity and quality of seston available to benthic suspension feeders. Estuar. Coast. Shelf Sci. 34: 393-412

Fischer HB, List JE, Koh RCY, Imberger J and Brooks NH (1979) Mixing in inland and coastal waters. Academic Press Orlando

Fréchette M, Bouget E (1987) Significance of small-scale spatio-temporal heterogeneity in phytoplankton abundance for energy flow in *Mytilus edulis*. Mar. Biol. 94: 231-240

Fréchette M, Grant J (1991) An in situ estimation of the effect of wind-driven resuspension on the growth of the mussel *Mytilus edulis* L. J. Exp. Mar. Biol. Ecol. 148: 210-213

Fréchette, M., Butman CA, Geyer WR 1989. The importance of boundary-layer flow processes in supplying phytoplankton to the benthic suspension-feeder, *Mytilus edulis* L. Limnol. Oceanogr. 34: 19-36

Gibbs MM, James MR, Pickmere SE, Woods PH, Shakespeare BS, Hickman RW, Illingsworth J (1991) Hydrodynamic and water column properties at six station associated with mussel farming in Pelorus Sound, 1984-85. N. Z. J. Mar. Freshw. Res. 25: 239-254

Grant J (1986) Sensitivity of benthic community respiration and primary production to changes in temperature and light. Mar. Biol. 90: 299-306

Grant J, Cranford PJ (1991) Carbon and nitrogen scope for growth as a function of diet in the sea scallop *Placopecten magellanicus*. J. Mar. Biol. Assoc. U.K. 71: 437-450

Grant J, Enright CT, Griswold A (1990) Resuspension and growth of *Ostrea edulis*: a field experiment. Mar. Biol. 104: 51-59

Grizzle RE, Lutz RA (1989) A statistical model relating horizontal seston fluxes and bottom sediment characteristics to growth of *Mercenaria mercenaria*. Mar. Biol. 95-105

Herman PMJ, Scholten H (1990) Can suspension feeders stabilise estuarine ecosystems? In Trophic relationships in the marine environment, (Barnes, M,Gibson RN eds) pp 104-116, Aberdeen University Press Aberdeen

Herman PMJ (1993) A set of models to investigate the role of suspension feeders in estuarine ecosystems. This volume

Hickman RW, Waite RP, Illingworth J, Meredyth-Young, JL, Payne G (1991) The relationship between farmed mussels, *Perna canaliculus*, and available food in Pelorus-Kenepuru Sound, New Zealand, 1983-1985. Aquaculture 99: 49-68

Incze LS, Lutz RA, True E (1981) Modeling carrying capacities for bivalve molluscs in open, suspended - culture systems. J. World Maricult. Soc. 12: 143-155

Johnson M (1990) The importance of particle-bound bacteria in the diet of the blue mussel *Mytilus edulis.* M.Sc. Thesis, Dalhousie University, Halifax

Jørgensen CB, Larsen PS, Riisgard HU (1990) Effects of temperature on the mussel pump. Mar. Ecol. Prog. Ser. 64: 89-97

Kautsky N (1982) Growth and size structure in a Baltic *Mytilus edulis* population. Mar. Biol. 68: 117-133

Knox, GA (1986) Estuarine ecosystems: a systems approach. CRC Press Boca Raton, Florida

Kremer JN, Nixon SW (1978) A coastal marine ecosystem: simulation and analysis. Springer-Verlag Berlin

Loo L-O (1992) Filtration, absorption, respiration and growth of *Mytilus edulis* L. at low temperatures. Ophelia 35: 123-131

Loo L-O, Rosenberg R (1983) *Mytilus edulis* culture: growth and production in western Sweden. Aquaculture 35: 137-150

Lucas MI, Newell RC, Shumway SE, Bally R (1987) Particle clearance and yield in relation to bacterioplankton and suspended particulate availability in estuarine and open coast populations of the mussel *Mytilus edulis.* Mar. Ecol. Prog. Ser. 36: 215-224

MacCaull WA, Platt T (1977) Diel variations in the photosynthetic parameters of coastal marine phytoplankton. Limnol. Oceanogr. 22: 723-731

MacDonald BA Thompson RJ (1986) Influence of temperature and food availability on the ecological energetics of the giant scallop *Placopecten magellanicus* III. Physiological ecology, the gametogenic cycle and scope for growth. Mar. Biol. 93: 37-48

Mallet, AL (1989) Culture of the mussel *Mytilus edulis*. In: Cold-water aquaculture in Atlantic Canada, (Boghen, AD ed) pp. 179-210, Canadian Institute for Research on Regional Development Moncton

Mallet AL, Carver CE (1991) An assessment of strategies for growing mussels in suspended culture. J. Shell. Res. 10: 471-478

Mallet A, Carver CE, Coffen SS, Freeman KR (1986) Winter growth of the blue mussel *Mytilus edulis* L.: importance of stock and site. J. Exp. Mar. Biol. Ecol. 108: 217-228

Moody JA, Butman B, Bothner MH (1987) Near-bottom suspended matter concentration on the continental shelf during storms: estimates based on *in situ* observations of light transmission and a particle size dependent transmissometer calibration. Cont. Shelf Res. 7: 609-628

Navarro E, Iglesias JIP., Camacho AP, Labarta U, Beiras R (1991) The physiological energetics of mussels (*Mytilus galloprovincialis* Lmk) from different cultivation rafts in the Ria de Arosa (Galicia, N.W. Spain). Aquaculture 94: 197-212

Newell CR (1990) The effects of mussel (*Mytilus edulis*, Linnaeus, 1758) position in seeded bottom patches on growth at subtidal lease sites in Maine. J. Shell. Res. 9: 113-118

Newell CR, Shumway SE, Cucci TL, Selvin, R (1989) The effects of natural seston particle size and type on feeding rates, feeding selectivity and food resource availability for the mussel *Mytilus edulis* Linnaeus, 1758 at bottom culture sites in Maine. J. Shellfish Res. 8: 187-196

Newell CR, Shumway SE (1993) Grazing of natural particulates by bivalve molluscs: a spatial and temporal perspective. This volume.

Newell RC, Branch GM (1980) The effects of temperature on the maintenance of metabolic energy balance in marine invertebrates. Adv. Mar. Biol. 17: 329-396

Officer CB, Smayda TJ, Mann R (1982) Benthic filter feeding: a natural eutrophication control. Mar. Ecol. Prog. Ser. 9: 203-210

Page HM, Hubbard DM (1987) Temporal and spatial patterns of growth in mussels *Mytilus edulis* on an offshore platform: relationships to water temperature and food availability. J. Exp. Mar. Biol. Ecol. 111: 159-179

Palanques A, Biscaye PE (1992) Patterns and controls of the suspended matter distribution over the shelf and upper slope south of New England. Cont. Shelf Res. 12: 577-600

Pérez Comacho A, Gonzalez R, Fuentes J (1991) Mussel culture in Galicia (N.W. Spain). Aquaculture 94: 263-278

Peterson CH, Black RW (1991) Preliminary evidence for progressive sestonic food depletion in incoming tide over a broad tidal sand flat. Estuar. Coast. Shelf Sci. 32: 405-413

Pieters H, Kluytmans JH, Zandee DI, Cadee GC (1980) Tissue composition and reproduction of *Mytilus edulis* in relation to food availability. Neth. J. Sea Res. 14: 449-361

Platt, T (1971) The annual production by phytoplankton in St. Margaret's Bay, Nova Scotia. J. Cons. 33: 324-334

Platt T, Prakash A, Irwin B (1972) Phytoplankton nutrients and flushing of inlets on the coast of Nova Scotia. Naturaliste Can. 99: 253-261

Rhoads DC, Boyer LF, Welsh BL, Hampson GR (1984) Seasonal dynamics of detritus in the benthic turbidity zone (BTZ); implications for bottom- rack molluscan mariculture. Bull. Mar. Sci. 35: 536-549

Riisgård HU (1991) Filtration rate and growth in the blue mussel *Mytilus edulis* Linnaeus, 1758: Dependence on algal concentration. J. Shellfish Res. 10: 29-35

Roden CM, Rodhouse PG, Hensey MP, McMahon T, Ryan TH, Mercer JP (1987) Hydrography and the distribution of phytoplankton in Killary Harbour: a fjord in Western Ireland. J. Mar. Biol. Assoc. U.K. 67: 359-371

Rodhouse PG, Roden CM (1987) Carbon budget for a coastal inlet in relation to intensive cultivation of suspension-feeding bivalve molluscs. Marine Ecology Progress Series 36: 225-236

Rodhouse PG, Roden CM, Burnell GM, Hensey MP, McMahon T, Ottway B, Ryan TH (1984) Food resource, gametogenesis, and growth of *Mytilus edulis* on the shore and in suspended culture: Killary Harbour, Ireland. J. Mar. Biol. Assoc. U.K. 64: 513-530

Rosenberg R, Loo L-O (1983) Energy-flow in a *Mytilus edulis* culture in western Sweden. Aquaculture 35: 151-161

Shortle JL (1991) The design of a recirculating flume and its use in the investigation of feeding rates in the mussel *Mytilus edulis*. M.Sc. Thesis, Dalhousie University, Halifax

Shumway S, Cucci TL, Newell RC, Yentsch CM (1985) Particle selection, ingestion, and absorption in filter-feeding bivalves. J. Exp. Mar. Biol. Ecol. 91: 77-92

Smaal AC (1991) The ecology and cultivation of mussels: new advances. Aquaculture 94: 245-261

Smaal AC, van Stralen MR (1990) Average annual growth and condition of mussels as a function of food source. Hydrobiologia 195: 179-188

Smaal AC, Prins TC (1993) The uptake of organic matter and the release of inorganic nutrients by bivalve suspension feeder beds: a mini-review. This volume

Smaal AC, Verhagen JHG, Coosen J, Haas H (1986) Interaction between seston quantity and quality and benthic suspension feeders in the Oosterschelde, the Netherlands. Ophelia 26: 385-399

Soniat TM, Brody MS (1988) Field validation of a habitat suitability index model for the American oyster. Estuaries 11: 87-95

Stenton-Dozey JME, Brown AC (1992) Clearance and retention efficiency of natural suspended particles by the rock-pool bivalve *Venerupis corrugatus* in relation to tidal availability. Mar. Ecol. Prog. Ser. 82: 175-186

Thompson JK, Nichols FH (1988) Food availability controls seasonal cycle of growth in *Macoma balthica* (L.) in San Francisco Bay, California. J. Exp. Mar. Biol. Ecol. 116: 43-61

Widdows J (1976) Physiological adaptation of *Mytilus edulis* to cyclic temperatures. J. Comp. Physiol. 105: 115-128

Widdows J (1978) Combined effects of body size, food concentration and season on the physiology of *Mytilus edulis*. J. Mar. Biol. Assoc. U.K. 58: 109-124

Widdows J, Bayne BL (1971) Temperature acclimation of *Mytilus edulis* with reference to its energy budget. J. Mar. Biol. Assoc. U. K. 51: 827-843

Widdows JP, Fieth P, Worrall CM (1979) Relationships between seston, available food and feeding activity in the common mussel *Mytilus edulis*. Mar. Biol. 50: 195-207

Wildish DJ, Kristmanson DD (1979) Tidal energy and sublittoral macrobenthic animals in estuaries. J. Fish. Res. Bd. Can. 36: 1197-1206

Wildish DJ, Miyares MP (1990) Filtration rate of blue mussels as a function of flow velocity - preliminary experiments. J. Exp. Mar. Biol. Ecol. 142: 213-219.
Wildish D, Kristmanson D (1993) Hydrodynamic control of bivalve filter feeders: A conceptual view
Wulff F, Field JG, Mann KH (eds) (1989) Network analysis in marine ecology: methods and applications. Springer- Verlag Berlin

A SET OF MODELS TO INVESTIGATE THE ROLE OF BENTHIC SUSPENSION FEEDERS IN ESTUARINE ECOSYSTEMS

Peter M.J. Herman
Netherlands Institute of Ecology
Center for Estuarine and Coastal Ecology
Vierstraat 28
4401 EA Yerseke
The Netherlands

INTRODUCTION

Benthic suspension feeders are an important component of estuarine ecosystems. They often represent a considerable biomass, and their filtration activities can be the key mortality factor for the phytoplankton. It is not unusual that the daily filtration capacity of the suspension feeders is in the order one fifth to one third of the total system water volume, thus imposing on the phytoplankton a mortality rate in the order of 0.20 to 0.33 d^{-1} (Dame et al 1991; Smaal 1991; Hily 1991). Coupled to this high grazing rate is the importance of the suspension feeders for nutrient recycling and for sedimentation/resuspension of particulate matter. Based on the relatively low mortality rates of benthic suspension feeders, it has been argued that their role in energy transfer pathways is minor, compared with their role in element cycling in the ecosystem (Kautsky and Evans 1987).

In ecosystem models of estuarine systems, where the aim is the description of major flows of carbon and nutrients, benthic suspension feeders are obviously an important component. In the Oosterschelde model SMOES, e.g., (Klepper 1989), benthic suspension feeders are responsible for the largest part of benthic-pelagic coupling. As was shown by Herman and Scholten (1990), benthic suspension feeders are the key factor in the robustness of SMOES to external pressure in the form of nutrient loading. It was argued in that study that shallow estuarine

NATO ASI Series, Vol. G 33
Bivalve Filter Feeders
in Estuarine and Coastal Ecosystem Processes
Edited by Richard F. Dame

systems may be stabilized by the activity of suspension feeders, because of the balance between grazing control and nutrient limitation control on the phytoplankton. Whenever nutrient limitation control would disappear, an increase in benthic suspension feeder biomass could increase grazing control, and counteract possible algal blooms in the model.

Contrasting with this approach that uses a large and difficult ecosystem model (with over 200 parameters) as 'black box tool' for the analysis, is the highly simplified approach by Officer et al. (1982). These authors made a theoretical analysis of a Lotka-Volterra type predator-prey model between benthic suspension feeders and phytoplankton. They argued that stability of this relationship is only possible when the water recycling time for the benthic suspension feeders is of the same magnitude as the time constant for phytoplankton growth. This is only attainable when the mortality rates of the benthos are low, leading to a high biomass with large filtration capacity, and when water depth is limited, leading to a low volume/surface ratio. Stability between phytoplankton and zooplankton is impossible due to the too high mortality rate of the zooplankton; stability between phytoplankton and benthic suspension feeders in deep waters is impossible due to the high volume of water per unit of surface.

An advantage of the approach followed by Officer et al. (1982) is that stability criteria can be expressed as a simple function of a few readily interpretable parameters. Drawbacks, however, are the difficulty to make the concept of 'stability' in a Lotka-Volterra system operational and measurable in the field, and the lack of any spatial and temporal variability in the system. The two points are linked to one another. In a seasonal environment, a relatively short period where the phytoplankton outgrows grazing control may suffice to build up a phytoplankton biomass on which the grazer can amply feed, providing a biomass reserve for the grazer sufficient to live on for several months in which it actually overgrazes the phytoplankton. Seasonality, in this way, may provide for persistent co-occurrence of what is in essence an unstable predator-prey pair. Lack of spatial resolution is most apparent in the analysis of the relation between water depth and control by suspension feeders. The conclusion that control is only effective in shallow waters depends heavily on the assumption that primary production (in g m^{-3} d^{-1}) is independent of water depth. This is obviously untrue in light-limited primary production.

The approach followed by Herman and Scholten (1990) does not suffer from these drawbacks. Resolution in time and space is very reasonable; the simulation model takes into account the particular settings of the Oosterschelde system. However, a major drawback of the approach, besides the complex nature of the model, is the fact that the biomass of suspension feeders is not actually modelled as a state variable in the model. Biomass of mussels, cockles and hard substrate suspension feeders is imposed on the model as a forcing function. Investigation of the effect on the model of various scenarios can be performed by checking the scope for growth of the suspension feeders. The biomass function can subsequently be adapted to maintain - on average - a sufficiently large scope for growth. The reason for choosing this solution is the lack of feedback mechanisms in an ecosystem simulation model, that act on the time scale of years, relevant to the suspension feeders. Because of the longevity of these organisms, relatively limited changes of some parameters can easily lead to chaotic (-like?) behaviour of the model. Such a destabilisation is no longer possible with the use of forcing functions; the parameters leading to critical destabilisation can consequently no longer be detected.

Other whole ecosystem models, such as GREWAQ (de Vries et al., 1988) used forcing functions for suspension-feeder biomass for the same technical reasons. Bacher (1989, 1991) and Bacher et al (1991) also followed in essence the same general reasoning. Growth of oysters is modelled in a realistic setting, where, however, no feedback of grazing on phytoplankton growth or on nutrient dynamics is present. The growth of oysters is related to their stock, which has been fixed by the modeller. The relation between stock and growth serves as a way to estimate the carrying capacity of the system for oysters. The Ems-Dollard model (Baretta and Ruardij, 1988), on the other hand, explicitly models the dynamics of suspension feeders. The formulations used are extremely complex, which limits their usefulness in different settings.

In the present contribution we present a set of suspension

feeder models that fit somewhere in between the extreme simplicity of a Lotka-Volterra predator-prey system, and the detail of a system simulation model. It has been envisaged that these simplified models could become part of an estuarine simulation model, or at least provide building blocks for use in such a model. In order to investigate problems of stability and influence on the system, we have, in a first instance, dynamically modelled an extremely simplified four-component system. Phytoplankton, nutrients (Si is taken as the example), benthic suspension feeders and bottom detritus are the four state variables in the model. Their dynamics are set in a seasonal context of light, day-length, temperature and water seston content. These dynamics are described as simply as possible.

The approach followed consisted of two phases. First, the basic model was developed, which is simple and fast, and therefore well suited for addition of different details of the physiology or ecology of the suspension feeders. Next, several of these details were added with two objectives: the model should be able to represent a set of field data reliably, and it should exhibit consistent and robust behaviour with respect to some changes in external forcing (notably different depth of the water system and increased nutrient concentrations in the system).

In general terms it has been described that the feeding capacity of an area is related to phytoplankton dynamics. Smaal and van Stralen (1990) demonstrate more precisely that mussel growth is related to annual primary production, not to chlorophyll *a* standing stocks. It is also known from selectivity studies that phytoplankton is selected from seston before ingestion (Kiørboe and Møhlenberg 1981). In the basic model, no feeding at all on detritus sources has been allowed. This has later been changed in a formulation where a (small) part of the detritus is considered edible, and part of the phytoplankton escapes predation when it is filtered.

Two different aspects of hydrodynamic forces are influencing suspension feeder biomass. Incomplete vertical mixing can be caused in areas with low tidal mixing, or can result from benthic boundary layer flow (Fréchette et al 1989; Fréchette and Grant 1991). As a result, local (partial) food depletion may occur at the height of water intake by the suspension feeders. This phenomenon can reduce the amount of food filtered from the water substantially (Fréchette et al 1989: factor 2 from model; Fréchette and Grant, 1991: factor 1.5 from data). The spatial scale of the vertical gradients is small: concentration differences of a factor 1.5 are observed and modelled over a vertical distance of tens of cm. Fréchette et al (1989) also demonstrate that the precise height of the water intake point of the suspension feeders has a material influence on the predicted vertical gradients and on food depletion. Monismith et al (1990) add to this that the precise spatial configuration of the siphons, and the characteristics of the siphon jets are of great importance for the steady-state formulation of Fréchette et al(1989). If one wants to incorporate these, obviously very important, mechanisms into an ecosystem-type of model, two alternatives exist. One is very unattractive: one can divide the spatial domain of the model into very small spatial compartments and attach a biological module to what is then essentially a multidimensional transport model. This requires very large computation times, and it is questionable whether the simulation of a full year is achievable. Besides, since very small temporal steps are taken, biological interactions can no longer be represented rather roughly as (partly parameterized) daily averaged rates. The other alternative is to parameterize these short-term, small-scale transport phenomena too, and grossly sketch their effect in the coarser ecosystem models. In the present modelling we have parametrized the vertical gradient in two different ways. In a first approach we have analyzed a water column model, with a number of vertical layers on top of each other. Daily- and tidally-averaged transport and mixing between the layers is parameterized as eddy-diffusion like dispersion. In an even more simplified approach, the effect of local food depletion is incorporated as a coefficient that makes part of the food unattainable by the benthic suspension feeders, and part of the nutrients unattainable for primary production.

On a larger horizontal scale, exchange of water with rivers and/or the sea may result in the advection of food utilizable by the estuarine suspension feeders. This phenomenon may be responsible for a considerable part of the carbon sources for these organisms (Rodhouse and Roden 1987; Smaal and van Stralen, 1990; Asmus and Asmus 1990). Asmus and Asmus (1990) remark that input of phytoplankton from deeper waters to a tidal flat system is an unstable and pulsating food source, which is most effectively used by the larger suspension feeders. An extension of the basic model with advective/dispersive transport in a simplified 2D universe with boundaries depicts this situation.

Questions asked with the aid of these models are focused on the potential stabilizing role of suspension feeders in ecosystem. Will a stable model result from a faithful description of the animals' physiology and environment? What are key processes determining the eventual biomass development? How do animal physiology and ecological and spatial setting interfere?

FIELD DATA

Although this paper aims at a relatively abstract representation of algae-suspension feeder systems, field data have been employed to depict realistic climatic forcing, and to check the performance of the models. All data used are taken from the data base of the Oosterschelde model SMOES (Klepper 1989; Klepper et al. in press). They apply to the 'central' compartment (compt. no. 2) of the Oosterschelde, a semi-enclosed sea-arm in the S-W Netherlands. Mussel data (*Mytilus edulis*) have been used as a reference for suspension feeders in the models. Reference is made to the mentioned papers for details of geographical position, etc., and for primary literature sources.

THE BASIC MODEL

The most basic version of the model depicts a vertically well mixed column of water, without any spatial gradients at all. Of the four state variables phytoplankton (gC m^{-3}), nutrients (g m^{-3}), suspension feeders (gC m^{-2}) and bottom detritus (gC m^{-2}), the latter two are fixed to the bottom. Silicate is taken as an example for 'nutrients', and wherever a conversion of mass from biological variables into nutrients (or the reverse) takes place, a Si/C ratio of 0.5 is assumed.

THE PHYTOPLANKTON SUBMODEL

Phytoplankton primary production is described with a parameterized function of light at the water surface, extinction coefficient, day-length, form parameter of the P/I curve, and average depth of the water column. Such a function was described by Klepper et al (1988) and Klepper (1989). Essentially, this integration is based on the following logic. At any particular moment, primary production at a particular depth in the water is described as a function of light intensity at this depth. This dependence of primary production on light intensity is described by the Eilers and Peeters formula (Eilers and Peeters, 1988). Light at a particular depth is a function of light at the water surface, and of light extinction following the Lambert-Beer relationship. Light at the surface is a function of time in the day, total day-length (= fraction of the day when sunlight is available) and total integrated light reaching the surface over the day. Finally, the relative amount of water at a particular depth is a function of the surface-depth relationship in a water body. For a great number of parameter combinations, the total primary production was integrated over time and depth, and the daily depth-integrated primary production divided over the volume of water, yielding the average daily depth-integrated primary production per unit of phytoplankton biomass available.

Parameterization has been performed for estuarine water bodies where the fraction S_d/S_0 of the total surface S_0 of the water body that has a depth d is given by a simple exponential expression:

$$\frac{S_d}{S_0} = e^{-t*d} \qquad (1)$$

Average water depth D in such a body is related to the parameter τ of this expression as:

$$D = \frac{S_0}{t} * \frac{\left(1 - e^{-t * d_m}\right)}{d_m} \qquad (2)$$

where d_m is maximal depth (taken as 30 m in the present parameterization).

Parameter	dimension	value
a	-	4.503
b	-	1.66
c	-	1.781
w	-	1
I_{10}	$W\ m^{-2}$	117
Q_I	-	0.055
P_{m10}	d^{-1}	4.392
Q_P	-	0.073
k_0	m^{-1}	1.37
ks	$m^2 g^{-1}$	0.047
D	m	4
Temp-FAVG	° C	10.8
Temp-AMPL	° C	8.04
Temp-PERIOD	d	365
Temp-reference time		79/12/31
Temp-phase	-	2.345
Light-FAVG	$W\ m^{-2}$	48.18
Light-AMPL	$W\ m^{-2}$	41.14
Light-PERIOD	d	365
Light-reference time		79/12/31
Light-phase		17.24
Seston-FAVG	$g\ m^{-3}$	21.7
Seston-AMPL	$g\ m^{-3}$	8.703
Seston-PERIOD	d	365
Seston-reference time	-	79/12/31
Seston-phase		-1.163
Day length-FAVG	-	0.5
Day length-AMPL	-	0.183
Day length-PERIOD	d	365
Day length-reference time	-	79/6/24
Day length-phase		0.75

Table I. Values for parameters in the phytoplankton submodel

Light intensity at the water surface during one day was described as a symmetric triangular function of time.

$$P_1 = \frac{a * P_{max} * L * \left(1 - e^{-b*U}\right) * (W+2)}{\left(k + \frac{c}{D}\right) * (W+5) * D} \qquad (3)$$

The parametrized function is given by:

where P_1 averaged daily depth-integrated primary production per unit of phytoplankton biomass, without nutrient limitation (d^{-1})

P_{max} maximal primary production per unit of phytoplankton biomass , realized at the optimal light intensity (d^{-1})

k extinction coefficient (m^{-1})

a, b, c dimensionless parameters

L day-length, i.e. fraction of day that the sun is shining (dimensionless)

D average water depth of an estuarine water body (m)

U relative (daily integrated) light intensity at the water surface : U=Light/Iopt, where Light is the integrated light intensity at the surface, and Iopt is the optimal light intensity of the P-I curve (dimensionless)

W shape parameter of the P-I curve (dimensionless)

Parameter values for the parameterized function are given in Table I.

Optimal light intensity I_{opt} was described as a function of temperature (Klepper, 1989) :

$$I_{opt} = I_{10} * e^{((T-10)*Q_I)} \qquad (4)$$

where I_{10} optimal light intensity at 10°C (W m^{-2})

Q_I a parameter describing temperature dependence (dimensionless)

P_{max} was a similar function of temperature

$$P_{max} = P_{m10} * e^{((T-10)*Q_p)} \tag{5}$$

where P_{m10} Optimal specific primary production at 10°C (d^{-1})

Q_P a parameter describing temperature dependence (dimensionless)

Values of the parameters used in this and the previous function are given in Table I. These parameter values were obtained by non-linear regression of the given functions on the data of the Oosterschelde.

k, the extinction parameter, was described as a function of seston content :

$$k = k_0 + k_s * S \tag{6}$$

where k_0 background extinction coefficient (m^{-1})

k_s dependence parameter on seston content ($m^2\ g^{-1}$)

S seston content ($g\ m^{-3}$)

Temperature, light, seston content of the water and day-length were described as harmonic functions fitted to the Oosterschelde data:

$$F(t) = F_{avg} + Ampl * sin\left(2\pi\left(\frac{1}{period} * (t - t_0) - phase\right)\right) \tag{7}$$

where $F(t)$ value of the forcing function at time t

F_{avg} average value of the function

$Ampl$ amplitude of the function

$period$ period of the function

$t-t_0$ time as difference from the reference time

$phase$ phase difference of the function

Parameter values for the four forcing functions are given in Table I.

Nutrient limitation of primary production was described by a simple Michaelis-Menten equation :

$$L_n = \frac{N}{(K_s + N)} \qquad (8)$$

where L_n dimensionless nutrient limitation factor
N nutrient concentration (gSi m^{-3})
K_s Michaelis-Menten constant (gSi m^{-3})

Gross production of the phytoplankton was described as :

$$P_g = A * P_l * L_n \qquad (9)$$

where P_g gross production of phytoplankton (gC m^{-3} d^{-1})
A Algal (phytoplankton) concentration (gC m^{-3})

Algal respiration is described as the sum of basal respiration and active (growth-dependent) respiration :

$$R_a = A * r_b * e^{((T-10)*Q_r)} + P_g * r_a \qquad (10)$$

where R_a Algal respiration rate (gC m^{-3} d^{-1})
r_b basic respiration rate parameter (d^{-1})
r_a active respiration rate parameter (dimensionless)

Algal excretion is dependent on gross production rate and on nutrient limitation :

$$E_a = P_g * E_m * (1 - L_n) \qquad (11)$$

where E_a Algal excretion rate (gC m^{-3} d^{-1})
E_m maximal fraction of gross production excreted (dimensionless)

The values of the parameters in eqs. (8)-(11) were determined by calibaration. Values in the model runs are given in Table II.

Parameter	Dimension	Value
K_s	gSi m^{-3}	0.033
r_b	d^{-1}	0.0099
r_a	-	0.094
E_m	-	0.080

Table II. Values for parameters in the phytoplankton submodel, obtained after calibration on Oosterschelde data.

THE SUSPENSION FEEDERS SUBMODEL

In the basic model only the biomass of the suspension feeders is described. Size or age structure in the population is neglected, and reproduction is not explicitly modeled, as it represents a transition of biomass from the suspension feeder component to itself. Because of this simple approach, all metabolic rates (e.g., respiration, clearance, etc.) are represented as linear functions of biomass. The usual representation of these functions with allometric relationships of body size is impracticable in this case, since individual body sizes are not known. Implicitly, the representation with linear functions assumes that the size or age structure of the population is stable (or at least in a steady state) in time. In that case the respiration per unit of body weight can be represented as a constant. Obviously, this assumption is a gross simplification of reality.

Clearance rate (volume of water swept free of particles per unit of time and per unit of suspension feeder biomass) is represented as a linear function of temperature. This is in accordance with the 'leaky, viscous, constant-force pump' model of Jørgensen et al (1990), where it was convincingly shown that clearance in mussels is essentially limited by the viscosity of the filtered water. It is assumed here, however, that some limitation of clearance occurs at high seston concentrations. This is represented as in Klepper (1989) with an exponential function.
Clearance rate is then given by:

$$C = (C_c + T^* C_T)^* e^{-C_s^* S} \qquad (12)$$

where C clearance rate (m^3 gC^{-1} d^{-1})

C_C clearance rate at 0°C (m^3 gC^{-1} d^{-1})

C_T dependence of clearance rate on temperature (m^3 gC^{-1} d^{-1} $\cdot C^{-1}$)

C_S dependence parameter of clearance rate on seston content of the water (m^3 g^{-1})

Grazing of suspension feeders is directly proportional to algal concentration and suspension feeder biomass. No local depletion phenomena, nor larger spatial gradients are taken into account.

$$G = A^* F^* C \qquad (13)$$

where G grazing rate of suspension feeders (gC m^{-2} d^{-1})

F biomass of suspension (Filter) feeders (gC m^{-2})

Part of the grazed material is rejected as pseudofeces. This process is simply modelled as a cut-off function, in accordance with Klepper (1989), who preferred it to more complicated models:

$$PS = G^* MAX\left(0, \left(1 - \frac{C_{ps}}{S}\right)\right) \qquad (14)$$

where PS rate of pseudofeces formation by suspension feeders (gC m^{-2} d^{-1})

C_{PS} cut-off seston concentration (g m^{-3})

Assimilation rate is proportional to the rate of ingestion of food material, i.e., to the difference between grazing rate and pseudofeces formation rate:

$$ASS = e_a^* (G - PS) \qquad (15$$

where ASS assimilation rate of suspension feeders (gC m^{-2} d^{-1})

e_a assimilation efficiency (dimensionless)

The non-assimilated material is turned into faces. The faces production rate is therefore:

$$FAEC = G - PS - ASS \qquad (16)$$

where $FACE$ rate of faces production by suspension feeders

Respiration rate is modelled as a temperature-dependent first-order process :

$$RES = F^{*} R_{c} {}^{*} e^{(T-10)^{*} Q_{m}} \qquad (17)$$

where $\boldsymbol{ERs}$ respiration rate of suspension feeders (gC m^{-2} d^{-1})

$\boldsymbol{R_C}$ basic metabolic rate at 10 °C (d^{-1})

$\boldsymbol{Q_m}$ parameter for temperature dependence of metabolic rate

In principle, one would model respiration as the sum of two processes: a basal respiration rate and an assimilation-dependent rate. However, in the present formulations (where oxygen is not modelled) the latter term introduces an extra parameter, while its effect is virtually indiscernible from a change in the assimilation efficiency.

Mortality rate is described as a (temperature-dependent) first-order process:

$$M = F^{*} C_{m} {}^{*} e^{(T-10)^{*} Q_{m}} \qquad (18)$$

where $\boldsymbol{M}$ mortality rate of suspension feeders (gC m^{-2} d^{-1})

$\boldsymbol{c_M}$ mortality coefficient (d^{-1})

Faces and pseudofeces are turned into bottom detritus, which is mineralized in a (temperature-dependent) first-order process:

$$MIN = B^{*} C_{MIN} {}^{*} e^{(T-10)^{*} Q_{MIN}} \qquad (19)$$

where $\boldsymbol{MIN}$ mineralizatio nrate of bottom detritus (gC m^{-2} d^{-1})

$\boldsymbol{B}$ bottom detritus mass (gC m^{-2})

$\boldsymbol{c_{MIN}}$ mineralization coefficient (d^{-1})

$\boldsymbol{Q_{MIN}}$ parameter for temperature dependence of mineralization (°C^{-1})

Finally, the rates of change of the state variables are:

$$\frac{dA}{dt} = P_g - R_a - E_a - \frac{G}{D} \qquad (20)$$

$$\frac{dN}{dt} + SiCratio^{*}\left(R_a + E_a - P_g + \frac{MIN + RES}{D} \right)$$

$$\frac{dF}{dt} = ASS - RES - M$$

$$\frac{dB}{dt} = PS + FAEC + M - MIN$$

where ***SiCratio*** is the Silicon/Carbon ratio of algae.

The model was implemented in SENECA 1.5 (de Hoop et al., 1992). It was integrated using Euler numerical integration with adaptive step size. The step was reduced whenever the change in one of the state variables per step exceeded 10% in the case when the change had the same sign as in the previous step, and 5 % when the change had the opposite sign. The precision of the numerical integration was confirmed by comparison with Runge-Kutta fourth-order integration with a one-day step size. Euler was preferred as routine integration method because it executes faster.

INCLUSION OF VERTICAL CONCENTRATION GRADIENTS.

In this approach, grazing by suspension feeders is reduced by a dimensionless factor ***Vr*** ($0 \le Vr \le 1$), which represents the effect of local food depletion. With this factor eq. (12) changes into:

$$C = (C_c T * C_T) * e^{-C_t * S} * Vr \tag{21}$$

Nutrient limitation of primary production is changed by the same factor. Therefore eq. (8) is changed into :

$$L_n = \frac{N * Vr}{(K_s + N * Vr)} \tag{22}$$

With these formulations the problem is reduced to the problem of calculating the vertical reduction factors. This was done with the following approach.

A vertical water column was considered, divided in 10 layers. The lowest layer was the thinnest. Thickness of subsequent layers increased in a logarithmic series, where each subsequent layer had a thickness of two times the previous one. Suspension feeding organisms take away algal cells near the bottom, from a water layer of finite thickness ***Di*** (intake depth). As the model should depict a steady state, the same amount of algae is immediately added to the surface layer. Vertical dispersion between the layers is represented by the formula :

$$\frac{\partial A}{\partial t}=\frac{\partial}{\partial X}\left(D_e(X)\frac{\partial A}{\partial X}\right) \tag{23}$$

where $D_e(x)$ (eddy) dispersion coefficient ($m^2 d^{-1}$)

x depth in the vertical column (m)

D_e was depth-dependent, as is the case in benthic boundary flows (Fréchette et al., 1989), and given by :

$$D_e = E_v * X * \left(1 - \frac{X}{D}\right) \tag{24}$$

where E_v constant (m d^{-1})

D total depth of the water column (m)

This formula was discretized and integrated using a semi-implicit Crank-Nicholson scheme (Press et al 1987) with a short time step. Depending on the values of the parameters, steady state was reached after at most half a day. For security one full day was calculated in each case.

The value of E_v was calibrated to represent the results in Fréchette et al (1989). Pumping rate at the bottom and total depth were set equal to those used by these authors, and E_v was adjusted until an algal concentration gradient similar to the one described by Fréchette et al (1989) was reached. This parameter is not directly comparable to the vertical eddy diffusion coefficient used in the model by these authors, since they described mixing in a horizontal flow. The effect of this horizontal flow is incorporated (parameterized) in our coefficient E_v . A typical result of the model is given in Fig. 1.

Parameter	dimension	value
C_C	$m^3\ gC^{-1}\ d^{-1}$	$2.8\ 10^{-4}$
C_T	$m^3\ gC^{-1}\ d^{-1}\ {}^{\circ}C^{-1}$	0.011
C_S	$m^3\ g^{-1}$	0.0071
C_{PS}	$g\ m^{-3}$	8.95
e_a	-	0.487
R_C	d^{-1}	0.005
Q_M	${}^{\circ}C^{-1}$	0.069
C_M	d^{-1}	$3.7\ 10^{-4}$
C_{MIN}	d^{-1}	0.0104
A_C	-	25
P_a	-	0.839
P_d	m^{-1}	0.029

Table III. Parameter values used in suspension-feeder model, obtained after calibration of the 'nominal model' on the Oosterschelde data.

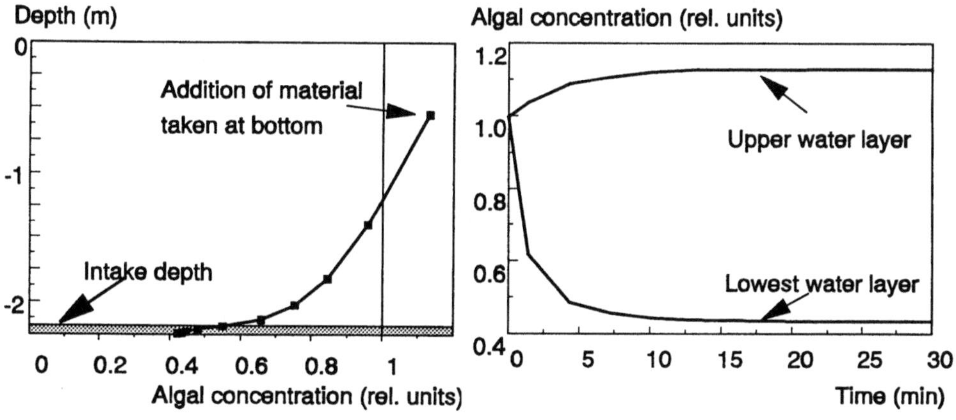

Figure 1 Algal concentrations in a structured steady state column: a) time course b) steady state concentration

Subsequently the model was run 175 times with different randomly drawn combinations of total depth ***D*** and clearance rate ***C*** at the bottom. At steady state, the ratio between algal concentration in the bottom layers (up to the intake depth D_i), to the average concentration over the water column, was calculated. For an intake depth of 5 cm this ratio could very well be described (r=0.94, n=175) by a simple linear regression on clearance rate and total depth :

$$V_r = 0.83 - 0.071 * log_{10}(D) - 0.0049 * C \qquad (25)$$

where ***C*** clearance rate ($m^3\ m^{-2}\ d^{-1}$)

It is apparent from eq. (25) that the vertical reduction factor is mainly dependent on the clearance rate. Especially at high clearance rates the reduction may be substantial. High local clearance rates are not uncommon, e.g., with the mussel *Mytilus edulis*, because of the clumped distribution of this species in space. It is therefore not sufficient to apply equation (25) to the average density of the suspension feeders. Rather, the local density in the clumps is relevant to the calculation of local food depletion. This has been, again roughly, taken into account by replacing the last term on the right-hand side of eq. (25) by:

$$0.0049 * C * A_c \qquad (26)$$

Spatial schematization. a) Depth (m) of compartments. b) relative flow intensity between compartments.

where A_c 'Aggregation factor' (dimensionless) representing the ratio between density in the clumps to average density.

FEEDING ON DETRITUS. THE 'NOMINAL' MODEL.

The basic model was further expanded by building in the possibility for the suspension feeders to feed not only on algae, but also on the detritus present in the system. This was achieved by a slight modification of eq. (13). Grazing is now described as the product of clearance, vertical reduction coefficient (see previous chapter) and food concentration. This food concentration is made up by algae and detritus :

$$FOOD = P_a * A + P_d * B$$

$$G = FOOD * C * Vr * F \tag{27}$$

The vertical column model (without layer structure) and with differential feeding is referred to as the 'nominal' model. This model was used to estimate the parameter values by automatic calibration (see below). Parameter values of the suspension feeder model after calibration are given in Table III.

MODEL WITH 2D (X-Y) WATER TRANSPORT

In order to represent spatial exchanges and possible phenomena of food import and accumulation related to the activity of suspension feeders, the nominal version of the model was mounted in a regular 2-dimensional square grid of 5x5 cells. This configuration represents an abstract space, and was not modelled after a real example. Each cell has a size of 1000m x 1000m. The depth of each of the different cells was fixed; the 25 cells had a depth distribution approaching eq. (2). Average depth over the entire system was 3.9 m. Dispersive exchange between adjacent cells was proportional to the exchange surface, independent of direction. Fig. 2a represents depth of the grid cells.

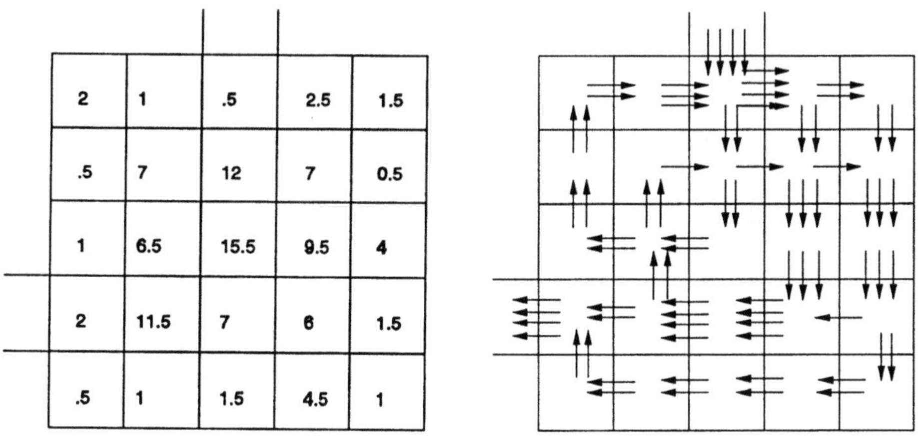

Figure 2. Spatial schematizations. a) Depth(m) of compartments, b) Relative flow intensity between compartments.

The grid also had two external boundaries, and advective flows (in a general clock-wise gyre) between the cells. Fig. 2b shows the relative intensity of these flows. By defining a value for the incoming flow $\boldsymbol{FADV_{in}}$, the intensity of all cell-to-cell flows is defined.

Transport of matter between cells is described by an advective-dispersive formulation, discretized using an upstream difference scheme:

$$V_{ij}\frac{\Delta C_{ij}}{\Delta t} = QX_{in} * C_{Xup} * Q_{Xout} * C_{Xdn} + QY_{in} * C_{Yup} - QY_{out} * C_{Ydn} +$$

$$E_{i+1j} * \left(C_{i+1j} - C_{ij}\right) - E_{ij} * \left(C_{ij} - C_{i-1j}\right) + \qquad (28)$$

$$F_{ij+1} * \left(C_{ij+l} - C_{ij}\right) - F_{ij} * \left(C_{ij} - C_{ij-1}\right)$$

where $\boldsymbol{QX_{in}}$ inflowing advective flow in X-direction (m^3 d^{-1})

$\boldsymbol{QX_{out}}$ outflowing advective flow in X-direction (m^3 d^{-1})

$\boldsymbol{QY_{in}}$ inflowing advective flow in Y-direction (m^3 d^{-1})

$\boldsymbol{QY_{out}}$ outflowing advective flow in Y-direction (m^3 d^{-1})

$\boldsymbol{C_{Xup}}$ upstream concentration X-direction (g m^{-3})

$\boldsymbol{C_{Xdn}}$ downstream concentration X-direction (g m^{-3})

$\boldsymbol{C_{Yup}}$ upstream concentration Y-direction (g m^{-3})

$\boldsymbol{C_{Ydn}}$ downstream concentration Y-direction (g m^{-3})

$\boldsymbol{C_{ij}}$ concentration in cell i,j (g m^{-3})

$\boldsymbol{E_{ij}}$ dispersive flux' into cell i,j along X direction (m^3 d^{-1})

$\boldsymbol{F_{ij}}$ dispersive flux' into cell i.j. along Y direction (m^3 d^{-1})

In the X-direction we have

$$E_{ij} = E * \frac{A_{i-1,i,j}}{\Delta X_{i-1,i,j}} :$$

where $\boldsymbol{E}$ dispersive coefficient (m^2 d^{-1})

$\boldsymbol{A_{i-1,i\,;\,j}}$ exchange surface between cells (i-1,j) and (i,j) (m^2)

$\boldsymbol{\Delta x_{i-1,i\,;\,j}}$ exchange distance between these two cells (m)

Parameter	dimension	value
E_d	$m^2\ d^{-1}$	10^4
Δx	m	10^3
Q_{in}	$m^3\ d^{-1}$	10^6

Table IV. Transport parameters used in horizontal transport model

and similarly for F_{ij} in the Y-direction. The same value for E is used in both directions.

The transport model was integrated using an alternate direction implicit (ADI) scheme (Press et al 1987). Values of the parameters used are given in Table IV. Two different boundary conditions in the upstream boundary were chosen, to model two radically different situations. In the first, the Oosterschelde data used throughout this study were used as boundary conditions, ensuring a high degree of synchrony between concentrations in the boundary and in the model domain. In the second, (time-) constant concentrations of nutrients and algae were used, thus minimizing the correlation between the boundary and the domain.

Results of the nominal model.

AUTOMATIC CALIBRATION OF THE PARAMETERS

For many parameters of the model, good estimates could be obtained by fitting the model relationships directly and separately to the Oosterschelde data. This applies, e.g., to light curves over the year, to P_{max} as a function of temperature, etc. Many other parameters, however, are only known within a broad range. Values for these parameters were obtained by calibrating the 'nominal model' to the Oosterschelde data. The automatic calibration routine of SENECA (de Hoop et al 1992) was used for this purpose. In this routine a possible range for every parameter is set; an algorithm that carefully avoids local minima then searches within the given ranges for the parameter set minimizing the weighted sum of squared deviations of modelled to observed time series. Details can be found in de Hoop et al (1992).

CALIBRATION RESULTS OF THE 'NOMINAL' MODEL.

The results of the nominal model, after calibration, are shown in Fig. 3. The model faithfully reproduces the timing of the decline in nutrient concentrations, the level of the winter nutrient concentrations, the timing and height of the first phytoplankton peak, and the evolution of the suspension feeder (mussel) biomass. The remarkable gap in the 'data' for suspension feeder biomass around April 1 is caused by the stocking of the Oosterschelde with young mussels imported from elsewhere; the strong decline from October-November onwards may be related to fisheries. Taking this into account, the model describes the suspension feeder data very reasonably.

The summer phytoplankton and silicate dynamics are not faithfully represented. The data provide strong evidence for a summer minimum followed by a - smaller - autumn peak. Most probably, limitation by more than one nutrient should be taken into account. In particular, nitrogen plays an important role in the Oosterschelde. It is depleted later than silicate, and its regeneration is probably faster. Klepper (1989) and Klepper et al (in press) give many details and modelling formulations for these nutrient dynamics in the Oosterschelde. Clearly the present model does not envisage to duplicate this effort. It is remarkable, however, how much of the dynamics of the system can apparently be represented by such a simple model as used here. It is a clear advantage of the present set of models, that their results can easily be interpreted in terms of one or a few parameters.

The parameter values leading to the results in Fig. 3 are very reasonable and well within the (admittedly often rather broad) ranges found in the literature. A Ks for silicate of 0.03 g Si m^{-3} is on the lower end of the range (0.02-0.13) given by DiToro (1980), and in the middle of the broad range (0.002-

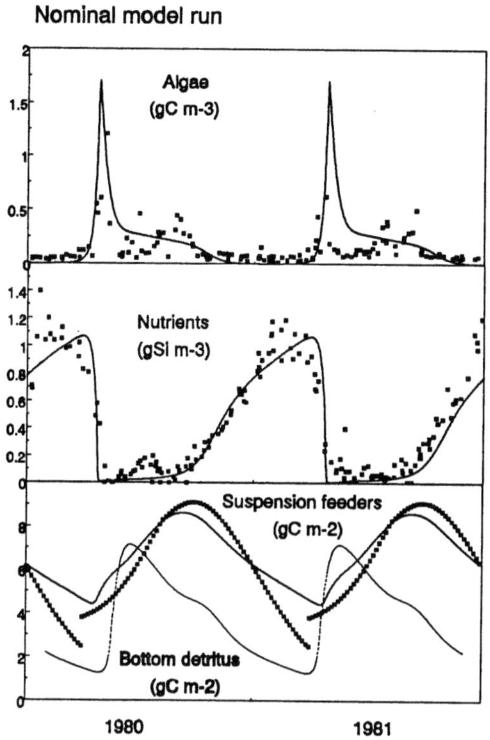

Figure 3. Results of nominal model.

0.1) given by Kremer and Nixon (1978). Respiration and excretion of algae are roughly between 10 % and 18 % d^{-1}. This is relatively low, but not unreasonable (Klepper 1989). It coincides well with the value of 15 % used by Rodhouse and Roden (1987), based on literature values from Burris (1980).

The limitation of clearance by seston concentration appears to be relatively high, compared with ranges given by Klepper (1989). However, the maximal depression with the modelled seston concentration is small; the model did not appear to depend critically on this parameter. Since both seston concentration and clearance are seasonally driven, some compensation between parameters is clearly possible.

Filtration rate is low at 0 °C, but increases rapidly at higher temperatures. At 12 °C it is about 120 l gC^{-1} d^{-1}. This is about 2 l h^{-1} for a 0.8 g AFDW animal. The latter value is comparable (although at the high side) to the values mentioned by Bayne (1976) and Thompson (1984). It is considerably lower than the values given by Møhlenberg and Riisgård (1979). Unfortunately, it is difficult to compare this value to the results of Jørgensen et al (1990). These authors do not give the weight of their experimental mussels. However, it is sure that the low intercept (clearance at 0 °C), as found in the calibration of the present model, is in disagreement with the model of Jørgensen et al (1990). This model also has a higher intercept than the data they showed as a test to the model.

Mortality and respiration rates of suspension feeders (both leading to approximately the same result and modelled similarly, and therefore only well calibrated as a group) are low, in accordance with the long life span of these animals.

The aggregation factor was not used for calibration. It was imposed so that mussels would occur in clumps with a (local) biomass of approximately. 500 g AFDW m^{-2}.

The assimilation efficiency of the suspension feeders is around 0.5, a reasonable figure. It is striking that an earlier calibration of the model, where detritus food was not taken up, resulted in a higher calibrated value for efficiency. In the present model a trade-off is observed between efficiency and fraction of detritus available as food: similar goodness of fit is found with higher efficiency and lower fraction of detritus used as food, or vice versa.

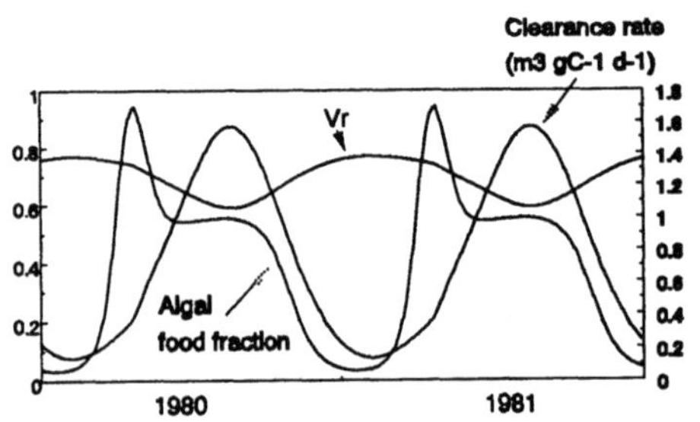

Figure 4. Vertical reduction factor, fraction of algae in food, and clearance rate of suspension feeders. Nominal model run.

The low value for the fraction of detritus taken up as food was similar to the results of Crosby et al. (1989), who described a low assimilation efficiency of 3 % for non-living POM in *Crassostrea virginica*. It is consistent with the results of Williams (1982) who failed to grow *Mytilus edulis* on detritus. The calibrated parameter in our model is also in the range indicated by Cranford and Grant (1990) for *Placopecten magellanicus* (Assimilation efficiency 9% for resuspended organic material). These authors' remark, however, the potential importance of this unpreferred and inefficiently used food in critical periods. Fig. 4 illustrates that this is also the case in our model. Despite the low availability of detritus as food, it makes up most of the total food in winter periods. Remark, however, that feeding rate is low in this period! Algae are the most important part of the food during the growth season. A comparison between the 'nominal' model and the 'simple' model has shown the importance of detrital food. Both models had very comparable time series for algae and nutrients. However, the nominal model had an almost double biomass of suspension feeders, and a considerably lower amount of detritus in winter. Extensive calibration exercises with the simple model have shown that this result is not due to peculiar parameter combinations. Within reasonable (but wide) ranges for all parameters, it was impossible to obtain a suspension feeder biomass near to the data. This was the main reason why feeding on detritus was incorporated into the nominal model.

The effect of the 'vertical reduction factor' on the magnitude of the state variables in the standard run (mean water depth 4 m) is rather limited. As shown in Fig. 4, this factor fluctuates between appr. 0.6 and 0.75. However, the formulation of these hydrodynamic effects had a dramatic effect on the stability of the model under different forcing.

MODEL BEHAVIOUR UNDER DIFFERENT FORCING

The 'simple model' was easily destabilised by deviations from the standard circumstances. Increasing the average depth of the system, or increasing the nutrient loading of the system, rapidly led to unstable and even chaotic model behaviour. With the nominal model, these manipulations still led to serious changes in the system, but they never provoked chaotic behaviour.

Increasing nutrient loading or increasing average depth of the system had comparable effects. In both cases the suspension feeders were able to build up a larger biomass, thereby increasing the grazing pressure on the system. This is not unexpected, since increase in average depth increases the volume to surface ratio : with the same algal concentration, more algae per m^2 of bottom surface are available. Only late in the season

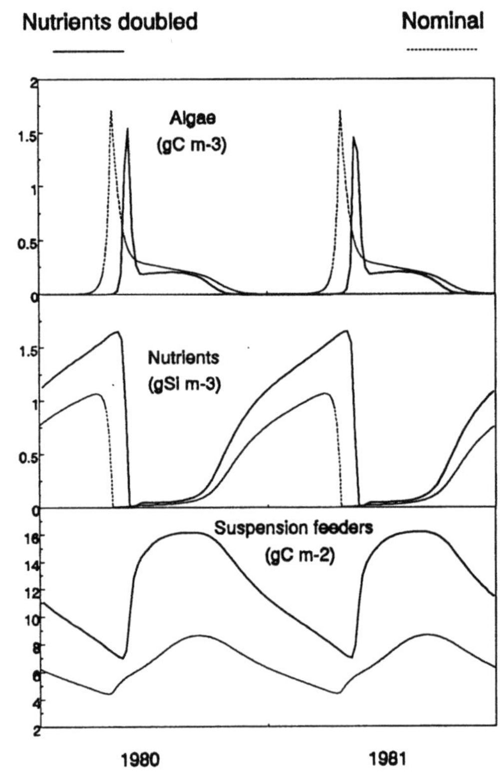

Figure 5. Comparison of the results of the nominal model with a run where initial nurient concentrations were doubled.

sufficient light is available to cause a phytoplankton growth rate exceeding the filtration rate of the suspension feeders. At that moment the phytoplankton blooms, and nutrient limitation takes over control. Later in the season control is again taken over by grazing. In the absence of correction for the creation of vertical gradients, and without any grazing on detritus, the short phytoplankton peak may suffice for an enormous increase in suspension feeder biomass. This biomass is then sufficient to prevent any bloom at all in the next season, and the outcome in the third season is dependent on very subtle ratios of suspension feeders and algae, hence the chaotic nature of the results. With the stabilisation following from the grazing on detritus, and with the moderation of grazing during peaks (correction for gradients), the model is far better behaved. Comparison of the results of the nominal model with a run where initial nutrient concentrations (1/1/79) were doubled.

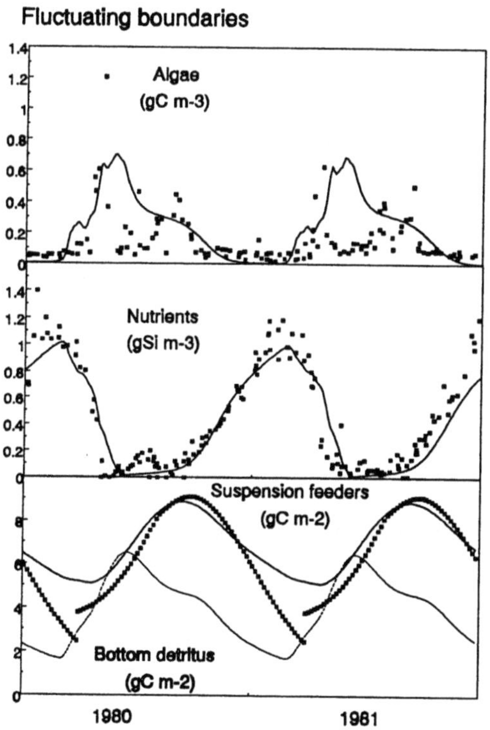

Figure 6 Result of the spatially structured model with (linearly interpolated) data as upstream boundary conditions.

The suspension feeder biomass, however, absorbs only a part of the nutrients added to the system (Fig. 5). Doubling of the initial value of nutrients at 1/1/1979 (from 0.77 to 1.5 g Si m⁻

3) was largely reflected in the subsequent winter concentrations and resulted in persistent effects on the functioning of the system. The role of nutrient limitation was reduced, and the system appeared as clearly more eutrophicated. This result is therefore in contradiction with the assertions on the stabilising properties of suspension feeders in estuarine ecosystems (Herman and Scholten, 1990).

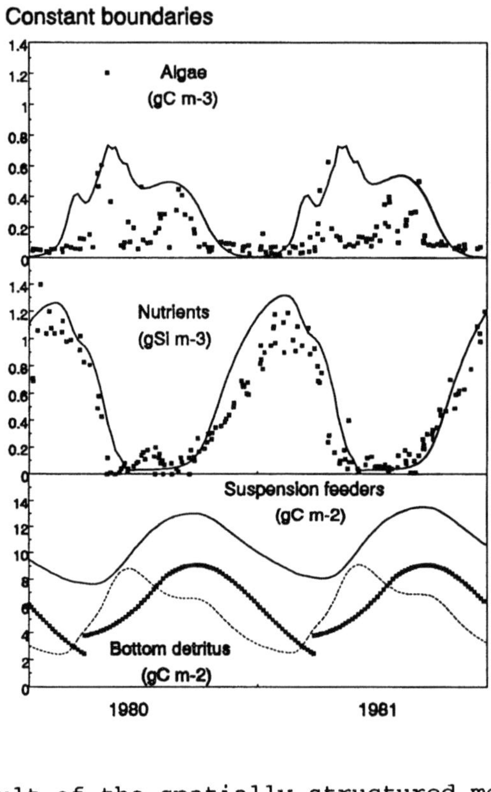

Figure 7 Result of the spatially structured model with constant upstream boundary conditions.

RESULTS OF THE SPATIALLY STRUCTURED MODEL

In several respects the spatially structured 2D (X-Y) model was the most interesting version. In the absence of advective transport, or with advective transport and a 'co-fluctuating' upstream boundary (Fig. 6), it provided excellent fit to the data. In particular the spring decline in nutrients and peak in algae are less sharp than in the nominal model. This is caused by the depth differences in the basin : at different depths the inherent timing of the model is different. Dispersive transport smoothes the differences to a certain extent, but the effect does never completely disappear.

The result with a constant upstream boundary (Fig. 7) sheds a sharper light on the 'stabilising' properties of suspension feeder populations. At the upstream boundary the concentrations of algae and nutrients were fixed at 0.4 and 0.8 g m^{-3}, respectively. Obviously, in summer important amounts of material were imported into the system. This effect led to the increase (by a factor 1.5 approximately.) of the suspension feeder biomass on average. In the upper right corner of the modelled universe this increase was much more important, since the imported material is first used here. Surprisingly, the constant upstream boundary led to limited changes in algae and nutrients. Equilibration may take several years, but by the end of the simulated period little change was left in algal blooms (Fig. 7), especially if one considers the large change in suspension feeder biomass. For the results of Fig. 7 the model simulations were started 4 years earlier than for the other simulations.

The situation depicted here essentially abstracts the homeostasis described in Herman and Scholten (1990). In the latter study too, eutrophication was simulated by increasing the nutrient concentrations at the upstream boundaries. This loading of the system with nutrients was not translated into algal biomass, even though it led to increased suspension feeders biomass. It was mentioned before that a similar homeostasis could not be found back in the nominal (closed column) model. The fact that it is found back here in the model with advective transport points to the importance of some form of 'flushing' of the system. As was noted by Herman and Scholten (1990), the presence of suspension feeders ensures that a possible weakening of nutrient limitation control is replaced by suspension feeder grazing control. However, this mechanism is dependent on the possibility to somehow lose again the imported amount of nutrients.

The role of the suspension feeders in this process is clear. Only after the establishment of a much increased suspension feeder biomass is the change in algal concentrations relatively limited. In the early years of the simulation depicted in Fig. 7, the algal bloom starts already early January. This date is gradually shifted later when suspension feeder biomass builds up.

Can suspension feeders stabilise estuarine ecosystems ? Apparently they can, but only in open systems. Openness can be realized, as in the present system, by hydrodynamic exchanges with fresh water and sea end members. If, however, other sinks such as denitrification for nitrogen nutrients, or burial for silicate are considered, even hydrodynamically closed systems could show some form of 'flow-through' nature. In this case the change induced by a heavier loading of the system essentially affects only the component with the longest time scale, and leaves the fastest component (algae in our case) but little affected.

CONCLUSIONS.

The models presented and analyzed in this paper point to a some important aspect of the dynamics of the suspension feeder-phytoplankton-nutrients interaction, to the role of suspension feeders in estuarine ecosystems, and to important aspects in modelling these interactions in the framework of estuarine ecosystem models. It was shown that even if suspension feeders largely prefer phytoplankton as food, and only take a few percent of the available detritus as food. This detrital food has a disproportionate effect on the dynamics of the interaction. The fact that detrital food is present in adverse periods may have a determining effect on survival in these critical periods.

The availability of detrital food also had serious influences on the stability of the models presented here. It is tempting to speculate about the role of this type of food in stabilising the interaction in natural systems.

With the aid of a closed, steady state, vertically structured system, local food depletion near the bottom as a consequence of suspension feeding could be parameterized in a 'vertical reduction parameter'. This simple formulation turned out to be a most important aspect of the 'nominal' model. As the importance of local food depletion increases with depth and with increasing suspension feeder biomass, the inclusion of 'vertical reduction' largely added to the stability of the feeding interactions modelled here. Obviously, it also added to the general applicability of the models, as it is the more untenable to assume complete and immediate vertical mixing, as one goes to deeper water.

Little homeostasis could be observed in the closed systems. The 'nominal model' was not easily destabilised by the increase of the nutrient loading e.g., but large and immediate reactions in all state variables were observed. This was different in the spatially structured open system. It was shown that increases in loading of the system could result in serious increases in suspension feeder biomass and productivity, without altering the algae and nutrients very much. The flow-through nature of the system was shown to be a very important characteristic.

ACKNOWLEDGMENTS

The author wishes to thank K. Soetaert, I. de Vries, J. Grant and C. Bacher for their valuable comments on the manuscript. This is publication no. 626 of the Netherlands Institute of Ecology, Center for Estuarine and Coastal Ecology.

REFERENCES

Asmus H Asmus RM (1990) Trophic relationships in tidal flat areas: to what extent are tidal flats dependent on imported food? Neth J Sea Res 27:93-99

Bacher C (1989) Capacité trophique du bassin de Marennes-Oléron: couplage d'un modèle de transport particulaire et d'un modèele de croissance de l'huître *Crassostrea gigas*. Aquat Living Resour 2:199-214

Bacher C (1991) Etude de l'impact du stock d'huîtres et des mollusques compétiteurs sur le performances de croissance de Crassostrea gigas à l'aide d'un modèle de croissance. ICES mar Sci Symp 192:41-47

Bacher C Héral M. Deslous-Paoli JM Razet D (1991) Modèle énergétique uniboite de la croissance des huîtres (*Crassostrea gigas*) dans le bassin de Marennes-Oléron. Can J Fish aq Sci 48:391-404

Baretta J Ruardij P (1988) Tidal flat estuaries. Simulation and analysis of the Ems estuary. Springer-Verlag Berlin

Bayne BL (1976) Marine mussels - their ecology and physiology. International Biological Programme 10

Burris JE (1980) Respiration and photorespiration in marine algae. In : Falkowsky PG (ed) Primary productivity in the sea. Plenum Press New York :411-432

Cranford PJ Grant J (1990) Particle clearance and absorption of phytoplankton and detritus by the sea scallop *Placopecten magellanicus* (Gmelin). J exp mar Biol Ecol 137:105-121

Crosby MP Langdon CJ Newell RIE (1989) Importance of refractory plant material to the carbon budget of the oyster *Crassostrea virginica*. Mar Biol 100:343-352

Dame R Dankers N Prins T Jongsma H Smaal A (1991) The influence of mussel beds on nutrients in the Western Wadden Sea and Eastern Scheldt estuaries. Estuaries 14:130-138

DiToro DM (1980) Application of cellular equilibrium and Monod theory to phytoplankton growth dynamics. Ecol Modell 8 :201-218

Eilers PHC Peeters JCH (1988) A model for the relationship between light intensity and the rate of photosynthesis in phytoplankton. Ecol Modell 42:185-198

Fréchette M Grant J (1991) An in situ estimation of the effect of wind-driven resuspension on the growth of *Mytilus edulis*. J exp mar Biol Ecol 148:201-213

Fréchette M Butman CA Geyer WR (1989) The importance of boundary-layers flows in supplying phytoplankton to the benthic suspension feeder, Mytilus edulis L. Limnol Oceanogr 34: 19-36

Herman PMJ Scholten H (1990) Can suspension-feeders stabilise estuarine ecosystems? In: M Barnes and RN Gibson (eds) Trophic relationships in the marine environment. Aberdeen University Press, Aberdeen pp 104-116

Hily C (1991) Is the activity of benthic suspension feeders a factor controlling water quality in the Bay of Brest ? Mar Ecol Prog Ser 69: 179-188

Hoop BJ de Herman PMJ Scholten H Soetaert K (1992) SENECA 1.5. A simulation environment for ecological applications. User Manual. Netherlands Institute of Ecology, Centre for Estuarine and Coastal Ecology. Yerseke

Jørgensen CB Larsen PS Riisgård HU (1990) Effects of temperature on the mussel pump Mar Ecol Prog Ser 64: 89-97

Kautsky N Evans S (1987) Role of biodeposition by *Mytilus edulis* in the circulation of matter and nutrients in a Baltic coastal ecosystem. Mar Ecol Prog Ser 38: 201-212

Kiørboe R Møhlenberg F (1981) Particle selection in suspension-feeding bivalves. Mar Ecol Prog Ser 5:291-296

Klepper O (1989) A model of carbon flows in relation to macrobenthic food supply in the Oosterschelde estuary (S.W. Netherlands). Ph D thesis, University of Wageningen, Wageningen, 1-270

Klepper O Peeters JCH van de Kamer JPG Eilers P (1988) The calculation of primary production in an estuary. A model that incorporates the dynamic response of algae, vertical mixing and basin morphology. in: Marani A (ed): Advances in environmental modelling, Elsevier:373-394

Klepper O van der Tol MWM Scholten H Herman PMJ (in press) SMOES, a simulation model for the Oosterschelde ecosystem. Part I: description and uncertainty analysis. Hydrobiologia

Kremer JN Nixon SW (1978) A coastal marine ecosystem - simulation and analysis. Springer Verlag, Berlin 271 pp

Møhlenberg F Riisgård HW (1979) Filtration rate, using a new indirect technique, in thirteen species of suspension feeding bivalves. Mar Biol 54:143-147

Monismith SG Koseff JR Thompson JK O'Riordan CA Nepf HM (1990) A study of model bivalve siphonal currents. Limnol Oceanogr 35:680-696

Officer CB Smayda TJ Mann R (1982) Benthic filter feeding: a natural eutrophication control. Mar Ecol Prog Ser 9: 203-210

Press WH Flannery BP Teukolsky SA Vetterling WT (1987)

Numerical recipes: the art of scientific computing. Cambridge University Press, Cambridge

Rodhouse PG Roden CM (1987) Carbon budget for a coastal inlet in relation to intensive cultivation of suspension-feeding bivalve molluscs. Mar Ecol Prog Ser 36:225-236

Smaal AC van Stralen MR (1990) Average annual growth and condition of mussels as a function of food source. Hydrobiologia 195:179-188

Thomann RV Mueller JA (1987) Principles of surface water quality modelling and control. New York, Harper and Row

Thompson RJ (1984) The reproductive cycle and physiological ecology of the mussel *Mytilus edulis* in a subarctic, non-estuarine environment. Mar Biol 79 :277-288

Smaal AC (1991) the ecology and cultivation of mussels: new advances. Aquaculture 94:245-261

de Vries I Hopstaken F Goossens H de Vries M de Vries H Heringa J (1988) GREWAQ: an ecological model for Lake Grevelingen. Rijkswaterstaat, Tidal Water Division report T 0215-03

Williams P 1982 Detritus utilization by *Mytilus edulis* Estuar Coastal Shelf Sci 12: 739-746

WHY CARRYING CAPACITY MODELS ARE USEFUL TOOLS FOR MANAGEMENT OF BIVALVE MOLLUSCS CULTURE

Maurice HERAL

Institut Francais de Recherche pour l'Exploitation de la Mer, Laboratoire de Biologie et d'Ecologie des Invertebres Marins, B.P. 133, 17390 LA TREMBLADE (France).

INTRODUCTION

In areas where oyster or mussel culture is very intensive, declines of growth rate and decreases of survival rate have occurred. For these reasons, plans have been proposed to regulate the cultivated biomasses in order to fit the carrying capacity of the different ecosystems (Heral et al 1990; Heral, 1991). In areas where new aquaculture of molluscs is beginning, oyster or mussel farmers need to know how large the extension of the culture could be and what the maximal densities should be in order to obtain the maximum economic benefit. Furthermore, as mollusc cultures are developed in coastal areas, they are very susceptible to changes in environmental conditions that can modify trophic relationships, or directly reduce growth rate, physiological functions, recruitment and mortality processes. To give responses to these three types of questions concerning regulation, development, and environmental impact, it is necessary to build models. These models should predict responses in terms of bivalve growth rate in relation to the different management strategies, taking into account biomasses, new species and environmental modifications that can be planned. Two types of models have been developed to achieve these goals, (i) general models based on long term series of population dynamics of the cultivated species, and (ii) trophic models that describe the main relationships that govern the major fluxes of energy, carbon or nutrients in ecosystems.

NATO ASI Series, Vol. G 33
Bivalve Filter Feeders
in Estuarine and Coastal Ecosystem Processes
Edited by Richard F. Dame

Locality	Total weight after 12 months (g)	Total weight after 24 months (g)
Emsworth Harbour (U.K.)	34	100
	(5) 60	
Newton Bay (U.K.)	52	
Linne Mhiurich (U.K.)	14	58
Emsworth Harbour (U.K.)	20	70
		(16 months)
Menai Straits (U.K.)	50	130
Rossmore (Ireland)	46	—
Carlinglard	16	—
Bullinakill	6	—
Carna	3	—
Flensburg		60
Baltic North Sea	8	
(Germany)	(12)60	
Flensburg Fjord (Germany)	24	—
Oualidia (Maroc)	(6) 43	120
Etel (France)	12—29	—
Marénnes-Oléron (France)	48	60
Marennes-Oléron (France)	35	60
Thau (France)	35—50	68
		116 (20 months)
Corse (France)	—	100 (17 months)
Arcachon (France)	15	58 (30 months)
	(18 months)	
Marénnes-Oléron (France)	8	27 (30 months)
	(18 months)	
Bretagne Sud (France)	28	99 (30 months)
	(18 months)	
Marénnes-Oléron (France)	1970 50	100
	1972 33	75
	1974 20	50
	1975—1981	40
	20	
	1984 15	30
Fish pond (Israel)	(4) 79	—
	92	
Lim Canal (Yugoslavia)	26	103

Note: Number in parenthesis represents weight at the beginning of the culture.

Table 1. Growth performance of the cupped oyster *Crassostrea gigas* on the European and Mediterranean coast. (from Heral and Deslous-Paoli (1991).

Substantial differences in growth rates were recorded for the same species in different areas for different countries, but also in the same bay at different levels of exploitation of the area (Table 1).

The main factor which can explain this variability is the available food which depends on two factors: first, the nutritional value of the bay for the mollusc which is a function of (i) the phytoplanktonic productivity, (ii) the estuarine organic matter inputs and (iii) the hydrodynamic characteristics of the bay in terms of current velocities, residence time of water masses and the time of immersion of the mollusc beds. Different bays, lochs, estuaries were tested for *C. gigas* growth and some sites showed ten times more growth than others. In these experiments the tested animals were in small quantities without any other large cultivation, e.g., mussels, scallops, clams, or abundant wild populations of molluscs in the area. The growth results indicate the level of the molluscan trophic capacity for each sector. The second factor affecting growth is the impact of density on the available food on two scales, (i) the local density, which is the density of the unit of culture. It is known that oyster or mussel growth is related to available settlement surfaces (Shafee and Sabatie, 1986; Berthome et al 1984; Boromthanarat and Deslous-Paoli, 1988). These observations for juveniles were also valid for adults. Several authors have shown that growth was a function of the density on the ground, in baskets, in racks or in ponds. This local interdependence is a function of food and space limitation (Frechette and Lefaivre, 1990; Frechette et al 1992). (ii) On a broad scale there is the impact of the total biomass of cultivated or wild populations. Heral et al (1986) found that for the bay of Marennes-Oleron there is a relationship (Fig. 1) between the decrease of the annual growth and the development of the total cultivated biomass of Pacific oyster for the last 15 years.

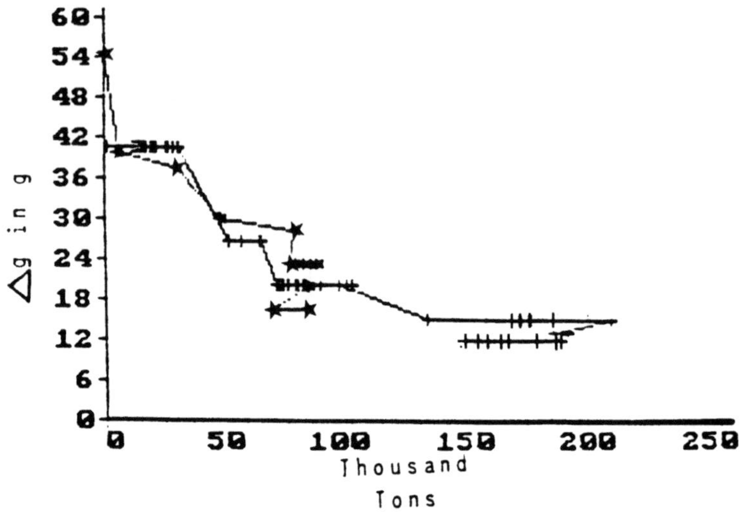

Fig. 1. Evolution of the annual growth rate of Portuguese oyster (+) and Japanese oyster (*) in relation to the total biomass of oysters in the bay of Marennes-Oleron (from Heral, 1991) .

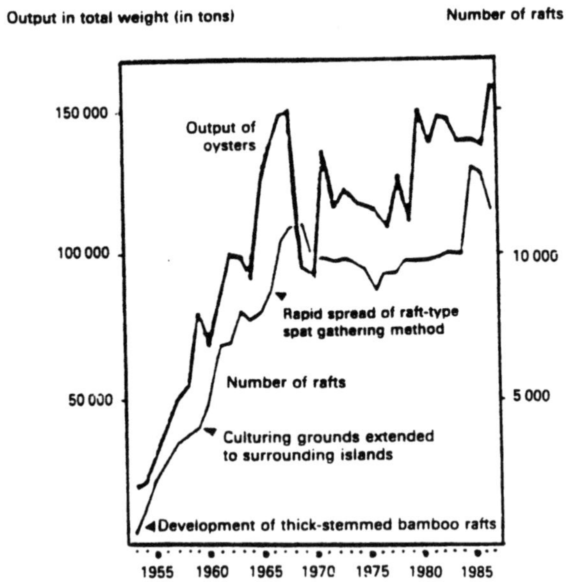

Fig. 2. Evolution of the annual production of oyster and the number of rafts in Hiroshima Bay (from Heral, 1991).

As to the collection of data on growth rate and mortality rate, they can be directly measured in *in situ* experiments or obtained from farmers. These methods gave different results because the farmers carried out all sorts of activities during the culture process (technical aspect, culture rotations). Annual production data also had to be collected. This data could be obtained by inquiries to the local administrations or from the professionals at the different fisheries or aquaculture operations. The data can also come from aerial observations of the intertidal areas as described by Bacher et al. (1986) and Heral (1991). Recording long term time series of oyster production produced clear trends of production which could be related to the management strategies that have been developed and to changes in environmental conditions.

HISTORICAL DATA SET

A number of examples of the major world oyster bays will be successively analysed:Hiroshima Bay, Chesapeake Bay, Marennes-Oleron Basin and Arcachon Basin.

Hiroshima Bay:The evolution of the number of rafts and the oyster production was monitored since 1950, before the boost of the production (Anonymous in Heral, 1991; Fig. 2). Until 1980, oyster growth reached market size in one year. After that date, a continuous increase of number of rafts and production (54,000 tons to 110,000 tons) induced a decrease in growth rate to 2 years that modified oyster turnover and profitability. For these reasons the oyster men associate in cooperative societies decided to reduce the number of rafts to re-obtain their initial growth rate. To maintain the biomass at the same level each year, they decided to fix the necessary number of spat collectors function of the exportation and the hardening processes. It is emphasized that this type of management has been possible only because Japanese oyster farmers are highly organized.

Chesapeake Bay: The landings from 1820 to the present show different trends of oyster production (Fig. 3; Heral et al 1990; Rothschild et al 1993). Three main periods were identified:(1) an increase in oyster fishery from 1840 to 1890 with a large overfishing and the destruction of oyster habitat caused by fishing gear, mainly dredging; (2) the decrease of landings from 1900 to 1980 due to the failure of the reseeding plan connected to the deterioration of environmental conditions, in particular, heavy sedimentation in relation with deforestation and anoxic conditions in summer; (3) the last decrease in production (1981-1988) was caused by high morality related to disease (MSX and *Perkinsus marinus*), predation and poor management practices which contributed to the spread of the diseases.

Marennes-Oleron Basin: The bay is the main European area of oyster production. Change in the production of the cupped oyster has been estimated since the beginning of the century with three difference sources of data (Heral et al 1986; Fig. 4). It is apparent that (1) there was low production from 1890 to 1927 related to the development of Portuguese oyster culture and to the fisheries practices which exploited natural oyster beds; (2) after 1927, there was an increase in production due to the beginning of oyster culture obtained by the control of recruitment using spat collectors and by the development of breeding techniques. Production increased until 1960, (3) from 1960 to 1969 severe declines of the growth rate were observed in relation to over exploitation. Oyster physiological condition was poor and densities were very high. All factors were converging to enhance the spread of disease. It is now known that an iridovirus caused the disappearing of *C. angulata*. It has been replaced by the Japanese oyster *C. gigas* (see Grizel and Heral, 1991). Oyster production reached previous levels quickly with the same overstocking consequences.

Arcachon Basin: The long term time series for this bay (Deltreil, personal communication; Fig. 5) showed the same pattern as Marennes-Oleron bay but with two severe environmental crises:(1) during 1950-1970, oyster production and recruitment deteriorated. This was caused by effluent of a pulp mill that discharged into the bay. The production of the oyster factory increased ten-fold between 1950 and 1960 from 1970 onwards the wastes have been collected in a sewage that discharges outside of the bay (Heral et al; 1990), (2) from 1977 to 1981: the decrease in production was related to contamination of the bay by Tributiltin (TBT) which reduced recruitment (His and Robert, 1980) and caused malfunction of the calcification processes of

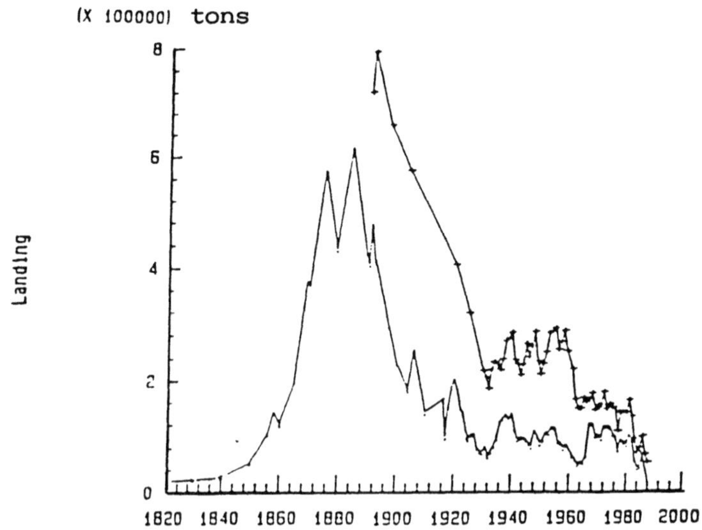

Fig. 3. Trends in the oyster production, in total weight Maryland part (-), Chesapeake Bay (+) (from Heral et al 1990)
-

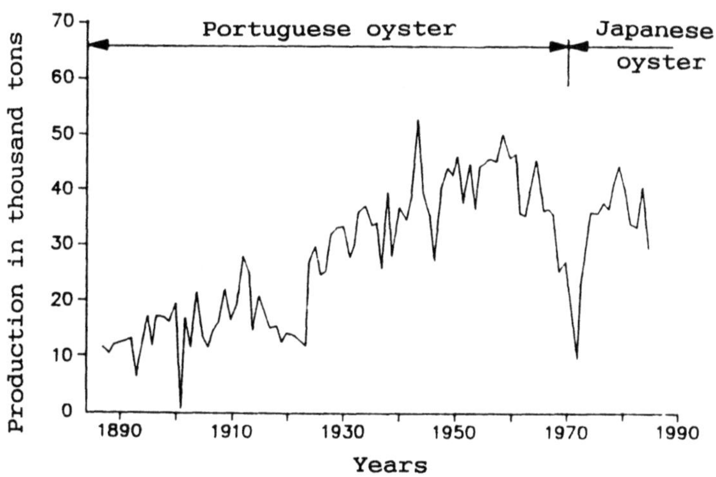

Fig. 4. Evolution of annual oyster production in Marennes-Oleron Basin (from Heral et al 1986).

the oyster shell (Alzieu and Heral 1983). After the ban of the TBT antifouling paint, recruitment and oyster growth returned to earlier levels (Alzieu et al 1985).

These different case studies clearly illustrated the three main factors that can largely alter mollusc production: diseases, environmental damages and overstocking.

GENERAL DYNAMIC POPULATION MODELS

Dynamic population models were designed to give management authorities an indication of the carrying capacity of a bay. They are based on the assumption that, for the given period, environmental factors have a constant mean, although a certain variation can occur around this mean. Thus, environmental parameters are not included in the regulations. Historical data from long time series allow the building of dynamic population models. For the bay of Marennes-Oleron, data sets have been collected for production, growth and mortality rates for the same years. These data showed that growth rates had decreased for both Portuguese and Pacific oyster while the mortality rate had increased (Fig. 6). The stocks in culture (biomass) have been calculated from the annual production, taking into account growth and mortality (Fig. 7). These calculations gave results comparable with the estimates of stock size obtained by sampling for the recent years.

The establishment of a relationship between the stock and production showed clearly an asymptotic value of 40,000 tons. This level corresponded to the maximum production capacity of the ecosystem (Fig. 8). This definition describes the carrying capacity of the bay for oyster cultivation during the period. Maximum production of the ecosystem can be determined by an equation of the same type as the one used for the growth of the populations. So the Von Bertalanffy equations $P = P_{max} (1-e^{-KB})$ fits the data well(P_{max} is the maximum production of the bay, B is the cultivated stock). For Marennes-Oleron for *C. angulata* $K = 0.026 \times 10^{-3}$ and P_{max} = 41,800 tons and for *C. gigas* $K = 0.028 \times 10^{-3}$ and P_{max} = 52,450 tons. The yield P/B in relation to the stock followed a negative exponential curve as the evolution of the annual growth rate in relation to the stock.

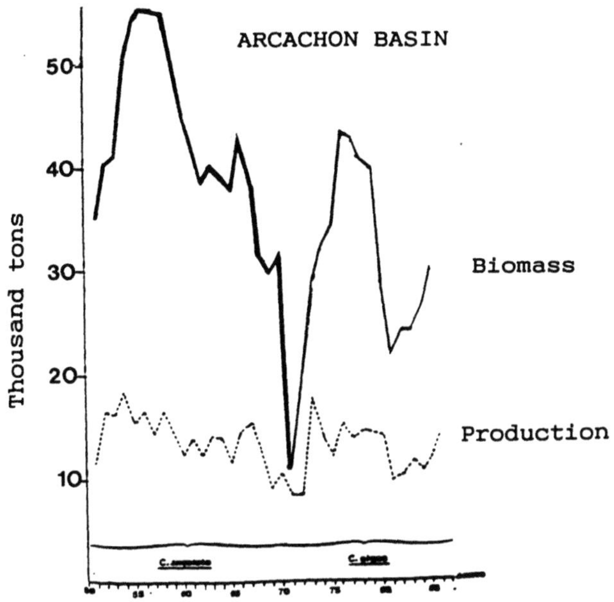

Fig. 5. Development of the oyster stock in Arcachon Bay (from Deltreil, personal communication).

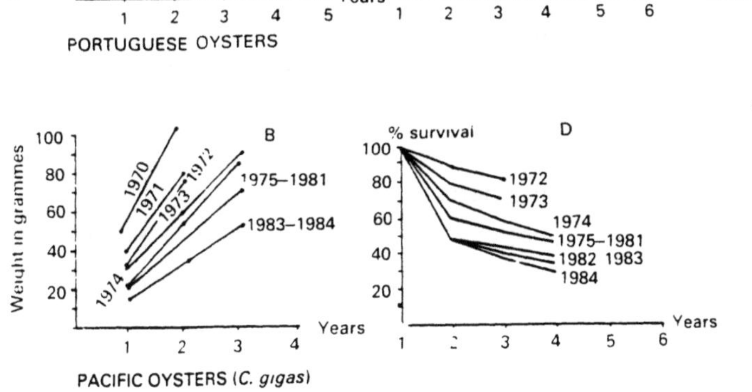

Fig. 6. Change in growth rate needed for an oyster to reach market size (A: Portuguese oysters; B: Pacific oysters) and survival rates after the first year of culture (C: Portuguese oysters; D: Pacific oysters) (from Heral et al 1986).

a stock of Portuguese oysters of 130,000 tons but also with a stock of Japanese oyster of 80,000 tons. This difference between the two species can be explained by the energetic demand of each oyster: the assimilation of food by the Japanese oyster is 1.7 times greater than by the Portuguese oyster (Heral et al 1986). For management and to estimate the impact of these two oysters, this transformation coefficient must be used.

These models showed that without management of cultivated oysters, the stocks tended to exceed the minimal biomass necessary to reach the maximum potential production. If stock regulation is applied, it gives oyster farmers the advantages of shorter breeding cycle and decreased mortality, which results in better profits.

ANALYTICAL MODEL

The quality of a model is dependent on its hypothesis, but its accuracy depends on the quality of the data sets that have been collected in the ecosystem. For these reasons, it is important to focus attention on the minimal data necessary to build a carrying capacity model.

(1) For hydrodynamic data, simultaneous records of tidal levels are necessary as well as precise and recent bathymetry. Some long time series of current velocity and direction could be of interest knowledge of wind action is necessary.

(2) Meteorological records, such as temperature, wind (strength and direction) and day light, have to be taken frequently over long time periods (10 years) to obtain mean values of different typical conditions with their probability of occurrence intensity.

(3) For sediment suspension and resuspension data, the area has to be mapped with physical characteristics of the sediment. This should be done several times a year because of seasonality of the meteorological conditions. Turbulence is one of the key factors controlling erosion in coastal areas. To study it, it is useful to have a data set on waves with their refraction-diffraction processes as well as measurements on wavelets created by local wind.

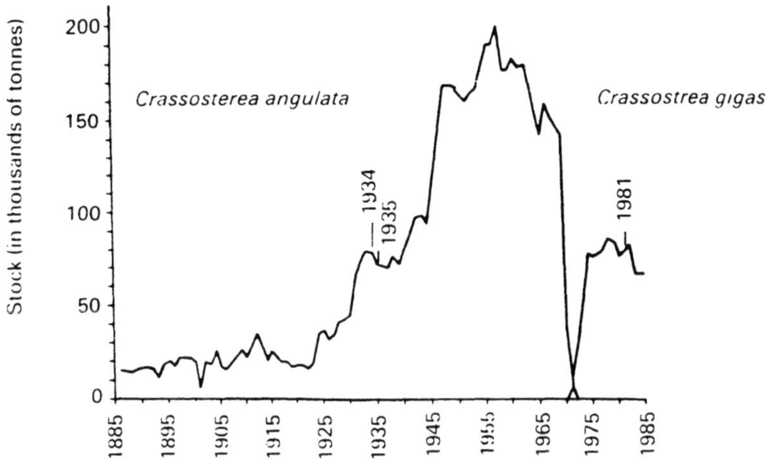

Fig. 7. Calculated evolution of the total biomass of cultivated oysters in the bay of Marennes-Oleron. (from Heral et al 1986)

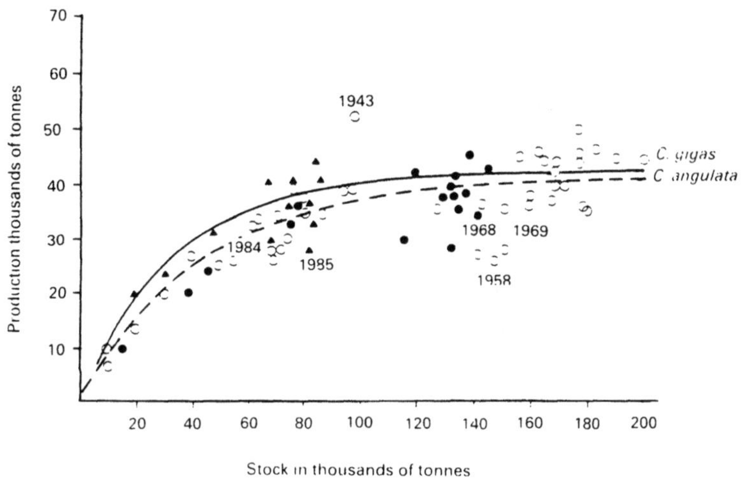

Fig. 8. Evolution of the annual production of the stock in culture for *Crassostrea angulata* (o), *Crassostrea gigas* (•) and for *Crassostrea gigas* converted in equivalent Crassostrea

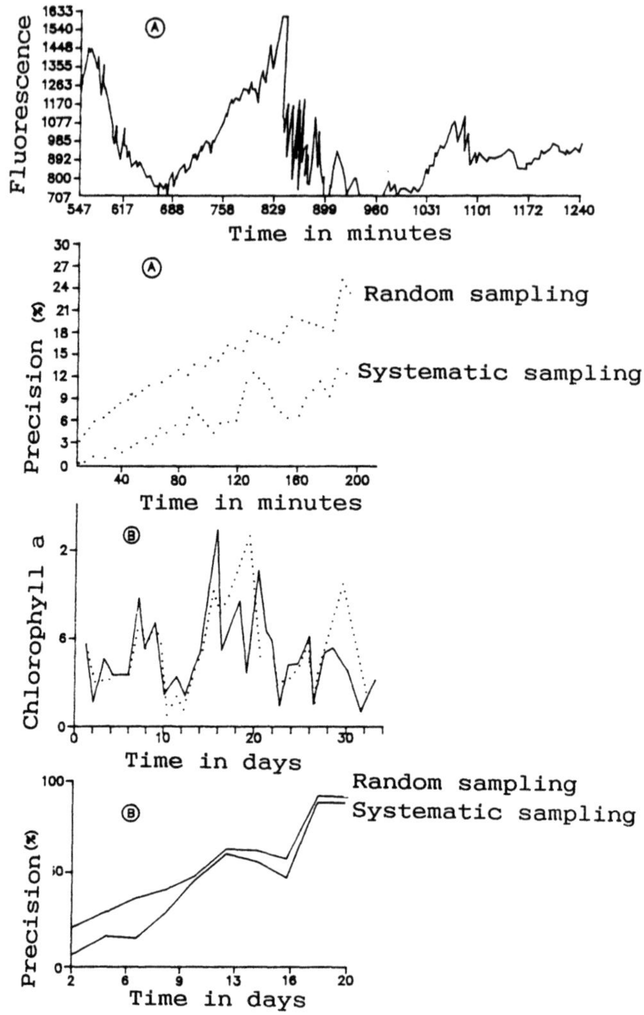

Fig. 9. Precision at the level of 95 % of the intradiurnal mean of the signal (A) as a function of the sampling time and the nature of the sampling (systematic or random sampling). Precision at the 95 % level for the interdiurnal monthly mean of the signal (B) as a function of the sampling time and the nature of the sampling (from Heral et al 1990).

For the water column, sampling strategy must be defined depending on the precision that will be introduced in the model and the time scale of the model. Figure 9 shows, for example, how intense the sampling must be to have good precision on phytoplankton biomasses on a daily or monthly basis. Furthermore, to validate advection dispersion models, continuous records of salinity are required in the different limit conditions. In parallel, continuous record of turbidity are very useful to validate a particular model of transport and sedimentation or erosion processes.

For primary production, for phytoplankton and phytobenthos, the use of photosynthetron techniques allows the determination of the different Michaelis-Menten parameters. A knowledge of nutrient loads from the river and fluxes from the sediment is important to estimate their respective contribution. Determination of limiting factors of phytoplankton production by test in bioassays will define the primary production model as either nitrogen or phosphorus based.

Trophic competitors, such as zooplankton and benthic populations, are estimated by adequate sampling strategies. For molluscs, stratified sampling based on granulometry and bathymetry with optimal allocation as a function of the variance could be a good compromise (Sauriau, 1992). Special techniques with the help of aerial observations or remote sensing are useful to contribute to the estimation of the cultivated biomasses. Population dynamics of the targeted cultivated species have to be followed in different areas to compare results given by the molluscs growth rate model and the observations.

After these intensive phases of acquiring a data set, which can take several years of multidisciplinary research, the main biomasses, production, and fluxes must be compared to identify which are the main processes that control the turnover of the ecosystem (Fig. 10).

It is only these main processes that will be retained for the ecological carrying capacity model. This preliminary work is crucial. Further validity of the model will depend on whether, at this stage, the crucial choices have been well made.

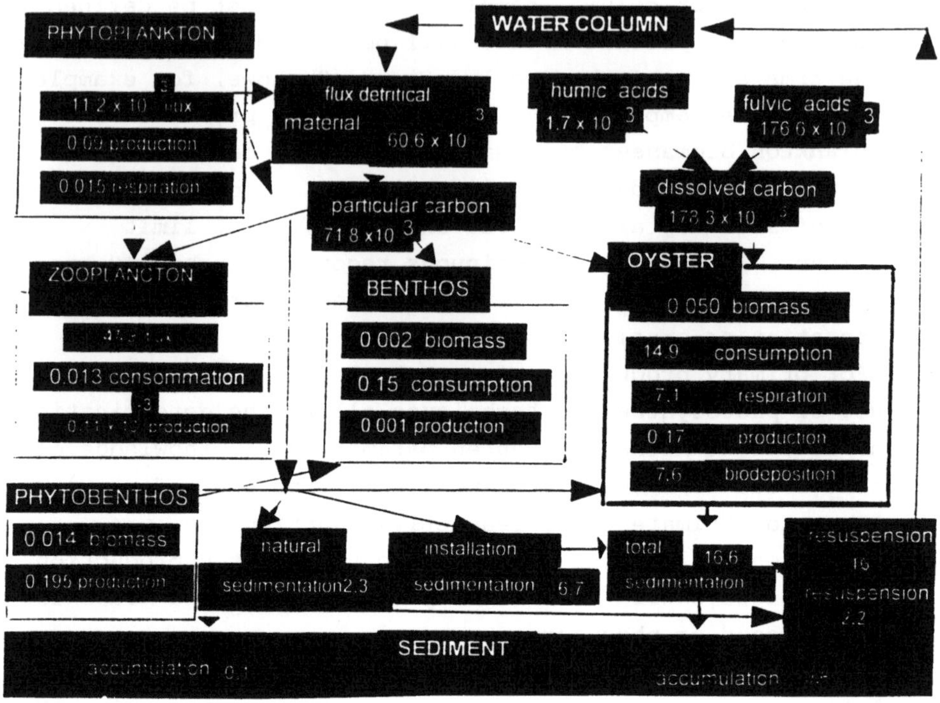

Fig 10. Fluxes of carbon in kg m^{-2} $year^{-1}$ in Marennes-Oleron (from Heral unpublished data).

Figure 10, for Marennes-Oleron bay, shows obviously that suspension and resuspension processes must be retained as well as the phytobenthos production. Detrital organic material is also a significant compartment. Trophic competitors, both benthic wild molluscs populations and zooplankton, play a secondary role in this ecosystem.

The following design could be retained for Marennes-Oleron bay (Figure 11.):

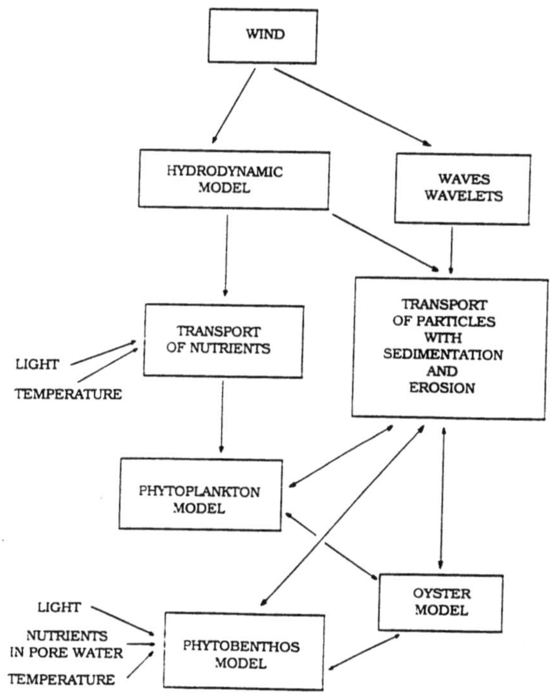

Fig. 11. Scheme of the main trophic interaction in Marennes-Oleron bay.

OYSTER MODEL

Food consumption

$Q_{[d,p,m]} = SES_{[d,p,m]} \cdot F$

$Qt = F / (Ws^{bf}) \cdot SES$

$F = F0 \cdot e^{(kf \cdot \min(0, Ts - SES))} \cdot Ws^{bf}$; $SES = SES_d + SES_p + SES_m$

Ingestion

$I_{[d,p]} = (1 - PF_o) \cdot Q_{[d,p]}$; $I_m = (1 - PF_m) \cdot Q_m$

$PF_{[o,m]} = PFX_{[o,m]} \cdot {}^{(1 - e(kp[o,m] \cdot \min(0, C1 - Qt)))} + (1 - PFX_{[o,m]}) \cdot {}^{(1 - e(kp2 \cdot \min(0, C2 - Qt)))}$

Absorption

$A_{[d,p]} = ae \cdot I_{[d,p]}$; $EA_{[d,p]} = A_{[d,p]} \cdot X_{[d,p]}$

$ae = aet \cdot TMP + ae0$; $X_d = (ES - SES_p \cdot X_p) / SES_d$

$ES = SES_L \cdot X_L + SES_G \cdot X_G + SES_P \cdot X_P$

Metabolic cost

$R = (art \cdot TMP + ar0) \cdot Ws^{br}$; $ER = R \cdot X_{O2}$

Spawning effort

$P = ap \cdot Wj^{bp}$; $p = P^{(1/dp)}$

Scope for growth

$dWj/dt = ((EA_d + EA_p) - ER) \cdot 10^{-3} \cdot pc \cdot TIM - p \cdot Wj$; $Ws = Wj \cdot Y_h$

PARAMETER	DEFINITION	UNITY
ae	Absorption efficiency	
ae0	Absorption efficiency at t=0°C	
aet	Slope of the absorption function of temperature	
ap	Constant for spawning effort	
ar0	Respiration at t=0°C	$mlO2\ h^{-1}\ gPs^{-1}$
art	Slope of the respiration function of temperature	$mlO2\ h^{-1}\ gPs^{-1}\ °C^{-1}$
$A_{[d,p]}$	Particular absorption	$d \rightarrow mgPs\ h^{-1}\ ind^{-1}$
bf	Allometric exponent for filtration	
bp	Allometric exponent for spawning	
br	Allometric exponent for respiration	
C1,C2	Level of pseudofeces production	$mgPs\ h^{-1}\ gPs^{-1}$
dp	Duration of the spawning period	j
$EA_{[d,p]}$	Energy absorbed	$J\ h^{-1}\ ind^{-1}$
ER	Metabolic cost	$J\ h^{-1}\ ind^{-1}$
ES	Total energy of the seston	$J\ l^{-1}$
F	Filtration rate	$l\ h^{-1}\ ind^{-1}$
F0	Standart filtration rate	$l\ h^{-1}\ gPs^{-1}$
$I_{[d,p,m]}$	Particular ingestion	$mgPs\ h^{-1}\ ind^{-1}$, $p \rightarrow \mu gChla\ l^{-1}\ ind^{-1}$
kf	Filtration exponent	
kp_{1o}, kp_{1m}, kp_2	Pseudofeces production exponents	
p	Daily ratio of the weight for the spawning effort	
P	Total ratio of the weight for the spawning effort	
pc	Percentage of absorption in energy for the flesh growth	
$PFX_{[o,m]}$	Thresold of pseudofeces production	
$PF_{[o,m]}$	Rate of pseudofeces production	
Qt	Total consumption	$mgPs\ h^{-1}\ gPs^{-1}$
$Q_{[p,d,m]}$	Particular consumption	$mgPs\ h^{-1}\ ind^{-1}$
R	Respiration	$mlO2\ h^{-1}\ ind^{-1}$
$SES_{[p,d,m,P,L,G]}$	Sestonic concentration of the water column	$mgPs\ l^{-1}$, $p \rightarrow \mu gChla\ l^{-1}$
TIM	Immersion ratio	
TMP	Water temperature	°C
Ts	Level of cloging	$mgPs\ l^{-1}$
W_j	Calorific content of the oyster flesh	$kJ\ ind^{-1}$
W_s	Dry weight of the oyster flesh	$gPs\ ind^{-1}$
X_{O2}	Energetic conversion for oxygen	$J\ mlO2^{-1}$
X_p	Energetic conversion for chlorophyl	$J\ \mu gChla^{-1}$
$X_{[d,P,L,G]}$	Energetic conversion of the weight of seston	$J\ mgPs^{-1}$
Y_h	Conversion in weight of the energetic value of the oyster flesh	$gPs\ kJ^{-1}$
Y_p	Conversion in dry weight of the chlorophyl	$mgPs\ mgChla^{-1}$

Table 2 : Processes presented in the model, variables and parameters used. The suffixes d, p, m, o, P, L, G, refer respectively to detrital organic matter, phytoplankton, particulate inorganic matter, particulate organic matter and particulate proteins, lipids, carbohydrates. The suffixes 1 and 2 refer to the two stages of the law of pseudofaeces production (from Raillard et al., 1993).

Oyster model

The objective of this model is to predict the growth rate in relation to environmental conditions (T, quantity and quality of food, mineral seston loads). These models cannot include all the physiological results, for example, ingestion rate in relation to the size of the particles. The general equation of the energy budget of oyster populations is established following the equation

$$\mathbf{P = A - R = C - (F + U) - R}$$

where **A** = assimilation, **R** = respiration, **F** = particulate excretion (feces and pseudofeces), **U** = dissolved excretion, C = consumption, **P** = production (**P = Pg + Pr + Ps** with **Pg** = production of the flesh, **Pr** = production used for the reproduction, **Ps** = production of the shell + mucus). Table 2 shows the different laws we have retained until this date for the *Crassostrea gigas* species (Raillard et al 1993) and the results of the model are shown in Fig. 12. Further physiological measurements are required to improve ingestion and assimilation function.

Connection of the different submodels

As time and spatial scales have to be consistent between physical models and biological ones, a box structure is applied to the oyster production area. The Lagrangian residuals of a tide are calculated and are the base for spatial box design. The time scale is the tidal cycle (Bacher, 1989). It has been demonstrated that this design does not alter physical characteristics, such as, residence time. Until now, resuspension and sedimentation terms have been calculated with the original hydrodynamic mesh and the mean values are calculated for each box within the grid (Raillard, 1992).

The ecosystem model is built with a stock of oysters of two age classes in each box, and the growth model is a function of the food that is transported by the physical model. The trophic molluscan shellfish competitors and their assimilation of food are introduced in each box as driving variables. This

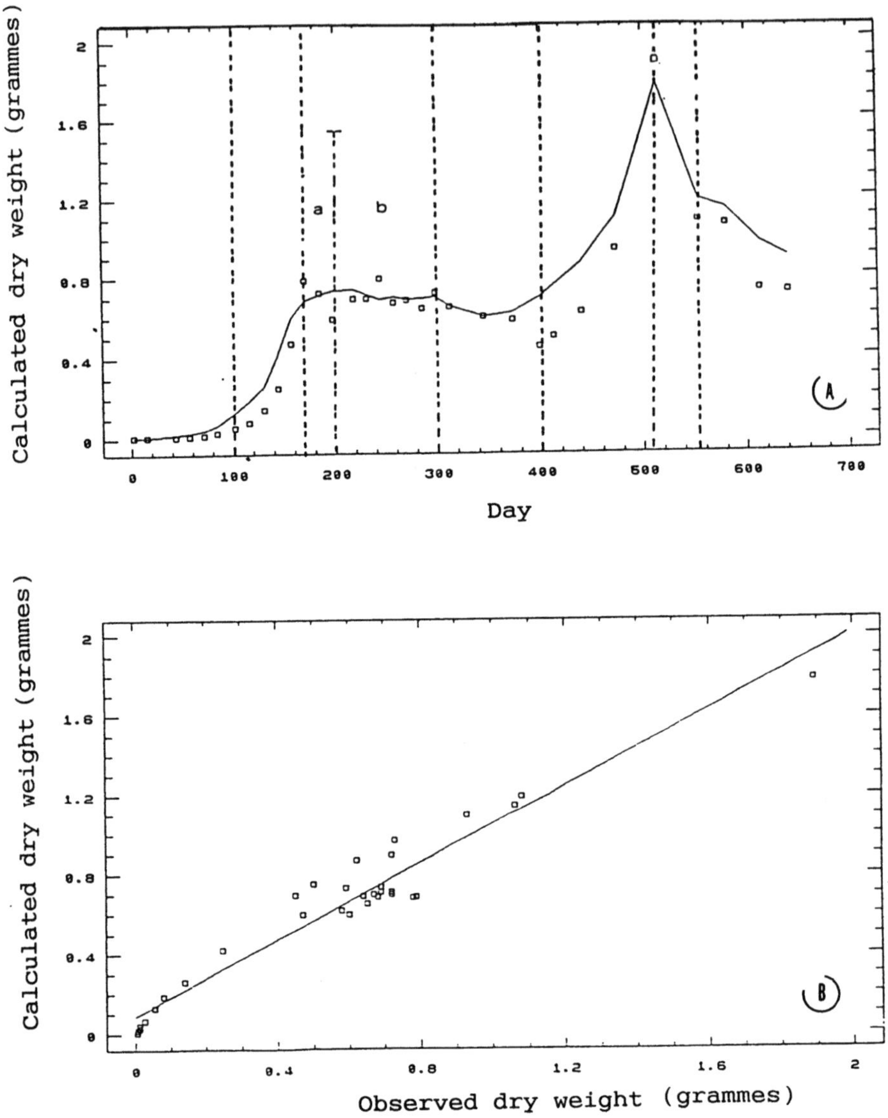

Fig. 12. Results of simulated and measured individual growth (O) of the flesh of *Crassostrea gigas* (KJ) during two years : (a) temporal distribution of S and M; (b) linear regression between simulation and observations (a = 0.96, b = 1.74).

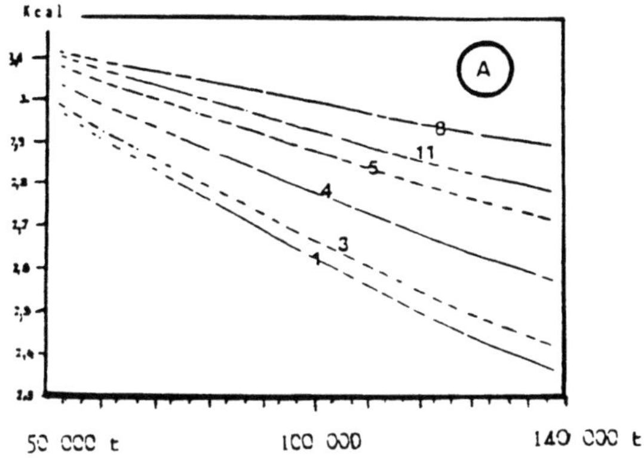

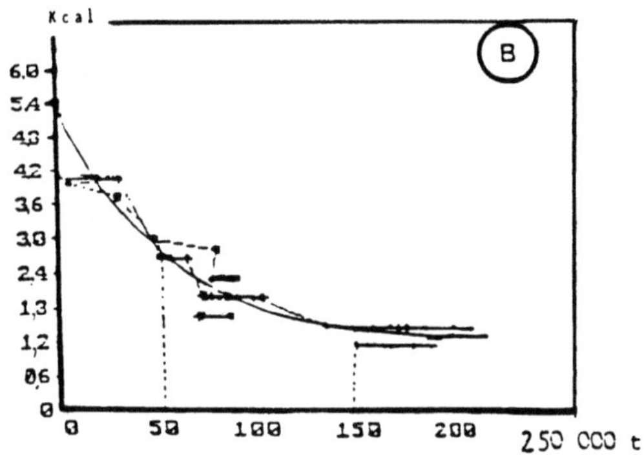

Fig. 13. (A) Simulation of annual production according to the analytical model (from Bacher 1991), (B) annual production of one oyster, results of the global model (from Heral 1986).

approach permits fluctuations in the cultivated stocks of oysters and allows predictions of growth rates in the different areas. The comparison between the results of the general model (Fig. 13B) and the simulation of the analytical model (Fig. 13A) demonstrates the same effects of the total cultivated biomass on the annual growth rate, but with more precision for the different areas with the analytical model. Sensitivity analysis was applied to the level of biomass, to seston load or to the nitrogen quantity (Bacher et al 1991; Bacher, 1991; Raillard and Menesguen 1991; Raillard 1992).

CONCLUSION

To date, these models use many simple hypotheses, but they indicate the future research needs to provide a useful management tool. They demonstrate that a multidisciplinary approach of biologists, physical and chemical scientist and sedimentologists could predict changes in the growth rate of the cultivated species as a function of the food and of all other factors, particularly pollution, that can modify the quality and quantity of the trophic requirements. It is evident that to be predictive and to be a management tool it is necessary to go further in the study of energy demands of the oysters for particulate and dissolved substances. Alternatively, a phytoplanktonic model, that simulates the variations of the input of nutrients from the estuary, would be helpful for the study of the consequences of the use of freshwater on oyster production.

With the results of the EEC TROPHEE programme, we hope we will be able to develop more deterministic laws for the physiology, recycling of nutrients and phytobenthos production and to increase the predictivity of the models.

REFERENCES

Alzieu C Heral M (1983) Ecotoxicological effects of organotin compounds on oyster culture, in Ecotoxicological testing for the Marine Environement, Vol 1, Persoone G., Jaspers E., Claus C. Eds Bredene Belgium 187

Alzieu C Sanjuan J Deltreil JP and Borel M (1986) Tin contamination in Arcachon Bay : Effects on oyster shell anomalities. Mar Poll Bull 17 (11):494-498

Bacher C (1989) Capacite trophique du bassin de Marennes-Oleron : couplage d'un modele de transport particulaire et d'un modele de croissance de l'huitre Crassostrea gigas. Aquat Liv Resour 2:199-214

Bacher C Baud JP Bodoy A Deslous-Paoli JM Dreno JP Heral M Maurer D Prou J (1986) A methodology for the stocks assessments of cultivated oysters along the French Atlantic coasts. ICES, Shellfish Committee K:36, 14

Bacher C Heral M Deslous-Paoli JM Razet D (1991) Modele energetique uniboite de la croissance des huitres (*Crassostrea gigas*) dans le bassin de Marennes-Oleron. Can J Fish Aquat Sci 48:391-404

Berthome JP Prou J Razet D Garnier J (1984) Premiere approche d'une methode d'estimation previsionnelle de la production potentielle d'huitre creuse *Crassostrea gigas* d'elevage. Haliotis, 14, 34

Boromthanarat S Deslous-Paoli JM (1988) Production of *Mytilus edulis* reared on bouchots in the bay of Marennes-Oleron : comparison between two methods of culture. Aquacul 72:255-263

Frechette M Lefaivre D (1990) Discriminating between food and space limitation in benthic suspension feeders using self thinning relationships. Mar Ecol Prog Ser 65:15-23

Frechette M Aitken AE Page L (1992) Interdependence of food and space limitation of a benthic suspension feeder : consequences for self-thinning relationships. Mar Ecol Prog Ser 83:55-62

Grizel H Heral M (1991) Introduction into France of the Japanese oyster (*Crassostrea gigas*). J Cons int Explor Mer 47:399-403

Heral M (1990) Traditional oyster culture in France. In Aquaculture Barnabe Eds Ellis Harwood, Chichester, voll:342-387

Heral M (1991) Approches de la capacite trophique des ecosystemes conchylicoles : synthese bibliographique. ICES mar Sci Symp 192:48-62

Heral M Bacher C Deslous-Paoli JM (1989) Capacite biotique des bassins ostreicoles. In L'homme et les ressources halieutiques. Troadec JP Eds IFREMER Brest:225-260

Heral M Deslous-Paoli JM (1991) Oyster culture in European countries in Estuarine and Marine Bivalve Mollusk Culture. W Menzel (Eds) CRC Press, Boca Raton:153-190

Heral M Deslous-Paoli JM Prou J (1986) Dynamique des productions et des biomasses des huitres creuses cultivees (*Crassostrea angulata* et *Crassostrea gigas*) dans le bassin de MarennesOleron depuis un siecle. ICES, Mariculture Committee F:41 p

Heral M Drinkwaard AC (1990) Management of oyster and mussel culture in Aquaculture Europe'89. Business joins Science. N de Pauw and R Billard (Eds) European Aquaculture Society, Bredene 12:147-162

Heral M Rothschild BJ Goulletquer P (1990) Decline of oyster production in the Maryland portion of the Chesapeake Bay : causes and perspectives. ICES Shelfish Committee K:20,37 p.

His E Robert R (1980) Action d'un sel organo metallique l'acetate de tributyletain sur les oeufs et les larves D de *Crassostrea gigas* (Thunberg). ICES Mariculture Committee F:27, 10 p

Raillard O (1991) Etude des interactions entre les processus physiques et biologiques intervenant dans la production de l'huitre *Crassostrea gigas* du bassin de Marennes-Oleron : essais de modelisation. These, Universite de Paris VI, 207 p

Raillard O Deslous-Paoli JM Heral M Razet D (1993) Modelisation du comportement nutritionnel et de la croissance de l'huitre japonaise *Crassostrea gigas*. Oceanologica Acta 16(1):in press

Raillard O Menesguen A (1991) Modelisation de la production de l'huitre *Crassostrea gigas* du bassin de Marennes-Oleron (

France) : Etude de l'influence des apports de nitrates du fleuve Charente sur la production primaire et ostreicole. ICES, Biological Oceanography committee L:74, 18 p

Rothschild B Ault JS Goulletquer P Heral M (1993) The deline of Chesapeake Bay oyster population : a century of habitat destruction and overfishing. Mar Ecol Prog Ser (in press).

Sauriau PG (1992) Les mollusques benthiques du bassin de Marennes-Oleron : estimation et cartographie des stocks non cultives, competition spatiale et trophique, dynamique de population de *Cerastoderma edule*. These Universite de Bretagne Occidentale, 309 p

Sauriau PG Mouret V Rince JP (1989) Organisation trophique de la malacofaune benthique non cultivee du bassin ostreicole de Marennes-Oleron. Oceanologica Acta, 12(2):101-112

Shafee MS Sabatie MR (1986) Croissance et mortalite des huitres dans la lagune de Oualidia (Maroc). Aquaculture 53:201

INTEGRATED ESTUARINE MANAGEMENT - OBTAINING A SUSTAINABLE YIELD OF BIVALVE RESOURCES WHILE MAINTAINING ENVIRONMENTAL QUALITY

Norbert Dankers
DLO Institute for Forestry and Nature Research
P.O. Box 167
1790 AD Den Burg/Texel
The Netherlands

INTRODUCTION

Coastal seas and estuaries have been used by mankind for many centuries and fulfilled many functions. In many parts of the world sustainable use of these areas is essential for survival of local people. Everywhere these areas are important for a number of reasons (Groot 1992). For example, they are important as nurseries for fish, function as shipping lanes, provide recreational facilities, feed many birds, act as a place to discharge sewage and provide for both commercial and recreational fishing. Over the last decades the commercial value of estuaries has shifted from fisheries production to a variety of interests among which recreation has increased significantly. The majority of estuarine functions are based on natural values; therefore it is essential to maintain these natural values. If one user is allowed to significantly alter the system, it may destroy the options for other or future users. In view of the changes in the past decades, one can conclude that it is difficult to predict what level of importance we will give to the different functions in twenty or more years from now. Therefore it is essential not to aim for a maximisation of one particular function but to aim for optimisation of a great number of functions. This places a burden on science to provide policy-makers with data that can be used to evaluate options.

One activity carried out in estuaries all over the world is fishing and culturing of bivalves. In Europe the most important

NATO ASI Series, Vol. G 33
Bivalve Filter Feeders
in Estuarine and Coastal Ecosystem Processes
Edited by Richard F. Dame

Fig. 1 The International Wadden Sea.

species are the blue mussel *Mytilus edulis*, the oysters *Crassostrea giga* and *Ostrea edule,* and the cockle *Cerastoderma edule*. There is a great variety in the ways culture or fishing of bivalves is performed. Mussel culture is practised using longlines, rafts, poles or bottom culture. Furthermore, there is fishery on wild, natural mussel beds. This is carried out primarily to obtain spat for the mussel culture, but also consumption size mussels are fished. This fishery occurs both in the intertidal and the subtidal. Oyster culture is practised as a bottom culture or on racks approx. 50 cm off bottom. Cockles are not cultured, but in the intertidal regions of the British and Dutch coastal waters there is an extensive fishery for cockles.

Depending on the way the culture or fishing is carried out, the impact on the natural ecosystem will be different. Some aspects of the impact such as shifts in carrying capacity, can be translated from one area to another. In this contribution I will limit myself to the culture and fishery of mussel and cockle and use the Wadden Sea as an example (Fig. 1).

CONFLICTS BETWEEN SHELLFISH PRODUCTION AND THE MANAGEMENT OF THE COASTAL ECOSYSTEM

In the different countries bordering the Wadden Sea, the approach towards the management of the system is different although the goals are similar (Dankers et al. 1992). All countries aim for a management of the area that will result in conservation of natural values. In the Netherlands, the protection of the Wadden Sea is based on a combination of physical planning instruments as laid down in a Wadden Sea memorandum and the Nature Conservation Act. The objective of the Wadden Sea memorandum is sustainable protection and development of the Wadden Sea as a nature area while providing for a number of human activities. It is not stated whether the activities should be restricted to specific areas or whether they are allowed to a certain impact level.

In Germany, the Federal Nature Conservation Act lays down provisions for national parks. These are defined as areas of eminent character, of which major parts fulfil the conditions for being designated a nature reserve, and which have not, or hardly, been influenced by man and will permit the preservation of as many native species of flora and fauna as possible. The major part of the German Wadden Sea has been designated as a national park by the federal states of Schleswig-Holstein, Niedersachsen and Hamburg. Within these national parks certain core zones have been designated. These zones are mainly intertidal areas and cover 30-60% of the area. Public admittance in the core zone is prohibited except on guided tours or assigned paths. Within the national park, actions are prohibited which destruct, damage or change the protected area or any part thereof, or that could lead to lasting disturbances. Because fishery existed before the national park was designated, a number of fishing activities and bivalve cultures are allowed within the park boundaries.

In Denmark the Danish Wadden Sea has been designated as 'Nature Reserve Wadden Sea' and 'wildlife reserve' (Miljoministeriet 1986). The nature reserve area is bordered by the mean-high water mark on the mainland and islands and by a line situated three nautical miles from the islands into the North Sea.

It can be concluded that mechanisms for protecting the natural values of the Wadden Sea are in place. On the other hand, there is still a considerable pressure of cockle and mussel fishery.

Cockle fishing is only carried out in the Netherlands. The biomass of cockles fluctuates widely from less than 10 million kg (1979, 1986, 1991) to more than 200 million kg (1980, 1981, 1988) (Fig. 2). In years of above-average biomass less than 10% of the stock was fished. In some low-stock years, fisheries took more than 25% and almost 100% of the dense beds. In 1991 the fishery was closed because recruitment failure in 1988, 1989 and 1990 caused a lack of cockles. Mussel recruitment failed in the same years and the effects of low stocks of these bivalves became apparant. Oystercatchers *Haematopus oestralegus* showed low reproductive success and eider ducks *Somateria mollissima* either starved or died because of parasite infections after having shifted to crabs as an alternative food source. Conservation groups, scientists and the national press urged for strong measures against the fishermen and even accused them of 'environmental vandalism'.

The mussel culture and fishery spoiled its image as a natural activity (Drinkwaard 1987). Since the 1950s, mussel culture has been practised in the Wadden Sea of the Netherlands, Niedersachsen and Schleswig-Holstein. The culture has some impact on the natural ecosystem but, in general, it has been considered acceptable within the Wadden Sea ecosystem. Seed mussels are collected from wild beds, both in the

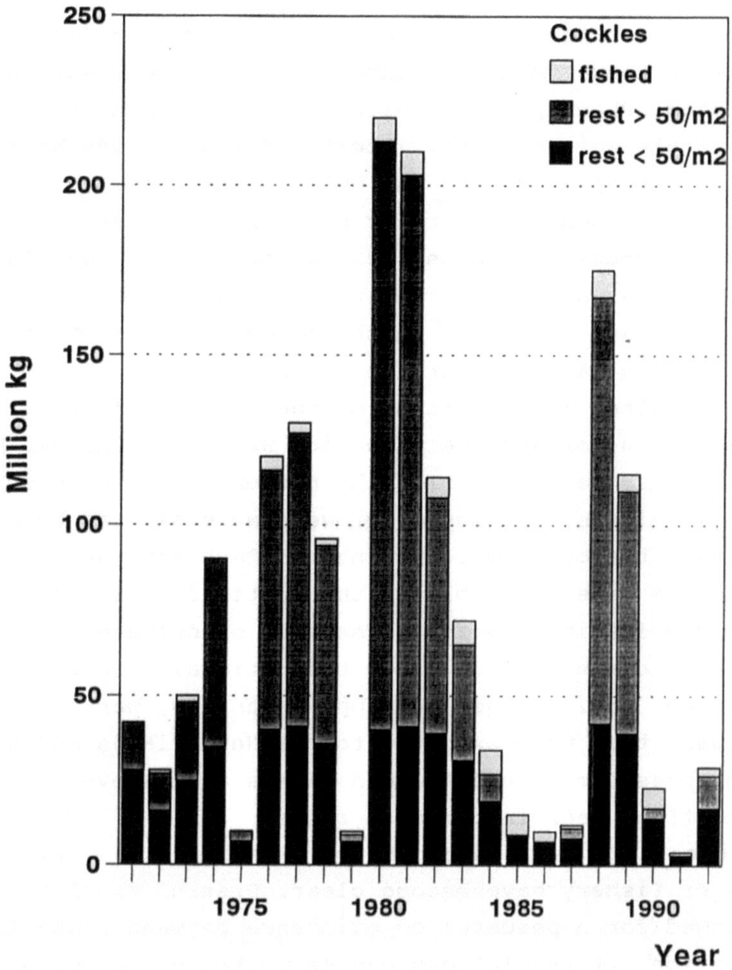

Fig 2. The biomass of cockles in the Dutch Wadden Sea in autumn. The amounts fished, dense beds and solitary cockles are presented separately. The figure is based on information by Dankers and Vlas (1992) and Stralen and Kesteloo (1992). Division between dense beds and rest calculated by M. van Stralen (pers. comm.)

intertidal and subtidal. Intertidal mussel beds can become very old (McGrorty et al. 1990; Dankers and Koelemaij 1989; Obert and Michaelis 1991) and form stable, complex communities (Kuenen 1942; Maas-Geesteranus 1942; Verwey 1952; Straaten 1965) of a number of year classes and an associated flora and fauna (Asmus 1987, Dittman 1990). Old mussel beds slowly disappear because of age, predation, storm, ice or human interference (Obert and Michaelis 1991). New spatfall on or in the vicinity of these beds is responsible for the continuous rejuvenation of these beds. Because of increased fisheries pressure on these juvenile mussels, the old beds were not rejuvenated. When spatfall failed since 1988 and remaining adult mussels were removed, intertidal mussel beds have virtually disappeared from the Dutch Wadden Sea. In the Danish part of the Wadden Sea there is no fishery for seed mussels but fishery for commercial sized mussels on natural beds is carried out when prices are high. This was the case in the period between 1984 and 1987 when Dutch and German production was lower than the market. Overfishing in combination with ice winters resulted in a collapse of the intertidal mussel stock in Denmark. In Germany the situation was more complex. Mussel culture occurs mainly outside the national park core zone, but due to old rights, the mussel growers were allowed to collect spat within the core zone. During the last ten years sufficient seed mussels were available in the subtidal region outside the Schleswig-Holstein core zone. The area of culture lots outside the core zone was doubled since the national park was implemented. In 1992 the seed supply was low, partly because large quantities were exported to the Netherlands and there was a strong pressure to use the old rights and remove mussels from the intertidal beds in the core zones.

The conflicts between nature conservation and bivalve culture or fishery have become clear. Present regulations have not allowed for a peaceful co-existence between fisheries and conservation. If regulations can be worked out which solve the problems in the Wadden Sea where nature conservation has priority, it should be possible to translate these regulations to other areas.

THE MUSSEL AND COCKLE IN THE COASTAL ECOSYSTEM

The development of dense assemblages

Both mussels and cockles form dense assemblages. After a good spatfall cockles initially occur in densities of up to 100,000 per m^2. The maximum density of one-year-old cockles is 2,000 per m^2 (Verwey 1952). Growth is possible only if space is created by mortality. Within a very dense cockle bed spatfall is reduced (Hancock 1973). Therefore beds often consist of one-year-class. Mortality is caused by low winter temperatures and heavy predation pressure and most cockle beds do not survive for more than 3-5 years. After that period, individual older animals may remain, but the density is too low for a bed to be recognised. Empty shells either remain in the sediment or are carried away by waves and currents. It can be concluded that cockle beds are not communities, but are simply high densities of individual cockles. There is no evidence of persistence of these beds nor of structure or a distinct assemblage being created. After the cockles in a bed have died, the remaining shells may provide a distinct habitat in the sandy surroundings.

Mussel larvae in general settle on hard substrate or on threadlike structures (Blok and Geelen 1958; Maas-Geesteranus 1942). Settled larvae have a greater chance for survival on the byssus threads of old mussels (McGrorty et al. 1990). Only in years with extraordinarily good spatfall, spat may settle in intertidal areas on all sorts of substrates including bare sand (Beukema 1982; Dare 1976). This was also observed in the Dutch Wadden Sea in 1987. When spat has fallen and a mussel bed develops, the mussels produce large quantities of faeces and pseudo-faeces which accumulate underneath them. If this causes the bed to rise high above the surrounding flat it becomes very susceptible for storm damage (Straaten 1965). Only mussel beds that have developed slowly over many years or even decades will form stable beds. These form on the pseudofaeces which is re-enforced with empty mussel shells. Because larvae seem to prefer byssus threads for settlement, there is a positive feed-back which keeps beds in existence and these mussel beds contain mussels of different age classes (McGrorty et al. 1990).

The Common Wadden Sea Secretariat (CWSS 1992) describes three different types of mussel beds. "In the lower reaches of the intertidal, beds are situated at the edges of tidal creeks and gullies. Such beds exist over long periods although the actual population size on such beds can vary extremely depending on the fishing intensity, natural mortality (mainly caused by ice, storm and predation) and the strength of the most recent recruitment. In periods without strong winters, these beds reach the status of mature beds characterised by small hills consisting of compressed biogenic mud and covered with mainly old, large mussels. Between those hills, tidal pools with a rich epibenthic fauna and flora can be found. The mussels are covered with brown algae *Fucus vesiculosus*. The hills will be swept more or less away by ice in the next strong winter, but these beds disappear finally only if the abiotic conditions of the site change drastically." The second type is "banks in the higher tidal region, far away from creeks. These are only formed in years with extraordinary strong recruitment. Suitable substrates for the settlement of juveniles in this region are mainly moderate silty regions covered with eelgrass *Zostera marina*, remains of old cockle beds or tubes of the polychaeta *Lanice* spec. Natural mortality, although caused by the same factors as in the lower regions of the intertidal is much higher and growth is slower. Natural mortality in combination with mortality caused by fishing leads to the disappearance of these beds within a few years of formation. This leads to restoration of the original situation of an area with aggregates of mussels within a *Zostera* bed until new strong recruitment leads to the formation of a new bed." The third group consists of mussel beds in the subtidal. In the Dutch Wadden Sea these are very common, possibly because nearby culture areas supply larvae and during storms clumps of mussels are flushed from the cultures. Mussel clumps are known to act as spat collectors. In the German part of the Wadden Sea, the subtidal beds only occur after a good spatfall. Predation by starfish, shrimp, crabs, fish and diving ducks is very high. Many of these beds are fished away each year. In the Netherlands they provide the majority of the seed mussels for the cultures. In a four-year-survey it was shown that on these beds less than 15% of the mussels were older than one year (Dankers et al. 1989). Also if they are not fished the lifetime of these beds is considered to be less than a few years (CWSS 1991).

The role of mussels and cockles in the ecosystem

Some aspects of the role of mussels and cockles are comparable. Both suspension feeders together often form the bulk of the benthic biomass (Beukema 1976; Dekker 1989; Smaal et al. 1986; Smaal and Nienhuis 1992; Wolf 1983). Both pump water and filter algae and suspended matter. Mussels and cockles occur in densities that can filter the volume equivalent to the whole water mass of an estuary within one week (Dankers and Koelemaij 1989; Smaal and Prins 1993; Smaal et al. 1986). They consume oxygen and produce excretory products, such as nitrogen compounds, which can stimulate primary production. In all these aspects their effects on the ecosystem can be added. For the Eastern Scheldt this has been done by Prins and Smaal (1990) and Dame et al. (1991). A review is given by Dame (1993). Mussels differ from cockles because they accumulate pseudo-faeces and they form morphological structures called beds. These beds form distinct communities that are important in the benthic pelagic coupling (Dame et al. 1980, 1991; Dame and Dankers 1988; Prins and Smaal 1990; Smaal and Prins 1993).

Mussels and cockles generate large amounts of dissolved nutrients (Smaal and Prins 1993) and a considerable biomass. Part of this biomass is produced as eggs and larvae, and the importance of these in marine ecosystems as consumers of phytoplankton or as prey is unknown. The juvenile and adult biomass consists of meat and shells. The meat is consumed by a variety of predators, including man, and the shells are either buried or deposited in gullies in a more or less intact form. Mussel shells break down in a matter of years, but cockle shells seem to be very resistant to decomposition. However, a considerable part of the Wadden Sea sediment consist of shell fragments of cockles which are crushed in the stomachs of eider ducks (Cadée, in press).

By consuming algae and producing dissolved nutrients suspension feeders influence the whole ecosystem. Zuidema et al. (in manuscript) showed experimentally that phytoplankton biomass and total primary production decreased when mussels were present, but that primary production per unit chlorophyll increased. This positive feed-back was also shown by Asmus and Asmus (1991). They reported that the potential primary production induced by the nutrient release from a mussel bed was higher than the uptake of the phytoplankton by the bed. On the basis of simulations with an ecosystem model (Smaal and Boeije 1991) it is shown that enlarging or decreasing the mussel biomass has a more profound influence on the more important ecosystem parameters than for example a reduction of the nutrient load by 50%.

This feedback and increased turnover may be the reason that, for example, in 1989 when the Wadden Sea contained an extremely large mussel population, an unusual autumn bloom of diatoms occurred (G.C. Cadée, pers. comm.). This bloom resulted in a high meat content of the mussel stock.

The role as a food source for birds

Mussels and cockles are important as a direct food source for birds. Mussel beds also provide a habitat which is favoured by many birds not feeding on mussels. Zwarts (1991) showed that mussel beds could contain more than 200 birds per hectare. Although mussel beds normally only cover 3-4% of the intertidal they contain about 25% of all waders at low tide. The majority of the birds consume the associated fauna or small mussels. Oystercatchers and eider ducks are important consumers of adult mussels and cockles. McGrorty et al. (1990) reported that oystercatchers can cause mussel mortality of more than 25% per year. Zwarts (1991) noted that 40% of the large mussels on a bed could be eaten by oystercatchers. This is the primary reason that few mussels will become older than five years.

Enough information is available about the number of birds and their diet in the Wadden Sea to calculate the amount of food required. Smit (1980) and Swennen (1991) calculated that the eider ducks in the Wadden Sea consume about three million kg ash-free dry weight per year. This amount consists of 1.2 million kg mussels, 1.3 million kg cockles and 0.5 million kg of crabs, starfish and other organisms. The majority of the cockles are consumed on the tidal flats, while 90% of the mussels are taken from the subtidal beds and cultures.

The oystercatchers consume yearly more than two million kg of food (calculated on the basis of information from Hulscher (1980), Smit (1980) and Lasiewski and Dawson (1967)) consisting of cockles (70%) and mussels (25%). When ash-free dry weights are calculated to fresh meat weights, Table 1 summarises the food requirements of the bird population.

It should be taken into account that cockles will be eaten only if they occur in densities of more than approx. 50 per m^2. This is the case for 30% of the stock in years with few cockles and 80% of the stock in years with abundant cockles (Stralen and Kesteloo 1992; Kesteloo and Stralen 1992). Partly due to

Table 1. Consumption of mussels and cockles in the Dutch Wadden Sea. All values in million kg fresh meat.

Species	Intertidal		Subtidal	
	cockle	mussel	cockle	mussel
Eider duck	6	0,6	0.5	5
Oystercatcher	8	3	-	-
Other birds	-	1	-	0.3

territorial behaviour, not more than 30% of the mussel bed are consumed by oystercatchers. The mussel population in the intertidal varies greatly (Dankers and Koelemaij 1989). Beukema (1976) calculated that the mussel stock in the intertidal of the Wadden Sea was approx. 30 million kg fresh meat. This means that each year approximately 15% are consumed by birds. For the cockle the average biomass was 74 million kg (RIN 1987; Stralen and Kesteloo 1992; Kesteloo and Stralen 1992). A consumption of 14 million kg means average predation of almost 20% per year.

COCKLE AND MUSSEL CULTURE AND FISHING

Cockle fishery

Cockle fishing has been practised in the Wadden Sea for many years. Already around the turn of the century cockles were exported to England. The fishery mechanised after 1950, but the total catch did not rise much above one million kg meat weight. After 1972 a spectacular increase took place (Fig. 2), and the annual yield now is approx. 7 million kg.

Within the Wadden Sea mechanical fishing on cockles is only carried out in the Dutch part. There is a small amount of fishing by hand, but the majority of cockles are caught by specialised dredges (Vlas 1987). The open season is set by the government, but vessel capacity is sufficient to remove all cockles from 120 km^2 within one season. This is more than the total surface area of high-density beds. The landings are normally determined by the market. In years with a low stock of cockles overfishing may occur (Vlas 1987). Since 1980, the cockle stocks were above average in six years, and only a small portion of the stock was removed. In four years almost all beds were removed, and in the two remaining years the fleet moved to the Eastern Scheldt. If the international market for cockle increases, the present fleet will be able to remove the majority of the dense beds in areas open for fisheries.

Mussel culture and fishery

Originally mussels were harvested from natural, intertidal beds. Most fishery and trade occurred in the Dutch part of the Wadden Sea. As early as 1880, 13,000 metric tons were exported to England. Due to a food shortage the landings increased to 40,000 tons in 1917. The absolute record was reached in 1918 when 124,000 tons were fished. The majority of this record amount was used as fertiliser because the potash trade had ceased during the war. Up to 1950 mussels on intertidal beds were mainly fished as fodder for duck breeders. Consumption mussels were cultured in the Delta region in the southern

Netherlands. Since 1949 mussel culture has developed in the Wadden Sea. First the culture beds were established in the Netherlands, later also in Germany and Denmark. Only a minor quantity (5%) of marketable mussels are now fished from natural beds. Mussel culture ceased completely in the Danish Wadden Sea. In Germany mussel culture was on a low level until 1980 but has increased considerably since then.

Mussel culture is based on the collection of small-sized seed from natural beds. The seed are subsequently spread on culture lots. Sometimes the mussels are transplanted several times, either to areas with better food, to prevent predation, or to areas which are less prone to winter storm damage. Mussels reach a marketable size (50-60 mm) two or three growing seasons after larvae have settled. When farmers have high-quality lots with low storm risk they obtain at least 3 tons of marketable mussels from about 1 ton of seed. In the Netherlands there are intense time restrictions on the seed fishery. This, plus the occurrence of culture lots in unfavourable areas and predation by starfish, crabs and eider ducks cause high mortality among the mussels. Dutch farmers have recently needed 1-2 tons of seed to obtain 1 ton of marketable mussels (Table 2). However, overall mortality on well-situated culture lots is much lower than on natural subtidal beds. Growth conditions are also better than on natural intertidal beds.

Table 2. The efficiency of mussel culture in the Dutch Wadden Sea (in million kg).

Year	Seeded	Production next year
1984	93	73
1985	68	39
1986	93	57
1987	70	37
1988	>100	80
1989	141	70
1990	33	c. 8

Over the last 25 years there has been no clear trend in mussel production from the Wadden Sea (Fig. 3) but the supply

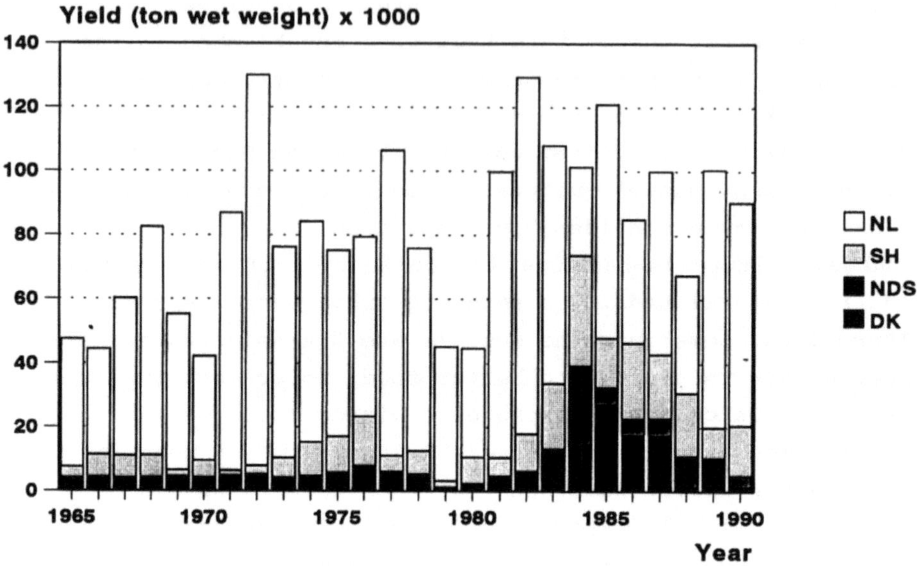

Fig. 3 Total landings - cumulative - of blue mussels in all regions of the Wadden Sea (from CWSS 1992).

was highly variable. The fishing on intertidal beds in Denmark only occurred for a few years around 1985 when prices were relatively high. This fishing resulted in an almost complete disappearance of the beds around 1987 (Laursen et al. 1992), and the Danes strictly regulated their mussel fishery industry.

ENVIRONMENTAL IMPACT

Cockle fishing

In 1979 and 1980 Vlas (1982, 1987) investigated the impact of cockle fisheries on the benthic fauna in the Netherlands. He concluded that the major impact was the removal of large cockles.

Benthic organisms living deeper than 5 cm in the sediment only experienced 1-4% mortality. Organisms in the upper sediment layer showed mortality rates of up to 30%. Böhme (1988) confirmed the conclusions of Vlas, but found morality rates for some species rising to 100%. Since the investigation of Vlas, cockle fishing increased and several years with a low stock occurred. For example in 1985, 1986, 1987 and 1990 more than 30% of the cockle stock were harvested. This means that almost 100% of the dense beds disappeared (Fig. 2).

In the last few years cockle spatfall has been extremely low. Natural mortality of old stock, and intense fishery resulted in a low stock of cockles in 1991. The same years also the stock of mussels was low. This combination resulted in mortality of eider ducks and low reproductive success of oystercatchers.

The cockle fishery has an impact on the Wadden Sea sediment. During fishing, fine material is suspended in the water and is transported by currents. Because large cockle shells are removed, settlement places for mussels and macroalgae disappear. Also potential seagrass settlement will be hindered if regular disturbance takes place (Jonge and Jong 1992).

Mussel fishery and culture

The most obvious impact of mussel fishing is the impact on the mussel beds. As has been described before, there are different types of mussel beds. When an unstable subtidal bed is removed the ecosystem level impact is not considered to be serious. If the bed had not been fished away, the mussels would not have lived for long. Because mussels are spread on culture lots, where crabs and starfish are combatted, the total biomass of mussels in an area is probably increased. Some mussel growers prefer seed from the intertidal because these mussels have shells which are about twice as thick as the subtidal mussels (Cadée, in press). Fishing on unstable intertidal beds results in disappearance or decrease of these beds. Small mussels serve as food for a number of bird species, and those larger than 2-3 cm are consumed by oystercatchers and eider ducks. Eider ducks also prey on mussels on culture lots in the subtidal. In many parts of the Wadden Sea mussel spat settles on seagrass leafs. When such a mussel bed develops further it often disappears because of natural causes (mainly predation and storms). The surrounding seagrass tends to spread again providing new possibilities for attachment of mussel spat. When these mussel beds are fished away, the seagrass is also destroyed even in the spots between the mussel ridges. Eventually this results in a loss of diversity for the ecosystem.

The most severe ecological effect of overharvesting is the destruction of stable mature beds. In Denmark these are fished away when commercial size mussels are present even though the prices paid for these are less than one third of the prices paid for cultured mussels. In the Netherlands and in Germany, these beds were harvested when there were almost no mussels on the culture lots. This fishery, combined with natural mortality caused the disappearance of the mature beds in the Netherlands in 1990 and 1991. Degeneration of mature beds is a normal phenomenon, but generally new spatfall keeps them 'alive'. Since the mid-1980s most spat has been harvested and no new mature beds could begin development. In the last few years spatfall has failed. The present situation is serious because there does not seem to be adequate substrate left on which mature beds can redevelop.

In Schleswig-Holstein the intertidal beds in the national park core zone are under threat. When the park was officially dedicated the mussel growers kept the right to collect spat from the core zone, if necessary. Culture lots were moved outside the core zone. Initially more than enough seed mussels were available in the gullies near the culture lots, and the culture area was doubled. In 1991 at least 40,000 tons of large seed was exported to the Netherlands, and new spatfall was low. Now there is strong pressure of the mussel growers to use their right to start fishing again on the intertidal mussel beds to supply mussels for a very large area of culture lots.

The disappearance of mussel beds with all associated species means a severe loss for the Wadden Sea ecosystem.

When mussels are cultured this means an increase in the amount of suspension feeding molluscs. They filter large amounts of water. The majority of the suspended matter is removed from this water. The mussels and the thick layer of pseudo-faeces and faeces provide an environment in which breakdown of organic matter is enhanced. This results in low oxygen conditions which sometimes induce mortality of a number of organisms, among which seed mussels, and in a production of nutrients, which in general are in low supply in summer. Therefore mussel culture may stimulate the primary production. On the other hand, the mussels may reduce the amount of phytoplankton to such an extent that primary production is reduced.

Sustainable management

In an area where management aims for wise and multifunctional use, the concept of sustainability should be broadened compared to the concept of sustainable yield as used in the sectoral fisheries approach. In the Wadden Sea sustainable management should consider:
- sustainable yield of target organisms;
- sustainable yield of organisms for other parties;
- acceptable impact on non-target organisms.

In the last case sustainable yield then means the yield which can be collected without impacting non-target species to a level higher level than acceptable for other parties, which often are aiming for nature conservation.

Cockle fishery

Natural mortality of cockles is high and unpredictable because it strongly depends on the severity of the next winter. From a fisheries point of view it would be wise to fish the maximum amount the market can take. Because spatfall in cockles is unpredictable and when the total stock drops below average it might be advisable to limit fishery in order to keep some reserves for the next years, taking the risk that they may disappear by natural causes.

As cockles are competitors for food with mussels, also for the mussel growers it is favourable if the cockle stock remains small. Because, in general, there is no stock-recruitment relationship, also from this point of view there is no reason to preserve large stocks of old cockles. Only in 1991, with an extremely low cockle stock in the Wadden Sea, there were indica-tions that settlement was low because of the low parent stock.

Dense beds of adult cockles for birds are important in sustainable management. In the Dutch Wadden Sea the birds required 14,000 tons of cockle meat occurring in beds of a density of at least 50 m^{-2}. Fig. 2 makes clear that this amount was only available in eleven years between 1971 and 1992. Therefore the yield for the fishermen in the other eleven years should be lower than the demand from the market.

Mussel culture and fishery

The majority of the mussels are produced by the mussel culture. Therefore, the sustainable yield of this culture must be considered. The yield is determined by a number of factors;

- the availability of seed mussels;
- the susceptibility for storms;
- the primary production in the estuary.

Since the late 1950s the quantity of seed mussels in the Wadden Sea was never a problem except in 1989, 1990 and 1991. As long as the seed mussels were available from subtidal beds, there was no conflict with aims of nature conservation. In general, enough subtidal mussel beds are available. When seed mussels are in low supply, careful fishing and seeding could improve mussel survival. If the seed-yield ratio (see Table 2) is only slightly improved, availability of seed mussels would not normally limit yields.

The large increase in cultures has established lots in places which are very susceptible to storms. The culture of marketable mussels takes at least 1.5 years after seeding and there is a large risk that a storm will occur. Wave action or high current speeds, flush mussels from the culture lots. In general they do not survive long in the gullies. When these mussels finish up in the intertidal areas outside the gullies, they may survive longer and act as anchor places for new spatfall. In the Dutch Wadden Sea storms have been the most important determinant factor for the mussel yield. From that point of view the culture lots should be concentrated in areas which lie sheltered from storms and where current speeds during storms remain below a certain threshold.

It has become clear that phytoplankton availability in an estuary can determine the yield and quality of mussels. It has been estimated that on average the western Dutch Wadden Sea contains 294 x 10^6 kg of mussels (fresh weight). This population would pump 920 x 10^6 m^3 of water each day. The western Wadden Sea contains 4,500 x 10^6 m^3 of water at low tide (Dankers et al. 1989). The ratio between the amount of water pumped and the total volume is such that the mussels have the potential to remove a considerable portion of the phytoplankton biomass. A large part of this is transformed into body tissue or respired, and a part is laid down with the faeces and pseudo-faeces. There are indications that the quality of mussels is limited by the availability of food. After the summer, when phytoplankton decreases, the mussels lose considerable weight (Fig. 4). This indicates food shortage.

In 1989 when the mussel biomass was extremely high due to good survival of the 1987 spatfall, the weight loss did not occur. In that year a large autumn bloom of diatoms occurred (G.C. Cadée, pers. comm.). Whether this autumn bloom was a result of the release of nutrients from mussel beds, when temperature and light were still sufficient for primary production, needs further investigation. Food limitation for bivalves has also been cited by Rodhouse et al. (1984), Peterson and Black (1987), Korringa (1956), Heral et al. (1986), and Stralen and Dijkema (1993). When studying food availability other suspension feeders should be taken into account. In the Eastern Scheldt where all trophic levels have been studied in detail, mussel quality (percentage of meat) was strongly correlated with the primary production in a particular year (Smaal and Stralen 1989). On the basis of the ecosystem studies, a mathematical model was developed. This model indicated that the ecosystem behaviour and potential mussel

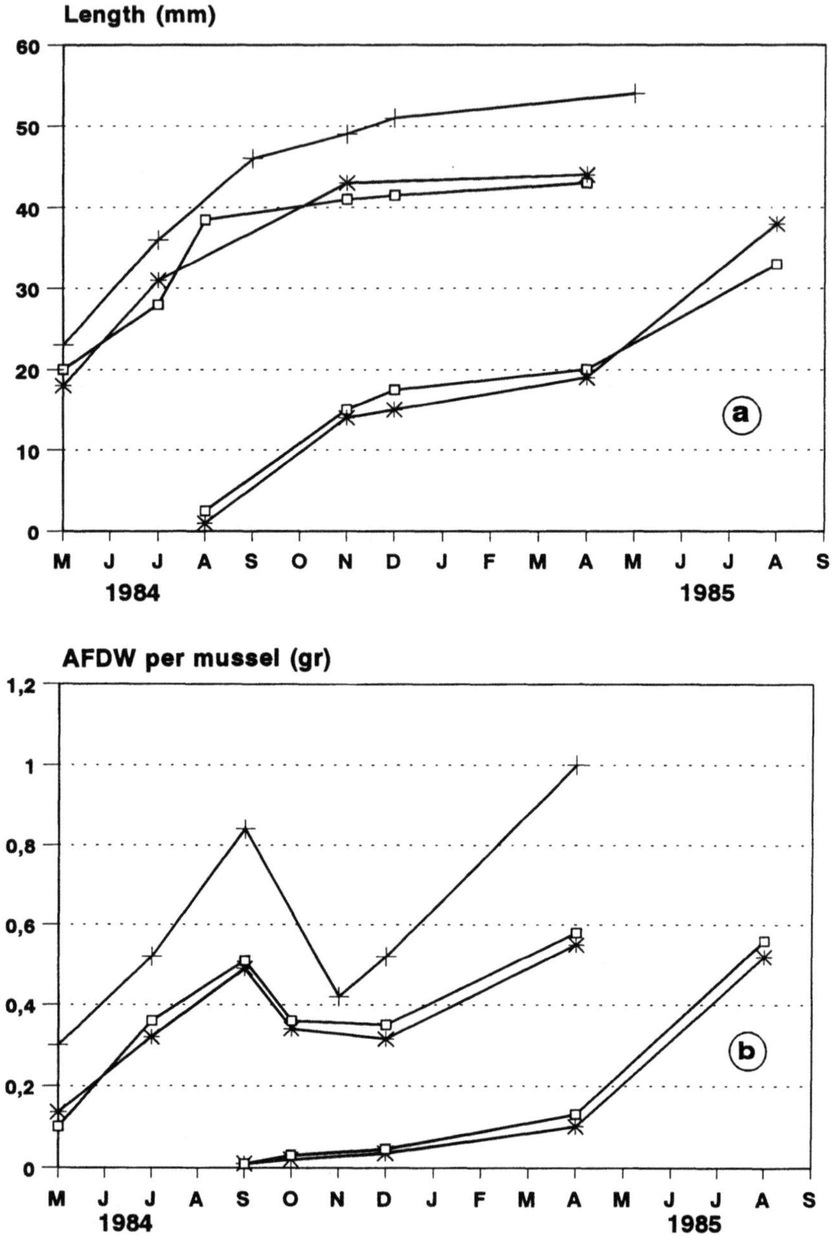

Fig 4. Variations in size (a) and ash-free dry weight (b) of mussels from two culture lots in the Dutch Wadden Sea. The different lines show the results of samples from different spots on the culture lots.

growth is strongly dependent on the biomass of mussels present (Table 3). Carrying capacity for mussel culture has been reached, and cultures should not be expanded (Smaal and Boeije 1991).

Table 3. Results of the SMOES-ecosystem model for the Eastern Scheldt (from Smaal and Boeije 1991). NUT- = reduction of the nutrient load by 50%; MOS- = reduction of the mussel biomass by 50%; MOS+ = doubling of the mussel biomass; COMB = combination of NUT- and MOS+.

	NUT-	MOS-	MOS+	COMB
Primary production	-8	+16	-14	-21
Consumption of zooplankton	-14	+31	-24	-33
Consumption of suspension feeders	-7	-19	+9	+1
Phytoplankton biomass	-4	+35	-20	-25
Zooplankton biomass	-9	+17	-19	-21
Suspension feeders biomass	0	-36	+41	+41
Phytoplankton turnover	-5	-14	+9	+5
Mussel growth	-12	+65	-32	-45

Experiences with other mathematical models (Klepper et al. 1993; Scholten and Tol 1993; Scholten et al. 1990; Tol and Scholten 1992; Vries et al. 1988, 1990) have indicated that these models can be an important tool in the management of coastal ecosystems.

THE INTERTIDAL MUSSEL BED AS A SUPER ORGANISM

As early as 1877, Möbius described an oyster bed in the Wadden Sea as a community and used the term biocoenosis. His characterization of a community is also valid for a mature mussel bed. A mature mussel bed contains mussels of different age classes. Growth speed decreases towards the centre of the bed because of food depletion and shorter submersion time. Incidental mortality causes the occurrence of open spaces giving room for other species. Often the mature bed prevents drainage, thereby creating pools. As mussel beds age and grow they increase not only their biological component, the living mussels, but also their physical component.Mussel beds can often attain considerable thickness and the living and dead shells also increase significantly the surface available for the settlement of associated organisms. A typical mussel bed community consists of three major components - a physical matrix of living and dead mussel shells interconnected by a complex mat of byssal threads, a bottom layer of accumulated sediments, mussel faeces and pseudofaeces, organic detritus and shell debris which accumulate in direct proportion to bed thickness and a taxonomically diverse assemblage of associated organisms (Seed and Suchanek 1992). Within the interstices of the mussel matrix, light, temperature and wave action are significantly reduced thus affording greater protection to those species with more restricted tolerance to these environmental parameters.

As structurally complex entities, mussel beds provide refuge and suitable habitat for a wide diversity of associated organisms. Such communities have been documented for several mytilids e.g. *Mytilus edulis* (Tsuchiya and Nishihira 1985, 1986; Asmus 1987), *M. galloprovincialis* (Tsuchiya and Ballan Santini 1989), *M. californianus* (Suchanek 1980, 1985; Seed and Suchanek 1992), *Septifer* spp. (Jackson 1976; Ong Che and Morton 1992), *Perumytilus purpuratus* (Ramirez 1965), *Brachidontes* spp. (Penchaszadeh 1973; Stephenson and Stephenson 1972) and *Modiolus modiolus* (Brown and Seed 1977; Comley 1978; Witman 1985).

The associated biota can be divided into three major functional categories albeit with some overlap between groups. Epibiotic organisms are sessile forms which grow on or even bore into the mussel shells themselves and include taxa such as algae, barnacles, bryzozoans and cnidarians. Mobile fauna move freely throughout the matrix and are typically represented by various small crustaceans and free-roving gastropods. Infaunal taxa are restricted to, and often dependent upon, the organic rich sediments or accumulated shell debris, and comprise those organisms that are typical of soft sediment environments such as polychaetes, sipinculids and ophiuroids. The specialisation that derives from the microhabitat differences encountered by the associated biota can result in a predictable vertical stratifi-cation throughout the matrix. The specific nature of preferred site selection within a mussel matrix may even occur on the scale of individual mussel shells (e.g. Laihonen and Furman 1986).

Both species richness and diversity are positively correlated with age and structural complexity of the mussel matrix (Tsuchiya and Nishihira 1985, 1986). Briggs (1982) recorded at least 34 species associated with *M. edulis* beds in northern Ireland, nearly half of which were crustaceans. Asmus (1987) found 41 species associated with a bed in the Wadden Sea and Dittmann (1990) recorded 96 species. In Japan Tsuchiya and Nishihira (1985, 1986) found 69 species within the *M. edulis* community.

By means of a series of field experiments Suchanek (1980) studied the colonisation of the associated biota on artificial mussels of varying degrees of structural complexity. The physical structure of the mussel matrix was the most important factor promoting biological diversity.

It has been shown (Dankers et al. 1989; Dame and Dankers 1988; Dame 1993) that the community metabolism of a mussel bed is significantly different from the metabolism calculated on the basis of individual mussels. Mussel beds are highly productive assemblages often rivalling the productivity of kelp beds and tropical rain forests (Leigh et al. 1987).

Like other congregates of living organisms that influence each other and create new habitats for other species or increase survival chances for themselves, such as oyster beds, coral reefs, and sublittoral seagrass beds, these biocoenosis are formed and maintained by biotic processes. Of course their survival depends, for a large part, on non-biological processes as is the case for all organisms. Individual mussels can be regarded as r-strategists. This makes them an interesting species for fishermen because in general fishing mortality does not influence the amount of spatfall in the next year. The fishermen's rule 'catch them before they die' is valid, even for the subtidal short living beds and beds in storm-sensitive places on the intertidal flats in a nature reserve like the

Wadden Sea. However, a mature mussel bed should be regarded as 'super-organism' of the K-strategy type. These beds persist for long periods because damage is easily repaired by the r-strategy of individual year classes (McGrorty et al. 1990). When mature beds disappear from an area it takes a long time before new stable beds re-develop.

Population dynamics of a mussel bed

As in general population dynamics the most important parameters for mussel beds are: 'birth rate', survival to maturity, and mortality. There are a number of factors important for a successful settlement. First there has to be sufficient supply of larvae and in most years this is not a problem. Sometimes, such as in May 1991, less than two larvae per litre were present in the Dutch Wadden Sea. It is not clear whether this was due to the extremely low stock of adult mussels that year. In May 1981, when there was a large stock of adult mussels more than 60 larvae were counted in each litre of seawater (C. de Vooys, pers. comm.). Mussel larvae need a good place to settle, and after settlement it is important that they have sufficient food and low predation pressure. When food is abundant young mussels grow fast and the larger specimens are less prone to predation by invertebrates. These factors, which are necessary for the 'birth' of a mussel bed, may not act simultaneously in each year and there are years without sufficient spatfall.

When a mussel bed has developed the mussels produce large quantities of pseudo-faeces. If these accumulate too fast the bed develops a high profile and becomes very susceptible to storm or ice damage. These beds do not often reach maturity. When a bed develops very slowly, for example, if there is high predation pressure which supplies the bed with empty shells and consecutive years of moderate spatfall stable, mature mussel beds may develop. These beds, although not containing the same individual mussels, may survive for decades. The 'stability' of the bed population is caused by the larvae that settle in open spaces and have high survival within the clumps of old mussels. Too many juveniles results in high mortality (McGrorty et al.

1990). Therefore one can conclude that the bed is in some way self-regulating its survival. Even when damaged by ice, the bed foundations are often left for new settlement. In more than 25 years of research on the Balgzand tidal flats, J.J. Beukema (pers. comm.) has only observed the development of a new mature mussel bed once. Also McGrorty and Goss-Custard (1991) do not mention the development of new mature mussel beds in their seven-year study on the population dynamics of mussel beds in the Exe-estuary.

It is clear that development of a mature mussel bed is a rare event. As in other organisms with low reproduction capacity it is of extreme importance to keep mortality low. Croxall et al. (1990) showed that a small increase in fisheries induced mortality in the wandering albatross, had a severe impact on the population size. Also in marine mammals with a low reproduction rate it has been shown that populations cannot sustain even moderate levels (only 4% of the population per year) of incidental mortality (Woodley and Read 1991). If a mature mussel bed is considered as a K-species, this implies that even moderate fisheries or damage resulting in less bed stability will cause the disappearance of these beds. If the less stable newly recruited beds are fished away, the possibility for formation of new mature beds is restricted.

WISE USE OF AN ESTUARINE ECOSYSTEM

Aims and goals of nature conservation

According to a report on wise use of the Wadden Sea which was supported by 59 NGOs (WWF 1991) the guiding principle for all Wadden Sea policy must be "to achieve a natural and sustainable ecosystem in which natural processes proceed in an undisturbed way, and in which for that reason the full variety of plant and animal species which should be in the Wadden Sea, encounters suitable living conditions." This is in line with the Dutch approach towards nature management which states that nature management in general should give as large as possible contribution to the conservation of the total diversity of natural processes, structures, biotic communities and species of plants and animals, both locally and in other parts of the world. The Dutch Wadden Sea memorandum states that there should be sustainable protection and, within the limits set by the aims of the conservation, human activities of an economical or recreational character should be possible. A similar view "nature will be protected and used at the same time" (Agger 1992) has been adopted in Denmark. The same is true in Germany according to articles 1 and 2 of the National Park Law (Andresen 1992; Helbing 1992). According to the German conservation societies, a zoning system should be introduced as a compromise between necessary conservation and traditional uses of the Wadden Sea (Bachert and Wesemüller 1992). The general view towards nature conservation in an area like the Wadden Sea is that natural processes which have shaped and maintain the system, should be allowed to act as much as possible undisturbed.

Aims and goals of fisheries

Fishermen are also interested in sustainable management, however, they emphasize a sustainable catch of their target organisms. They want a stable supply based on catches in good years. Often the aims and goals of the fishermen run parallel with the conservationists. Problems arise when target organisms occur in low quantities and if stock-recruitment relations do not exist. In that case the fishermen will continue to harvest their target organisms and may cause disruption of the system. Fishermen also aim for specific size classes. In the cockle fishery large (3 years old) cockles are favoured, while for the mussel culture, young (1-2 years old) seed beds are selected over mature beds with many empty shells. It is only when expectations for sustainable harvest are based on levels that do not account for normal ecosystem variance of stocks, that problems become severe.

Solutions

Several aspects should be taken into account when designing a management plan aiming at wise use. The carrying capacity of the system should be studied. Mathematical models have proved to be useful for this. It is important to study the carrying capacity for target organisms but it is also essential to examine the effect of competition for food or space on other species. It is also important to consider natural 'biocoenosis' such as mature mussel beds. Sustainable fishery on these beds has not been possible. On the other hand, fishery on subtidal beds has proven to be sustainable. In order to maintain the mussel production the subtidal culture lots should be spread more evenly over the area in order to avoid local food competition and to spread the risk during storms. Care should be taken not to situate culture lots in gullies which have high current speeds during storms. In the initial stages of the culture care should be taken to prevent high mortality of the seed mussels. Because of the importance of intertidal mussel beds from a conservation point of view, these beds should not be used for seed fishery nor for fishing on mussels of consumption size. There will always be conflicts between cockle fishery and nature conservation. In years with an average amount of cockles, the present fisheries will leave enough cockles for the birds, but in some years there will be 'food competition' between birds and fishermen. As fisheries aim for a quantity the market can take, this means that in years with few cockles problems will arise. In the case of the cockle fishery the best solution would be to close the majority of the intertidal flats on a permanent basis. When cockles are in short supply the fishing target cannot be reached. The fishing industry should be organised in such way that they can survive several years with a low catch.

From experience it is clear that rules should be set by agencies dealing with sustainable use based on nature conservation. When fishery agencies set the rules there will be no problems in years when stocks are abundant, however, when the stock is low there will be overexploitation. Fishery agencies typically only consider the sustainability of the target species but not the other aspects of the ecosystem.

ACKNOWLEDGEMENTS

I wish to thank the participants of the NATO workshop for the stimulating discussions. Earlier discussions with Jaap de Vlas helped me to develop the ideas worked out in this chapter. Ray Seed provided information and parts of the text for description of mature mussel beds. I am indebted to John Kraeuter who refereed this chapter. His great effort improved the readability greatly. The critical approach of Michaela Scholl must be acknowledged. She was responsible for the metamorphosis of a draft manuscript into a camera-ready copy including figures and literature.

REFERENCES

Agger P (1992) Aims and Goals for the future management of the Danish Wadden Sea. In: Proceedings of the 7th International Wadden Sea Symposium of the 7th International Wadden Sea Symposium, Ameland, The Netherlands 22-26 October 1990 (eds) Dankers N Smit CJ Scholl M pp 69-72 Neth J Sea Res Publ Ser No 20

Andresen FH (1992) Future management of the Schleswig-Holstein National Park In: Proceedings of the 7th International Wadden Sea Symposium of the 7th International Wadden Sea Symposium, Ameland, The Netherlands 22-26 October 1990 (eds) Dankers N Smit CJ Scholl M pp 73-75 Neth J Sea Res Publ Ser No 20

Asmus H (1987) Secondary production of an intertidal mussel bed community related to its storage and turnover compartments. Mar Ecol Progr Ser 39: 251-266

Asmus RM Asmus H (1991) Mussel beds: limiting or promoting phytoplankton? J Exp Mar Biol Ecol 148: 215-232

Bachert S Wesemüller H (1992) The future of conservation in the German Wadden Sea - policy and management proposals from NGOs In: Proceedings of the 7th International Wadden Sea Symposium of the 7th International Wadden Sea Symposium, Ameland, The Netherlands 22-26 October 1990 (eds) Dankers N Smit CJ Scholl M pp 77-78 Neth J Sea Res Publ Ser No 20

Beukema JJ (1976) Biomass and species richness of the macro-benthic animals living on the tidal flats of the Dutch Wadden. Sea Neth J Sea Res 10: 236-261

Beukema JJ (1982) Annual variation in reproductive success and

biomass of the major macrozoobenthic species living in a tidal flat area of the Wadden Sea. Neth J Sea Res 16: 37-45
Blok de JW Geelen HJFM (1958) The substratum required for the settling of mussels (*Mytilus edulis* L) Extrait des archives Neerl Zool Tome XIII I Suppl 1958: 446-460
Böhme B (1988) Auswirkungen der Herzmuschelfischerei auf die Bodenfauna des Wattenmeeres. Nationalparkverwaltung Niedersächsisches Wattenmeer Wilhelmshaven
Briggs RP (1982) Community structure and growth of *Mytilus edulis* L. in Lough Foyle. Proc R Ir Acad 82: 245-259
Brown RA Seed R (1977) *Modiolus modiolus* (L.) - an autecological study In: Biology of benthic organisms (eds) Keegan BF O'Ceidigh P Boaden PJS pp 93-100 Pergamon Press Oxford
Cadée GC (1993) Eider, shellduck and other predators, the main producers of shell fragments in the Wadden Sea. Paleontological implications. Paleontology (in press)
Comley CA (1978) *Modiolus modiolus* (L.) from the Scottish west coast. I Biology Ophelia 17: 167-193
Coosen J Smaal SC (1985) Jaargemiddelde biomassa en activiteit van de dominante bodemdieren in de Oosterschelde Balansrapport 1985-12 Deltadienst Middelburg
Croxall JP Rothery P Pichering SPC Prince PA (1990) Reproductive performance, recruitment and survival of wandering albatrosses *Diomedea exulans* at Bird Island, South George. J Anim Ecol 59: 775-796
CWSS (Common Wadden Sea Secretariat) (1992) Mussel fishery in the Wadden Sea Working Document 1991-2, revised version 1992 Wilhelmshaven
Dame R (1993) The role of bivalve filter feeding material fluxes in estuarine ecosystems. (this volume)
Dame R Dankers N (1988) Uptake and release of materials by a Wadden Sea mussel bed. J Exp Mar Biol Ecol 118: 207-216
Dame RF Zingmark R Stevenson H Nelson D (1980) Filter feeding coupling between the estuarine water colomn and benthic subsystems In: Estuarine perspectives (ed) Kennedy VS pp 521-526 Academic Press New York
Dame R Dankers N Prins T Jongsma H Smaal A (1991) The influence of mussel beds on nutrients in the western Wadden Sea and Eastern Scheldt estuaries. Estuaries 14: 130-138
Dankers N Koelemaij K (1989) Variations in the mussel population of the Dutch Wadden Sea in relation to monitoring. Helgoländer Meeresunters 43: 529-535
Dankers N Vlas de J (1992) Multifunctioneel beheer in de Waddenzee - integratie van natuurbeheer en schelpdiervisserij. RIN-rapport 92/15 DLO-instituut voor Bos- en Natuurbeheer Texel
Dankers N Dame R Kersting K (1989) The oxygen consumption of mussel beds in the Dutch Wadden. Sea Scient Mar 53: 473-476
Dankers N Koelemaij K Zegers J (1989) De rol van de mossel en de mosselcultuur in het ecosysteem van de Waddenzee. RIN-rapport 89/9 Rijksinstituut voor Natuurbeheer Texel
Dankers N Smit CJ Scholl M (eds) (1992) Proceedings of the 7th International Wadden Sea Symposium Ameland The Netherlands 22-26 October 1990 Neth Inst Sea Res Publ Ser 20

Dare PJ (1976) Settlement growth and production of the mussel *Mytilus edulis* L in Morecambe Bay. England Fishery Invest Lond (Ser 2) 28 (1): 1-25
Dekker R (1989) The macro-benthos of the subtidal western Dutch Wadden Sea: 1 Biomass and species richness. Neth J Sea Res 23: 57-68
Dittmann S (1990) Mussel beds - amensalism or amelioration for intertidal fauna. Helgoländer Meeresunters 44: 335-352
Drinkwaard B (1987) Musselculture a natural culture In: Proc 5th Int Wadden Sea Symp, Esbjerg, Denmark, 29 Sept-3 Oct 1986 (eds) Tougaard S Asbirk S pp 185-193 The National Forest and Nature Agency and The Museum of Fishery and Shipping Esbjerg
Groot de RS (1992) Functions of nature; evaluation of nature in environmental planning, management and decision-making. Wolters-Noordhoff Groningen
Hancock DA (1973) The relationship between stock and recruitment in exploited invertebrates. In: Fish stocks and recruitment (ed) Parrish BB pp 113-131 Rapp P-v Réun Cons perm int Explor Mar 164
Helbing C (1992) Future management in the Niedersachsen part of the Wadden Sea In: Proceedings of the 7th International Wadden Sea Symposium of the 7th International Wadden Sea Symposium, Ameland, The Netherlands, 22-26 October 1990 (eds) Dankers N Smit CJ Scholl M pp 83-86 Neth J Sea Res Publ Ser No 20
Heral M Deslous Paoli JM Prou J (1986) Dynamiques de production et des biomasses des huîtres creuses cultivées (*Crassostrea angulata* et *Crassostrea gigas*) dans le bassin de Marennes Oléron depuis un siecle. ICES CM 1986/F: 41 Mariculture Committee
Hulscher JB (1980) Oystercatcher (*Haematopus ostralegus* L) In: Birds of the Wadden Sea. (eds) Smit CJ Wolff WJ Balkema Rotterdam
Jackson LF (1976) Aspects of the intertidal ecology of the east coast of South Africa. S Afr Ass Mar Biol Res Invest Rep 46: 72 pp
Jonge de VN Jong de DJ (1992) Role of tide, light and fisheries in the decline of Zostera marina L. in the Dutch Wadden Sea In: Proceedings of the 7th International Wadden Sea Symposium Ameland The Netherlands 22-26 October 1990 (eds) Dankers N Smit CJ Scholl M pp 161-176 Neth Inst Sea Res Publ Ser 20: 161-176
Kesteloo JJ Stralen van MR (1992) Het kokkelbestand in de Oosterschelde, de Westerschelde en de Waddenzee in 1992. RIVO-rapport AQ 92-06 Rijksinstituut voor Visserijonderzoek IJmuiden
Klepper O Tol van der MWM Scholten H Haman PMJ (1993) SMOES: a simulation model for the Oosterschelde ecosystem. Part I: description and uncertainty analysis. In: The Oosterschelde estuary (The Netherlands): case study of a changing ecosystem. (eds) Nienhuis P Smaal RC Kluwer (in press)
Korringa V (1956) Oyster culture and biological productivity. Rapp P-v Réun Cons perm int Explor Mar 140: 30/31, 59-61
Kuenen DJ (1942) On the distribution of mussels on the intertidal sand flats near Den Helder. Arch Neerl Zool 6

(2/3): 117-160
Laihonen P Furman ER (1986) The site of settlement indicates commensalism between blue mussel and its epibiont Oecologia 71: 38-40
Lasiewski RC Dawson WR (1976) A re-examination of the relation between standard metabolic rate and the body weight in birds. Condor 69: 13-23
Laursen K Asberg KS Frikke J (1992) The impact of mussel fishing in the Danish Wadden Sea on numbers, distribution diet and weight of eiders (*Somateria mollissima*). National Environmental Institute Denmark
Leigh EG Paine RT Quinn JF Suchanek TH (1987) Wave energy and intertidal productivity. Proc Nat Acad Sci (USA) 84: 1314-1318
Maas-Geesteranus RA (1942) On the formation of banks by *Mytilus edulis* L Arch Neerl Zool 6 (2/3): 283-326
McGrorty S Goss-Custard JD (1991) Population dynamics of the mussel *Mytlius edulis:* spatial variations in age-class densities of an intertidal estuarine population along environmental gradients. Mar Ecol Progr Ser 73: 191-202
McGrorty S Clarke RT Reading CJ Goss-Custard JD (1990) Population dynamics of the mussel *Mytilus edulis*: density changes and regulation of the population in the Exe estuary Devon Mar Ecol Progr Ser 67: 157-169
Miljoministeriet 1986. Naturreservat Vadehavet, Bekendtgarelse nr. 328, af 15. Juli 1985 og Landbrugsministeriets bekendtgarelse af 5. Juli 1984 op 21 August 1984.
Möbius K (1877) Die Auster und die Austernwirtschaft. Wiegund Hempel and Parey Berlin
Obert B Michaelis H (1991) History and ecology of the mussel beds (*Mytilus edulis* L.) in the catchment area of a Wadden Sea tidal inlet. In: Estuaries and coasts: spatial and temporal intercomparisons (eds) Elliot M Ducrotoy JP pp 185-194 Olsen and Olsen Fredensburg Denmark
Ong Che RG Morton B (1992) Structure and seasonal variations in abundance of the macro-invertebrate community associated with Septifer virgatus (Bivalvia: Mytilidae) at Cape d'Aguilar, Hong Kong. Asian Mar Biol 9: (in press)
Penchaszadeh PE (1973) Ecologia de la comunidad del mejillin (*Brachidontes rodriguezi* d'Orb) en la mediolitoral rocoso de Mar del Plata (Argentina): el proceso de recolonization. Physis (Buenos Aires) 32: 51-64
Peterson CH Black R (1987) Resource depletion by active suspension feeders on tidal flats: Influence of local density and tidal evaluation. Limnol Oceanogr 32 (1): 143-166
Prins TC Smaal AC (1990) Benthic-pelagic coupling: The release of inorganic nutrients by an intertidal bed of *Mytilus edulis* In: Trophic relationships in the marine environment (eds) Barnes M Gibson RM pp 89-103 Proc 24th Eur Mar Bio Symp Aberdeen Univ Press
Rijksinstituut voor Natuurbeheer (1987) Effecten van de kokkelvisserij in de Waddenzee. RIN-rapport 87/18
Rodhouse PG Roden CM Burnell GM Hensey MP McMahon T Ottway B Ryan TH (1984) Food resource, gametogenesis and growth of *Mytilus edulis* on the shore and in suspended culture:

Killary Harbour, Ireland. J Mar Biol Ass UK 64: 513-529
Scholten H Tol van der MWM (1993) Responses of the food web of the Oosterschelde to a changing environment. Part II: calibration and validation. Neth J Sea Res 30 (1) (in press)
Scholten H Klepper O Nienhuis PH Knoester M (1990) Oosterschelde estuary (SW-Netherlands): a self-sustaining ecosystem? Hydrobiologia 195: 201-215
Seed R (1993) Invertebrate predators and their rol in strcturing coastal and estuarine bivalve populations (this volume)
Seed R Suchanek TH (1992) Populationa and community ecology of *Mytilus:* In: The mussel *Mytilus:* Ecology, physiology, genetics and culture (ed) Gosling EM pp 87-169 Elsevier Press Amsterdam
Smaal AC Boeije RC (1991) Veilig getij, de effecten van de waterbouwkundige werken op het getijdemilieu van de Oosterschelde. Nota GWWS 91088 DGW/directie Zeeland Middelburg
Smaal AC Nienhuis PH (1992) The Eastern Scheldt (The Netherlands) from an estuary to a tidal bay. A review of responses at the ecosystem level. Neth J Sea Res 30 (1), (in press)
Smaal AC Prins TC (1993) The uptake of organic matter and the release of inorganic nutrients by bivalve suspension feeder beds (this volume)
Smaal AC Stralen van MR (1989) Average annual growth and condition of mussels as a function of food supply. Hydrobiologia 195: 179-188
Smaal AC Verhagen JHK Haas HA Coosen J (1986) Interactions between seston quantity and quality and benthic suspension feeders in the Oosterschelde (The Netherlands) Ophelia 26: 385-399
Smit CJ (1980) Production of biomass by invertebrates and consumption by birds in the Dutch Wadden Sea In: Birds of the Wadden Sea (eds) Smit CJ Wolff WJ pp 290-301 Balkema Rotterdam
Straaten van LMJU (1965) De bodem der Waddenzee. Het Waddenboek Thieme Zutphen
Stralen van M Kesteloo J (1992) Het kokkelbestand en de broedval van kokkels in de Oosterschelde en de Waddenzee in 1991. RIVO-rapport AQ 92-05 Rijksinstituut voor Visserijonderzoek IJmuiden
Stralen van MR Dijkema R (1993) Mussel culture in a changing environment: the effects of a coastal engineering project on mussel culture in the Oosterschelde In: The Oosterschelde estuary (The Netherlands): case study of a changing ecosystem. (eds) Nienhuis P Smaal RC Kluwer (in press)
Suchanek TH (1980) Diversity in natural and artificial mussel bed communities of Mytilus californianus. Am Zool 20: 807
Suchanek TH (1985) Mussels and their role in structuring rocky shore communities In: The ecology of rocky coasts (eds) Moore PG Seed R pp 70-96 Hodder and Stoughton Sevenoaks
Swennen C (1991) Ecology and population dynamics of the common eider in the Dutch Wadden Sea. Thesis Rijksuniversiteit Groningen

Tol van der MWM Scholten H (1992) Responses of the food web of the Oosterschelde to a changing environment. Part I: adaptation or functional. Neth J sea Res 30 (1), (in press)
Tsuchyia M Nishihira M (1985) Islands of *Mytilus* as habitat for small intertidal animals: effect of island size on community structure. Mar Ecol Prog Ser 25: 71-81
Tsuchyia M Nishihira M (1986) Island of *Mytilus* as habitat for small intertidal animals: effects of *Mytilus* age structure on the species composition of the associated fauna and community organisation. Mar Ecol Prog Ser 31: 171-178
Tsuchyia M Bellan-Santini D (1989) Vertical distribution of shallow rocky shore organisms and community structure of mussel beds (*Mytilus galloprovincialis*) along the coast of Marseille, France. Mesogée 49: 91-110
Verwey J (1952) On the ecology and dsitribution of cockle and mussel in the Dutch Wadden Sea. Their role in sedimentation and the source of their food supply With a short review of the feeding behaviour of bivalve molluscs. Arch Neerl Zool 10: 171-239
Vlas de J (1982) De effecten van de kokkelvisserij op de bodemfauna van de Waddenzee en Oosterschelde. RIN-rapport 82/19 Rijksinstituut voor Natuurbeheer Texel
Vlas de J (1987) Effects of cockle fisheries on the macrobenthos in the Wadden Sea. Fishery Museum Esbjerg Biologiske Meddelser 31: 215-228
Vries de I Vries de M Hopstaken F (1990) Development and application of VEERWAQ for a management analysis of Lake Veere, WL, DGW dir. Zeeland (in Dutch)
Vries de I Hopstaken F Goossens H Vries de M Vries de H Heringa J (1988) GREWAQ: an ecological model for Lake Grevelingen. Documentation report T 0215-03 WL, DGW Delft
Witman JD (1985) Refuges, biological disturbance and rocky subtidal community structure in New England. Ecol Monogr 55: 421-445
Wolff WJ (1983) Estuarine benthos In: Ecosystems of the world Vol 26 Estuaries and enclosed seas (ed) Ketchum BH pp 151-182 Elsevier Amsterdam
Woodley TH Read AJ (1991) Potential rates of increase of a harbour porpoise (*Phocoena phocoena*) population subjected to incidental mortality in commercial fisheries. Can J Fish Aquat Sci 48: 2429-2435
WWF (1991) The common future of the Wadden Sea. WWF Husum
Zwarts L (1991) Mosselbanken: wadvogels op een kluitje. Vogels Jan/Feb 1991: 8-12

PRELIMINARY OBSERVATIONS ON THE HYDRODYNAMICS OF FILTER FEEDING IN ZEBRA MUSSELS

Josef Daniel Ackerman
Mechanical Engineering
University of Toronto
Toronto, Ontario
CANADA M5S 1A4

Zebra mussels (*Dreissena* spp.) have rapidly colonized North American freshwater ecosystems causing severe economic and ecological impacts due to their epibyssate nature and life history (Nalepa and Schloesser 1993). While these traits are more similar to marine mussels than to freshwater ones, zebra mussels exist in an environment where energy inputs are reduced relative to marine systems (Nixon 1988). An examination of zebra mussel filter feeding within these environments would be useful to assess their impact on the seston.

To examine the hydrodynamics of zebra mussel filter feeding, a rock (8x20x7 cm) from Lake Erie was examined in a flume (31x38 cm x 10 m). The mussels (1-20 mm in length) covered the rock at a density of $\approx 2.3 \cdot 10^4$ m^{-2}. Flow visualizations, using fluorescein-labeled *Chlorella*, were conducted under no flow, slow flow ($\approx$ 2 $cm \cdot s^{-1}$), and fast flow ($\approx$ 5 $cm \cdot s^{-1}$) conditions.

In the no flow condition, the "marker" remained distinct above a height of 4-5 cm, below which, plumes of dilute marker were ejected from exhalant siphons in a spiral-like fashion (arrow in Fig. 1A). Under both slow and fast flows the marker was transported downstream, except close to the rock where it remained between the mussels (Fig. 1B). The exhalant plumes were less distinct, and were obscured in the mean flow above the mussels (Fig. 1B).

These results indicate that zebra mussel filter feeding affects the surrounding fluid on the order of centimeters (Monismith *et al.* 1990). The ability of individuals to filter large volumes of water (Reeders *et al.* 1989) has led to estimates of seston

NATO ASI Series, Vol. G 33
Bivalve Filter Feeders
in Estuarine and Coastal Ecosystem Processes
Edited by Richard F. Dame

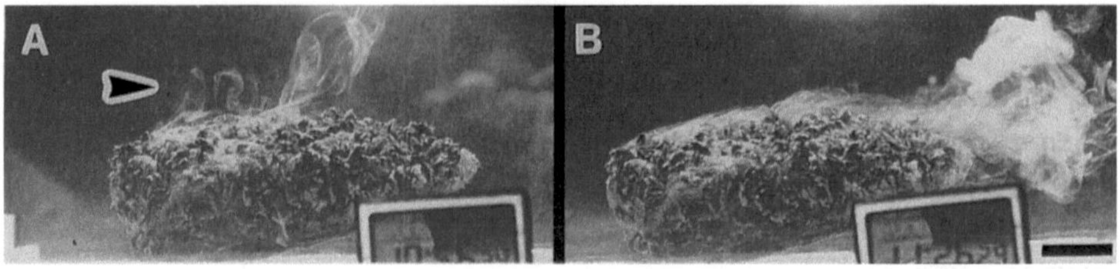

Figure 1. Plumes of dye from zebra mussel exhalant siphons extend several cm above a rock under no flow (arrow in A) and much less under 5 $cm \cdot s^{-1}$ flow (B; flow from left). Bar = 4 cm.

utilization that are unrealistic because they fail to account for the reduced mixing and mass transfer within these systems. Under certain conditions (slow flow, depth, etc.) the hydrodynamic forces may be insufficient to replenish the local depletion of seston, dissolved gases, etc. (Wildish and Kristmanson 1984; Fréchette *et al.* 1989). These mussels may be limited by transport processes. However, mussels located in well mixed areas (surface waters, intakes, etc.) will benefit from these transport processes. Future laboratory and field studies are planned to confirm these observations.

I would like to thank Ernesto Morala, Anthony Corriveau and the Ontario Ministries of the Environment and Natural Resources.

REFERENCES

Fréchette M Butman CA Geyer WR (1989) The importance of boundary-layer flows in supplying phytoplankton to the benthic suspension feeder, *Mytilus edulis* L. Limnol Oceanogr 34:19-36

Monismith SG Koseff JR Thompson JK O'Riordan CA Nepf HM (1990) A study of model bivalve siphonal currents. Limnol Oceanogr 35:680-696

Nalepa TF Schloesser DW (1993) Zebra Mussels. Lewis Boca Raton

Nixon SW (1988) Physical energy inputs and the comparative ecology of lake and marine ecosystems. Limnol Oceanogr 33:1005-1025

Reeders HH bij de Vaate A Slim FJ (1989) The infiltration rate of *Dreissena polymorpha* (Bivalvia) in three Dutch lakes with reference to biological water quality management. Freshwater Biol 22:133-141

Wildish DJ Kristmanson DD (1984) Importance to mussels of the benthic boundary layer. Can J Fish Aquat Sci 41:1618-1625

PELAGIC - BENTHIC INTERACTIONS IN AN ENCLOSED COASTAL ECOSYSTEM: THE IMPACT OF *MYTILUS EDULIS* POPULATIONS ON WATER QUALITY IN THE SOUTH DOCKS, LIVERPOOL.

Janette R. Allen & Stephen J. Hawkins.
Port Erin Marine Laboratory
University of Liverpool
Isle of Man.

MYTILUS EDULIS POPULATIONS AND WATER QUALITY CONTROL IN THE SOUTH DOCKS

The South Docks, Liverpool are one of many dock systems in the U.K. which, after a period of disuse, have been redeveloped under urban regeneration schemes. The South Docks are an interconnected chain of dock basins, with water depths of 3 to 10m. The water in the Docks is brackish (26 - 28 ^{o}/oo) and rich in nutrients. Water quality problems experienced included dense phytoplankton blooms (sometimes of potentially toxic dinoflagellate species), and periodic oxygen depletion. These problems were an impediment to the development scheme and solutions were required. Various methods of water quality control were examined in the South Docks. These included the introduction of a population of mussels (*Mytilus edulis* L.) to an experimental dock as a biological filter. Improvements in water clarity and benthic oxygen concentrations, due to reduced phytoplankton blooms, were associated with greatly increased mussel densities from both introductions and natural settlement. Reduced phytoplankton biomass occurred despite generally longer hours of bright sunshine, reduced zooplankton populations and similar nutrient concentrations in the years after mussel settlement. The changes in water quality have been described previously (Allen & Hawkins in press, Hawkins *et al* 1992).

BENTHIC - PELAGIC INTERACTIONS AND MAINTENANCE OF THE CLEAR WATER STATE.

The sudden switch from the turbid (high phytoplankton / few filter feeders) to the clear water (low phytoplankton / abundant filter feeders) condition in the South Docks may represent a transfer between two alternative stable states. Several buffering mechanisms exist within each state which will tend to enforce the existence of the alternative states and to resist switches from one to the other. Benthic oxygen concentrations are considered to be an important feedback mechanism in the South Docks. In

NATO ASI Series, Vol. G 33
Bivalve Filter Feeders
in Estuarine and Coastal Ecosystem Processes
Edited by Richard F. Dame

considered to be an important feedback mechanism in the South Docks. In the clear water state the filter feeders reduce algal blooms and associated depletion of benthic oxygen concentrations during microbial decay, ensuring the continued survival of benthic fauna at all depths. In the turbid state low hypolimnetic dissolved oxygen concentrations during bloom die-off inhibits colonisation by benthic filter feeders. Abundant phytoplankton in surface waters may also enhance thermal stratification (Sathyendranath *et al*, 1991) leading to further oxygen depletion. No obvious trends in dissolved nutrient concentrations were seen over the period of monitoring and nutrients were not considered to play a major part in the changes in water quality that were observed. The effects of oxygen saturation and mussel populations on nutrient cycling may influence phytoplanktonic production however, and these interactions should be considered.

Water quality management by manipulation of filter feeding populations aims to push the system from the turbid to the clear water state in the hope that feedback mechanisms will ensure its continuation. Benthic filter feeders will provide a useful tool for water quality management only if a degree of stability of populations can be ensured. Benthic communities and water quality continue to be monitored in the South Docks.

REFERENCES

Allen JR & Hawkins SJ (1993) Can biological filtration be used to improve water quality? In: Urban Waterside Regeneration, problems and prospects. KN White, EG Bellinger, AJ Saul, M Symer & K Hendry (eds) Ellis Horwood series in Environmental Management, Science and Technology, Ellis Horwood Press U.K. pp 377-385

Hawkins SJ, Allen JR, Russell G, White KN, Conlan K, Hendry K & Jones HD (1992) Restoring and managing disused docks in inner city areas. In: Restoring the Nations Marine Environment,.GW Thayer (ed). National Oceanic and Atmospheric Administration, Maryland Sea Grant College publication, Maryland. pp 473-542

Sathyendranath S, Gouveia AD, Shetye SR, Ravindran P, & Platt T (1991) Biological control of surface temperature in the Arabian Sea. Nature 349:54-56

ECOSYSTEM MODELLING: SENSITIVITY OF PHYSICAL CHARACTERISTICS TO SPATIAL BOX DESIGN

C. Bacher
IFREMER
rue de l'Ile d'Yeu
BP 1049
44037 Nantes Cedex 01
France

The residence time of the water in different parts of the Marennes-Oleron Bay is one important characteristic of the biological system (Raillard 1991). For instance, the primary production potentiality of a given area depends on the local turbidity and the residence time of the waters in that area. Further, global assessment of the nitrogen flow between the biological compartments and inflows from the ocean or the Charente river as derived by the Raillard (1991) ecosystem model, showed the predominance of ocean inputs. A new hydrological model of the Marennes-Oleron Bay yielded two dimensional velocity fields for a single mean tidal level (Lazur, unpublished). The computations were the basis for the calculation of the transport of conservative substances throughout the Bay. For this work, the instantaneous values of the bidimensional flows and the volume of each cell were used to estimate the residence time of water inside the Bay. Eularian residual values were computed for the sake of simplicity. The simplification of the equations allowed the formulation of transport processes in a probabilistic fashion and the use of matricial computation methods to estimates the water residence time.

The calculation of the residence time was also solved as before for the single box model (Raillard 1991). The goal of this model was to simulate the growth of the oyster population in subtidal areas in the southern part of the Bay over a two year period. Biological and hydrodynamical processes were coupled with a 15 compartment model. Considering the transport processes

NATO ASI Series, Vol. G 33
Bivalve Filter Feeders
in Estuarine and Coastal Ecosystem Processes
Edited by Richard F. Dame

only, the computation of the transport terms and volumes yielded a transition matrix, the components of which represented the transition probabilities between the compartments. The residence time results were compared to ones obtained from the more precise calculation with the initial grid. The area residence time was 10.3 days. It was, however, highly cell dependent, ranging from 0 for cells close to the boundaries, to more than 30 days elsewhere. Near Oleron Island there was a steep gradient of residence times in a southwestern direction and was thought to be connected to velocity vorticities in the area. There was a similar residence time gradient north of Oleron Island, while the central portion of the bay was relatively homogenous with respect to residence time.

The residence time estimation was plotted against the local values averaged on each compartment and ranged from a few hours to 6 days for boxes 14 and 15 near the north-eastern boundary. When compared to the mean cell values, true differences between different sectors seemed to be smoothed by the box calculations. Thus, in any system with complex bidimensional circulation resulting from gyres, over- and under-estimation of residence time might result. Since many of the compartments concerned with oyster cultivation are near these gyres, inadequate representation of water residence time could result in a poor estimate of biological production.

REFERENCE

Raillard O (1991) Etude des interactions entre les processus physiques et biologiques interventant dans la production de l'huitre *Crassostea gigas* du bassin de Marennes-Oleron: essais de modelisation. These Universite Paris VI.

OYSTER FARMING IMPACT ON THE ENVIRONMENT OF A MEDITERRANEAN LAGOON (THAU) (Preliminary results of the OXYTHAU programme)

DESLOUS-PAOLI JM, MAZOUNI N, SOUCHU P, LANDREIN S, PICHOT P, JUGE C
IFREMER, Laboratoire Ecologie,
1, rue J.Vilar 34200 Sète FRANCE.

The lagoon of Thau is a shallow microtidal system exposed to terrestrial and marine influences. The residence time and the vertical mixing of water-mass are controlled by winds. Periodical absence of wind in Summer can lead to strong anoxia in sediments and bottom waters. Moreover, low inorganic nitrogen and chlorophyll-a levels confer seasonal oligotrophic features to the ecosystem. Nevertheless, a large productive stock of oysters (*Crassostrea gigas*) cultivated on suspended lines and colonized by epibiota (invertebrates, algae, etc...) covers 20 % of the total area. The aim of the present programme (**OXYTHAU 1991-1995**) is to study effects of oyster farming on the environment of the lagoon of Thau. Preliminary results obtained are presented here.

The spatial distribution of both chlorophyll-a and oyster growth are closely linked (fig 1). Since no chlorophyll-a spatial distribution was drawn up before the establishment of oyster farming, it is difficult to assert that the occurence of oyster farming depletes chlorophyll-a distribution. Even though oyster growth is clearly limited by the availability of phytoplancton, it is demonstrated that shell farming continues to act as a sink for phytoplankton (Kautsky 1982; Frechette and Bourget 1985). Moreover, filter feeders size select suspended matter (Bayne and Newell 1983). Therefore, we can expect qualitative shifts in the microbial communities to result from this selection (Baines and Pace 1991).

NATO ASI Series, Vol. G 33
Bivalve Filter Feeders
in Estuarine and Coastal Ecosystem Processes
Edited by Richard F. Dame

Biodeposition can represent up to 75 % of the total matter ingested (Widdows *et al.*1979; Deslous-Paoli *et al.* 1990). In this lagoon, with an increase of the sedimentation rate of 40 10^6 kgDW $year^{-1}$, biodeposition leads to a threefold increase in the amount of organic nitrogen in sediment beneath the lines and a twofold in that of organic carbon (fig 1).

Anoxia conditions produce hydrogen sulphide which diffuses into the water column and is oxidized immediately to sulphate. This oxygen uptake associated to the strong oxygen demand by the culture itself enhances the depletion in the water column (fig 1). Consequently anoxic conditions developing in the water column lead to the periodic mortality of oysters in large number.

This situation raises the problem of sustainable intensive marine farming in coastal ecosystems undergoing dystrophic crisis. The results obtained in the **OXYTHAU** programme will be used to set up a biological modelisation. This modelisation will contribute to the prevention of irreversible deterioration in the ecosystem.

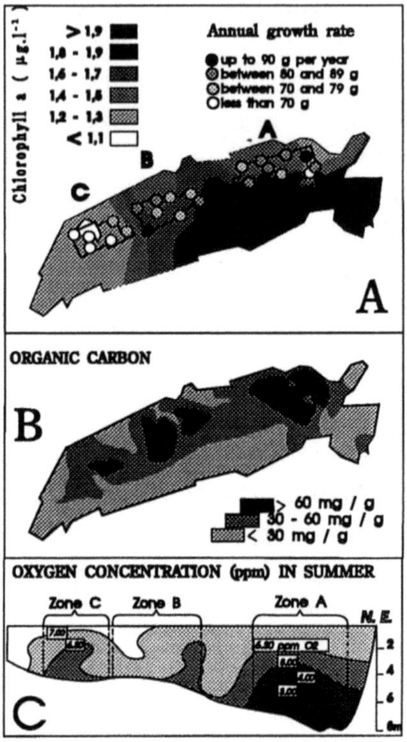

Fig. 1 A- Chlorophyll and rearing distribution and annual oyster growth.
B- Organic carbon in upper sediment.
C- Oxygen distribution in water column under oyster rearing.
For more explanations see text

REFERENCES

Baines SB Pace ML (1991) The production of dissolved organic matter by phytoplanKton and its importance to bacteria: patterns across marine and freswater systems. Limnol Oceanogr 36(6): 1078-1090

Bayne BL Newell RC (1983) Physiological energetics of marine molluscs. In "The Mollusca", Wilburg KM, Saleuddin ASM (eds), Academic press, London 4(1): 407-515

Deslous-Paoli JM Boromthanarat S Héral M Boromthanarat W Razet D (1990) Energy budget of a *Mytilus edulis* L. population during its first year on bouchots in the bay of Marennes-Oléron. Aquaculture 91: 49-63

Frechette M Bourget E (1985) Food limited growth of *Mytilus edulis* L. in relation to the benthic boundary layer. Can J Fish Aquat Sci 42: 1166-1170

Kautsky N (1982) Growth and size structure in a Baltic *Mytilus edulis* population. Mar Biol 68: 117-133

Widdows J Fieth P Worrall CM (1979) Relationships between seston available food and feeding activity in the common mussel *Mytilus edulis*. Mar Biol 50: 195-207

BIODEPOSIT PRODUCTION IN MEDITERRANEAN SHELLFISH FARMS : NON-LINEAR RELATIONSHIPS WITH THE ENVIRONMENT

Christian Grenz
Centre Océanologie Marseille - CNRS UA 41
rue de la batterie des lions - F13007 Marseille

INTRODUCTION

In intensive shellfish cultures, biodeposit production leads to several kinds of problems : on one hand, with regard to the establishment of the carrying capacity of the site, and on the other hand, concerning the organic matter accumulation on the bottom and the consecutive risks of pollution (Haven and Morales-Alamo 1966, Bernard 1974, Kautsky and Evans 1987). Thus, an estimation of the biodeposition rate of mussels (*Mytilus galloprovincialis*) and oysters (*Crassostrea gigas*) was performed during a two year field study using an *in situ* sediment trap design, moored at mean depth (2 m). In parallel to these vertical flux measurements, seston concentrations and current speeds were recorded in order to quantify the upstream horizontal food fluxes.

MATERIELS AND METHODS

Experiments were performed at two different microtidal sites : an estuarine bay (Carteau Bay in the Gulf of Fos) and a semi-enclosed lagoon (Thau lagoon). Details of these sites are given elsewhere (Grenz 1989, Grenz et al. 1991, Deslous-Paoli et al. this vol.). Between 1986 and 1988, at 12 successive one week field experiments, biodeposit production was measured during a 24 hours period, using a funnel (open mouth 60 cm) on top of which nearly 50 mussels or 12 oysters were placed. Traps without shellfish were used as controls. Particulate Suspended Matter (PSM) and Particulate Organic Matter (POM) were sampled (standard methods) and water flow recorded using RCM7 Aanderaa current meters.

NATO ASI Series, Vol. G 33
Bivalve Filter Feeders
in Estuarine and Coastal Ecosystem Processes
Edited by Richard F. Dame

RESULTS AND CONCLUSIONS

Compared to the controls, biodeposit production enhances from 2 to 6 times the sedimentation flux inside the farming zones (25 up to 100 gPOM m^{-2} d^{-1} according to site and season). When considering the horizontal seston flux defined as PSM multiplied by current speed for a given section of a culture structure (i.e. 25 m^2), the relationship between biodeposition and seston flux is "dome-shaped".

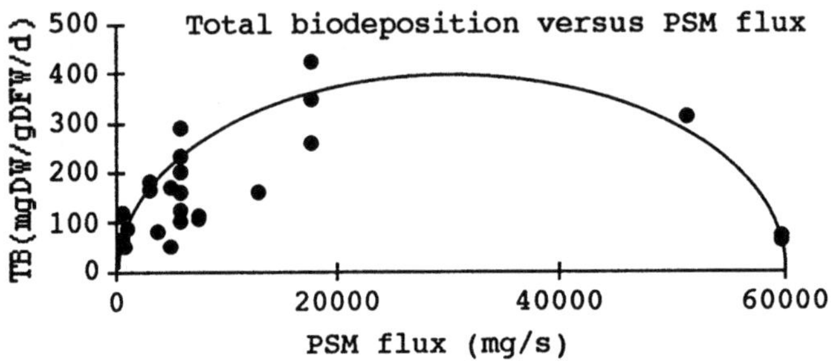

Biodeposit production is non-linearely related to seston flux. The negative effect of PSM flux on total biodeposition after an environmental optimum is probably due to turbidity enhancement in the water column as a result of sediment resuspension events which stress the filtration activity of the filter feeders. The ACE computation was applied to seston flux and to a mean wind speed over 1-4 days before biodeposit flux measurement. Transformation of the biodeposit production produces a positive linear relationship that explains 52% of the variance. Transformation of the seston flux was also "dome-shaped" with an optimum at 30-40 gDW s^{-1} section^{-1}. Considering wind intensity, the transformations describe a negative effect until 3 m s^{-1} and then a beneficiel effect of increasing short term mean wind speed with an optimum at 6 m s^{-1}. This second effect can be the consequence of vertical turbulent diffusion which is replaced by horizontal advective dispersion after an environmental optimum (Millet and Cecchi 1992). Thus, food availability for filter feeders depends on physical variables that interact together with the suspended particules, on filtration activity and consequently on biodeposit production.

REFERENCES

Bernard FR (1974) Annual biodeposition and growth energy budget of mature pacific oysters *Crassostrea gigas*. J Fish Res Bd Can 31:185-190

Grenz C (1989) Quantification et destinée de la biodépostion en zones de production conchylicole méditerranéenne.Thèse Doct Univ Aix-Marseille II

Grenz C Massé H Morchid A Parache A (1991) An estimate of the energy budget between cultivated biomass and the environment around a mussel-park in the northwest Mediterranean Sea. ICES mar Symp 192:63-67

Haven DS Morales-Alamo R (1966) Aspects of biodeposition by oysters and other invertebrate filter feeders. Limnol Oceanogr 11:487-498

Kautsky N Evans S (1987) Role of biodeposition by *Mytilus edulis* in the circulation of matter and nutrients in a baltic coastal ecosystem. Mar Ecol Prog Ser 38:201-212

Millet B Cecchi P (1992) Wind-induced hydrodynamic control of the phytoplankton biomass in a lagoon ecosystem. Limnol Oceanogr 37:190-196

BIOLOGICAL AND PHYSICAL PROCESSES IN MEDITERRANEAN COASTAL ECOSYSTEMS : FIELD MEASUREMENTS AND MODELING.

Christian Grenz
Centre Océanologie Marseille - CNRS UA 41
rue de la batterie des lions - F13007 Marseille

INTRODUCTION

Numerical modeling is more and more considered as a complementary but useful tool in marine ecosystem studies. Since Riley (1965) or Steele (1962), simulation models were developped for many different purposes : eutrophication studies (Jorgensen 1986), microbiotique ecology (Salomon and Pommepuy 1991), North Sea ecology (Fransz et al. 1991) or oyster culture (Bacher 1989). The aim of this study is to describe the application of a vertical microbiological processes model to the Gulf of Fos (France), an estuarine system under freshwater input influence, and to make some conclusions that has been drawn from the model regarding the sources and sinks of nitrogen and organic carbon in the Gulf.

METHODS

Study site

The Gulf of Fos (43.25°N, 4.56°E) situated in southern France, is a semi-enclosed shallow neritic area, influenced by fresh-water supplies from the Rhône river (10 m^3 s^{-1} mean inflow). Strong wind regimes from the Rhône Valley or the open sea, and the flat surroundings generate for the system drastic physical energy inputs. Maximum depth is 14 m at the limit between the gulf and the shelf. The mean depth is 8 m and the approximate area is 45 10^6 m^2.

Model description

As described in Tett (1990) the model embodies (1) the effects of surficial heat exchange, and stirring at the sea surface and bed, on water column stratification and vertical exchange ; (2) the suspension and deposition of inorganic, light absorbing, and organic sediments ; (3) diffusional exchange between the water-column and

NATO ASI Series, Vol. G 33
Bivalve Filter Feeders
in Estuarine and Coastal Ecosystem Processes
Edited by Richard F. Dame

pore water in superficial oxic sediments ; (4) the growth of phytoplankton, and associated microheterotrophs, according to cell-quota, threshold limitation theory, with nitrogen as the limiting nutrient ; and (5) the mineralization of detritical organic material. The model consists of sets of differential equations solved numerically by using simple forward finite difference approximations with timesteps 1 day or less.

RESULTS AND CONCLUSIONS

A standard simulation was run for 1990 for which high frequency chlorophyll a data were available. Simulated and observed phytoplankton biomass were of the same order of magnitude.

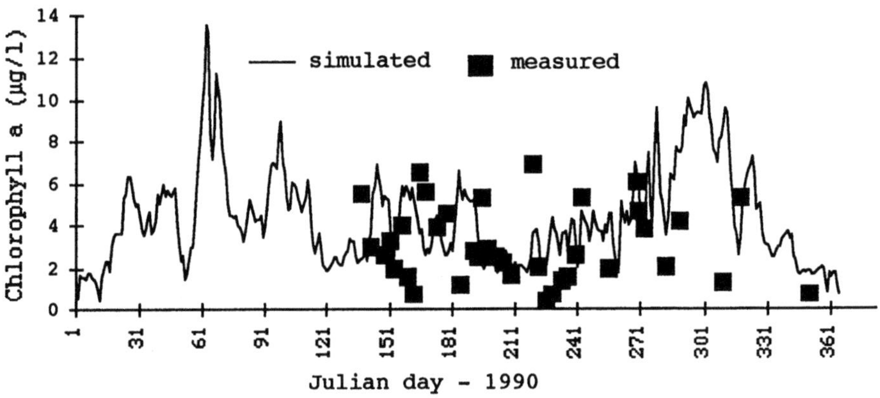

High frequency variability could be explained by alternating conditions of strong wind mixing and resuspension, and periods of calm leading to stratification of the water column, enhanced light intensity and a rapid development of phytoplankton blooms. During windy periods, blooms cease either due to increased turbidity or to nutrient limitation, grazing and export reducing phytoplankton biomass. An annual budget of nitrogen and organic carbon showed that freshwater inputs much nitrogen to the Gulf and contribute to the overall requirements for primary production more than the benthic remineralization.

REFERENCES

Bacher C (1989) Etude de la capacité trophyque du bassin de Marennes-Oléron : utilisation d'un modèle couplé de transport particulaire et de croissance de l'huître *Crassostrea gigas*. Thèse Doct Univ Bretagne Occidentale
Fransz HG Mommaerts JP Radach G (1991) Ecological modelling of the North Sea. Neth J Sea Res 28:37-140
Jorgensen SE (ed) (1986) Application of ecological modelling in environmental management. Elsevier
Riley GA (1965) A mathematical model of regional variations in plankton. Limnol Oceanogr 10:R202-R215
Salomon JC Pommpuy M (1991) Modèle mathématique de dispersion/décroissance des bactéries rejetées dans l'estuaire de Morlaix. Oceanol Acta 11:71-79
Steele JH (1962) Environmental control of photosynthesis in the sea. Limnol Oceanogr 7:137-150
Tett P (1990) A 3-layer vertical microbiological processes model for shelf seas. Report N 14 UCNW Bangor - Proudman Oceanographic Laboratory UK

DELAWARE BAY OYSTER POPULATIONS: EFFECTS OF SEED MOVEMENT, HARVESTING, AND DISEASES

J.N. Kraeuter, S.R. Fegley[1], S.E. Ford, and H.H. Haskin
Haskin Shellfish Research Laboratory
Rutgers University
Port Norris, NJ 08349 USA

ABSTRACT

Oysters are not directly harvested for market from Delaware Bay natural beds. They have been moved from upper Delaware Bay seed beds to planting grounds, located in lower Delaware Bay, since the mid 1800's. The States of Delaware and New Jersey have retained control of the seed beds, but lease bottom to private individuals in the planting grounds. Oysters placed on leased areas become the property of the lease holder and may be harvested at any time. Depending on the size of the oyster moved to the planting grounds (oysters from the uppermost seed beds are smaller on average than those from lower bay seed beds), one to three or more years are required for the oysters to reach market size. This mixture of direct harvest from the seed beds and subsequent grow-out on the planting grounds prior to final harvest has, in the presence of oyster disease and predation, changed the patterns of oyster abundance in Delaware Bay.

Historically, the greatest production of oysters probably occurred on the lower seed beds, and with so many oysters transported into the planting grounds, the greatest oyster population was in the lower bay. This distribution pattern was enhanced in the late 1940's and early 1950's by depletion of seed oysters in the upper bay by overharvesting. Although seed had been imported into Delaware Bay from other regions since the 1880's, larger proportions of seed were imported, principally

[1]Maine Maritime Academy, Castine, ME 04420

NATO ASI Series, Vol. G 33
Bivalve Filter Feeders
in Estuarine and Coastal Ecosystem Processes
Edited by Richard F. Dame

from Chesapeake Bay and the Virginia seaside bays, to supplement the relatively poor Delaware Bay seed bed harvests of the mid 20th century. Consequently, the oyster fishery was able to maintain itself in Delaware Bay even as the indigenous oyster population was being greatly depressed.

The oyster disease MSX was first noticed in Delaware Bay in 1957. Over the following three years approximately 90% of the oysters on the planted grounds and 60% of the oysters on the lower seed beds died. Very little mortality of oysters was observed on the upper seed beds. After the initial epizootic, MSX remained enzootic in Delaware Bay, but disease related mortality varied from year to year. Because of the disease intensity, lower bay seed beds were the primary source of oysters to be planted. The goal of the harvesters was to market oysters from the planted grounds in one year or less.

MSX disease has had little measurable effect on early life-history stages of oysters in the Bay. For example, monitoring of oyster spat settlement onto collectors has been done continuously since 1953. There are large interannual fluctuations in abundance of oyster spat, as is typical of most bivalve settlement records. However, these fluctuations do not correlate with the large changes in the adult oyster population over the past 40 years.

Relative abundance of oysters on the seed beds has been monitored continuously since 1953. Generally, numbers of oysters per bushel are greater on upper seed beds than on lower seed beds. This is due, in part, to the smaller average size of oysters on the upper seed beds. Long-term patterns of harvest and the distribution of predators and MSX also contribute to the differences in abundance between upper and lower seed beds. Prior to 1971, over 80% of the seed harvest was from upper seed beds. From 1971 to 1981 the average contribution of upper seed bed oysters to the planted total was 37%. After 1981 the portion of the harvest from the upper seed beds increased to a high of 62% in 1986. All seed beds were closed to harvest from 1987 to 1989. As a result of lower harvest intensity, predation and disease prevalence on the upper beds, the abundance of oysters currently exceeds pre-MSX levels. Abundances of oysters on the lower beds are not above pre-MSX levels because of continued, relatively high, predation, disease mortality, and harvest pressure. Consequently, the present Delaware Bay oyster distribution is strongly skewed toward the upper seed beds.

Although oyster abundance on the upper seed beds now exceeds that present in the years immediately preceding the MSX epizootic, the total of quantity of oysters baywide is greatly reduced. The effects of overharvesting and disease on the oyster population is likely producing ecosystem level changes. Comparison of 1880 and 1960 estimates of oyster abundance (a 91% decrease in abundance occurred over the interval) indicates that the 1880 populations would have been able to filter the entire volume of Delaware Bay in 10 to 20 days. The 1960 levels represent the lowest oyster populations in the bay, and would require 190 to 250 days to do the same. Further, the 1880 oyster population would have consumed about 17% of the bay's daily primary production and would have deposited 42 to 82% of the suspended solids present. The 1960 population could only utilize about 1% of the daily primary production and would have deposited only 4 to 5% of the suspended solids. These estimates are based on estimated standing stocks of oysters, annual averages of primary production, oyster carbon requirements, Delaware Bay volume, sediment loads, etc. and can only be considered as indicators of change. However, the differences in the indicators are substantial and suggest that management programs in estuarine ecosystems must consider more than just pollution and land-based anthropomorphic factors.

ACKNOWLEDGEMENTS

This is contribution No. 92-58 from the Institute of Marine

and Coastal Sciences, Rutgers University and New Jersey Agricultural Experiment Station Publication No. K-32406-1. Support for some of this work has been provided by NJ State funds and the NJ Commission on Science and Technology.

PHYSIOLOGY AND GRAZING WORKING GROUPS JOINT REPORT

BL Bayne[1] and S Shumway[2]

[1]Plymouth Marine Laboratory
Prospect Place, West Hoe
Plymouth PL1 3DH
United Kingdom

[2]Bigelow Laboratory
McKown Point
Boothbay Harbor, Maine 04575
USA

Our objective was to identify areas of ignorance where further research is required. We addressed this objective under three subheads with no priority order.

Topics of Fundamental Physiological Interest

We include here topics considered important for furthering our understanding of the bioenergetics of bivalve molluscs, with particular reference to feeding, digestion and assimilation of nutrients.

Uncertainty still exists concerning the fundamental mechanisms of particle retention and selection by bivalve suspension feeders. The classical model of the bivalve gill as a sieve is challenged by the concept of micro-scale hydrodynamic forces determining the properties of filtration; there is a controversy over the true role of mucous in the processes of particle trapping and selection. We know relatively little of the processes that determine the efficiency of selection amongst different particle types, both by the gills and by the labial

NATO ASI Series, Vol. G 33
Bivalve Filter Feeders
in Estuarine and Coastal Ecosystem Processes
Edited by Richard F. Dame

palps. Recent evidence due in large part to the application of analytical flow cytometry, suggests an impressive capacity for selection both below and above the threshold for pseudofeces production. The empirical support for variable filtration capabilities is growing; we need a truly mechanistic understanding of the processes involved.

There is a need for more research on integrations amongst physiological traits as they respond to complex suites of environmental factors, feeding behavior, which consists of many related traits (shell gape, pumping, filtration, clearance, particle sorting, ingestion, absorption) cannot be understood as a simple, maximizing engine subject only to modulation by disturbance; interactions and feedbacks between traits play a large part in determining net energy intake. These traits are also differently affected over time, and by various exogenous factors such as temperature, salinity and food quantity and quality. We need also a better understanding of the role of endogenous factors (such as reproductive condition, genotype) in determining the nutrient demand that is to be met by feeding.

Ecological Aspects

In this category we considered a variety of topics relevant to understanding how bivalves feed within natural populations in the field.

The role of water flow in determining the rates of filtration both of individuals and of populations has become a topic of considerable interest, and future studies both in the laboratory and the field should not ignore this important variable. Experiments that deal directly with the flow regime as a variable have highlighted the importance of interactive effects amongst individual in natural populations (down-stream particle depletion, self thinning), but further research is necessary to explore these effects so as to become predictive of particle filtration by populations of bivalves, from studies on individuals. More experiments are needed in carefully selected field sites, where the flow regimes can be fully characterized, and in laboratory flumes under controlled conditions of flow.

Discussions at this workshop highlighted differences in feeding behavior between species, e.g., oysters, mussels, cockles and scallops, but, however, we know little of the bases of these differences, with respect to morphology, physiology or ecology. Reasonable hypotheses exist concerning micro-habitat differences such as infaunal/epifaunal distributions, high flow/low flow habitats, or differences in the source of particulate material as food. There is a need for more comparative studies, undertaken under realistic conditions of flow and seston distribution, to elucidate further species differences and their functional basis.

The topic of "food quality" was the subject of much discussion. We recognized the need for various measures of quality depending on the specific purpose of the study. Nevertheless, it has become clear that, to understand mechanisms of feeding, we need to deal with particle volume, since processes of digestion and adsorption are clearly constrained by volume (gut capacity, surface area:volume ratios of particles). When modelling feeding in ecological and/or management models, however, food quality is treated in units of mass. More research is needed to calibrate between particle volume and mass, related to measures of quality with total organic content, specific biochemical composition, chlorophyll or other pigment content.

The Experimental/Modelling Interface

There was good agreement between the physiology and modeling working groups regarding the priority areas for study. We considered models of two types, those concerned with the feeding process, and models designed to facilitate effective management of bivalve stocks.

Optimal foraging models have already proved successful in representing bivalve feeding. The need here is for more interaction between empirical studies and the models; the models should be structured around measurable features of feeding behavior, and the model predictions challenged by appropriate experiments.

For ecological/management models, a number of challenges to the physiologists emerged. A paramount concern is to understand the variance amongst individuals in their feeding behavior and net energy intake; how much of observed variance is fixed within the genotype, how much is phenotypic, and what is the genetic basis for observed "plasticity" in feeding behavior.

It is important that physiologists consider the time-scales of feeding behavior with respect to environmental cycles, and that they describe theses in terms suitable to modelling; questions of endogenous or exogenous "forcing" of feeding are relevant here. The topic of food quality is also important at this interface between physiology and modelling, as discussed above. Specifically, we need to disassemble the term "detritus" into measurable units that are then relevant to studies of physiological response. Another topic requiring further study concerns the sensitivity to temperature change of different traits of feeding, and how these may be represented in models; specifically, what are the appropriate Q_{10} values to use? The physiologists' interests in understanding species differences in feeding behavior was reflected also by the modellers.

Finally, presentations at the Workshop showed an impressive ability for successful modeling of bivalve growth. This is encouraging, suggesting that the general features of feeding physiology are reasonably well understood. There is, however, a concern that good agreement between modeled and measured growth rates may be spurious, based perhaps on a canceling-out amongst model formulations and parameterizations. The modellers would therefore like to employ intermediate tests of model validity, prior to using a model to predict growth. Therefore, the physiologists are challenged to represent the processes of food capture, sorting, digestion and absorption in terms appropriate to the model testing stage that takes place prior to the calculation of the energy balance.

A final comment from these two working groups. Participants were inspired by the challenges posed to physiologists by the increased interest in benthic suspensions feeding, both as an important ecological process within the dynamics of coastal ecosystems, and with respect to the needs of the aquaculture industry. These challenges are seen as providing both a welcome new focus and a stimulus to new research on the physiology of benthic bivalves.

PREDATION AND POPULATION DYNAMICS WORKING GROUP REPORT

Raymond Seed
University of Wales Bangor
School of Ocean Sciences
Menai Bridge
Gwynedd LL55 5ey
United Kingdom

Following a wide ranging discussion there was general agreement that more fundamental research into coastal bivalves, their major predators, and the complex interactions between predator and prey is required. Whilst some bivalve predators have been extensively studied others are only poorly documented. Little is know about micropredators or the importance of predation on the vulnerable planktonic and early juvenile stages of even the most common bivalve species. Manipulative experiments should test the effects of varying the density of both predator and prey since at higher population densities many predators are known to spend proportionately more time on intraspecific interactions than on foraging. Future work should also consider the energetic costs of foraging, the way in which prey items are evaluated, and the extent to which predators make valued judgments of the risks involved in selecting particular prey types. Feeding history should also be considered since this may be important in determining whether a predator remains within a specific locality or moves to an alternative site where feeding opportunities are expected to be more favorable.

Mathematical models, particularly those based on diet optimization, would help conceptually in formulating the most appropriate questions regarding the ecological impact of predators. However, whilst comparatively simple models might adequately describe single predator-prey interactions,

NATO ASI Series, Vol. G 33
Bivalve Filter Feeders
in Estuarine and Coastal Ecosystem Processes
Edited by Richard F. Dame

predictive models explaining the effects of predation at the ecosystem level would be far more challenging. A major difficulty in developing such models would undoubtedly be the stochastic and variable nature of many of the parameters involved. Temporal and spatial heterogeneity in the abiotic environment may influence profoundly the abundance and foraging behavior of the predators; the controlling processes are thus likely to vary from one bivalve species to another. Feeding efficiency in many bivalve predators increases with experience or particular prey types and such learned behavior would also need to be incorporated into any realistic model.

Longer time series monitoring programs will be required if models are to incorporate aspects of seasonal variability. Severe winters are often followed by successful bivalve recruitment but the reasons for this are still largely unknown.

Increased mortality of major predators may be partly responsible, though egg quality in some bivalves seems to be enhanced by severe winter conditions. Mild winters may favor sustained growth of the prey enabling them to attain more rapidly a larger body size thus narrowing the window when most bivalves are especially vulnerable to predation.

Bivalve population density is often highly variable over time and even a small change in predation intensity could have significant effect on prey abundance. Annual fluctuations may reflect predation intensity on the planktonic or recently settled stages, though larval production in most bivalves is unlikely to be a limiting factor. Recruitment processes of predators are probably more important amongst invertebrates that typically have a high reproductive output; such predators may thus be able to respond more rapidly to changing densities of their prey and exploit bivalve populations to an extent that is unlikely amongst vertebrate predators such as birds, which have a lower reproductive output and extended periods of parental care.

Thus, despite a growing literature relating to the foraging behavior of predators that feed extensively on coastal and estuarine bivalves, surprisingly little is yet know about the impact of these predators on natural prey populations. Moreover, much of the information that is available is of an anecdotal nature and restricted to only a few common species.

This working group concluded that multidisciplinary approach, bringing together the disciplines of ecology, physiology, oceanography, and mathematical modelling would greatly assist our understanding of the complex interactions that exist between predators, bivalve populations, and the abiotic environment.

MATERIAL PROCESSING
WORKING GROUP REPORT

Richard F. Dame
Coastal Carolina College
Conway, South Carolina 29526
USA

Historically, the material processing studies concerning bivalve filter feeders have focused on uptake processes like particle filtration and biodeposition. In reality, bivalve filter feeders both take up and release materials to the surrounding system and this duality has only been addressed by researchers in the past 10-years. The working group recognized that from an ecosystem perspective bivalve filter feeders exhibit both functional and structural controls on material processing with in coastal and estuarine systems. Although numerous mechanisms and processes were discussed, only those which are poorly understood and which need to be investigated in future studies are outlined below.

Functional Controls

1. Scale

Previous studies of bivalve filter feeders (Cohen et al 1984; Cloern 1982) have generally limited the influence of these systems to the overlaying water column. In some recent studies, it is evident that there is insufficient food in the overlaying water column to support the bivalve community (Bernard 1974) and the additional needs must be met from other sources or from other locations through the influx of material on tidal currents. On a meters to kilometers scale, the hydrodynamics of a given system determines the extent of influence of bivalve

NATO ASI Series, Vol. G 33
Bivalve Filter Feeders
in Estuarine and Coastal Ecosystem Processes
Edited by Richard F. Dame

filter feeders on nutrient processing. At scales less than 10m, filter feeders may actually influence the hydrodynamics in and around bivalve communities.

There is a continuum of coastal systems in which the influence of bivalve filter feeders in relation to spatial scale changes in relation to water currents and geomorphology (Dame et al 1992). At one end of the continuum are the no flow systems typical of isolated embayments and lagoons. These systems are often depleted of suspended materials and elevated in inorganic nutrient concentrations as a result of the dense bivalve filter feeder communities inhabiting them. Intermediate in this continuum are coastal systems that are often coupled to the near shore and deeper ocean, i.e., Spanish lagoons on the Atlantic coast. In these systems, upwelled materials are often the driving force for new nutrients that support increased phytoplankton productivity which in turn supports increased densities of bivalve filter feeders. Finally, the tidal system common to estuaries constantly import and export materials to the coastal ocean. In many cases, these shallow estuarine systems support bivalve filter feeding communities far larger than the overlaying water column can support.

2. Stimulation of Primary Production

To date, a number of studies of natural systems have through indirect scaled-up estimates implicated bivalve filter feeders in the depletion of phytoplankton (Cloern 1982; Cohen et al 1984; Alpine and Cloern 1992). Filter feeders have been attributed to (a) keeping phytoplankton in a younger population age that enhances log growth; (b) quickly recycling the nutrients which phytoplankton need for growth; (c) keeping phytoplankton densities low so more light would be available to the living cells; and (d) shifting the species of phytoplankton within the natural community. There appear to be no direct observations at the system level that show either depletion or enhancement of primary production (Dame this volume). All of these are interesting questions with potentially large impacts on coastal ecosystems. In the future, these ideas need to be addressed at the ecosystem level utilizing mesocosms, ecosystem level field experiments and detailed observations on a wide variety of natural systems.

3. Selective Pressure on Phytoplankton

Due to the size of the bivalve's filter and mouth, these organisms have the potential to differentially remove certain size classes of organisms from the plankton (Prins et al 1991). This ability of bivalve filter feeders to shift the phytoplankton community by species or size class had only been observed in one natural system, Thau Lagoon, France (Deslous-Paoli et al this volume). In this system, areas with oysters had smaller phytoplankton cells and areas without oysters had larger cells. Because this aspect deals with the community of phytoplankton interacting with the bivalves, these questions should be addressed through field and mesocosm studies.

4. Mineralization

Mineralization processes in the sediments surrounding bivalve filter feeder beds have been deduced from increased material concentrations in the vicinity of the sediments (Dame et al 1989 and 1991; Prins and Smaal 1990) and from studies on general benthic substrates. As these sediments surrounding beds of bivalve filter feeders are highly enriched with organic matter and organisms direct observations to elucidate their biogeochemistry should be conducted in conjunction with studies of the microbiota and meiofaunal communities in these sediments.

5. Eutrophication Control

Bivalve filter feeders have been touted as bio-filters to control eutrophication in estuaries and coastal waters (Officer et al 1982). Although there have been some studies to investigate this potential, there is a need for long-term field and large mesocosm studies to address the feasibility and application of bivalves in such a task.

Structural Controls

1. Geomorphology and Sediments

The control of channel and estuarine geomorphology and sediment structure (Keck et al 1973; Letzsch and Frey 1980; Stevens 1983) by bivalve filter feeders is poorly known. The available sedimentary and geomorphological information needs to be reevaluated in a filter feeder and ecosystem context to identify areas of fruitful research.

2. Food Web Theory and Structure

Food web theory and structure as described earlier in this volume (Dame 1993) offer a potentially promising approach to synthesizing the interaction of nutrients, phytoplankton filter feeders and filter feeder predators. On a larger ecosystem scale, the structural dynamic ecosystem models exemplified by Ulanowicz (1986) and applied to shellfish studies on a total system scale (Ulanowicz and Tuttle 1992) offer great promise in furthering our understanding the workings of estuarine ecosystems and their large bivalve communities. This work should conducted with both field, laboratory and modeling aspects to provide a realistic ecosystem perspective.

REFERENCES

Alpine AE Cloern JE (1992) Trophic interactions and direct physical effects control phytoplankton biomass and production in an estuary. Limnol Oceanogr 37:946-955

Bernard FR (1974) Annual biodeposition and gross energy budget of mature Pacific oysters, *Crassostrea gigas*. J Fish Res Bd Canada 31:185-190

Cloern JE (1982) Does the benthos control phytoplankton biomass in South San Francisco Bay? Mar Ecol Prog Ser 9:191-202

Cohen RH Dressler PV Phillips EJ Cory RL (1984) The effect of the Asiatic clam, *Corbicula fluminea*, on phytoplankton of the Potomac River, Maryland. Limnol Oceanogr 29:170-180

Dame RF (1993) The role of bivalve filter feeder material fluxes in estuarine ecosystems. This volume

Dame RF Childers D Koepfler E (1992) A geohydrologic continuum theory for the spatial and temporal evolution of marsh-estuarine ecosystems. Neth J Sea Res 30:63-72

Dame RF Dankers N Prins T Jongsma H Smaal AC (1991) The influence of mussel beds on nutrients in the Western Wadden Sea and Eastern Scheldt estuaries. Estuaries 14:130-138

Dame RF Spurrier JD Wolaver TG (1989) Carbon, nitrogen ad phosphorus processing by an oyster reef. Mar Ecol Prog Ser 54:249-256

Deslous-Paoli JM Mazouni N Souchu P Landrein S Pichot P Juge C (1993) Oyster farming impact on the environment of a Mediterranean lagoon (Thau) This volume.

Keck R Maurer D Watling L (1973) Tidal stream development and its effect on the distribution of the American oyster. Hydrobio 42:369-379

Letzsch WS Frey RW (1980) Deposition and erosion in a Holocene salt marsh, Sapelo Island, Georgia. J Sed Petrol 50:529-542

Officer CB Smyda TJ Mann R (1982) Benthic filter feeding: a natural eutrophication control. Mar Ecol Prog Ser 9:203-210

Prins TC Smaal AC (1990) Benthic-pelagic coupling: the release of inorganic nutrients by an intertidal bed of *Mytilus edulis*. In Barnes M and Gibson RN (eds) Trophic Relationships in the Marine Environment Aberdeen U Press, Aberdeen pp 89-103

Prins TC Smaal AC Pouwer AJ (1991) Selective ingestion of phytoplankton by the bivalves *Mytilus edulis* and *Cerastoderma edule*. Hydrobiol Bull 25:93-100

Stevens SA (1983) Ecology of intertidal oyster reefs: food distribution, carbon and nutrient flow. Ph.D. Dissertation, University of Georgia, Athens, Georgia, 195pp

Ulanowicz RE (1986) Growth and development: ecosystem phenomenology. Springer-Verlag New York

Ulanowicz RE Tuttle JH (1992) The trophic consequences of oyster stock rehabilitation in Chesapeake Bay. Estuaries 15:257-265

WORKING GROUP REPORT: MODELLING

Jon Grant
Department of Oceanography
Dalhousie University
Halifax, Nova Scotia B3H 4J1 Canada

INTRODUCTION

Although there is a vast literature on ecosystem modelling, models which focus on bivalves and their role in energy flow in natural or aquaculture systems are now proliferating. It is possible to separate bivalve-oriented models into two categories relative to their goals. The first of these are models which consider the ability of suspension-feeding bivalve populations to stabilize coastal ecosystems with regard to phytoplankton production (Herman 1993). These models also allow assessment of how wild or cultured populations impact coastal environments via biodeposition, nutrient release, and other processes. The second (though not mutually exclusive)class of models addresses questions of carrying capacity of the environment for bivalve culture (Grant et al. 1993). The latter may also be referred to as management models, since this describes their purpose more accurately. Models in both categories are based on flows of material or energy, and include a scope for growth equation for individual bivalves or for total bivalve biomass. In addition, they are generally more focused on the role of bivalve populations in coastal environments as opposed to large and cumbersome holistic ecosystem models which tend to include bivalves only as incidental components.

NATO ASI Series, Vol. G 33
Bivalve Filter Feeders
in Estuarine and Coastal Ecosystem Processes
Edited by Richard F. Dame

RESEARCH DIRECTIONS IN MODELLING

The management of cultured bivalve stocks does not descend directly from the tradition of stock/recruitment models used in conventional invertebrate fisheries ecology (e.g. Caddy 1989). For cultured populations, the importance of suspended food as a limiting resource for optimal bivalve growth results in an emphasis on ingestion and energetics of the individual rather than on population dynamics. Consequently, several aspects of population ecology are neglected in many energy-based models including recruitment, predation, and emigration. While some of these factors are less important in hatchery-derived and/or contained culture (e.g. suspended culture), they are essential in models of natural populations, some types of bottom culture, or culture operations which rely on natural spat settlement and collection. There are several limitations in quantifying spatfall and recruitment in coastal environments because it is difficult to assess larval dispersion, supply rates to adult settlement sites, substrate selection, early post-settlement mortality (e.g. adult-larval interactions), and subsequent movement of spat (Beukema and de Vlas 1989; Roegner 1991; Gaines and Bertness 1992). Mortality of juveniles and adults is also a complex issue due to the variety of predators (Seed 1993) and environmental factors (e.g. temperature stress), which contribute to loss of individuals from a population. There are few empirical observations on some of these phenomena such that their inclusion in predictive models is not easily accomplished. However, the value of such information and its integration into energy flow models is recognized as being important to future attempts to forecast population production in both wild and cultured populations.

In a related context, competitive interactions are not explicitly included in many energy-based models although they are implicit in food limitation resulting from seston depletion. Spacing of individuals may be indicative of food competition (e.g. mussel beds; Newell and Shumway 1993), but in other cases, such as bottom distribution of scallops, food resource limitation is not obvious. In general, the distinction between space and food (space in which to feed?) as a limiting resource has long been ambiguous in benthic ecology. Models based on concepts such as self thinning (Fréchette et al. 1992) provide an approach to this problem which could be utilized in the wider application of models to management of cultured species.

One aspect of modelling which continually arose in the Working Groups involved the enormous variation in coastal environments (e.g. temperature, food) and the response of bivalves to this variation. Variables such as suspended food can change on time scales of minutes to months. Seasonal changes in seston are readily incorporated into models (e.g. seasonal phytoplankton cycles), but modelling of higher frequency variation due to tidal and wind forcing requires a better understanding of the mechanisms which regulate seston quality and quantity. Resuspension is perhaps the least understood of these processes; neither sediment resuspension nor its effects on suspension feeder nutrition are easy to simulate in a model. Although sedimentological aspects of resuspension are understood, the determination of organic particle fluxes from the bottom to the water column is difficult due to its dependence on sediment type, organic content, and the relationship of microflora to sediment particles. In dense populations of bivalves, feces and pseudofeces can also have a large impact on particle fluxes (Kautsky and Evans 1987). Moreover, biodeposits may foster an active microbial community in the sediments (Grenz et al. 1990). Resuspension of biodeposits and other sediment particles, and its effect on microbial activity on particles are clearly in need of more field and laboratory research (Ritzrau and Graf 1992). Empirical progress in this area will foster improved submodels of seston as a bivalve food resource. Modelling of the linkage between bivalves and their food supply is also limited by an understanding of the time scale of physiological responses to fluctuations in particle fields as well as other environmental variables (e.g. temperature). Interspecific differences in this regard are also of interest. This topic is discussed further in the report of the Physiology Working Group.

One area of research which has provided substantial insight into the role of bivalves in coastal ecosystems has been the use of *in situ* tunnels and flumes placed over mussel beds, oyster reefs, and other bivalve communities (see Asmus and Asmus 1993; Dame 1993; Newell and Shumway 1993). This approach has successfully elaborated the role of bivalves in regulating nutrient budgets and the functioning of coastal ecosystems. However, there are few dynamic models which attempt to simulate these dense aggregations. Most of the emphasis in these studies has been on seston depletion and nutrient regeneration by the bivalves. Further work on the interaction of bivalves with the adjacent sediments and benthos would be useful in understanding the larger system. The physical effect of the enclosures on processes such as resuspension and total particle fluxes at the bed also needs more attention. An example of the utility of modelling these bivalve communities arises from suggestions by Newell and Shumway (1993) regarding the placement of seed in bottom mussel culture. Although seston may be depleted along the length ($\underline{x}$-axis) of a mussel bed, seeding width could be manipulated to optimize exchange in the $\underline{y}$-direction such that seston renewal would be enhanced by cross-stream flux. Clearly, hydrodynamic modelling could have major input into the arrangement of bottom or suspended culture, interacting with biological models of species-specific responses to food quality/quantity. These ideas reiterate the role of modelling in aquaculture management.

Throughout this discussion, we cannot emphasize strongly enough the importance of understanding and quantification of physical processes in models of bivalve populations. Factors such as tidal flushing may completely dominate particle exchange (Herman 1993) and override the local influence of suspension feeding on particle fields. Wildish and Kristmanson (1979) developed this concept previously in contrasting turbulent renewal versus suspension feeder depletion as a measure of food limitation. It is apparent from the research described in this volume (e.g. Grant et al. 1993; Herman 1993; Newell and Shumway 1993) that physical oceanographic studies are being more fully integrated into models of bivalve populations and their food resources.

The NATO Workshop has been an important step in integrating various approaches to the study of bivalves in marine ecosystems. An important next step is a modelling workshop wherein a common data set can be applied to various models constructed by several research groups. Boundary conditions of variables could be distributed prior to the meeting so that participants would be prepared to run simulations and sensitivity analyses during workshop. This exercise would provide a rigorous means of comparing different modelling approaches to bivalve populations. At present, energy-based modelling of bivalve populations is still in a research phase rather than in a widespread application phase. The power of modelling tools to predict growth rates in bivalve culture and wild fisheries is substantial; these models are on the verge of becoming an important implement in fisheries management. A workshop as suggested would facilitate the utility of models as a management tool for wild and cultured bivalve populations.

REFERENCES

Asmus H, Asmus R (1993) Phytoplankton-mussel bed interactions in intertidal ecosystems. This volume

Beukema JJ, de Vlas J (1989) Tidal-current transport of thread-drifting postlarval juveniles of the bivalve *Macoma balthica* from the Wadden Sea to the North Sea. Mar. Ecol. Prog. Ser. 52: 193-200

Caddy JF (1989) A perspective on the population dynamics and assessment of scallop fisheries, with special reference to the sea scallop, *Placopecten magellanicus* Gmelin. In: Marine invertebrate fisheries: their assessment and management (Caddy JF ed.) John Wiley and Sons

Dame R (1993) The role of bivalve filter feeder material fluxes in estuarine ecosystems. This volume

Frechette M, Aitken AI, Page L (1992) Interdependence of food and space limitation of a benthic suspension feeder - consequences for self-thinning relationships. Mar. Ecol. Prog. Ser 83: 55-62

Gaines SD, Bertness MD (1992) Dispersal of juveniles and variable recruitment in sessile marine species. Nature 360: 579-580

Grenz C, Hermin M-N, Baudinet D, Daumas R (1990) In situ biochemical and bacterial variation of sediments enriched with mussel biodeposits. Hydrobiologia 207: 153-160

Grant J, Dowd M, Thompson K, Emerson CW, Hatcher A (1993) Perspectives on field studies and related biological models of bivalve growth and carrying capacity. This volume

Herman PMJ (1993) A set of models to investigate the role of suspension feeders in estuarine ecosystems. This volume

Kautsky JN, Evans S (1987) Role of biodeposition by *Mytilus edulis* in the circulation of matter and nutrients in a

Baltic coastal ecosystem. Mar. Ecol. Prog. Ser. 38: 201-212

Newell CR, Shumway SE (1993) Grazing of natural particulates by bivalve molluscs: a spatial and temporal perspective. This volume

Ritzrau W, Graf G (1992) Increase of microbial biomass in the benthic turbidity zone of Kiel Bight after resuspension by a storm event. Limnol. Oceanogr. 37: 1081-1086

Roegner GC (1991) Temporal analysis of the relationship between settlers and early recruits of the oyster *Crassostrea virginica* (Gmelin). J. Exp. Mar. Biol. Ecol. 151: 57-69

Seed, R (1993) The role of invertebrate predators in structuring coastal and estuarine bivalve populations. This volume

Wildish DJ, Kristmanson DD (1979) Tidal energy and sublittoral macrobenthic animals in estuaries. J. Fish. Res. Bd. Can. 36: 1197-1206

Wildish DJ, Kristmanson DD (1993) Hydrodynamic control of bivalve filter feeders: A conceptual view. This volume

PHYSICAL FACTORS
WORKING GROUP REPORT

Marcell Fréchette
Institut Maurice-Lamontagne
Ministère des Pêches et des Océans
C.P. 1000
Mont-Joli, Québec
Canada G5H 3Z4

The discussions of the working group covered a number of topics:

- the problems of scale in the context of linking physical and biological processes;
- site selection for bivalve aquaculture;
- topographic aspects and their consequence for the relevant physical phenomena.

The discussions were held in a management and modeling context. The group recognized that the physical environment, that is, current speed and wind, is a dominant factor in the control of energy flow in bivalve systems by its action on processes such as advection, diffusion, and resuspension.

SCALE

The importance of choosing the right spatial scale was recognized. Spatial scale of biological objects -- whether one is looking at intra-animal, individual, population or geographical patterns -- has implications for the relevant physics. This aspect is overviewed in the Wildish and Kristmanson paper. The group, however, dealt with the problem from the perspective of a modeler: the present thrust in ecological and physiological research is to produce mechanistic models of biological processes at the individual level, to be

NATO ASI Series, Vol. G 33
Bivalve Filter Feeders
in Estuarine and Coastal Ecosystem Processes
Edited by Richard F. Dame

included in physical models of transport at large scale. This approach raised some concern precisely on the necessity and difficulty of extrapolating from small scale to large scale especially in the context of linking physical aspects of ecosystems with its biology, especially considering the feed-backs, may have proper physics and spatio-temporal scales. In short: how do we link scales and processes?

This concern was not shared by all members of the group. It was argued that models at small scales need not be linked with models of processes operating at large scale. Parameterization of small-scale processes is sufficient to account for them in large-scale models. Others pointed out, nevertheless, that it was necessary to have a clear idea of the scales relevant to any given situation, because they would indicate the proper physical processes (and associated variables) operating, which bears consequences for experimental designs, model specifications, etc. For example, let us consider the paradox presented by Wildish and Kristmanson -- populations profit from increases in current speed, while individuals suffer from high current speed. It appears that the response of individuals and populations to current speed is not the same because the control is not the same at the scale of the individual and of the population.

Other participants noted that much of the present research efforts implied small-scale physics, and that advection at larger scales (km) should also be considered.

Site Selection And Management

Physical models were suggested to be necessary components for aquaculture site selection and management. There was disagreement on this assertion, with other participants feeling that in practice, site selection and sowing density problems should be addressed using simple, predictive, relationships.

Topography

Topography was pointed out as an important determinant of physical processes. Two examples were given. First, it was reported that in Lake Grevelingen, mussel beds are found on gently sloping bottoms. Present models of the effect of hydrodynamics on the food supply to bivalve beds assume a flat topography. It is possible that in Lake Grevelingen, horizontal advection and vertical turbulent transport are inadequate to provide sufficient resources for growth on flat bottoms. With sloping bottoms, additional components, such as vertical advection, would be added to the supply of food to a bivalve bed. Second, the presence of non-tidal water planes in places like Etang de Thau may be related to topography. Transport in these systems is strongly governed by winds that are unpredictable in frequency, duration and direction. In such situations, the shape and orientation of the water plane will have something to do with the system's response to wind because of the varying fetch.

The group concluded that our understanding of the relationship between current speed and other physical forcing and bivalve growth was in its early stage. Future research should include: the study of wind effects, aspects linked to topography, and the direct effects of current speed on bivalves.

THE ROLE OF SCIENTISTS IN THE MANAGEMENT OF COASTAL ECOSYSTEMS

Norbert Dankers
DLO Institute for Forestry and Nature Research
P.O. Box 167
1790 AD Den Burg/Texel
The Netherlands

It is clear that politicians and managers should make decisions, and scientists should provide background information that informs decision-makers.

Scientists experience a great lack of specific and detailed knowledge among policy-makers and politicians. These decision-makers have to deal with numerous special interest groups with often very diverse interests. These groups provide the policy-makers with, often biased information. Therefore the decision-makers need objective background information and knowledge. Often this information is of a descriptive nature, but scientists also have to be prepared to answer detailed questions about activity-effect relations.

In the discussion of the working group, some examples were given of the interrelationships of policy and decision-makers with bivalve scientist. Dutch policy-makers have greatly benefited from the graphical representation of the status of the environment, the so-called AMOEBE approach. It should be stressed to policy-makers that environmental predictions, for example the biomass of bivalves are difficult to make and also therefore a continuous consultation between policy-makers and scientists is essential in order to take decisions based on the latest available information.

RESPONSIBILITIES OF POLICY-MAKERS

It is the responsibility of the policy-makers to make their goals clear. If the main goal is nature conservation in which other uses have to be fitted, as is the case in the Wadden Sea,

NATO ASI Series, Vol. G 33
Bivalve Filter Feeders
in Estuarine and Coastal Ecosystem Processes
Edited by Richard F. Dame

the questions asked will be different from when the main goal is oyster culture. There the main activity might conflict with yachting, agriculture, industry, tourism and waste treatment as is the case in the Bay of Oléron.

Policy-makers should also be clear about their vision of management. Do they act from a natural development point of view, a classical view on nature conservation, or do they take a functional approach, i.e. the use of natural resources as a starting point (Dankers et al. 1991).

If these two points, the goals and the vision are not clear, the questions will not be clear. In practice, the scientists have to phrase or rephrase the questions in order to make answers possible, and this is difficult if the goals and vision of the demanding party are not clear.

RESPONSIBILITY OF SCIENTISTS

There are two main tasks for scientists in the management of coastal ecosystems. First of all they have to provide information to decision-makers. This information consists of ecological, socio-economical and technical knowledge. Secondly they have to answer specific questions in the order of: If you do this, you will get that result, or such and such cannot be predicted. The effects of alternative policies may be discussed stressing uncertainties. Scientists should avoid being pushed into the role of being decision-maker although policy-makers often try this.

MAJOR ISSUES IN MANAGEMENT OF LARGE COASTAL ECOSYSTEMS

There is at least one question which always comes up when management plans have to be developed for large areas:
What is best? Do we concentrate all users in some areas and leave the rest of the system as a low impacted nature reserve or do we aim for a multifunctional use where we strive for an optimisation of many users in peaceful co-existence.

When the first option is chosen the questions asked to scientists are relatively simple. In general we only have to indicate which parts of the ecosystem are the most important.

It is much more difficult to advise on the sustainable co-existence of sometimes conflicting uses. We need in depth knowledge of all sort of activity-effect relationships such as disturbance by recreational activities, impact of fisheries, impact of dredging and harbour works, impact of pollution, etc. It should also be made clear that it is sometimes necessary to regulate the numbers of species in extreme cases even birds or seals. It also should be accepted that natural events 'destroy' natural things of beauty. For example extreme weather conditions may cause animals to die or islands to disappear. It certainly is also a task of scientists to make this clear to a broad public.

On a complex issue, i.e. the management of a coastal ecosystem ecological, knowledge is not enough. Experiences on economical, planning and sociological issues have to play a role. It is doubtful that we have the right tools to join these very different sciences in order to give good advice to politicians. For this reason, it is most important to develop models that couple social, socio-economic, carrying capacity and ecological models. Making decisions based on only one model is often not acceptable for one of the parties. A clear example can be found in Fig. 1. The figure shows the relation between yearly oyster production and oyster stock in the Bay of Oléron. A large oyster stock means a large effort by the fishermen, pollution by thousands of outboard engines and the environmental impact by the aquaculture, etc. From a fisheries point of view, a large oyster stock is preferred because it provides the highest yearly production. From an ecological point of view it would be advisable to reduce the stock to approximately 60-80 thousand tons. The yearly production would be close to 90% of the maximum, at a considerably reduced environmental impact. On the other hand, such strategy would probably result in large-scale unemployment and high incomes for those that stay in business. The decision to allow a large stock is taken without the availability of a comprehensive tool to evaluate different alternatives.

Scientists are necessary for informed management decisions,

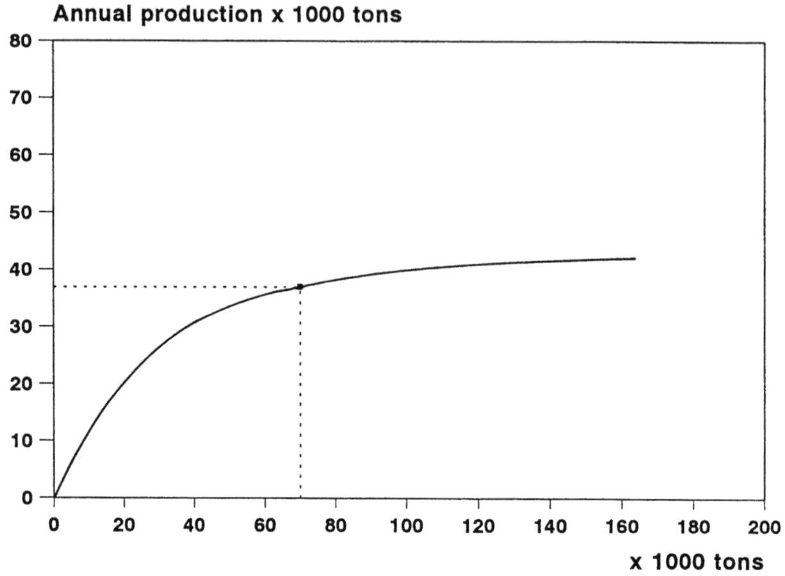

Fig. 1 Relation between biomass and annual production (from Héral et al. 1986)

but it is doubtful whether they will ever play a real role when short term achievements are politically more interesting and expedient to policy-makers.

REFERENCES

Brink ten BJE Hosper SH Colijn F (1991) A quantitative method for description and assessment of ecosystems: the AMOEBA-approach. Mar Poll Bull 23: 265-271

Dankers N Dijkema KS Reijnders PJH Smit CJ (1991) The Wadden Sea in future - why and how to reach. RIN-contribution to Research on Management and Natural Resources 1991-1 Research Institute for Nature Management Texel

Héral M Deslous Paoli JM Prou J (1986) Dynamiques de production et des biomasses de huîtres creuses cultivées (*Crassostrea angulata* et *Crassostrea gigas*) dans le bassin de Marennes Oléron depuis un siecle ICES CM 1986/F: 41 Mariculture Committee

BIVALVE FILTER FEEDERS AND ESTUARINE AND COASTAL ECOSYSTEM PROCESSES: CONCLUSIONS

Richard Dame
Coastal Carolina College
Conway, SC 29526
USA

Bivalve filter feeding mollusks are important components of coastal ecosystems because they remove large quantities of suspended material from the water column and excrete abundant amounts of reactive nutrients. These animals are also major prey for numerous predators including birds, fish, mammals, and invertebrates. They are also significant food resources to humans. While organismic and population level studies have dominated bivalve ecology, the recent focus on the ecosystem roles evidenced by filter feeding systems has led to even larger scale investigations. This workshop ties the various scales of study together using the process/functional philosophy of the ecosystems approach. Thus, the specific topics of physiology, grazing, predation, nutrient cycling, physical environment, computer simulation modeling and environmental management are combined into a meaningful whole.

Because many models and ecosystem level estimates are based on organismic rate observations, physiological mechanisms incorporating the role of physical factors, particularly hydrodynamics, in filter feeding need to be determined. Physiologist must devise the simplest and most realistic representation of filter feeding by mollusks (Bayne; Navarro and Inglesias). New techniques such as endoscopy and flow cytometry may prove fruitful when applied to whole organisms and beds. In defining mechanisms, the question of particle selection by bivalve filter feeders lacks adequate resolution both at the organismic and ecosystem levels. Finally, the physiology of nitrogen and phosphorus excretion

NATO ASI Series, Vol. G 33
Bivalve Filter Feeders
in Estuarine and Coastal Ecosystem Processes
Edited by Richard F. Dame

must become a focal point for future temporal studies in order to understand the role of bivalve filter feeders in ecosystem nutrient recycling.

Even though predators are thought to change the entire character of certain coastal ecosystems their role in ecosystems dominated by bivalves is not well known. Predation of bivalves by invertebrates and birds is well studied, but translation to ecosystems is lacking (Seed; Meire). Understanding of predation by large rays and marine mammals is poor and there appears to be a difference in dominant predator type between Europe and North America. The role of micro-predators on the newly settled larvae of bivalves and how they influence the older age classes of mollusks is unknown.

The development of flumes, chambers, tunnels and other techniques to directly measure fluxes between bivalve beds and the adjacent water mass has opened an entirely new aspect of research and technology (Dame; Smaal and Prins). Bivalve filter feeding systems may play important roles in the cycling of nutrients in estuarine and coastal ecosystems. Filter feeders remove large amounts of particulate material and release equivalent amounts of dissolved materials into the surrounding water. Thus, they appear to function as material processors and nutrient retention mechanisms within ecosystems. In addition to the rapid consumption and remineralization of materials through feeding, metabolism and excretion, dense populations of bivalves change the biogeochemistry of the sediments surrounding the beds. Unfortunately, there are few direct observations of nutrient release from the organically enriched sediments of bivalve beds. The sources and fates of materials entering and leaving bivalve filter feeding beds are often kilometers from the beds themselves and imply an ecosystem coupling role far beyond the immediate surroundings.

The influences of water flow on the processes of bivalve growth and filter feeding is critically important (Wildish and Kristmanson). Vertical profiles of seston and current velocity are of paramount importance for unbiased estimates of food flux to bivalve communities (Frechette et al). Because most estimates of hydrodynamic influences are for semi-mobile mussels, future observations should include sessile oysters and infaunal bivalves. Recent efforts to cross calibrate the various methods at the same time and place are to be applauded and encouraged (Asmus et al 1992). Future studies should test different techniques simultaneously, i.e., up-stream/down-stream, flumes, chambers, tunnels, etc., and should continue to be developed though cooperative international experiments. It is only through such endeavors that understandable progress in observation and measurement can be achieved.

Simple physical-biological models utilizing time series data offer a strong empirical approach to representing bivalve filter feeder trophic relationships (Grant). By using ecosystem models of the interactions between filter feeders, phytoplankton and nutrients in a 2-dimensional framework, a hither to unknown estuarine feature, homeostasis in the phytoplankton, is suggested (Herman). It was also observed by the modellers that their models are not stable unless detritus in small quantities are provided. This later finding supports the physiological observations by other groups. One very provocative and potentially productive suggestion by the modelling group is to test a number of existing models on the same data sets and to due this in a special modelling workshop.

This workshop indicates that although there is an immense body of knowledge concerning bivalve filter feeders, there are large gaps in our understanding of how these organisms and communities function within estuarine and coastal ecosystems. Wise decisions on the use and management of waters dominated by bivalve filter feeders is highly dependent on understanding how these ecosystems work and how they interact with the human species (Dankers; Heral). The carrying capacity or sustained yield population model approach offers one avenue to bivalve filter feeder's management, while the whole system or "super organism" concept of Dankers (this volume) offers another. The latter idea involves many new and unique interrelations between environmental, sociological and practical concerns involved in the cultivation and management of systems with large populations of bivalve filter feeders. As many of the issues raised by this workshop can only be answered by examining large coastal ecosystems, comparative studies between Europe and North America are strongly suggested to help elucidate differences and ecosystem level mechanisms in estuarine and coastal systems dominated by bivalve filter feeders.

REFERENCES

Asmus H Asmus R Phytoplankton-mussel bed interactions in intertidal ecosystems. This volume

Asmus H Asmus R Prins TC Dankers N Frances G Maass B Reise K (1992) Benthic-pelagic flux rates on mussel beds: tunnel and tidal flume methodology compared. Helgoland Meeresunters 46:341-361

Bayne BL Feeding physiology of bivalves: time-dependence and compensation for changes in food availability. This volume

Dame RF The role of bivalve filter feeder material fluxes in estuarine ecosystems. This volume

Dankers N Integrated estuarine management - obtaining a sustainable yield of bivalve resources while maintaining environmental quality. This volume

Frechette M Lefaivre D Butman CA Bivalve feeding and the benthic boundary layer. This volume

Grant J Dowd M Thompson K Emerson C Hatcher A Perspectives on field studies and related biological models of bivalve growth and carrying capacity. This volume

Heral M Why carrying capacity models are useful tools for managementof bivalve molluscs culture. This volume

Herman P A set of models to investigate the role of benthic suspension feeders in estuarine ecosystems. This volume

Meire P The impact of bird predation on marine and estuarine bivalve populations: a selective review of patterns and underlying causes. This volume

Navarro E Iglesias JIP Infaunal filter feeding bivalves and the physiological response to short-term fluctuations in food availability and composition. This volume

Newell CR Shumway SE Grazing of natural particulates by bivalve molluscs: a spatial and temporal perspective. This volume

Seed R Invertebrate predators and their role in structuring coastal and estuarine populations of filter feeding bivalves. This volume
Smaal AC Prins TC The uptake of organic matter and the release of nutrients by bivalve suspension feeder beds. This volume
Wildish DJ Kristmanson DD Hydrodynamic control of bivalve filter feeders: a conceptual view. This volume

INDEX

Springer-Verlag and the Environment

We at Springer-Verlag firmly believe that an international science publisher has a special obligation to the environment, and our corporate policies consistently reflect this conviction.

We also expect our business partners – paper mills, printers, packaging manufacturers, etc. – to commit themselves to using environmentally friendly materials and production processes.

The paper in this book is made from low- or no-chlorine pulp and is acid free, in conformance with international standards for paper permanency.

NATO ASI Series G

Vol. 1: **Numerical Taxonomy.** Edited by J. Felsenstein. 644 pages. 1983. (out of print)

Vol. 2: **Immunotoxicology.** Edited by P. W. Mullen. 161 pages. 1984.

Vol. 3: **In Vitro Effects of Mineral Dusts.** Edited by E. G. Beck and J. Bignon. 548 pages. 1985.

Vol. 4: **Environmental Impact Assessment, Technology Assessment, and Risk Analysis.** Edited by V. T. Covello, J. L. Mumpower, P. J. M. Stallen, and V. R. R. Uppuluri. 1068 pages.1985.

Vol. 5: **GeneticDifferentiationand Dispersal in Plants.** Edited by P. Jacquard, G. Heim, and J. Antonovics. 452 pages. 1985.

Vol. 6: **Chemistry of Multiphase Atmospheric Systems.** Edited by W. Jaeschke. 773 pages. 1986.

Vol. 7: **The Role of Freshwater Outflow in Coastal Marine Ecosystems.** Edited by S. Skreslet. 453 pages. 1986.

Vol. 8: **Stratospheric Ozone Reduction, Solar Ultraviolet Radiation and Plant Life.** Edited by R. C. Worrest and M. M. Caldwell. 374 pages. 1986.

Vol. 9: **Strategies and Advanced Techniques for Marine Pollution Studies: Mediterranean Sea.** Edited by C. S. Giam and H. J.-M. Dou. 475 pages. 1986.

Vol. 10: **Urban Runoff Pollution.** Edited by H. C. Torno, J. Marsalek, and M. Desbordes. 893 pages. 1986.

Vol. 11: **Pest Control: Operations and Systems Analysis in Fruit Fly Management.** Edited by M. Mangel, J. R. Carey, and R. E. Plant. 465 pages. 1986.

Vol. 12: **Mediterranean Marine Avifauna: Population Studies and Conservation.** Edited by MEDMARAVIS and X. Monbailliu. 535 pages. 1986.

Vol. 13: **Taxonomy of Porifera from the N. E. Atlantic and Mediterranean Sea.** Edited by J. Vacelet and N. Boury-Esnault. 332 pages. 1987.

Vol. 14: **Developments in Numerical Ecology.** Edited by P. Legendre and L. Legendre. 585 pages. 1987.

Vol. 15: **Plant Response to Stress. Functional Analysis in Mediterranean Ecosystems.** Edited by J. D. Tenhunen, F. M. Catarino, O. L. Lange, and W. C. Oechel. 668 pages. 1987.

Vol. 16: **Effects of Atmospheric Pollutants on Forests, Wetlands and Agricultural Ecosystems.** Edited by T. C. Hutchinson and K. M. Meema. 652 pages. 1987.

Vol. 17: **Intelligence and Evolutionary Biology.** Edited by H. J. Jerison and I. Jerison. 481 pages. 1988.

Vol. 18: **Safety Assurance for Environmental Introductions of Genetically-Engineered Organisms.** Edited by J. Fiksel and V.T. Covello. 282 pages. 1988.

Vol. 19: **Environmental Stress in Plants. Biochemical and Physiological Mechanisms.** Edited by J. H. Cherry. 369 pages. 1989.

Vol. 20: **Behavioural Mechanisms of Food Selection.** Edited by R. N. Hughes. 886 pages. 1990.

NATO ASI Series G

Vol. 21: **Health Related Effects of Phyllosilicates.**
Edited by J. Bignon. 462 pages. 1990.

Vol. 22: **Evolutionary Biogeography of the Marine Algae of the North Atlantic.**
Edited by D. J. Garbary and G. R. South. 439 pages. 1990.

Vol. 23: **Metal Speciation in the Environment.**
Edited by J. A. C. Broekaert, Ş. Güçer, and F. Adams. 655 pages. 1990.

Vol. 24: **Population Biology of Passerine Birds. An Integrated Approach.**
Edited by J. Blondel, A. Gosler, J.-D. Lebreton, and R. McCleery. 513 pages. 1990.

Vol. 25: **Protozoa and Their Role in Marine Processes.**
Edited by P. C. Reid, C. M. Turley, and P. H. Burkill. 516 pages. 1991.

Vol. 26: **Decision Support Systems.**
Edited by D. P Loucks and J. R. da Costa. 592 pages. 1991.

Vol. 27: **Particle Analysis in Oceanography.** Edited by S. Demers. 428 pages. 1991.

Vol. 28: **Seasonal Snowpacks. Processes of Compositional Change.**
Edited by T. D. Davies, M. Tranter, and H. G. Jones. 484 pages. 1991.

Vol. 29: **Water Resources Engineering Risk Assessment.**
Edited by. J. Ganoulis. 551 pages. 1991.

Vol. 30: **Nitrate Contamination. Exposure, Consequence, and Control.**
Edited by I. Bogárdi and R. D. Kuzelka. 532 pages. 1991.

Vol. 31: **Industrial Air Pollution. Assessment and Control.**
Edited by A. Müezzinoğlu and M. L. Williams. 245 pages. 1992.

Vol. 32: **Migration and Fate of Pollutants in Soils and Subsoils.**
Edited by D. Petruzzelli and F. G. Helfferich. 527 pages. 1993.

Vol. 33: **Bivalve Filter Feeders in Estuarine and Coastal Ecosystem Processes.**
Edited by R. F. Dame. 584 pages. 1993.

FSC
www.fsc.org
MIX
Papier aus verantwortungsvollen Quellen
Paper from responsible sources
FSC® C105338

If you have any concerns about our products,
you can contact us on
ProductSafety@springernature.com

In case Publisher is established outside the EU,
the EU authorized representative is:
Springer Nature Customer Service Center GmbH
Europaplatz 3, 69115 Heidelberg, Germany

Printed by Libri Plureos GmbH
in Hamburg, Germany